Prestressed Concrete

Other Books of Interest

Advanced Mechanics of Solids and Structures (81-7319-066-6)
N. Krishna Raju and D.R. Gururaja

Applications of Random Vibrations (81-85198-58-6)
N.C. Nigam and S. Narayanan

Boundary and Finite Elements (1-84265-013-0)
J. Ramachandran

Computers in Urban Planning and Urban Management (81-7319-216-2)
P.K. Sikdar et al

Design of Foundation Systems (81-7319-028-3)
Nainan P. Kurian

Fly Ash Disposal and Deposition (81-7319-286-3)
U. Dayal et al

Grillage Analogy in Bridge Deck Analysis (81-7319-153-0)
C.S. Surana and R. Agarwal

Groundwater Evaluation (1-84265-077-7
P.K. Garg and C.S. Agarwal

Hydrology and Water Resources Engineering (1-84265-023-8)
K.C. Patra

Industrial Solid Waste Management and Landfilling Practice (81-7319-317-7)
M. Datta et al

Management of Ash Ponds (81-7319-338-X)
S.R. Gandhi et al

N. Rajagopalan

Prestressed Concrete

Alpha Science International Ltd.
Pangbourne England

Dr. Ing. N. Rajagopalan
Professor (Retd.)
Formerly, Head of Department of Civil Engineering & Dean
Indian Institute of Technology, Madras
Chennai 600 036, India

Alpha Science International Ltd.
P.O. Box 4067, Pangbourne RG8 8UT, UK

ISBN 1-84265-073-4

Printed in India.

Foreword

It is a proud privilege and pleasure offered to me to write the Foreword for the book on "Prestressed Concrete" authored by Prof. N. Rajagopalan as the subject is quite close to my heart.

The innovation of the idea of prestressing of concrete by Eugene Freyssinet in France is a landmark in the history of civil engineering. It is our good fortune in India that we are not lagging far behind in employing the techniques of prestressing in various construction projects in India. The Palar Bridge near Chennai is by far the fourth application of prestressed concrete in highway bridges in the world. Prestressed concrete by combining high strength concrete has enabled large span structures to be built reducing the effect of deadweight of structures. Concrete, being strong in compression and reinforced concrete being strong in countering bending moments the only shortcoming is its weakness in tension. This has been taken care of very effectively by precompressing the concrete. Tensile stresses in the members in service condition are nullified by compressive stresses induced by prestressing in advance. Thus prestressed concrete has proved to be cost-effective when compared to steel and reinforced concrete structures in numerous applications.

The journey of its use and application to concrete structures has had contributions from numerous practitioners and theoreticians. Dr. Rajagopalan has painstakingly gone about assimilating information from various sources and reflecting his own experience in the field to write a book that is technical in nature and contemporary in its application. He has brought forth the history of development of this art that would interest the reader in delving deeper into the subject.

Prof. Rajagopalan has covered the various forms of prestressing, the materials that constitute prestressed concrete in the context of their behaviour, durability and its effects in various stages of performance of the structure out. Prestressing systems and anchorages in various structures are elaborately explained. Some non-conventional prestressing methods are also mentioned. The concepts of analysis for flexure and types of prestressing strands are also dealt with. Losses in prestress are also explained in an elaborate manner. Various codal provisions for checking the ultimate strength in flexure are also covered. Approaches to deisgn philosophy and design for flexure have also been touched upon.

Prestressed concrete members are rarely stand-alone structures. Composite construction lends various additional complexities to the analysis and design. These are also addressed along with relevant clauses of various codes. Statistically indeterminate structures are also dealt with in detail. Cylindrical structures such as pipes and tanks are also covered. Partial prestressing for economy and optimising structural behaviour are also discussed.

Prestressed concrete structures bring to mind only girders and beams. However, other structures like flat slabs, skew slabs and cylindrical shell structures whose performances could be enhanced with prestressing are also dealt with.

As a complement to the many books on prestressed concrete I believe this book, written by an Indian professor who has taught students in Indian Institute of Technology Madras, will, besides being a text book on prestressed concrete, serve as very useful reference material for practising civil engineers in the Indian subcontinent. This is due to the fact that Prof. Rajagopalan understands the environment in which civil engineers in India and surrounding countries have to work including the cost of prestressed concrete vis-à-vis other alternatives. It is my firm belief that prestressed concrete has a place much more in our region than probably in the advanced world and, therefore, the best tribute to the author of this book is for more and more people to practise and apply prestressed concrete in the numerous opportunities India and the neighbouring countries will provide for economical application of the techniques of prestressing in the structural engineering practice.

A. Ramakrishna
President (Operations) & Deputy Managing Director
M/s. L&T-ECC, Chennai

Preface

The off quoted passages such as "Offence is the best form of Defence", "Prevention is better than Cure" are good preambles for prestressed concrete.

The man made material concrete is first thought as equivalent to stone. Unlike stone, concrete can be given any shape one wants because of the facility of casting. But the tensile capacity is found to be far too low in comparison to stone. Hence it requires to be strengthened by reinforcement. Even then it is susceptible for disintegration by cracking. Instead of **curing** the deficiency in tensile capacity by introduction of steel, pre-compression by prestressing is thought of as **preventive measure**. Similarly providing reinforcement is considered as **defence** against tension and that could be resisted by an opposite step of prestresing **as an offence measure** before the external loads are applied on it.

To facilitate the civil engineers with this tool of protection from failure, the subject is compulsorily introduced for the graduates and postgraduates of various Universities in India and abroad around 1960..

Bruggeling in his book on "Structural Concrete" has said that prestressed concrete is no new material but falls in the spectrum of concrete with a built-in preventive measure as prestressing. This concept was also expressed in the conference of CEB/FIP in 1970. It is now followed in FIB Codes and Euro Codes.

With introduction of limit state concepts as a realistic method for design, revised version of various Indian and International Codes for structures came into existence.

This book encompasses the principles of prestressing as well as its applications to structural concrete as envisaged by Bruggeling. Limit state concepts for failure and serviceability are also covered to go with the modern design trends. This book is intended for civil engineering students of graduate and post graduate level and also for practising engineers for understanding the basic concepts in formulation of code provisions and their applications.

Various Indian Standard Codes such as IS-1343; IS-456 and IRC etc., which are meant for structural design have been well covered in this book. International systems of units (SI) have been used throughout. But, for specific cases where some examples are worked out relating to other national code provisions, units related to those code provisions have been used.

The salient features of this book are as follows :

a) It gives a good follow-up of reinforced concrete in as much that it makes the student feel that prestressed concrete is not a different material and it is one end of the spectrum of concrete structures. Hence every chapter speaks about some aspects of reinforced concrete and ***how prestressed concrete improves the behaviour of reinforced concrete. structures*** (such as in Flexure, Shear, Torsion etc.,)

b) ***Enough worked out examples*** are given covering various types of problems that one would face, starting from analysis, design and execution.

c) References have been widely given.

d) ***The design provisions in relevant Indian Standard Codes (IS-1343, IS-456 and IRC etc.,)*** have been elaborated from fundamentals so that any reader (Student and also Practising Engineer) could understand the basis for all expressions in IS code and how they have been derived from the fundamentals.

e) The corresponding expressions of ***International codes like BS-8110, ACI-318, Euro code ENV 1992, DIN 4227, SIA Norms 162, FIP specifications etc.,*** are also given with basis for such expressions. This aspect will also be useful for practising engineers when they are required to design any structure based on other codes.

f) A few design problems (based on all codes) have also been worked out so that a student/ professionalist can understand how the quantities of material can only marginally vary if the philosophy is understood.

g) Certain aspects of ***detailing*** have also been given such as reinforcement ***at indirect supports, anchorage zones, continuity portion in structures etc.,*** Since detailing forms a part of reinforced concrete structure, in general, a full chapter has not been given and it is considered that it is beyond the scope of this book.

h) ***Prestressing of indeterminate structures*** have been dealt with in detail. Analysis and design of continuous beams and portal frames using fundamental approach have been explained with examples inclusive of choice of best profile for a given structure for given loadings. Efficient ways of ***prestressing of skew slabs*** are also covered. These will be very useful for Practising Engineers also.

The basic concepts of prestressing and how it is developed from reinforced concrete to remedy the defects are covered in the first chapter of Introduction. The materials used in prestressed concrete are covered in chapter 2. The various prestressing systems and devices are described in chapter 3 while analysis for flexure, concrete in prestressing, ultimate strength in flexure are discussed in chapters 4,5 & 6. The philosophy of design namely the limit state of collapse and limit state of serviceability as compared to the earlier approaches based on stresses at working loads are described in chapter 7. The chapter 8 discusses about the design for flexure. Chapter 9 deals with shear and torsional behaviour of prestressed concrete structures. Chapter 10 deals with the analysis and design of anchorage zone in prestressed concrete structures. Behaviour of composite structures is dealt in chapter 11. The concept of statically indeterminate prestressed structures both in the form of analysis and designs are given in chapter 12. The space structures like cylindrical pipes and tanks are briefly dealt in chapter 13.

With the open spectrum for structural concrete the smooth leading into reinforced concrete from prestressing can be had from the principles of partial prestressing. The principle, analysis and design concepts of partial prestressing are dealt with in chapter 14. The use of prestressing is not limited only to a particular type of structure such as beams and frames or a planar structure. Other types of structures like surface structure, folded plates etc., could well be prestressed and they are discussed in chapter 15, enthusing the reader for higher planes in the use of prestressed concrete.

Illustrative examples are given for each concept explained in various chapters. To facilitate the reader for greater depths in understanding if he desires so, references have been given widely.

The author believes that :

- No created work is perfect
- Perfection is an ambition and never achievable. An attempt towards the same is lot to be appreciated.

In general the author has attempted to present a reasonably good book for a student who has gone through a basic course in concrete structures and the fundamentals have been explained. The professionalists will find their use in this book to find clarifications on code provisions, be it Indian standards or other International standards and also some detailing aspects and choice of profiles for special type of structures.

The author will be pleased to receive constructive criticisms and suggestions, which will be immensely useful for incorporating in the text in further editions.

N. RAJAGOPALAN

Acknowledgements

The author has used the published information from Institutions, Associations and Journals in this book while reproducing certain data, code regulations etc. The Organisations and Institutions are American Concrete Institute, Australian Standards Association, American Water Works Association, Bureau of Indian Standards, British Standards Institution, European Concrete Committee, Cement and Concrete Association, German Standards, Institution of Civil Engineers, International Association for Bridge and Structural Engineering, Indian Concrete Journal and Structural Engineer, SIA norms, etc. The author gratefully acknowledges the same.

Author acknowledges the use of lecture notes of Prof. F. Leonhardt and Prof.R.Walther at University of Stuttgart when the author was working as a Research Scholar under DAAD scheme and also the information published as short-term course materials conducted by SERC and National Council for Building Materials.

This book is an outcome of the teaching material of the author while he was offering courses to Graduate students at the Indian Institute of Technology Madras, Chennai for number of years. The students and faculty colleagues have been a great source of strength and knowledge to the author and hence their contribution though not visible is gratefully acknowledged.

The author wants to re-call with a sense of gratitude the guidance of Prof. F.Leonhardt of University of Stuttgart and Prof. P.C. Varghese, former Head of the Department at the Indian Institute of Technology Madras, Chennai, who have guided the author to choose this area of specialisation for professional career and this has gone a long way in preparing this book.

The support received by the author from the project staff working under this guidance during his stay at the Indian Institute of Technology Madras, is gratefully acknowledged. He also acknowledges the financial support received from the Indian Institute of Technology Madras, Chennai (Centre for continuing education) for preparing camara ready copy of the manuscript. Special mention of thanks have to be made to Mr. U. Manohar, Project Associate, Ms. D. Savithri of M/s. Allied Publishers who helped in getting the manuscript of this book to proper shape.

Last but not the least the author records his thanks to Mr. A. Ramakrishna, President (Operations) & Deputy Managing Director of M/s. L&T-ECC, Chennai who has found this book worthy of writing the "Foreword". I am grateful to him for the same.

Notations

The list of symbols and notations given here are only for general guidance. In places where notations are independently given for any formula in the text, those notations govern.

A	Cross-sectional area of member
A_{br}	Bearing area
A_c	Area of concrete section
A_{eff}	Area of effective concrete section
A_s	Area of non-prestressed tension reinforcement
A'_s	Area of compression reinforcement
A_{sp}	Area of prestressing reinforcement
A_{sv}, A_{st}	Area of stirrup reinforcement
A_t	Area of transverse reinforcement or torsional reinforcement
A_{tl}	Area of longitudinal reinforcement for torsion
A_{tv}	Area of strip reinforcement for torsion
A_i	Area of the transformed concrete section
A_{pun}	Punching area
B	Bredth of beam
B_f	Bredth of flange of flanged section
D	Total depth of the beam
E	Modulus of elasticity in general
E_s	Modulus of elasticity untensioned steel
E_{sp}	Modulus elasticity of prestressing steel
E_c	Modulus of elasticity of concrete
F	Total tension transverse to the axis of beam
F_{bst}	Bursting tension
F_k	Characteristic load
G	Torsional moment of inertia
I	Second moment of area of section
I_e, I_{eff}	Effective second moment of area for computation of deflections
I_t	Second moment area of transformed section
I_r	Second moment of area of cracked concrete section
J	Polar moment of inertia of the section
L	Span
L_{eff}	Effective span
L_b	Bond length
L_t	Transmission length
M	Bending moment

M_{cr}	Cracking moment
M_g	Bending moment due to dead loads
M_q	Bending moment due to live loads
M_t	Total bending moment
N	Force normal to a section
N_s	Force along the interface
N_d	Design tensile load
P	Prestressing force
P_{inf}, P_∞	Prestressing force at time ∞
P_i	Initial prestressing force
P_0	Prestressing at the jacking end
P^0	Prestressing force at the jack
P_f	Prestressing force after friction loss
P_x	Prestressing force at a distance x from the jack
Q	Imposed load
Q_k	Characteristic concentrated imposed load
R	Radius of the curvature
S	First moment of area of concrete section
T	Torsional moment
T_u	Ultimate torsional moment capacity of section
V	Shear Force
V_c	Shear resistance provided by concrete
V_{cw}, V_{co}	Shear capacity of concrete in web shear crack zone
V_{cr}	Shear capacity of concrete in flexure shear crack zone
Z	Section modulus
Z_b	Section modulus of bottom fibre of beam
Z'_b	Section modulus of the bottom fibre of the composite section
Z_{in}	Section modulus at the interface
a	Deflection
a_w	Spacing between cracks
b_w	Breath of web in a beam
c	Cover to steel
d_f	Thickness of flange
d_e	Effective depth of beam
d_w	Depth of web
e	Eccentricity of prestressing shell
e_c	Eccentricity of centre of compression
f_c	Compressive stress in concrete
f_{ck}	Characteristic cube strength of concrete
f_{cy}	Characteristic cylinder strength of concrete

f_{ct}	Allowable compressive stress in concrete at initial transfer of prestress
f_{cp}	Compressive stress at the centroidal axis due to prestress after all losses
f_{cr}	Flexural tensile stress of concrete
f'_p	Permissible ultimate tensile stress in tendon
f_{pu}	Characteristic strength of prestressing tendons
f_y	Characteristic strength of reinforcement
g	Distributed dead load or acceleration due to gravity
h	Overall depth of section
h_{max}	The larger dimension of section
h_{min}	The smaller dimension of section
i	Radius of gyration
k	Friction coefficient for wave effect of cable profile
k_1	Constant
k_2	Constant
k_3	Constant
n	Neutral axis depth factor
q	Distributed live load
r_t, r_b	Ratio of gyration of section for top fibre and bottom fibre
s	Spacing of stirrups of links
t	Time
u	Perimeter or Radial force due to curved prestressed tendon
w	Crack width
x_1	Smaller dimension of transverse reinforcement
y	Vertical distance of a point from centroid of concrete section
z	Lever arm between the compressive force in concrete and the tensile steel force
α	Angle, ratio or dimensionless coefficient
β	Coefficient to represent wobble effects
γ	Partial safety factor for concrete, steel depending on suffixes
ε	Strain in concrete/steel depending on suffixes
η	Reduction factor for loss in prestress
θ	Rotation of the beam at supports
μ	Friction coefficient
τ	Bond stress/shear stress
ρ	Reinforcement ratio in terms of cross section
ϕ	Initial curvature
ψ	Dimensionless coefficient
Δ	Difference of increment
Φ	Diameter of reinforcing bar

Contents

1. Introduction

1.1 Development of Prestressed Concrete

Prestressing, itself means, an intentional application of a pre-determined force on a system, for resisting the internal stresses that may be developed in the system, due to external loads. It is meant for improving the behaviour and capacity of the system. If the system is expected to have tensile stresses due to external loads, pre-compression can be given to the system to a desired level even before the application of the external loads. Then the tension developed due to the external loads, will get largely nullified by the initially applied compression. This principle was used long back in our civilization. It is reflected in the design of barrels to contain liquid (Fig 1.1) or wheel of a bicycle or guyed ropes of mast or inflation of tyres of a vehicle.

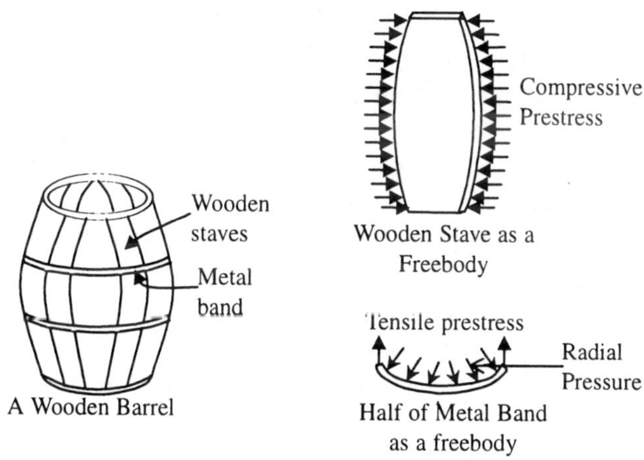

Fig.1.1. Principle of Prestressing

Metal bands wound around wooden staves of barrel are tightened to create precompression on the barrels. When liquid is stored in the barrels, tension is generated on the wooden staves and the already provided precompression nullifies this tension. Similar is the effect of spokes of the bicycle wheel, which has an initial tension to resist subsequent compression under loading. These are the cases of uniform prestressing of the material, in an element, to counteract the uniform stresses due to external loads. It will be of the same nature (say tension or compression) throughout the material. This philosophy of prestressing is extended later to materials, which are subjected to different types of stresses, namely tension or compression within the element itself due to external loads. Doehring of Germany first applied this concept of prestressing for concrete to take care of tensile stresses ***alone*** in concrete. Concrete which is weak in tension can be prestressed with initial compression in those zones where tensile stresses occur due to external loads. Doehring secured a patent (1888) of a concrete slab reinforced with steel with initial tension applied to it in the tension zone before the slab was loaded externally.

The other pioneers who used this concept are P.H. Jackson Sanfrancisco (1872) and C.R. Steiner (1908) of U.S.A. The application of precompression in concrete is based on the fact that concrete is weak in tension. In the zones where tension is caused by external loading, precompression is introduced. This precompression can be applied by an external system (say by jacks) unconnected with the structural element or internally from inside the element itself. But an external force system needs a reaction frame to take up the reaction. It is not always possible to set up a facility to take the reaction of the external prestressing system, which applies force on the structural element throughout its lifetime. **Hence prestressing by an external system unconnected with structural element** is not a possible preposition. By necessity, the prestress on concrete should be given from inside the structural element itself. This is achieved by using steel element with a fixed profile along the length of the element, concentric or eccentric to the axis of the element. Steel is pretensioned by stretching and then anchoring it to the concrete either by bond or by external anchorages. Because of the introduced elongation, steel tries to get back to its original length and presses the concrete introducing a concentrated force at the point of the contact. This is distributed through the full crosssection either by bond or by an anchor plate over a length of the element.

For this type of prestressing, mild steel was first used. It could be elongated only up to its permissible stress limit of 140 N/mm^2 namely upto a strain of .00067. This force in steel is transferred to large concrete area, which also gets compressed elastically. The elastic compression of concrete reduces the effective elongation in steel, and consequently reduces the effective stress on steel. Further, with time, creep and shrinkage strain in concrete, also reduces the available tensile strain on steel and thus the compression on concrete (Fig 1.2).

The permanent strain due to shrinkage and creep is normally of the order of .0006. Hence the left over effective tensile strain in mild steel (originally stressed upto 140 N/mm^2) is only of the order of .00007 and corresponding stress is only 14 N/ mm^2 which is 10% of the applied stress. So also the prestressing force transferred is only 10% of the original prestressing force,

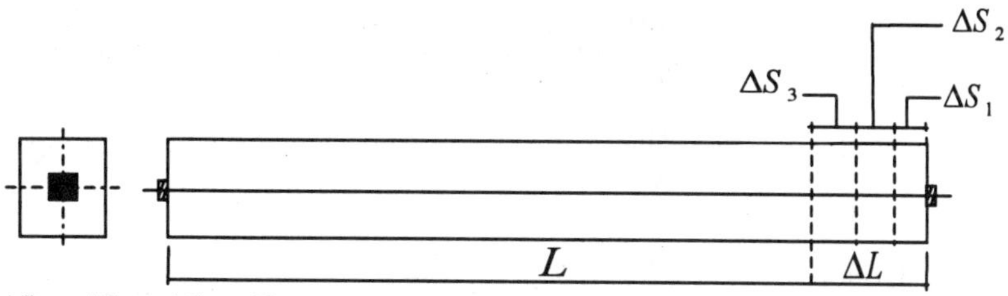

$\Delta S_1 + \Delta S_2 + \Delta S_3 = \Delta L$

$\Delta S_3 = \Delta L - \Delta S_1 - \Delta S_2$

For mild steel :

$\Delta S_3 = 0.0067\text{-}0.006$
$= 0.0007$

$\Delta S_3 / \Delta L = 0.1$ i.e 10%

ΔS_1 = Elastic deformation of concrete due to prestressing force transmitted by steel

ΔS_2 = Creep and shrinkage deformation of concrete at $t = \infty$

ΔS_3 = Left over elongation which results in P_∞

Fig. 1.2. Residual Prestress in Prestressed Elements

which is negligible. Even if the steel is stressed to a very high level nearing to its yield limit of 210 N/mm^2, the corresponding strain gets nullified by elastic, creep and shrinkage strains of concrete to a extent of 80%. Hence the prestressing effect on the concrete cannot be maintained for long. In 1908 Stainer in USA recognised the importance of loss in prestressing force due to creep and shrinkage and suggested retensioning the steel after creep and shrinkage strain in concrete had occurred. This led to practical difficulties. In 1923 F. Emperger [1] of Vienna developed a method for prestressing reinforced concrete pipes by winding high tensile steel wires on pipes at stress levels from 160 to 800 N/mm^2. In 1926 - 1928 Eugene Freyssinet [2] recognised the necessity of having the facility of large initial strain in steel, and correspondingly high tensile capacity of steel, so that the reduction in tensile strain in steel due to the elastic, creep and shrinkage strain in concrete would be only marginal, say upto 20%, giving a effective prestress of 80% of initial stress on concrete. This led to the development of High strength steel, also called as High tensile steel.

Prestressed concrete could also be looked at as an advancement of material ***concrete***, which passed through the stage of ***reinforced concrete*** before it was used in profession as prestressed concrete.

The advantages of concrete as a material are well known. It can be cast in a semi solid state in any particular shape desired and a monolithic jointless structure can be obtained. After hardening, it gets the properties of a stone. It is durable and involves very little maintenance cost. But its permissible strength properties per unit weight when compared with other structural material like steel and timber is not very good.

Steel : $\frac{140}{78.5} \times 10^6 = 1.8 \times 10^6$ N / mm². mm³ / N

Timber : $\frac{10}{9} \times 10^6 = 1.1 \times 10^6$ N / mm². mm³ / N

Concrete : $\frac{10}{23} \times 10^6 = 0.4 \times 10^6$ N / mm². mm³ / N

Though concrete is not as efficient as timber or steel as structural material as seen above, still it is sought after, as a construction material, because of its versatility of use in getting the desired form at the place of construction. But the strength of concrete in tension is very poor. To overcome this disadvantage of low tensile strength, steel was introduced as reinforcement. Thus ***reinforced concrete*** came into picture. The reinforcement can take care of the tensile stress and it acts monolithic with concrete to resist the forces like flexure and shear. Though Steel reinforcement is provided to resist tensile forces, still the concrete in the tension zone of the flexural member cannot be avoided. It is necessary to transfer shear forces and to provide continuity between compression and tension. It is also necessary to protect the steel from atmospheric action leading to corrosion. Thus the concrete is necessary in the tension zone. It has no structural use, but adds to the self-weight of the structural element. Moreover large strain in steel cannot also be permitted. That will develop wide cracks in concrete, which is not acceptable under serviceability conditions. An attempt to overcome these disadvantages led to the development of ***prestressed concrete***. Here the concrete is kept fully under compression and hence useful in resisting the structural effects. The evolution of achieving the prestress in concrete is explained earlier.

1.2 Advantages of Prestressed Concrete

By prestressing the concrete (introducing precompression) the structural element, which is subjected to tensile stresses due to external load, can be kept under compression during its service time. The tensile stresses generated by external loads is nullified by the precompression and there could also be a residual compression. The amount of precompression is accordingly decided. Hence the structure remains uncracked and the full cross-section is effective in transferring the load. The concrete in the tension zone, which only unnecessarily adds to the self-weight of the structure in reinforced elements, now becomes effective. By the choice of position of prestressing element namely steel, an initial camber opposite in nature to that caused by an external load can also be given to the structural element. Hence on application of external loads, the resultant deflection can be controlled to a large extent. This leads to a lighter and more slender structure and improves the performance of the structural element under service load conditions. Remember that **in prestressed concrete element, the entire load is resisted only by concrete at service load stage. Prestressing steel contributes very little in resisting the load unlike in reinforced concrete. Its main role is only to transfer prestress to concrete.**

Further prestressed concrete also functions more efficiently in resisting shear stresses because of precompression.

If precompression is introduced in concrete, the principal tensile stresses at any cross section become smaller. The shear carrying capacity accordingly is improved. Considering a reinforced/prestressed concrete beam subjected to flexure (Fig 1.3).

Principal stress at any point in the beam is calculated as

$$\sigma_{\mathrm{I}}, \sigma_{\mathrm{II}} = \frac{(\sigma_x + \sigma_y)}{2} \pm \sqrt{\frac{(\sigma_x + \sigma_y)}{4} + \tau_{xy}^{\ 2}}$$

In case of reinforced concrete beams, in tension zone, stresses acting on an element are $\sigma_x = 0$ since concrete cannot take any tensile stress and $\sigma_y = 0$ by assumption in the theory of bending.

On application of the above, principal stress σ_{I} can be calculated as equal to τ_{xy}. The same can be seen from Mohr's diagram also (Fig 1.3). Hence, in the reinforced concrete beam, when the shear stress at any point reaches the permissible tensile stress in concrete, the beam would fail. But, in case of prestressed concrete structures, in the tension zone $\sigma_x = \sigma_{x1}$ (a precompressive stress), $\sigma_y = 0$ and σ_{I} becomes less than τ_{xy} as per the above equation. Hence in case of prestressed concrete beams, even when shear stress (τ) reaches a value of the permissible tensile stress in concrete F_t, the principal tensile stresses will be still less than permissible tensile stress for concrete. Hence the beam ***will not*** fail. Accordingly the capacity of beam under shear becomes larger. This is due to the axial compression present in the beam. The shear stresses can further be reduced by draping of tendons, which introduces an upward force.

Bond stresses are important in Reinforced Concrete elements for transfer of forces from steel to concrete in an uncracked zone. Continuity is maintained between tension and compression zone, which is necessary to resist moments. In prestressed concrete bond stresses do not play such an important role in flexure zone since, concrete is uncracked throughout the section during service loads.

High strength steel can be used and higher strain in steel can be allowed since the strain results only in a compressive force in concrete. There is no problem of cracking in concrete and its related ill effects. Hence the advantages of using prestressed concrete elements can be summarised as follows:

- Concrete remains uncracked which makes the structure weather resistant, durable and it protects steel from corrosion.

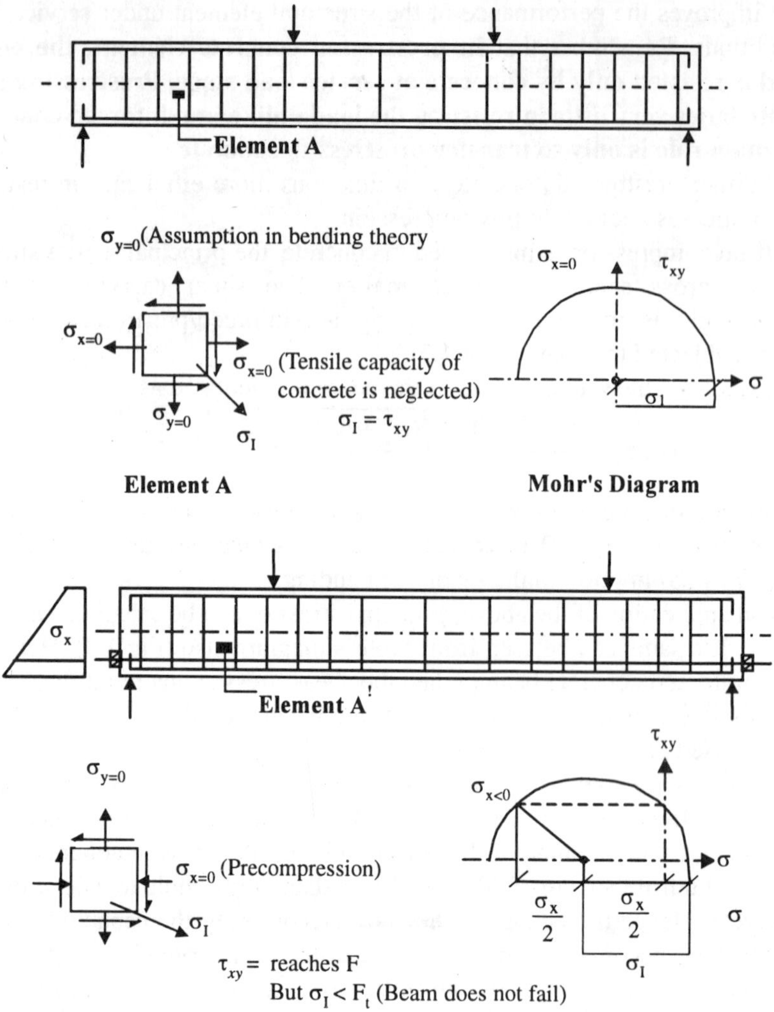

Fig. 1.3. Shear Resisting Capacity in Prestressed Concrete Beams

- Because of the uncracked conditions such structures can be used as liquid retaining structures more effectively.
- Because of uncracked conditions of concrete, the stiffness of the structure is more, which results in a reduced deflection or deformation.
- Because of full use of concrete in tension zone also, slender sections are achieved when designed with prestressing. This results in reduction of dead weight and thus results in economy.
- Because of precompression, shear-resisting capacity is also increased.

Economy, the structural efficiency, resilience for impact and dynamic loads, aesthetics and elegance etc have been achieved using prestressing techniques in concrete, world over in past few decades. Few examples are shown as Photographs in Annexure to this chapter.

1.3 Classification of Types of Prestressing

The prestressing of concrete can be achieved in number of ways and hence can be classified depending on the way in which precompression is transferred on concrete. Accordingly, the classification is identified in the following four groups [3]:

- External or internal prestressing
- Linear or circular prestressing
- Pretensioning or post tensioning
- Partial prestressing, limited prestressing or full prestressing

1.3.1 External or internal prestressing

This is based on the transfer of precompression on concrete from outside or from inside the structural element. External prestressing can be achieved by using systems like jacks for applying force from outside or by elongated steel elements kept ***outside*** the concrete element. Internal prestressing is done by elongated steel elements kept ***inside*** the concrete elements. This is achieved in two ways namely pretensioning or post tensioning explained later.

1.3.2 Linear or circular prestressing

In case of beams, slabs, columns, piles etc prestressing is applied in straight alignment and this is called linear prestressing. In cylindrical structures like silos, circular tanks, pipes etc, hoop tension is generated due to the applied loads. Hence pre hoop compression is be applied by circular prestressing. Here the prestressing elements are curved. It will be mostly following the shape of the structure. The prestressing steel elements can be continuously wound round the cylindrical structural elements with initial tension and anchored at ends after number of windings. This is also achieved by prestressing the steel elements in sectors of 120° and anchored at pilasters monolithically constructed with cylindrical structural elements.

1.3.3 Pretensioning or post-tensioning

Pretensioning is a method of prestressing in which the high tensile steel is stretched between two bulkheads in a prestressing bed or within a self-straining frame called a stress bench. First, moulds are placed in position between the bulkheads or within the self-straining frame. The high tensile steel is threaded through the moulds at the required levels with or without eccentricity to the elements to be cast. The steel is then stretched and anchored on the bulkheads or on the booms of the self-straining frames. Afterwards concrete is poured in the mould and allowed to set.

When the concrete is well set and has attained required strength, the high tensile steel is released from the bulkheads or the self-straining frames. The steel, not being able to get back to normal position because of the bond established between steel and concrete during setting, transfers the force on concrete by bond. No special extra anchorage facilities are provided. Here the force is transferred by bond between steel and concrete. For a small distance from the end of the element, called the ***transmission length***, the force is not felt throughout the cross-section. The force is felt in the full cross-section only after the transmission length. A typical pretensioning bed and a stress bench are shown in Fig 1.4. This facility can be established only in a factory or a laboratory.

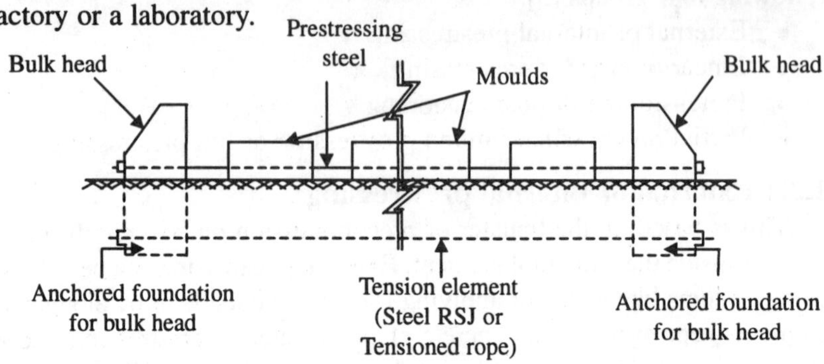

(a) Pretensioning Bed

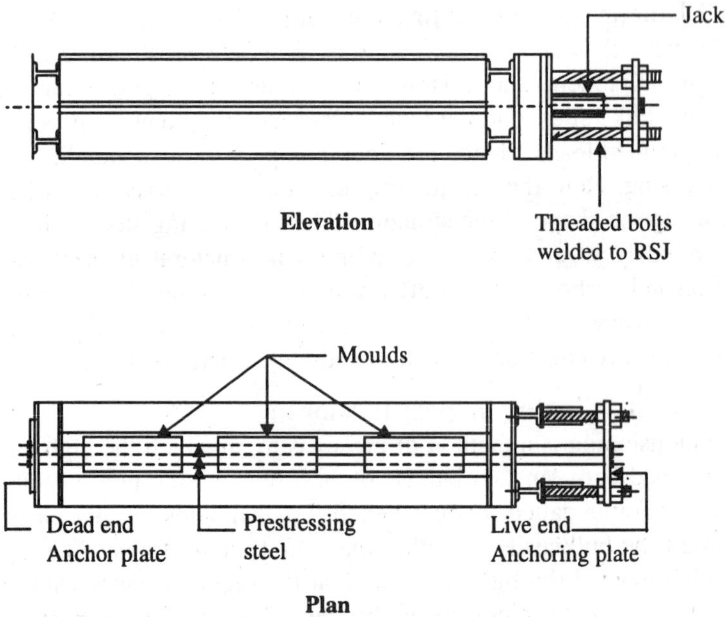

(b) Stress Bench - Self Straining Frame

Fig. 1.4 Pretensioning systems

The disadvantages in this system are

- High tensile cannot be given any profile under normal conditions
- The prestressing cannot be transferred immediately on to concrete

and one has to wait till the concrete acquires sufficient strength to take the precompression generated by the prestressing.

- The prestressing can be done only where there is a prestressing bed or a self-straining frame namely stress bench.
- The bond properties between concrete and steel should be good.
- The prestressing force is transferred gradually from the end and there is a transmission distance before the prestress is felt on the full cross-section of the structural elements.

In case post tensioning, structural elements are cast first with holes left for prestressing elements to run through. The holes are provided with metal sheathing ducts placed in proper position and they are called cable ducts. The prestressing steel elements are threaded through cable ducts later. After the concrete attains the required strength, the prestressing steel elements are pulled against the concrete element and anchored at the ends by means of special type of patented external anchorage systems. The force is transmitted through the external anchorage system at the ends of the structural concrete element. The cable ducts may be grouted with grouting material such as neat cement grout, mastic asphalt etc.

Grouting material such as neat cement grout cannot be removed later, whereas grouting material like mastic asphalt could be removed by increasing the temperature and pumping it out. In the second case there is a possibility of restretching the High Tensile steel elements afterwards.

In case of grouted prestressed elements, the bond between concrete and the element is generated and hence it is called bonded tendons. If grouting is not done, and the tendons are connected to the concrete elements only at the ends where it has anchorages, tendons are called un-bonded tendons. In case of bonded tendons the strain compatibility between concrete and the steel element at any cross-section has to be maintained, whereas in case of un-bonded tendons, strain compatibility between concrete and steel need not be maintained at any cross section. But the total elongation of the structural element should be compatible with the elongation of the pretensioned steel element at that level.

In case of post tensioning, any cable profile can be chosen for the prestressing steel. The force is transferred by external anchorages at ends. For a small distance from the anchored ends there is the effect of concentrated force flowing into the concrete. The zone within which the force gets uniformly distributed on to the structural element is called the anchorage zone.

Pre-tensioning method is preferred when

- the structural element to be prestressed is small and can easily be transported
- more numbers of similar prestressed structural elements are to be produced, such as prestressed concrete sleepers

- labour at site is costly
- number of external anchorages and the corresponding costs are high

Post tensioning method of prestressing is preferred when

- The structural element to be prestressed is very heavy
- The product is unique and not repeated many times
- Labour at site is not very costly
- Number of external anchorages and the corresponding costs are not high

1.3.4 Partial prestressing, limited prestressing or full prestressing

This type of classification was introduced in CEB/FIP recommendations in 1970 [4] depending on the level of prestress introduced in the structural element to nullify the stress that may develop during applications of external load. IS 1343–1980 [5] has also introduced this approach of classification of prestressing under "limit state of serviceability" and designated the structural elements as Type 1, Type 2, Type 3 structures. Type 1 refers to a fully prestressed concrete and Type 2 to limited prestressing of concrete where tensile stresses do not exceed the cracking stresses of concrete. Type 3 structures are those which allow tensile stresses to occur leading to limited crackwidths and that too under extraordinary loading conditions which may occur rarely on the structure. The cracks may get closed as soon as that extraordinary loading moves out from the structure. The permissible tensile stresses for such conditions are related to permissible crackwidths. Once the section is cracked, the tensile stresses cannot be calculated using the section modulus of the uncracked section. Hence the tensile stresses calculated using the section modulus of uncracked section are called hypothetical tensile stresses. IS 1343–1980 [5] recommends limits on such hypothetical tensile stresses for limited crack widths. The limitations on crackwidths are based on exposure conditions of the structural element. These values are given in the next chapter on "Materials" for use of the readers.

The aim of prestressing the concrete is to get the desired stress conditions in concrete when external loads are applied and the structural element will be the same stressed conditions during large portion of time in its life. The external loads coming on the structure could be of different levels of magnitude say 50 kN to 1000 kN. The frequency of occurrence of heavier loads may be very small. But the frequency of regularly passing loads (say 50 kN to 500 kN) may be large. In such a case, it is not necessary to prestress the structure for the rarely occurring load of 1000 kN and the prestressing force can be designed for regularly occurring loads say upto 500 kN. Under extraordinarily (heavy) loads (say 1000 kN) the structure may be subjected to tension and cracking. These occasions are far fetched. The cracks also close down as soon as such loads leave the element. In such cases the structures are partially prestressed so that the structural element is under desired levels of stress during major portion of service time. Hence the classifications are based on the level of prestress imparted to the structural elements. CEB/FIP (1970) [4] has classified them as Class I (fully prestressed), Class II (limited prestressed) and Class III (partially prestressed) and Class IV (Reinforced Concrete structures). In Class III type, the level upto which precompression is done can be achieved (a) by prestressing the High tensile steel upto desired level (much below the permitted level of stress for prestress for such steel) and allowing the same steel to act of reinforcement for taking care of extra tensile force that may occur due to occasional large loads or (b) and by prestressing the high tensile steel near to permitted level of stressing and providing untensioned steel to take care of the extra tensile force that may occur due to occasionally occurring large loads. The first is called partially prestressed structures and the second is called prestressed reinforced concrete structure [6]. These are also explained later on chapter on partial prestressing.

Class IV type namely Reinforced Concrete structural elements do not have any precompression or prestress in concrete before the external loads are applied.

1.4 Tendons and Cables

The prestressing steel in a prestressed concrete element is normally referred to as a tendon. A tendon may consist of one or more individual wires, one or more number of rods, number of wires 2, 3 or 7 twisted together to form an individual strand and number of strands grouped together. Depending on the force to be transferred and depending on the facilities/necessities, stretching the prestressing elements individually or in a group is identified. The tendon may be straight, inclined or curved. The profile of the tendon is based on the design requirements so that the prestressed force transmitted to concrete as compression effectively resist or counteract the super imposed design forces due to external loads. The performance characteristics of the tendons should satisfy the requirements laid down in relevant statutory norms. Bureau of Indian Standards has laid down norms for various types of tendons (prestressing elements) and the same has been discussed in Chapter 2.

The number of tendons grouped together is called a cable. The tendons grouped together in a cable will be anchored with one external anchor. The external anchorage can be pulled with all the tendons together by jack and this is called bulk pulling. Individual tendons in the cable can also be pulled and anchored. The tendon is accommodated in a cable duct. The duct has a sheathing of a minimum thickness governed by relevant codes. The duct is also to be tested for containing the cables and grouting materials. The tests are stipulated under various statutory organisations like Indian Roads Congress [7].

The cables will be spaced with minimum spacing as specified in various construction codes like Indian Roads Congress [7], Euro Code ENV 1992–Part 2 [8] etc. Sometimes they are also grouped together though grouping of cable may as far as possible be avoided.

The cables will be grouted permanently with cement grout or removable grouts like mastic asphalt or some organic materials. Grout materials shall also satisfy specifications laid down by statutory organisations. The methods of grouting should also be known earlier at the design stage itself and should be incorporated in the construction procedure manual.

1.5 Concepts of Prestressing

Prestressing the concrete is meant to transfer precompression (compressive stresses) on the concrete. When a prestressed structural member is subjected to tensile stresses due to external loads, the initial precompression (compressive stresses) is nullified and the structural member is under the desired state of stress conditions as explained earlier. This way of considering prestressing force transmitted to concrete as initial internal stress to counteract the internal stress developed due to external loads is called ***stress concept***. This concept is propagated by Eugene Freyssinet [2]. The effect of external loads is also converted as stresses on the material.

The combination of the effect of external loads and of prestressing are studied together as equivalent stresses and compared with the permissible levels of stresses in the material. Hence it is called the ***stress concept.***

1.5.1 Stress concept

In this approach, in a given structural element the stresses due to the self-weight and the external loads across any cross-section are first calculated. Undesirable stresses, which the concrete cannot take, namely tensile stresses have to be counteracted by introducing compressive stresses. The quantum of prestressing force P and the location of such prestressing force with respect to the centroidal axis of the element (defined as eccentricity e) and the shaping of the prestressing element along the length of the element, called profiling within the concrete cross-section are then decided. The eccentricity can be zero or can be up to a limit, based on the facility available in the cross-section. The stress generated by the prestressing force across the depth of any cross-section is calculated as

$$f_{cb_p}, f_{ct_p} = \frac{P}{A} \pm \frac{P_e}{I} \times y$$

where P is the prestressing force, A the area of cros-section, I the moment of inertia cros-section and y is the distance of the fibre under consideration from the centroidal axis of the structural element.

The stress so developed along with the stress due to dead load should be within the permissible limits. This stage of checking is called stage 1. The prestressing never acts alone. The minute prestressing starts acting on the structure, due to accidental eccentricity a moment is generated which moves the element from the contact of surface. This immediately brings in the effect of the dead load and hence prestressing and dead load are always taken together at stage I.

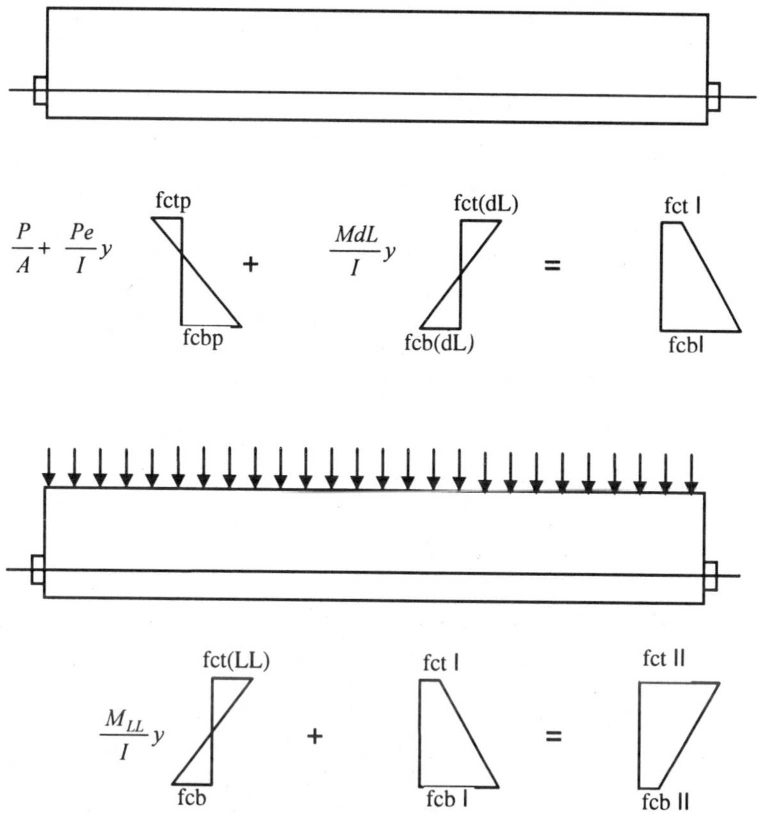

fct I, II = Stress in concrete at top fibre in stage I, II
fcb I, II = Stress in concrete at bottom fibre in stage I, II

Fig. 1.5 Stress Concept Approach

At the service load stage, the stresses caused by prestressing, dead or live load and other external loads are calculated altogether through the depth of the cross-section. The stresses should be within the permissible limits. This is considered as stage 2. In the theory of bending, it is assumed that plane sections remain plane before and after the moments are applied. Hence the top and the bottom fibre of the structural element are subjected to maximum stresses. They are considered as critical levels for the checking of stresses both in stages 1 and 2. Fig. 1.5 gives the details of the stress method of approach and its application to the analysis and design of prestressed concrete elements.

It is well known that deformation is a reality and stresses generated are only conjecture. It is based on the stress strain relationship of the material. Hence the stress concept is also based on the knowledge on the stress strain relationship of the materials.

Cracking is a deformation phenomenon. In the stress concept cracking it is considered as exceeding the limit of flexure tensile stress. The permissible stresses under compression and tension in the materials concrete and steel do have a major role to play in analysis and design of prestressed concrete structure based on ***stress concept.***

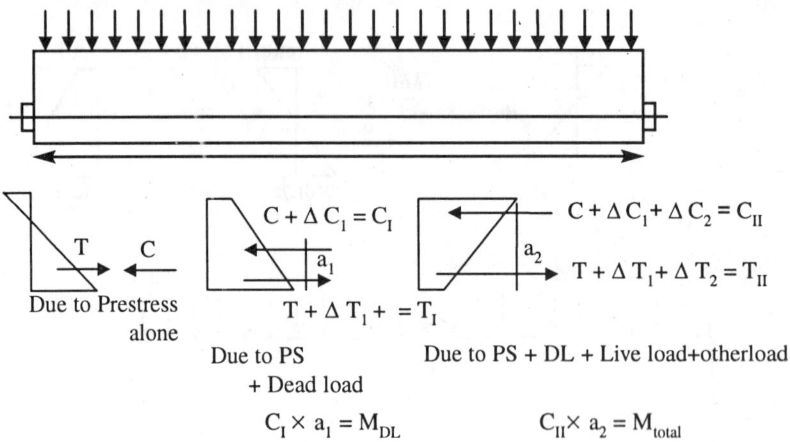

Fig. 1.6. Force Concept

1.5.2 Force concept

There is another concept called Force concept. In this approach the structural element is considered as if it is a reinforced concrete element. The total prestressing force given is taken as tensile force acting on any particular cross section and the effect of the prestressing force which was described as stress in the material concrete in the stress concept now shall be described as total compressive force acting on the element as the resultant of all the stresses. Hence the structural element at any cross-section will be subjected to the tensile force in the prestressing element and a compressive force in the concrete which is the resultant force of all the compressive stresses acting on that cross section. If, only prestressing force is considered, both the tensile force and the compression force will be collinear to keep the element in equilibrium. If any additional load (say even the dead load) is considered the tensile force in prestress element will be modified a little and the center of compression will also be shifted. For a usual case of positive sagging bending moment applied on the structure due to external loads the tensile force in the prestressed steel element is marginally increased and the compressive force which is the resultant of the compressive stresses caused by the prestressing and by the loading will be shifted upwards from the line of action of tension. The tensile force or the compressive force multiplied by the shift between these two forces will be the external moment. This concept is made use of by Lyn [9] for the design of structures to arrive at the cross section and also to arrive at the moment resisting capacity on the cross section at any point of time inclusive of the time at failure. Since the capacity of the section is decided based on the total tension and compression it carries, this approach is called the ***force concept*** [9].

1.5.3 Load balancing concept

The concept of prestressing can also be explained from fundamentals as introduction of opposite type of loads in structural element opposite in nature to the external loads. If the external loads cause a sagging curvature in the beam element, any load which introduces the hogging curvature on to the beam, equal and opposite in nature to that caused by external loads, is also called prestressing and this method of the prestressing or preload is recognised as ***load balancing concept*** [9]. In this approach the external loads are treated only as loads and not converted as stress on the structure. Prestressing is also converted as equivalent load and this equivalent load must counteract or balance the external loads. The combination of loads caused by external system and the prestressing system should keep the structure in a desirable state condition.

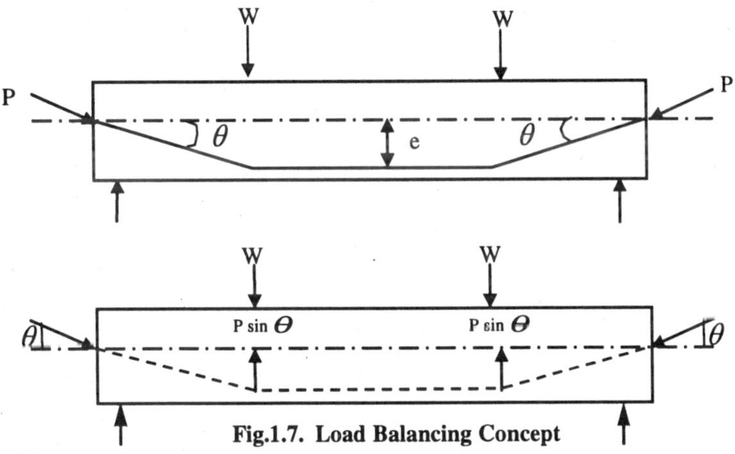

Fig.1.7. Load Balancing Concept

For example the parabolic profile of the prestressing cable with the prestressing force *P*. (Fig 1.7) can be considered equivalent to the upward force of $P.\left(\frac{8\times e}{l^2}\right)$,where *e* is the dip of the parabolic profilefrom the line joining the end points. If the external load acting on the structural element is *W* the choice of *P* and *e* can be decided, so as to counteract this external load *W*. Similarly, if there is a straight profile with a kink at specific points that cable prestressed with the force P, at the kink point there will be a upward force equal to $2P \sin \theta$, where θ is the change of angle of the profile at the point of kink. If the structural element is subjected to the concentrated load of W at any point, prestressing can be so designed with two straight profiles and a kink at the point of application of the load so that $2P \sin \theta$ will cause an equivalent upward load. Similarly, any kind of external loading on a structural element can be counteracted by a prestressing force and a predetermined cable profile. This approach is very useful in solving indeterminate prestressed concrete structure, specially continuous beams.

While the stress concept is widely used for checking the designs and the permissibility of stress at various stages of construction, during service load, just the other two concepts are used for specific conditions. The Force concept is used for designing the specific dimensions of the structure and for checking the failure capacity of the section. The load-balancing concept is used for analysis of indeterminate prestressed concrete structures, and complicated analysis where the effect of prestressing cannot be easily depicted.

Example 1.1

A simply supported prestressed concrete beam of cross-section 400 mm × 600 mm covers a span of 10 m. It is subjected to an uniformly distributed load of 30 kN /m in addition to its self-weight and is prestressed with a force of 1740 kN with a prestressing cable of parabolic profile. The cable is anchored at the center of gravity of the cross section at support section and has a dip of 160 mm at the middle cross section. Analyze the beam for the effects of prestressing and the loads at mid cross section, using the philosophy of ***stress concept.***

Solution

Span of the beam	= 10 m
Cross-section	= 400 × 600 mm (rectangular)
External Load	= 30 kN/m
Unit weight of Concrete	= 24 kN/m^3
Prestressing Force	= 1740 kN
Cable Profile	= Parabolic
Eccentricity at mid cross-section	= 160 mm
Eccentricity at support section	= 0 mm

Properties of the section

Area of Cross -section $A = 0.4 \times 0.6 = 0.24 \text{ m}^2$

$$\text{Moment of Inertia } = I = \frac{0.4 \times 0.6^3}{12} = 7.2 \times 10^{-3} \text{ m}^4$$

$$\text{Modulus of section } Z_t = Z_b = \frac{7.2 \times 10^{-3}}{0.3} = 0.024 \text{ m}^3$$

Self weight of the beam $w/m = 0.24 \times 24 = 5.76$ kN/m

Stress Concept

Bending moment at mid span section due to

a) Self weight = $5.76 \times \frac{10^2}{8}$ = 72 kN m (bottom tension)

b) External loads = $30 \times \frac{10^2}{8}$ = 375 kN m (bottom tension)

Prestressing force = P = 1740 kN

Eccentricity at mid span section = e = 160 mm

c) moment due to prestressing force = 1740 × 160

= 278.4 kN m (top tension)

All causes and effects are converted as stresses in ***stress concept*** for evaluation.

The two salient levels where stresses are to be calculated are at top fibre and at bottom fibre.

Stresses at top fibre due to

a) Prestress $= \frac{P}{A} - \frac{P \times e}{I} \times y = \left[\frac{1740}{0.24} - \frac{1740 \times 0.16 \times 0.30}{7.2 \times 10^{-3}}\right]$

$= 7250 - 11600 = -4350$ kN/m^2

$= -4.35$ N/mm^2 (tension)

b) Dead load $= M._{D.L.} \times \frac{y}{1} = \frac{72 \times 0.3}{7.2 \times 10^{-3}}$

$= 3000$ kN/m^2 $= 3.0$ N / mm^2 (compression)

c) External loads $= M._{L.L} \times \frac{y}{1} = \frac{375 \times 0.30}{.2 \times 10^{-3}}$

$= 15625$ kN/m^2 $= 15.625$ N / mm^2 (compression)

Stresses at bottom fibre due to

a) Prestress $= \frac{P}{A} - \frac{P \times e}{I} \times y = 7250 + 11600$

$= 18850$ kN / m^2 $= 18.85$ N/ mm^2 (compression)

b) Dead Load $= M._{D.L} \times \frac{y}{I} = -3.0 N / mm^2$

$= 3.0$ N / mm^2 (tension)

c) Live Load $= M._{D.L.} \times \frac{y}{I} = 15.625$ N / mm^2 (tension)

Stresses at stage I (Prestress + Dead Load)

Top fibre level $-4.35 + 3.0 = -1.35$ N/ mm^2 (tension)

Bottom fibre level $+ 18.85 - 3.0 = 15.85$ N/ mm^2 (compression)

Stresses at Stage II (Prestress + Dead Load + Other Loads)

Top fibre level $-14.35 + 3.0 + 15.625 = 14.275$ N / mm^2 (compression)

Bottom fibre level $+18.85 - 3.0 - 15.625 = 0.225$ N/ mm^2 (tension)

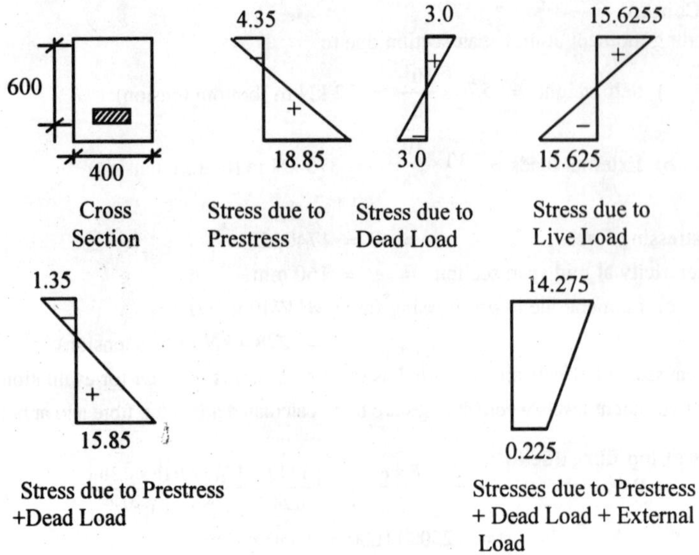

Example 1.2

Analyze the beam described in Example 1.1 using *force concept*.

Solution

Span of the beam = 10 m

Cross Section = 400 × 600 mm (rectangular)

External Load = 30 kN/m

Unit weight of Concrete = 24 kN/m^3

Prestressing Force = 1740 kN

Cable Profile = Parabolic

Eccentricity at mid cross-section = 160 mm

Eccentricity at support section = 0 mm

Properties of the section

Area of cross-section $A = 0.4 \times 0.6 = 0.24$ m^2

$$\text{Moment of Inertia } I = \frac{0.4 \times 0.6^3}{12} = 7.2 \times 10^{-3} \text{ m}^4$$

$$\text{Modulus of section } Z_t = Z_b = \frac{7.2 \times 10^{-3}}{0.3} = 0.024 \text{ m}^3$$

Self weight of the beam w/m = 0.24 × 24
= 5.76 kN/m

Force concept

In *force concept* all causes and effects are considered as forces for evaluation.

Bending moment at mid span due to

a) Self weight = 72 kN m (bottom tension)
b) External loads = 375 kN m (bottom tension)

Total Bending moment = 447 kN m

Prestressing Force = 1740 kN

When the prestressing force of 1740 kN (tensile) in the cables alone is acting, the stresses generated in concrete will lead to a resultant compression of equal value (1740 kN) and the compression also acts at the same level of prestressing force. The forces are collinear.

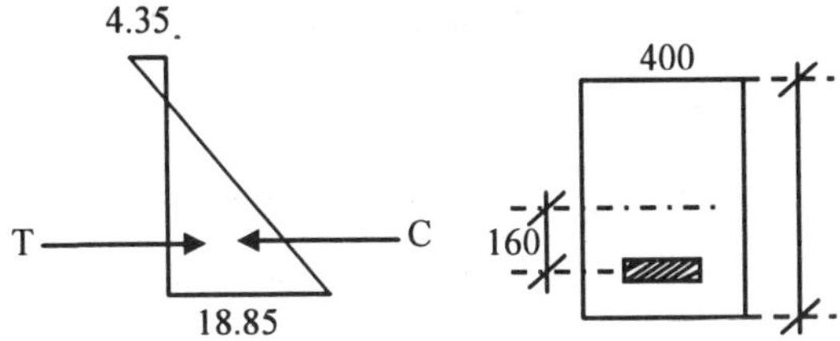

Stage I

When self-weight starts acting (which is immediately after prestress) there will be a small increase in the tensile force in the cable. But this is neglected. This aspect is discussed later in the chapter on analysis.

Hence the total tensile force in the cable = 1740 kN

Total compressive force = 1740 kN (for keeping the section in equilibrium)

But the resultant compression will act at a different level, so that the compression and the Tension will form a couple to resist the dead load bending moment.

Dead load bending moment= 72 kN m

Distance between tensile force (cable position) and the center of resultant compression = $\dfrac{72}{1740}$ = 0.04138 m

Position of center of compression from the center of gravity of cross-section = 0.16 – 0.04138 = 0.11862 m = 118.6 mm

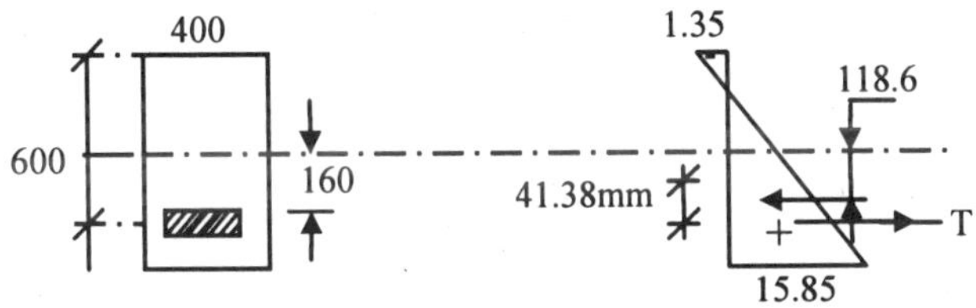

Due to prestress + DL

The resultant compression will act at 118.6 mm from center of gravity only for a given stress distribution.

The stress distribution can be evaluated as detailed below

$$\text{Stress at top} = \frac{1700}{0.24} - \frac{1740 \times 118.62 \times 10^{-2}}{0.024} = 7250 - 8599.95$$

$$= -\ 1349.95\ \text{kN/ m}^2 = -\ 1.35\ \text{N / mm}^2\ \text{(tension)}$$

$$\text{Stress at bottom} = 7250 + 8599.95 = 15849.95\ \text{kN/ m}^2$$

$$= 15.849\ \text{N/ mm}^2 = 15.85\ \text{N / mm}^2\ \text{(compression)}$$

Stage II

When the external load also starts acting the resultant (final) bending moment shall be resisted by the couple of total compression and total tension with a lever arm.

The total tension will slightly increase in the cable under uncracked conditions the increase in tensile stress will be small. That multiplied by the area of steel (which is also small compared to total cross sectional area of the section) will lead to very small increase in force, which is neglected.

Hence total tension = 1740 kN

Total moment to be resisted = Dead load bending moment + bending moment due to other loads

$$= 72 + 375 = 447 \text{ kN m}$$

$$\text{Lever arm required} = \frac{447\text{kNm}}{1740 \text{ kN}} = 0.2569 \text{ m} = 256.9 \text{ mm}$$

Resultant center of compression will be located at 256.9 mm from the center of tension i.e. cable position.

Hence position of center of compression from center of gravity of cross - section = 256.9 – 160

= 96.9 mm (upward)

= (–) 0.0969 m

This resultant compression (1740 kN) will act at 96.9 mm above center of gravity of section only for one particular stress distribution across the section.

That stress distribution is evaluated as follows

$$\text{Stress at top} = \frac{1740}{0.24} + \frac{1740 \times 0.0969}{0.024} = 14275 \text{ kN / m}^2$$

$$= 14.275 \text{ N / mm}^2 \text{ (compression)}$$

$$\text{Stress at bottom} = \frac{1740}{0.24} - \frac{1740 \times 0.0969}{0.024} = 225 \text{ kN / m}^2$$

$$= 0.225 \text{ N / mm}^2 \text{ (tension)}$$

The stresses are the same as got in the **stress concept**.

Example 1.3

Analyse the prestressed concrete beam described in Example 1.1 using ***Load balancing concept.***

Solution

Span of the beam = 10 m

Cross section = 400 × 600 mm (rectangular)

External load = 30 kN/m

Unit weight of concrete = 24 kN/m ^{3}

Prestressing force = 1740 kN

Cable profile = Parabolic

Eccentricity at mid cross-section = 160 mm

Eccentricity at support section = 0 mm

Properties of the section

Area of Cross-section $A = 0.4 \times 0.6 = 0.24 \text{ m}^2$

$$\text{Moment of Inertia} \quad I = \frac{0.4 \times 0.6^3}{12} = 7.2 \times 10^{-3} \text{ m}^4$$

$$\text{Modulus of section} \quad Z_t = Z_b = \frac{7.2 \times 10^{-}}{0.3} = 0.024 \text{ m}^3$$

Self weight of the beam $w/m = 0.24 \times 24 = 5.76$ kN/m

Load balancing concept

In this concept all the causes and the effects will be considered as loads and the member will be analyzed.

Self-weight and external loads put together will be acting downwards. Hence the total downward load acting on the beam = 30 + 5.76 = 35.76 kN / m

Prestressing force is applied along a parabolic profile with maximum dip of 160 mm at center section.

The equivalent upward uniformly distributed load provided by prestress

$$= \frac{8 \times P \times e}{l^2} = \frac{8 \times 1700 \times 0.16}{10^2} = 22.272 kN / m$$

The net downward on the beam = 35.76 − 22.272 = 13.488 kN / m

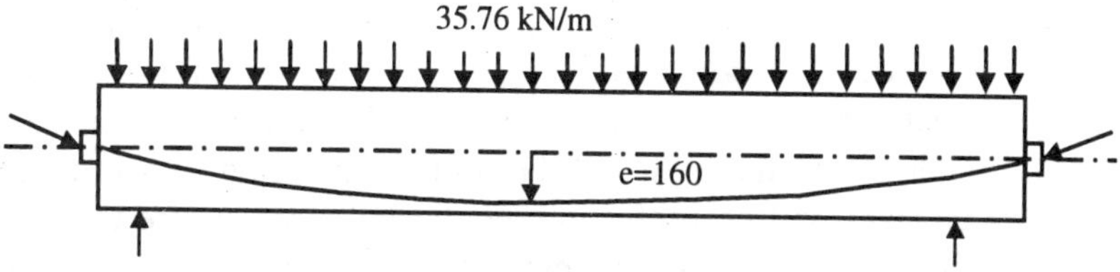

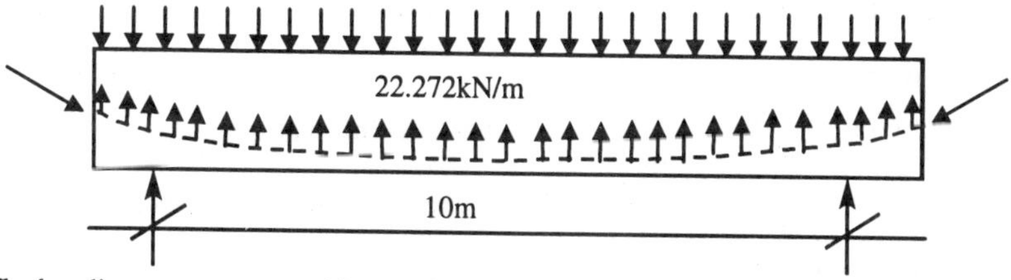

The bending moment caused by resultant downward force at center section

$$= 13.488 \times \frac{10^2}{8} = 168.6 \text{ kN / m}$$

This would cause stress on concrete section (mid section)

Stress at top fibre $= \dfrac{168.6 \times 0.3}{7.2 \times 10^{-3}}$

at bottom fibre = 7025.0 kN / m^2

= 7.025 N / mm^2 (tension at bottom compression at top)

At the anchorage end prestressing force is acting at the centroid of the section.

Stresses caused by that force at mid section = $\dfrac{P}{A}$ = 7250 kN/ m^2

= 7.25 N/ mm^2 (compression)

Hence net stresses :

at top fibre = 7.25 + 7.025 = 14.275N/ mm^2 (compression)

at bottom fibre = 7.25 - 7.025 = 0.225 /m^2 (compression)

The stresses are the same as obtained in the ***stress concept*** as well as ***force concept*** at Stage II.

These examples show that the prestressed concrete beam can be analyzed by any one of the methods. But certain methods are convenient for certain conditions of analysis.

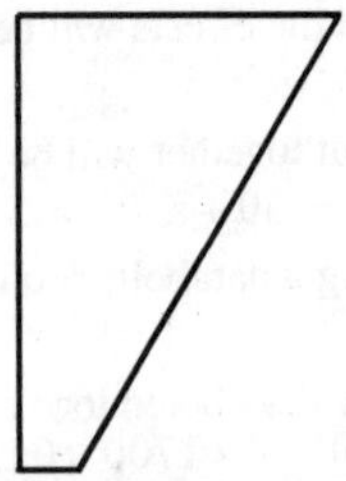

Net Stresses

As mentioned earlier, stress method is preferred to check the adequacy of section during service loads. Force method will be useful for arriving at the cross-sectional area of concrete and also steel required to resist a given loading conditions. That is, it will be useful for design of section. Load balancing concept will be useful in complicated cases where the effect of prestress will be very difficult to be evaluated. At such places the prestressed force will be converted as equivalent loads.

This will be better understood as we or one in the preceding chapter.

References

1. EMPERGER, F. (1923), "**Concrete tubes for internal pressure**", **Concrete and Constructional Engineering,** Vol.18.
2. FREYSSINET, E. (1926), "**Une revolution dans les techniques du beton**". Librairie de l' Enseignement Technique, Editeur Leon Eyrolles, Paris.
3. LEONHARDT F. (1964), **Prestressed Concrete Design and Construction**, Wilhelm Ernst and Sohn, Berlin,.
4. C.E.B - F.I.P. (1970), **Internationale Richtlinien zur Beruchnung und Ausfuh rung von Betonbauwerken. Prinzipien und Richtlinien**.2., Auflage, 6 congress der F.I.P., Prag.
5. IS 1343 (1980), Indian Standard **"Code of Practice of Prestressed Concrete",** Sixth reprint, Bureau of Indian Standards, New Delhi.
6. WALTHER, R. BHAL. N.S. (1973), **"Teilweise Vorspannung"**, Deutscher Ausschuss Fuer Stahlbeton HEFT 223, Wilhelm Ernst & Sohn, Berlin.
7. IRC 18 (1983), **"Design Criteria for Prestressed Concrete Road Bridges (Post tensioned concrete)"**, Indian Roads Congress, New Delhi.
8. ENV 1992-1,2 (1991), **Europen Standard for. Design of Concrete Structures Part I - General rules and rules for Buildings.** European committee for Standardisation - B - 1050 Brussels.
9. LIN, T.Y. (1963), **"Design of Prestressed Concrete Structures",** John Wiley & Sons, Inc. New York.

Seletar Sewage Treatment Digesters – Singapore : Design and installation of horizontal post-tensioning for six egg-shaped digesters

Jess H. Jones Memorial Bridge, Houston – Texas/USA

The Millennium dome under construction on the Green-wich Peninsula, London

Grand Central Tower – South Africa : Post-tensioning of this 40-m high unusal cone-shaped structure

Black Mountain Tower, Canberra. Special Support Bearings in Stainless Steel and Hercuflon

2. Materials

2.1 Introduction

In prestressed concrete works, the main materials used are high strength concrete and high strength steel. In the present day context one should aim at high performance concrete compatible with high strength.

High strength high performance concrete is necessary for prestressed concrete works for many reasons. As mentioned in the earlier chapter, only high strength steel can be used for prestressing concrete because the losses in prestress in steel due to creep and shrinkage of concrete can be kept low compared to the applied stress level. For full utilization of the higher capacity of steel, high strength of concrete is necessary. Further, nowadays heavy prestressing forces are transferred as concentrated forces in concrete by using special type of anchorage systems. These anchorage systems are designed based on the use of higher strengths of concrete. Weaker concrete would lead to failures specially at anchorages due to shear and bond and is liable to more shrinkage leading to greater loss in stress in the pretensioned steel. These aspects will be dealt with in greater detail later in this book. With varying environmental conditions, deterioration of the cementing properties of concrete is imminent. It is reflected in the carbonization of concrete and the consequent development of cracks leading to corrosion of reinforcement, thus affecting the performance level of the structural elements. A well-designed high performance concrete would have minimum shrinkage and creep and high durability. Thus would protect against the possibility of early deterioration. From the details discussed, it can be clearly seen that high strength steel and high strength high performance concrete are necessary for prestressed concrete works.

For achieving high strength high performance concrete, the ingredients like cement, coarse aggregate, fine aggregate, water and admixtures (mineral or chemical) have to be properly identified. These individual ingredients have to satisfy the minimum strength and performance requirements specified in relevant codes. The concrete mix should be designed using an approved procedure. Acceptance tests for concrete have to be followed and the quality of concrete should conform to the requirements specified in relevant statutory bodies for general design and construction [viz IS 1343 - 1980 (1), ACI 318 - 1998 (2), BS 8110 - 1985 (3), Euro Code ENV 1992 (4), FIP 1984 (5)]. For special structures like Bridges, Chimneys etc. there are further statutory guidelines to define the requirements of materials to be used in concrete, which are also to be followed.

Regarding the steel to be used in prestressed concrete works, it is a factory made product. As such it is expected to fully meet the codal specifications regarding the requirements of strength and other properties. The various aspects relating to the use of concrete and steel in Prestressed Concrete works inclusive of codal provisions are discussed in greater length in the next sections.

2.2 Concrete

2.2.1. Ingredients of concrete

The main ingredients of concrete are cement, coarse aggregate, fine aggregate, water and/or admixtures, both mineral as well as chemical. Pozzolonic materials like fly ash, blast furnace slag, silica fumes etc are called as ***mineral admixtures*** whereas the admixtures which are used to increase the slump, workability for a given water cement ratio etc, are identified as ***chemical admixtures***. The chemical admixtures should be compatible with the cement used. Cement and pozzolanic materials called mineral admixtures are nowadays grouped as ***cementitious materials*** and they are referred so in the updated version of National and International Codes. The following types of cement can be used for prestressed concrete construction.

- Ordinary Portland cement
- Portland slag cement but with not more than 50 percent slag content
- Rapid-hardening Portland cement
- High strength Ordinary Portland cement

These cements should also confirm to the approved the standards of various countries. IS 269, IS 455, IS 8041 and IS 8112 detail out the specifications for the above cements in India.

Under special circumstances other cements like super sulphated cement confirming to IS 6906 - 1973 can also be used with the approval of engineers in charge. IS 10262-1983 (6) gives some guide lines on concrete mix design.

All aggregates got out of igneous (granite, basalt) and metamorphic (gneics), formations may be used as coarse aggregates. Sediments of the above as finer particles (sand) may be used as fine aggregates, but they should satisfy the requirements laid down in statutory provisions like IS 383 confirmed by specific tests. It is preferable that the maximum size of the aggregates is restricted to 10 mm less than the minimum clear spacing between individual reinforcements or individual cable ducts or 10 mm less than the minimum cover for the reinforcements whichever is smaller [7].

A nominal size of 20 mm coarse aggregates may mostly satisfy these requirements in prestressed concrete works. Coarse aggregates shall be proportionately gap graded for proper adhesion of material. The fine aggregates shall be free from dust, lumps, soft flaky materials, mica and other deleterious material and their presence, if at all, should be restricted so that they do not reduce the strength and durability of concrete or attach the embedded steel. Water used in preparation of concrete should be potable and the permissible limits for certain ingredients in water which could be used as guidelines, is given in Table 2.1.

Table 2.1 Permissible Limit for Solids in Water [7]

Organic solids	200 mg/*l*
Inorganic solid	3000 mg/*l*
Sulphates (SO_4)	400 mg/*l*
Chlorides (as Cl)	2000 mg/*l* for plain concrete work and 500 mg/*l* for reinforced concrete work
Suspended matter	2000 mg/*l*

The pH value shall generally the not less than 6. Whenever necessary tests should be done as per IS: 3025/1964. Mixing and curing with seawater shall not be permitted.

For Indian conditions, admixtures used should conform to IS 9103–1979 and should be with the approval of engineers in charge. The compatibility of the chemical admixtures with the cement used should be tested before the same is used for construction [8]. The testing should check against any adverse effects of these admixtures on the basic properties of concrete in hardened state. Use of admixtures containing chloride in any form is totally prohibited.

2.2.2 Concrete mix and its requirements

For prestressed concrete construction works, properly designed controlled concrete alone can be used. The concrete mix proportions should be such that strength, durability, workability for placing the concrete, etc are acceptable for the construction for which the concrete mix is designed.

While designing the mix, the type and quality of materials available, placement, requirements, vibration, compaction, the type of construction procedure followed etc have to be kept in mind. If the volume of concrete to be placed is large, additional quality control has to be assured. In such a case, mixing plant (a weigh batching plant) may be required for which the design of concrete mix may be different.

While high strength is desired in concrete for prestressed concrete works, it is necessary to keep creep and shrinkage to a minimum. These properties depend on the quantity of cement used, water cement ratio and the curing procedure adopted. Low cement content and a low water cement ratio are preferred. But, for obtaining high strength concrete, larger cement content is required. Hence a minimum and a maximum limit for the quantity of cement to be used in high strength concrete is normally specified. ***Indian Standards specify that the cement content in the mix should be preferably not greater than 540 kg/m³ [7]. The minimum cement content required in cement concrete to ensure good durability under different conditions of exposure is given in the Table 2.2.*** BS 8110 (3) gives the upper limit of cement content as 550 kg/m³ and the lower limit is based on exposure conditions and it can go down upto 330 kg/m³. The details are given in Table 6.1 of BS 8110 - 1985 (3).

The concrete should have a low initial heat of setting (heat of hydration). Especially in thick concrete members, heat gets initially locked up inside, causing differential stresses while setting. The cement must therefore not emit considerable amount of heat during hydration process. This brings in a restriction on the use of rapid hardening cement for prestressed concrete works.

Concrete mixes of consistent quality are obtained only if various aggregate sizes and proportions are batched by weight and not by volume. Hence weigh batching is necessary for Prestressed Concrete works.

The strength of concrete is substantially dependent on the amount of water in the mix. Lower water cement ratio leads to higher strength. But to have a good workability, a minimum quantity of water is required. This minimum could be reduced by using vibrators for compaction or using suitable admixtures. Hence, for prestressed concrete works, water cement ratio of as low as 0.32 to 0.42 and compaction using high frequency vibration varying from 4000 to 9000 rpm are preferred.

Table 2.2 Minimum Cement Content Required in Cement Concrete to Ensure Durability Under Specified Conditions of Exposure [1]

Exposure	Minimum Cement Content kg/m³
Mild - For example, completely protected against weather, or aggressive conditions except for a brief period of exposure to normal weather conditions during construction	300
Moderate - For example, sheltered from heavy and wind driven rain and against freezing, whilst saturated with water, buried concrete in soil and concrete continuously under water	300
Severe - For example, exposed to sea water, alternate wetting and drying and to freezing whilst wet subject to heavy condensation or corrosive fumes.	360
NOTE: The minimum cement content is based on 20 mm nominal maximum size. For 40 mm aggregate, minimum cement content should be reduced by about 10 percent under severe exposure condition; for 12.5 mm aggregate, the minimum cement content should be increased by about 10 percent under moderate and severe exposure conditions only.	

The gain in strength of concrete with time is of particular importance in prestressed concrete since prestress could be transferred only after the concrete attains a minimum strength. Early gain in strength could be achieved by use of high early strength cement or by introduction of accelerators in concrete mix and by steam curing. The use of accelerators has to be carefully done after studying their adverse effects on other properties of concrete. For example, calcium chloride, though enhances the rate of gaining strength, leads to more of shrinking characteristics and also corrosion in steel. Hence, the use of accelerators in concrete have to be carefully decided by experts in the field. A comparison of gain in strength of concrete of a given value but prepared using three different cements namely with ordinary portland cement, high early strength portland cement and portland pozzolonic cement is given in Table 2.3.

It could be noted that the concrete prepared with high early strength cement gains strength faster than that prepared with Ordinary Portland Cement. But in 28 days the strength obtained by both types of concrete is the same and then after the gain of strength in both the cases is of the same order. But the concrete prepared with Portland Pozzolonic Cement gains strength at a slower rate earlier and reaches the designed strength in 28 days. All these concretes gain further strength with age. The additional strength gained beyond its 28 days strength with age is defined by a term called age factor. This age factor is more for concrete prepared using Portland Pozzolonic Cement than for that prepared with Ordinary Portland Cement. While the age factor for the concrete prepared with Ordinary Portland Cement and high early strength concrete is of the order of 1.2, the same for the concrete prepared with Pozzolonic Cement is 1.25. This could be observed in Table 2.3. ***But the IS 456 - 2000 [10] has constrained the use of this age factor in design.***

Table 2.3 Values of Ration f_c/f_{ck} at Various Ages

Age of Concrete in Days (T)	1	3	7	14	28	90	180	360
Ordinary Portland Cement	-	0.5	0.66	0.8	1.0	1.1	1.15	1.20
High early strength Portland Cement	0.3	0.6	0.75	0.85	1.0	1.1	1.15	1.20
Portland Pozzolana Cement	-	0.3	0.5	0.7	1.0	1.15	1.20	1.25

Gaining of strength with time for concrete of different strengths, but prepared using the same type of cement namely Ordinary Portland Cement is shown in Fig 2.1. High strength concrete picks up strength faster than low strength concrete. But the age factor for one year for high strength concrete is of the order of 1.20 whereas the same for low strength concrete is 1.25 as could be seen in Figure 2.1.

Another important point, which plays a role in the gain in strength of concrete, is curing. The concrete should be kept moist and warm at least for

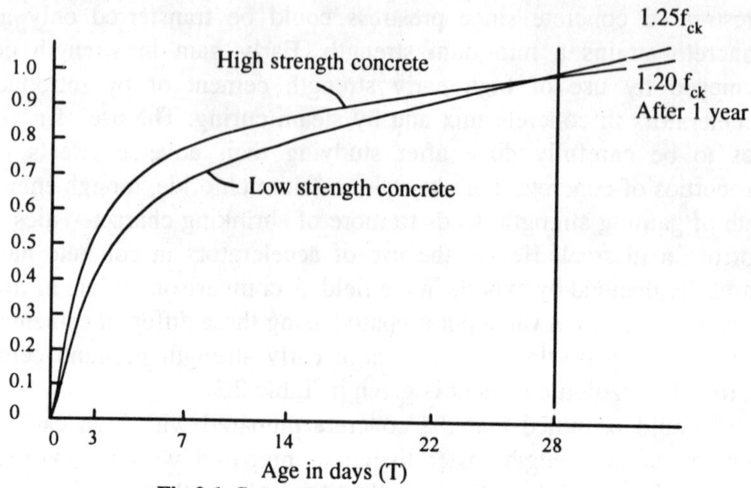

Fig.2.1 Strength of Concrete Vs Age

the first 8 days and preferably for a greater length of time. Heat of hydration should be dissipated slowly and uniformly. For this purpose curing with wet gunny bags is preferable upto atleast 7 days when 70 percent to 80 percent of the strength is achieved.

Steam curing leads to early gain in strength. Moist steam at 60°C to 70°C temperature, if used for curing the concrete, leads to a gain in strength of 70 to 80 percent within 10 hrs. For this, a properly designed steam curing cycle is to be adopted. Fig. 2.2 [9] shows a typical steam curing cycle adopted for pretensioned products. Afterwards, water curing can be done by sprinkling water on concrete products continuously. It is to be noted that raising the

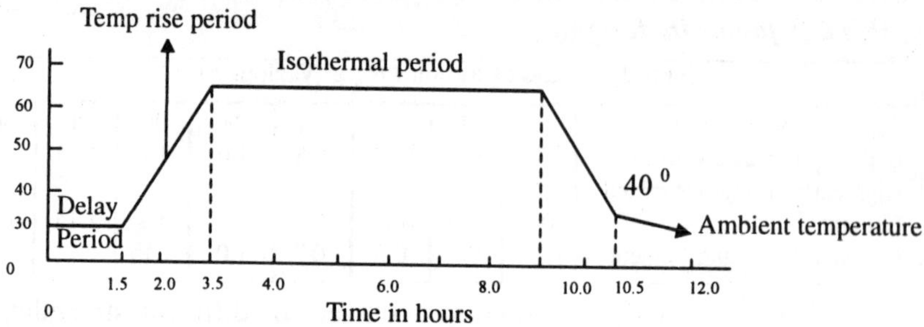

Fig.2.2 Typical Steam Curing Cycle

temperature leads to early gain in strength but this leads to increased evaporation and hence the moist condition on the top of the specimen has to be maintained.

While attention is to be given for achieving consistent, good and early strength in concrete, equal care has to be given for prevention against possible deteriorations of the durability of concrete. To minimise the chances of deterioration of concrete, a check on the harmful salts and their level in the concrete has to be maintained, though such a check is already made on the ingredients. Accordingly, it should be total water-soluble sulphate content of concrete mix expressed, as SO_3 kept less than 4% by mass of cement used in the mix and the total chloride content in concrete expressed chloride ion should not exceed 0.06% of mass of cement. [1]

In conclusion a design mix should specify

- ***the designation***
- ***the type of cement used***
- ***maximum nominal size of aggregates***
- ***minimum and maximum cement content***
- ***the type of compaction***
- ***desired workability***
- ***details of admixtures to be used, and their properties***
- ***the type of curing etc.***

In any case the control tests to be performed before the designed concrete mix is adopted in the field.

IS 10262 - 1982 [6] provides standard guidelines for design of concrete mixes, with different variable parameters mentioned above.

2.2.3 Hardened Concrete, its properties and requirements

2.2.3.1 Compressive strength and permissible stresses

Most of the Statutory Codes like ACI, BS, IS, Euro Code all over the world define the characteristic strength of concrete (f_{ck}) as the strength of concrete based on 95% probability of the test results passing the minimum strength defined. Trial mixes shall be made using samples of approved materials and tested for acceptance criteria defined in Statutory Codes, before final mix is decided. As per IS 456 - 2000 [10] the minimum target mean strength for a given designated concrete mix shall be more than the f_{ck} by 1.65 times the standard deviation on sample test results taken form 40 tests. For want of information of test results, one could take the values from IS 1343–1980 [1] or assume a standard deviation of 6.6 MPa for concrete mixes upto M 40 and 7.8 MPa for mixes upto M 60. All international regulatory bodies limit the minimum and maximum characteristic strength of concrete to be used in pretressed concrete.

The permissible stresses for such concrete are defined by various National Codes (1,2,3,5) differently. The comparison of the same is given in Table 2.4. ***The permissible stresses in concrete in prestressed concrete construction in India are governed by IS 1343–1980 and these are discussed below elaborately.***

IS 1343–1980 [1] limits the minimum characteristic strength to 30 MPa for post tensioned works and 40 MPa for pre-tensioned work and the maximum characteristic strength to 60 MPa for both pre-tensioned and post tensioned works. The higher strengths beyond 60 MPa is not at present recognized as design information for checking the safety of the structure.

The strength is determined based on the test on 15 cm cube specimens after 28 days (with proper curing) from the date of casting. The methods of tests are prescribed in IS 516–1959.

Indian Standards Code IS 1343 [1] recommends:

- *a minimum strength for concrete at the time of transfer of prestress as 0.5 times the characteristic cube strength. The allowable stresses at the time of transfer and service loads are defined by certain coefficients varying from 0.54 to 0.37 for post-tensioned products and 0.51 to 0.44 for pretensioned products as shown in Fig.2.3. These coefficients should be multiplied by the strength of concrete on the day of transfer of prestress for transfer conditions.*
- *For service load conditions coefficients varying from 0.41 to 0.27 should be used with characteristic strength as could be seen from Fig.2.4*
- *The stresses under direct compression is limited to 0.8 times the permissible stresses obtained from the details shown above for flexural conditions.*

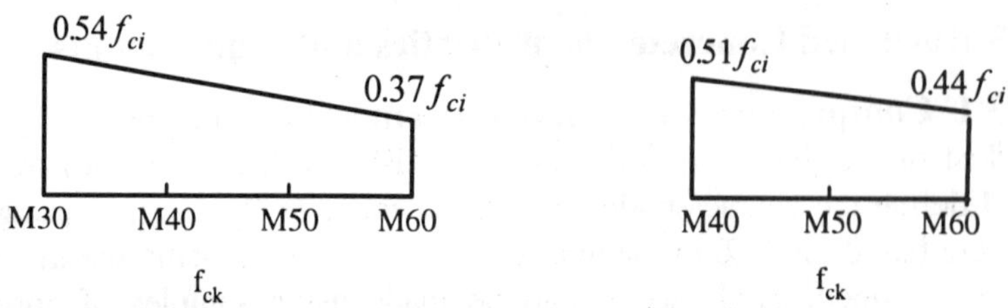

a. Post Tensioned work b. Pre Tensioned Work

Fig.2.3. Premissible Compressive Stress in Flexure at Transfer for Various Concrete Strengths[1]

Limitation of compressive stresses behind the anchorages is different as detailed below.

For prestressed concrete structural elements, depending on the type of force transfer, anchorage zones are to be separately designed. At those places the anchorage unit bears on the concrete directly causing heavy concentration of stresses which are compressive in nature just behind the anchorage zone. The bearing stresses in anchorage zone is limited to $0.48\, f_{ci}\sqrt{\frac{A_{br}}{A_{pu}}}$ or $0.8\, f_{ci}$ whichever is smaller. A_{br} and A_{pu} are the areas in bearing and punching of the anchor plate. is the cube strength of concrete at the time of transfer of force. The punching area is the area of contact of the anchor devices, which if circular in shape, shall be replaced, by square equivalent area. The bearing is the maximum area around the anchor plate, which could

Table 2.4 Maximum Permissible Stresses in Concrete (N/mm^2)

1	2	IS: 1343 - 1980	ACI : 318 - 1998 4	BS: 8110 - 1985 5	FIP : 1984 6
At transfer	Compressive stress	Varies linearly from 0.54 to 0.37 f_{ci} for post-tensioned work and from 0.51 to 0.44 f_{ci} for pre-tensioned work depending on the strength of concrete	0.60 f_{ci}	0.5 f_{ci} for bending 0.4 f_{ci} for axial loads $\not<$ 25 N/mm^2	---
	Tensile Stress	---	(a) At the ends of simply supported members: $6\sqrt{f_{ci}}$ (b) At others places: $3\sqrt{f_{ci}}$	Class 1 structures: 1 N/mm^2 Class 2 and 3 structures: Pre-tensioning: $0.45\sqrt{f_{ci}}$ Post-tensioning: $0.36\sqrt{f_{ci}}$	
At Service load	Compressive stress	Varies linearly from 0.41 to 0.35 f_{ck} depending upon the strength of concrete for loads which are not acting in a sustained manner and it varies from 0.34 to 0.27 f_{ck} in case of elements with sustained loads acting on them with time	$0.45\ f'_c$ for cases of prestress + sustained load $0.6 f'_c$ for cases of prestress + total load (not sustained)	0.33 f_{cu} in flexure, which may be increased to 0.4 f_{cu} in the the region of support moments in statically indeterminate structures. 0.25 f_{cu} in direct compression. 50% increase is permitted in prestressed units of composite construction	50 percent of the charecteristic compressive strength (f_{ck}) of concrete —

Table 2.4 Maximum Permissible Stresses in Concrete (N/mm^2) (contd)

At service load	Tensile stress	*Type 1 members:* None *Type 2 members:* Tensile stresses not to exceed 3 N/mm^2 *Type 3 members:* Hypthetical tensile stresses vary from 3.2 N/mm^2 for M-30 to a maximum of 7.3 N/mm^2 for M-50 grade concrete depending upon the limiting crack - width. Ref Table 2.5	Uncracked sections: $6\sqrt{f'c}$ Cracked transformed sections $12\sqrt{f'c}$	*Class 1 members:* Nil *Class 2 members:* Pre-tensioning 0.45 $\sqrt{fcu}$ Post-tensioning: 0.36 $\sqrt{fcu}$ These stresses may be increased by upto 1.7 N/mm^2 in specified cases. *class 3 members:* Hypothetical tensile stresses should not exceed a limiting value depending on the permissible crack width. strength of concrete and depth of members) Ref Table 4.1 & 4.3 of BS of 8110 - 1985	*Class 1 members:* (Full prestress) : Limiting Tensile stress at extreme fibre = 0 *Class 2 members:* (limited-perstress): Limiting Tensile stress at extreme fibre = 75 per cent of *Class 3 members:* (partial-prestress): No limit imposed on tensile stress. However, degress of prestress controlled by limiting deflection and crack width.

f_{cu}'f_{ck} - Characteristic strength of Concrete (N/mm^2)
f_{ci} - Strength of Concrete at the time of load transfer (N/mm^2)
f'_c - Strength of Concrete (cylinder) in pounds/sq inch

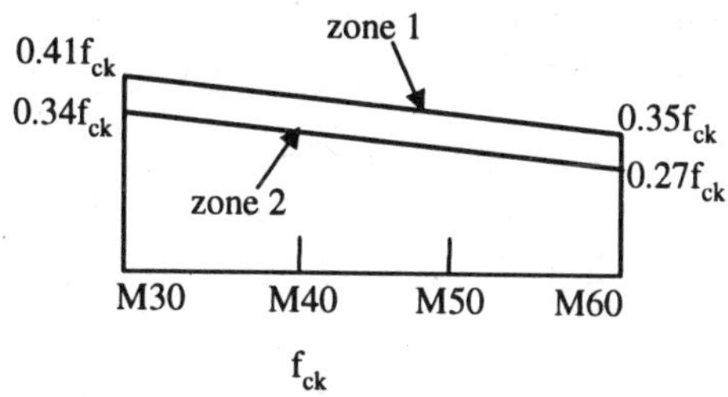

Fig.2.4. Premissible Compressive Stress at Service Load for Concrete Strength [1]

Zone I : and Zone II are defined as follows

Zone I : The places where the compressive stresses are not likely to increase in service

Zone II: The places where the compressive stresses are likely to increase in service

be located concentric to the effective punching area. Anyhow the diameter of the bearing area shall not be greater than twice the corresponding dimension of the punching area. During tensioning, allowable bearing stress may be increased by 25% provided the temporary value (with the increase) does not exceed f_{ci}.When more number of anchorages are used the bearing area my get overlapped. In such a case the overlapped portion of the area will be omitted. This aspect of permissible bearing the stress is discussed in greater detail later in the chapter on the design for anchorage of prestressing forces on concrete.

2.2.3.2 Tensile strength and permissible stresses

Tensile strength of concrete is defined under two categories. The maximum tensile stress that the material can withstand when it is subjected to flexure (bending) is called flexural tensile strength f_{cr} (also called modulus of rupture). The maximum tensile stress that the material can withstand when it is subjected to direct tension is referred as f_{ct}. The flexural tensile strength will be decided by a specified bending test on specimen of 15 cm × 15 cm cross-section 70 cm long with 60 cm span (center to center, distance between supports) subjected to two point loading at 1/3rd points. The direct tensile strength will be decided by a split tension test on a cylindrical specimen of 15 cms dia, 30 cms long. These tests have to be carried as per the specifications laid down in IS 516/1959. ***IS 1343/1980 [1] specifies a value f_{cr} as equal to .7 $\sqrt{f_{ck}}$ in N/mm².*** No definite value for direct tension has been specified in this code, though in certain international codes some relationship between flexural tension and direct tension is given. However it could be safely assumed to be 60% to 70% of capacity under flexural tension. Euro Code ENV, 1992 (4) has given the value of axial tensile strength of concrete of different strength as given in Table 2.5.

Tension, either as flexural tension or direct tension, is rarely permitted in prestressed concrete structures under service loads. However British Code, Euro Code and IS Code define different types of structures wherein limited tensile stresses are permitted. The details are furnished in Table 2.4. IS 1343 defines the types of structures as follows:

Table 2.5 Concrete Strength Classes, Characteristic Compressive Strengths fcy (Cylinders) Mean Tensile Strength fctm ENV 1992

Strength Class of concrete	C12 /15	C16 /20	C20 /25	C25 /30	C30 /37	C35 /45	C35 /40	C45 /55	C50 /60
fcy	12	16	20	25	30	35	40	45	50
fctm	1.6	1.9	2.2	2.6	2.9	3.2	3.5	3.8	4.1

Table 2.6 Hypothetical Flexural Tensile Stresses for Type 3 Members [1]

Type of tendons	Limiting Crack Width	Tensile stress in Concrete for Grade				
		M 30	M 35	M41	M45	M 50 and above
Pre-tensioned tendons	0.1 0.2	- -	- -	4.1 5.0	4.4 5.4	4.8 5.8
Grouted Post - tensioned tendons	0.1 0.2	3.2 3.8	3.6 4.4	4.1 5.0	4.4 5.4	4.8 5.8
Pre-tensioned tenders distributed in the tensile zone and the positioned close to the tension faces of concrete	01 0.2	- -	- -	5.3 6.3	5.8 6.8	6.3 7.3

When additional reinforcement is distributed within the tension zone and positioned close to the tension face of concrete, the hypothetical tensile stresses may be increased by an amount, which is proportional to the cross-sectional area of the concrete. For 1 percent of additional reinforcement, the stress may be increased by 4 N/mm^2 for the members with pre-tensioned and grounted post-tensioned tendons and by 3 N/mm^2 for other members. For other percentages of additional reinforcement the stress may be increased in proportion excepting that the total hypothetical tensile stress shall not exceed 0.25 ztimes the characteristic compressive strength of concrete.

a) Type 1 - No tensile stresses.

b) Type 2 - The tensile stress shall not exceed 3 N/mm^2. However, where part of the service loads is temporary in nature, this value may be exceeded by 1.5 N/mm^2, provided under the permanent component of the service load the stress remains compressive.

c) Type 3 - For type 3 members in which cracking is permitted, it may be assumed that the concrete section is uncracked, and the hypothetical tensile stresses for use in these calculations for members with either pre-tensioned or post-tensioned tendons are given in the Table 2.6, to be modified depending on the depth of member in mm as shown in Fig. 2.5. In members made up of precast units, no tension shall be allowed at any stage at mortar or concrete joints. For a member, which is free of joints, the tensile stress shall not exceed the values specified for class 3 types members.

Hypothethical tensile stresses are multiplied by the factors varying from 0.7 to 1.1 depending on the depth of the member as shown in Fig 2.5. It is well known that prestressed/reinforced flexural members are normally T sections or I (bulb) sections. In such cross sections, the center of gravity will always be nearer to the top fibre. The ratio of the distance of center of gravity from bottom fibre and to that of top fibre (greater than one) will be going on reducing for sections of larger depth. Hence in the sections of larger depths the tensile zone will increase and the cracked concrete section area will be more.

Crackwidth is the function of distance at extreme bottom fibre from the centre of gravity of the section. For the same stress level at the bottom fibre,

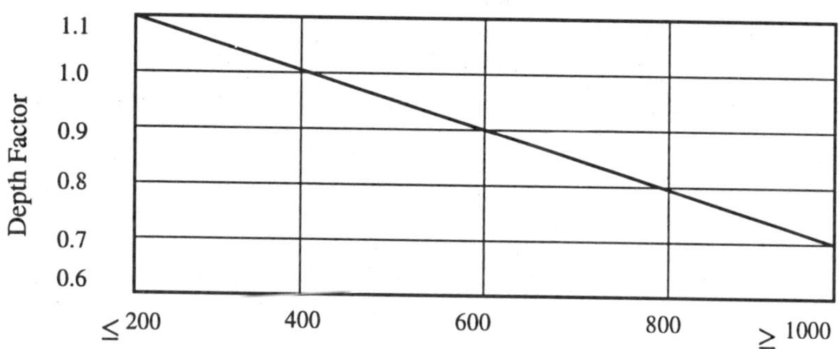

Fig. 2.5. Depth Factors for Tensile Stresses for Type 3 Members

the crackwidth will be more for deeper sections than for shallower sections. For the given crackwidth, the stress level at the bottom fibre will be less in case of deeper sections than in shallow sections. This aspect has been considered while fixing up hypothetical tensile stress for a given crackwidth. Accordingly the reduction factors for the hypothetical tensile stress calculation for deeper sections has also been given as could be seen from Fig 2.5.

Principal tensile stress is another term, which has definite relevance in the design of prestressed concrete structures specially for shear resistance.

- *Principal tensile stress resulting from direct tension or direct tension and shear force is to be calculated at the centroid of the section or at sections where the width of section changes.*
- *The principal tensile stress (f_{ct}) at those points shall not exceed $0.24\sqrt{f_{ck}}$.*
- *In case of flanged section, this principal tensile stress will be evaluated at the junction between the flange and the web.*

This value will be made use of in the design for shear for Prestressed Concrete sections.

2.2.3.3. Deformation behaviour

The deformation behaviour of concrete is highly complex even under working stress conditions. The load deformation characteristics are identified by the stress strain relationship. Indian Standard Code specifies that the stress strain diagram should be drawn by carrying out a compression test on a standard cylinder of 15 cms dia, 30 cms long. The procedure for the conduct of test is defined in IS 516–1959. The typical stress strain curves for the concrete of different strengths are given in Fig. 2.6 a [6]. It may be noted that the linearity of the stress strain curve is limited up to a very low stress level and the curve turns to be nonlinear showing that even after complete unloading there could be certain amount of deformation left over as permanent set. Hence the deformation at any load stage (specially at higher load stages) consists of an elastic (retrievable) part and an inelastic (non-retrievable) part. The inelastic deformation component is higher at higher stress levels. It can also be seen that the maximum stress does not occur at maximum strain, but corresponds to a strain ε_q, which is lower than the strain at failure. At the time of failure, the stress level is little lower than maximum stress that the material can stand.

The drooping portion of the curve shown in Fig 2.6 a is very difficult to be traced in the experimental procedure unless the strain-controlled experiment is conducted. RôS, et.al, [12] have suggested a formula for defining the stress strain characteristics of concrete. Their works out reasonably coinciding with experimental curves inclusive of the descending part

$$\sigma = \frac{E \in}{1+\left(\frac{\in}{\in_0}\right)^2}$$

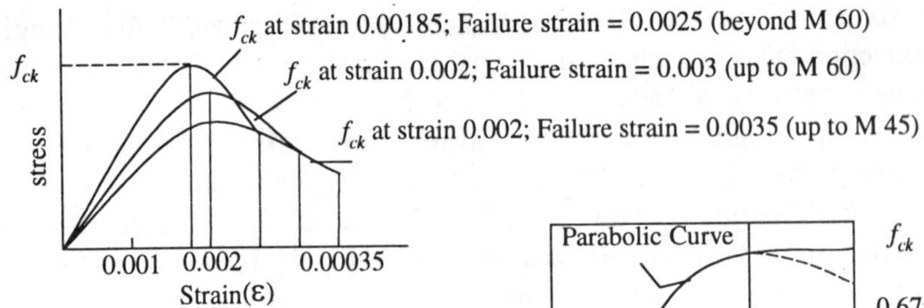

Fig.2.6 a. Typical Stress-Strain Curve for the Concrete

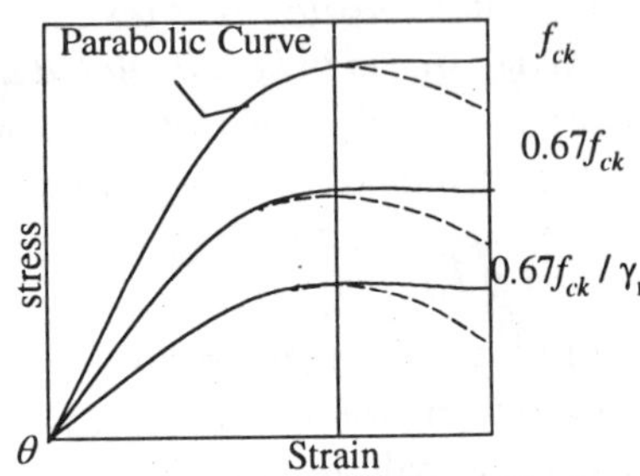

Fig.2.6 b. Idealised Stress Strain Curve for Concrete

$$\in_{total} = \frac{\sigma_c}{E_c} + 0.1\frac{1}{1000(\gamma f_{ck} - \sigma_c)} = \in_{el} + \in_{pl}$$

γ = 1.15 to 1.20 for M 40 to M 60
γ = 1.10 to 1.15 up to M 40
σ_c = the stress level at which strain is calculated

Indian Standard 1343/1980 [1] and also IS 456–2000[10] suggest/ identify the stress strain curve for concrete as shown in Fig. 2.6 b.

It could be seen that, though the stress strain curve for concrete is non-linear, the deviation from linearity is minimal up to the stress level of 0.30 to 0.35 f_{ck}. The permissible stress limits in compression at service loads under sustained loads are also of this order as per many of the National Codes (Ref Table 2.4). Hence linearity of the stress strain relationship is assumed for analysis and design at service load stages. This linearity is expressed with a constant modulus of elasticity E_c upto to 0.35 f_{ck}, which is initial tangent modulus. The E_c is also defined as a function of characteristic strength of concrete in various national and international codes. ***IS 1343 limits the permissible stress in prestressed concrete element to a maximum of 0.34 f_{ck} for service load conditions and recommends the modulus of elasticity of 5000 $\sqrt{f_{ck}}$ in . It is also assumed that the stress strain curve in tension shall be similar to that in compression up to the cracking stress, and the cracking stress is taken as 0.7 $\sqrt{f_{ck}}$.***

The other National Codes permit either initial tangent modulus of elasticity or secant modulus of elasticity at a particular stress level for analysis and design. AC 318 of 1998 recommends a value of E_c of 5050 $\sqrt{f'_c}$. If the stress strain curve is available. E_c is defined as secant modulus corresponding to 0.45 f'_c (f'_c cylinder strength). BS 8110 of 1985 [3] and FIP 1984 [5] gives the range of values for the tangent modulus E_c depending on the grade of concrete as shown in Table 2.7.

Table 2.7 Typical Range of the Static Modulus of Elasticity at 28 days of normal-weight concrete BS:8110 - 1985/FIP 1984

Cube strength at 28 days (N/mm^2)	BS 8110	Modulus of elasticity kN/mm^2	FIP	Modulus of elasticity kN/mm^2
	Mean value	Typical range	Mean value	Typical range
20	24	18 to 30	29	20 to 38
25	25	19 to 31	30	22 to 39
30	26	20 to 32	32	23 to 42
40	28	22 to 34	35	24 to 45
50	30	24 to 36	37	26 to 48
60	32	26 to 38	38	27 to 49

Euro Code ENV 1992 [4] defines modulus of elasticity of concrete as secant modulus at the stress level of 0.4 f_{ck} . At this stress level the secant modulus and tangent modulus of elasticity are practically the same, since the stress strain curve is linear up till this point. This Code defines E_c as equal to $9.5(f_{ck}+6)^{\frac{1}{3}}$ in *N, mm* units. These values are given in Table 2.8. Euro Code also gives the maximum strain (failure strain) for concrete as a function of strength of concrete as shown in Table 2.8. It may be noted that the failure strain is decreasing with the increase in strength of concrete. It could also be seen in Fig 2.6 a.

Table 2.8 Values of Secant Modulus of Elasticity E_{cm} (*in kN/mm²*) at 0.4 f_{ck} and E_{cu}

Strength Class C	C12 /15	C16 /20	C20 /25	C25 /30	C30 /37	C35 /45	C40 /50	C45 /55	C50 /60
E_{cm}	26	27.5	29	30.5	32	33.5	35	36	37
E_{cu}	-3.6	-3.5	-3.4	-3.3	-3.2	-3.1	-3.0	-2.9	-2.8

Further all these values, namely the modulus of rupture, the direct tensile strength and the modulus of elasticity are defined as the functions of $\sqrt{f_{ck}}$. This is based on the statistical information collected all over the world till recently. With advent of high strength high performance concrete, these expressions are not fully valid for concrete of strength beyond 60 MPa. For such high strength concrete, the exact values have to be decided by tests on control specimens.

2.2.3 Inelastic properties

Though concrete shows inelastic behaviour even under short term loading, in literature, whenever mention is made about inelastic properties, it is attributed to the long time-dependent inelastic behaviour only. Such an inelastic property of concrete under a load is also called creep of concrete. Creep is defined as a variation of deformation of concrete with time for the same applied load. This is of great importance in prestressed concrete works since it causes losses in applied prestress to the structural element with time. There is one more inelastic property of concrete called shrinkage and it is not considered as function of applied stress. But both of them are dependent on atmospheric conditions.

2.2.4.1. Creep of Concrete

Concrete when subjected to the constant stress undergoes irrecoverable deformation mostly due to plasticity and flow conditions. Such a permanent deformation is called creep. It ***is associated with time, applied stress level, density of the concrete, cement content in concrete, water cement ratio, atmospheric conditions like the humidity and temperature, thickness of the elements etc.*** These variables, with certain assumptions, ***can be incorporated in a single factor φ called creep coefficient.*** Then the deformation due to creep can be evaluated as creep coefficient times elastic deformation. But a detailed study of creep and shrinkage is necessary for predicting the time dependent deformation of special prestressed concrete structures like prestressed concrete cantilever bridges, where a precamber has to be given to keep the profile of the deck at appropriate level. It may also be necessary in some cases where composite construction of prestressed concrete elements and reinforced concrete elements are involved.

Tests on creep of concrete have shown that creep strain is linearly dependent on the elastic strain at ***lower stress levels.*** Hence creep strain is defined as $\in_{cr} = \in_{cl}\varphi$ where φ is called creep coefficient. The factor φ is constant for lower levels of applied stress upto 0.4 f_{ck} and is of the order of 1.0 to 4.0 depending on the age of concrete at the time of loading, exposure condition (relative humidity) notional thickness of elements etc. Fig 2.7 a. shows the variation of creep strain with time for different applied stress levels for a concrete of $f_{ck} = 50N/mm^2$ for mild exposure with relative humidity of 50% and notional size 150 $\left[\frac{2 \times area}{permeter}\right]$. The applicability of constant value of φ (the linear relationship between $\in_{cr}$ and $\in_{el}$) upto 0.4 f_{ck} is also seen in Fig 2.7 a. Beyond the value of 0.4 f_{ck}, the co-efficient φ is an increasing function.

Further it is noted that creep strain increases with time, at a decreasing strain rate [8]. Fig. 2.7 b. shows the results of series of creep tests on concrete specimens of a given strength of $20N/mm^2$ and for a specified stress level of $10N/mm^2$. But the age of concrete on the date of application of stress are different. The humidity and temperature were maintained at 35% and 20°C respectively.

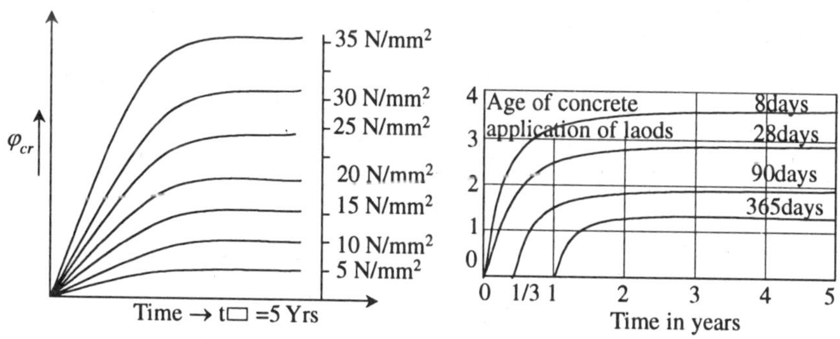

$\in_{cr} = \in_{el} \times \varphi_{\infty}$

$\varphi = \frac{\in_{cr}}{\in_{cr}}$, Humidity 35%, Applied stress 10 N/mm²

φ is a very nearly same for lower applied stress Level say upto 0.4 fck =20N/mm²

Fig. 2.7a. & b. Creep Time Curves for Concrete [13]

Since creep is caused mostly due to plasticity and flow properties of concrete, recovery of creep deformation after the removal of the stress is small. For a well set concrete, creep recovery may be nil. The effect of certain factors on creep are listed as follows.

- Creep is more in dry environment than in wet environment.
- Creep is more in case of specimens loaded early and to higher level.
- Higher cement content causes a large creep deformation.
- Well-compacted (high density) concrete with gap graded aggregate will have minimum creep.

For design purposes creep strain is considered at infinite time (5 years is a limiting time as found by experience).

The National Building Code [14] recommends creep strain per unit applied stress as

$$C = 1.75\left[\frac{1.25-h}{h}\right]\left[\frac{\sqrt{40}}{f_{ck}}\right]\left[\frac{f_{ck}}{f_{ci}}\right]10^{-5}\,mm^2/N$$

The effect of age of concrete at the time of the application of load is reflected in the ratio of concrete strength f_{ck} at 28 days to strength f_{ci} at the time of transfer. Factor 'h' takes care of humidity and other atmospheric conditions.

h = 1.0 in water
= 0.7 to 1.0 for moisty atmosphere
= 0.7 to 0.9 for highly humid temperature
= 0.5 to 0.75 for average humidity
= 0.2 to 0.5 for dry atmosphere

IS 1343–1980 recommends simplified version for calculating the creep strain as shown in Table 2.9 a. It is shown to be dependent only on the age on which the stress is applied. As per IS Specification the maximum permissible applied stress itself is of the order of only 0.34 f_{ck}. The strength of concrete at the time of applied stress shall not be less than $0.5f_{ck}$. Adopting these limiting values and an average humidity factor of 0.6 etc, creep coefficients have been defined in IS 1343/1980 for various ages of loading. Ultimate creep strain can be evaluated as creep coefficient multiplied by elastic strain for applied loading.

Table 2.9 a. Creep Coefficient as per IS 1343 - 1980

Age at loading	Creep Coefficient
7 days	2.2
28 days	1.6
1 year	1.1

NOTE: The ultimate creep strain estimated as per above does not include the elastic strain.

These recommendations are only for simple structures where prestressing is done at one stretch and hence would not form adequate information if the prestressing is done in stages. If stagewise prestressing is carried out, more detailed calculations for creep have to be done, considering the effect of delayed elasticity and plasticity [8,15].

BS 8110 (3) and ENV 1992 (4) have given the creep coefficient and shrinkage as function of age at loading, atmospheric condtitions (exposure) and notional size defined by 2 times the cross sectional area of the element divided by perimeter of the cross section. There values are given in Fig 2.8 and Table 2.9 b respectively. But these codes also take applied stress level to be constant and less than 0.4 *fck*

It may be noted that the applied stress in prestressed concrete element is varying with time due to creep losses. Hence creep coefficients defined based on one particular applied stress cannot be considered for arriving at the creep loss. There will be a reduction in creep rate with time due to reduction in applied stress also. This aspect has been discussed by Trost [16] as creep relaxation and the same can be advantangeously used in the design of prestressed concrete elements.

Table 2.9 b, Final creep coefficient as per EuroCode ENV 1992 (φ_∞, t_0) of normal weight concrete

Age at loading t_0 (days)	National size $2A_c/u$ (in mm)					
	50	150	600	50	150	600
	Dry atmospheric conditions (inside) (RH-50%)			Humid atmospheric conditions (outside) (RH-50%)		
	5.5	4.6	3.7	3.6	3.2	2.9
7	3.9	3.1	2.6	2.6	2.3	2.0
28	3.0	2.5	2.0	1.9	1.7	1.5
90	2.4	2.0	1.6	1.5	1.4	1.2
365	1.8	1.5	1.2	1.1	1.0	1.0

2.2.4.2 Shrinkage in concrete

Shrinkage is shortening of the member due to drying and removal of moisture. Excess of water in concrete evaporates and the cement matrix around the aggregates contracts. Similar to creep, shrinkage also depends on the atmospheric conditions such as humidity, the temperature difference between the concrete surface and inside called temperature gradient and notional size as defined earlier. The other factors which affect shrinkage are w/c ratio used, age of concrete, amount of cement used per unit volume of concrete and the quality of cement. Use of large quantities of cement also increase the shrinkage since the heat energy generated will be more. Different notional codes detail the values of shrinkage strains in different ways (Table 2.10 a, 2.10 b). ***IS Codes recommends the values of shrinkage strains to be adopted as given in Table 2.11.*** Hence IS 1343–1980 recommends a maximum content of cement per unit volume of concrete for prestressed works. Shrinkage is also to some extent dependent on the applied stress level.

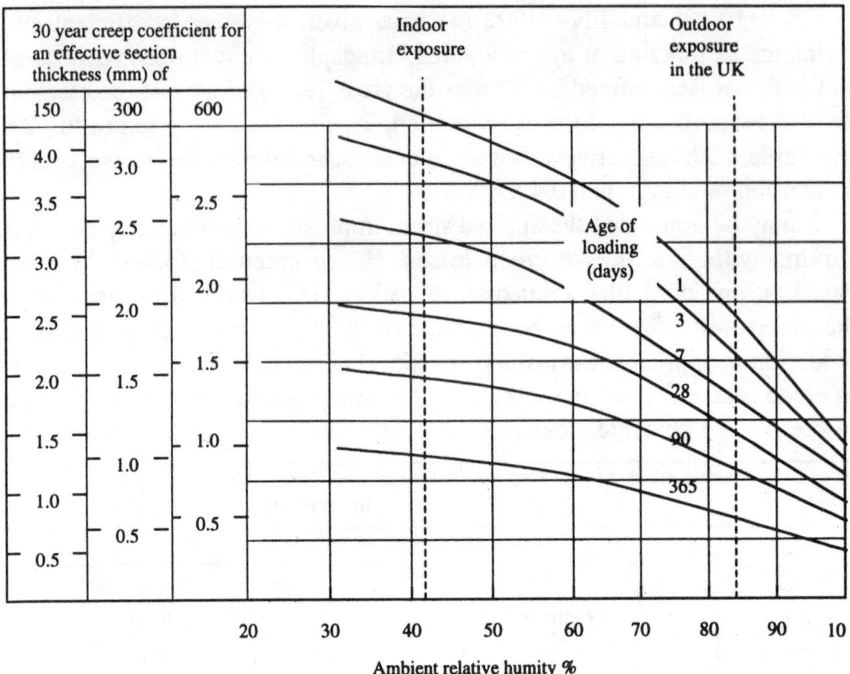

Fig.2.8 Effects of Relative Humidity Age of loading and section thinkness upon creep factor [3]

Table 2.10 a. Final shrinkage strains ecs (in o/oo) as per Euro Code for normal weight concrete

Location of the member	Relative humidity (%)	Notional size $2A_c/u$ (mm) ≤ 150	600
Inside	50	-0.60	-0.50
Outside	80	-0.33	-0.28

where: Ac cross-sectional area of concrete
u perimeter of that area

Shrinkage in prestressed concrete is unrestrained and hence it is more in prestressed concrete members than in reinforced concrete for similar conditions. Shrinkage is restrained by steel in reinforced concrete. Here the steel will be in compression and the corresponding bonded concrete will be in tension. This causes shrinkage cracks. But due to creep these cracks close down. But in prestressed concrete, the tendons do not have any restraining effect on concrete and hence the creep and shrinkage produce the same type of stresses i.e. compression in concrete.

Table 2.10 b. Final Values of Shrinkage and Creep Coefficient (F.I.P) [5]					
		Atmospheric Conditions			
		Humid (RH = 75%)		Dry (RH=55%)	
Equivalent thickness ($2A_c/u$)		Small 200 mm	Large 600 mm	Small 200 mm	Large 600 mm
Shrinkage * $\in_{cs}(t_\infty - t_0) \times 10^3$	Fresh (3 - 7 days)	0.26	0.21	0.43	0.31
	Medium (7 - 60 days)	0.23	0.21	0.32	0.30
	Mature (> 60 days)	0.16	0.20	0.1	0.28
Creep	Fresh (3-7 days)	2.7	2.1	3.8	2.9
	Medium (7-60 days)	2.2	1.9	3.0	2.5
	Mature (> 60 days)	1.4	1.7	1.7	2.0

A_c denatoes the area of the concrete section
u is the perimeter in contact with atmosphere (which includes the interior perimeter of a hollow section only, if there is a connection between the inetiro and the free atmomphere)

* t_o: age of concrete at the instant from which the shrinkage effect is being considered

\+ t_o : age at the commencement of loading

Table 2.11 Shrinkage Strains as per IS 1343 - 1980 [1]
a) For pretensioned members $\in = 3 \times 10^{-4}$
b) For post - tensioned members $\in_s = \dfrac{2 \times 10^{-4}}{\text{Log}_{10}(t+2)}$ (where t is the age in days at which stress is transferred.)
This value for case (b) can be increased by 50 percent in dry atmospheric conditions but not to exceed .003

2.2.5 Durability of Concrete

A durable concrete is one that is designed and cast to perform satisfactorily while responding to loads for which it is designed and in the environment in which it is placed over the life time of the structure. The durability of the concrete is assured by

- Minimum permeability for ingression of water, oxygen, carbon dioxide and other potentially deleterious material.
- Low free water cement ratio, sufficient hydration of cement through proper curing giving leading to low permeability
- Limiting the chloride in concrete to avoid the risk of corrosion (0.06% by mass of cement)
- Limiting the Sulphates to 4% of SO_3 by mass of cement to minimize expansion and disruption of concrete
- in the mix can be considered as upper limit.
- Limiting the Na_2O and K_2O available in cement to a minimum to avoid Alkali silica reaction.

More details can be had from specialized literature or books on Concrete Technology. Indian Institute of Bridge Engineers, Indian Roads Congress and Indian Railways have published special reports on Durability of Concrete and Prestressed Concrete Structures, which could also be referred to.

2.3 High Tensile Steel

As explained in the first chapter, the loss in stress in steel (loss in given strain/elongation) due to time dependent material properties of concrete such as creep and shrinkage should be preferably limited to 10 to 15% of the applied force. There is always an upper limit on the applied stress in steel in terms of its ultimate stress. Various national codes limit the level of applied stress as a percentage of the tensile strength of steel at the time of initial tensioning.

Hence, to keep the loss in stress in steel due to creep and shrinkage of concrete to a low level, the steel used in prestressed works should have high 0.2% proof strength and ultimate strength. The stress strain relationship should also be well defined. It should have good elongation property so that any loss in elongation (strain) may not affect the stress level considerably. This means, the steel should have a low modulus of elasticity. At the anchorage, the steel have to bend a little to accommodate the anchorage units. Accordingly the steel should have a high bendability without leading to surface cracks. When the force in steel is released there should not be any permanent set, in deformation (elongation). This is possible only if plastic flow property of steel is minimum. The steel should be corrosion resistant. Considering all these desirable qualities in prestressing steel, the requirements of steel for prestressed concrete works can be listed as follows:

- Steel for prestressed concrete works should have high strength (yield stress, 0.2 percent proof stress and ultimate strength). The failure stress or ultimate stress shall not be less than around 1400 N/mm^2.
- ***For steel, which does not have a definite yield point, the elastic limit is defined at 0.01 percent proof stress.*** Upto the elastic limit linear stress strain behaviour is assumed to be valid. This elastic limit should be as high as possible ssso that the inelastic deformation in steel can be kept as a minimum.
- Good ductility is expected in prestressing steel so that at anchor points when cold working is necessitated or inherent, the steel should not abruptly rupture for want of ductility.
- Bendability of steel for prestressed concrete works should be good. The specimen should be subjected to a bend test (reverse bend) where it is bent through 90° twice (in two directions) around a mandrel of a specified diameter for a given diameter of the wire and there should not be any crack seen on the surface. The requirement also is to take care of possible bending of steel at anchorage points.

- The dimensional tolerances should be as small as possible so that when they are used with patented anchorage system, no problem is encountered.
- The strain at the time of failure measured over a gauge length of 600 mm should be as large as possible.
- High bond properties are preferred in high tensile steel so that they can be confidently used in pretensioned products where the force is transmitted by means of bond. High bond properties keep the transmission length to a minimum.
- Corrosion is one of the major causes of damage in steel. Hence it should be seen that the metallurgical properties are such that the possibility of corrosion of the steel is kept to a minimum
- Low relaxation properties are preferred in high tensile steel, so that the loss in prestressing due to relaxation of steel is minimum.

The above requirements are specified with different limiting values by the different national statutory bodies responsible for code formation. Indian Standards Institution stipulates the above requirements of steel in various codes like IS 1785–1983 [17]. IS 6003–1983 [18], IS 6006–1983 [19], IS 14268–1995[20] for prestressing steel. A few relevant points will be discussed later in this chapter. International Codes specify the property of steel in BS 5896, BS 4486, ASTM, A 416, A 421, A 722, ENV 10138.

Most of the Indian National Codes are to be revised because of updated Technology. As such, the industry uses only strands 3 x 3 ply, 7 ply and 7 mm and 8mm wires for pretressing. 3mm, 2mm wires are used only for stranding purposes.

Though different types of steel such as, Hard drawn steel (stress relieved), cold worked steel, Alloy steels, Heat treated wires etc are available for prestressing works, only Hard drawn steel is permitted in Indian Standard specifications for use in Prestressed Concrete works. Anyhow brief information on other types of steel is given here as background material.

2.3.1 Types of Steel

2.3.1.1 Hard drawn stress relieved wires

Rolled steel wires are patented i.e. heated upto 900° to 1000°C and then cooled to 550° C to 500 in lead or salt bath to get a favourable crystal structure for cold working. These patented wires are drawn through dies and the diameter is slowly reduced in stages. For each time drawing through 'dies', the strength increases but the ductility reduces. By experience it has been found that a limiting lower diameter of 3 mm and an upper diameter of 8 mm gives the required strength and ductility for the material and hence IS 1785/1966[17] recommends only wires of diameter varying from 8 to 3.0 mm wire. The limits of certain ingredients are as follows [13]:

Carbon 0.7% to 0.9%
Manganese 0.5% to 0.9%
Silicon 0.12% to 0.15%
Phosphur and sulphur upto 0.05%

The hard drawn wires normally have a smooth surface. For improving the bond characteristics, sometimes indentations are made on the surface and they are called indented wires. The hard drawn wires of small diameter say 3 to 4 mm are also used to make strands with 3,5,7 plys. Artificial ageing of this steel is possible by a mild heat treatment under 150 to 420°C. By this process the proof stress/ the tensile strength is improved by 20 percent to 40 percent and percentage elongation is improved by 5 percent to 10 percent. Cold working introduces internal energy in the material, which is released by a mild heat treatment. Such wires are called stress relieved wires. Indian Standards for prestressed concrete works (IS 1343/1980) permits only hard drawn wires with or without indentation and with or without stress relief in

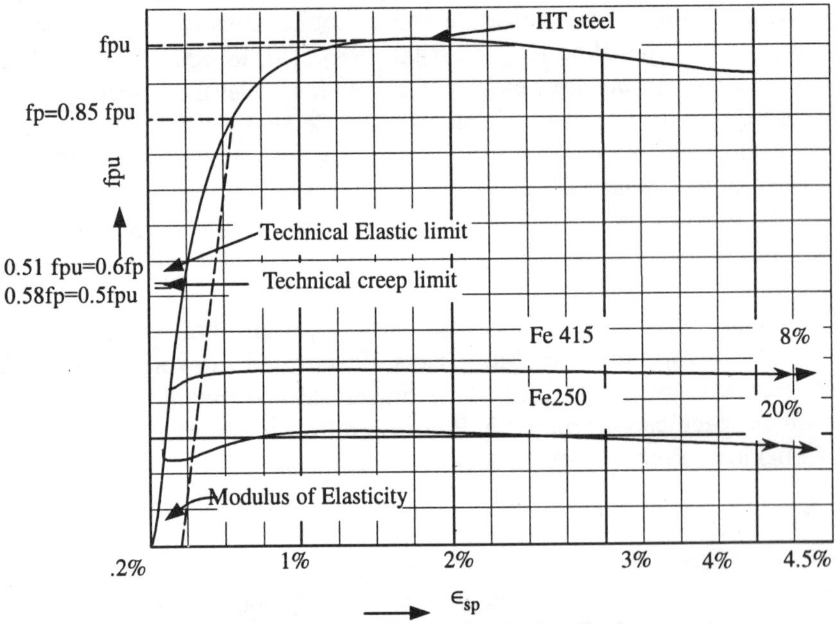

Fig. 2.9 Typical Stress-Strain for Steel

the form of wires or strands. The minimum performance requirements are specified in relevant IS Codes already listed. Certain specific information such as stress strain properties, ultimate strength, percentage elongation etc are given in Fig 2.9 [13] .

2.3.1.2 Heat - treated wires

Rolled steel bars of suitable composition are heated upto 810° C and quenched in oil bath to room temperature and then another heat treatment is carried out in a molten lead bath at 450° C. By this treatment a finer structure is obtained. The steel manufactured is in the form of 5 mm to 8 mm round or oval wires. It can be provided with ribs to improve bond properties.

2.3.1.3 Alloy Steel wires/ rods

High strength can be obtained by alloying some other metal like manganese with steel in a suitable proportion and combination. They are not drawn and are available at higher diameters of 9 mm to 32 mm. These steel have good ductility and definite yield.. Because of alloying material the cost of this steel becomes high. These are also used as pure tension members in Foundation Structures, Bowstring Girders etc.

2.3.1.4 Cold worked steel

Bars, which are hot rolled originally, can be cold worked by twisting or stretching. The strength is increased in this process. They can also be used for prestressing purposes. One has to be very careful about the ductility of such steel.

2.3.2 Mechanical Properties

The Mechanical properties of the steel are characterized by its stress strain relationship. Fig 2.9 gives typical stress-strain curves for prestressing steel. The ultimate strength, 0.2 percent proof stress, 0.01 percent proof stress (technical elastic limit) percentage elongation, maximum strain at failure etc are the details which could be evaluated from a mechanical tensile test. The point at which the stress strain diagram deviates from the linearity is defined as ***technical elastic limit.*** This normally happens corresponding to ***0.01 percent proof stress***. If the specimen is loaded upto this proof stress and released, there will not be any permanent left over deformation.

It may be noted that the tensile strength of the material can vary upto 2 percent in the manufacturing process itself, change in section, difference in composition etc. For hard drawn steel the ultimate strength will be in the order of 1400 N/mm^2 to 2000 N/mm^2 depending on the diameter of the wires. Smaller diameter wires have higher strengths. Heat-treated wires normally have a definitc yield point. The ductility will be good. Alloy steels, which are hot rolled, has a better ductility and good yield pattern. But the ultimate strength will be low. For all these types of steel 0.2 percent proof stress can be evaluated form the corresponding stress strain curve. It is necessary that there is a sufficient margin between 0.2 percent proof stress and ultimate stress. ***Indian Standard permitts only hard drawn steel to be used as prestressing steel and the typical stress-strain curve for such steel as idealised diagrames are given in 1S 1343 and shown here in Fig 2.10 a & b.***

2.3.3 Permissible stresses in Steel

High tensile steel is used mainly to apply Prestress (precompression) in concrete, as a reactive force. When steel is tensioned against the concrete and anchored, the reactive force on concrete is the precompression. Hence the tensile properties of the High tensile steel and the limitation of the tensile properties while using the steel as a prestressing element are to be defined. These limitations on the stress level on steel to be used for prestressing the

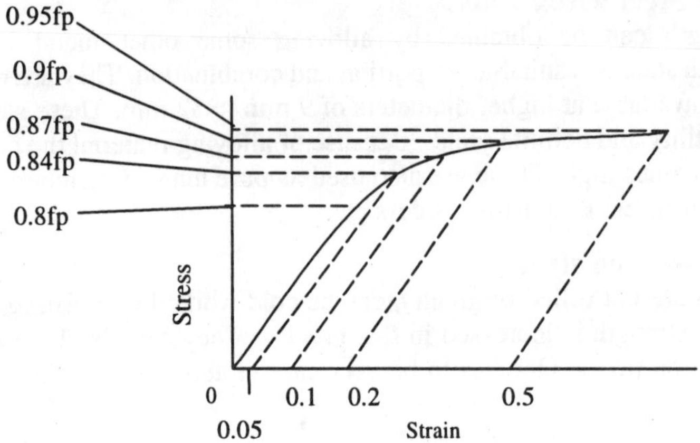

Fig. 2.10.a. Wires (Stress relived), Strands and Bars [1]

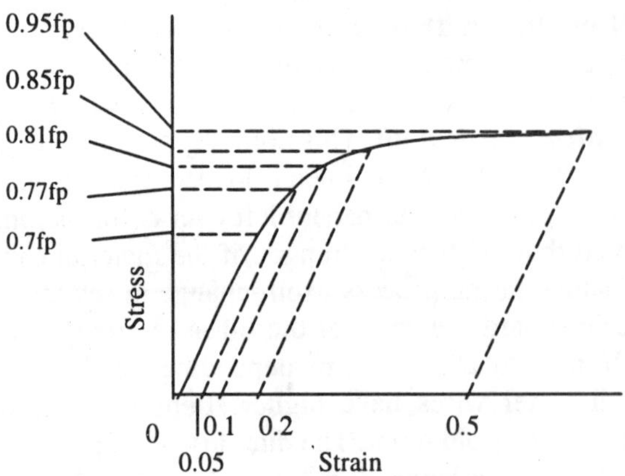

Fig. 2.10.b. Wires (As-Drawn)

concrete are expressed as a function of (a) of ultimate strength of steel (b) 0.2% proof stress of steel and (c) 0.1% proof stress of steel and (d) equivalent compression generated on concrete or (e) ultimate strain in steel. The limitations are also defined in two different stages namely at the time of transfer of force to concrete and at the service load stages. Various national and international codes define the permissible stress in steel at 2 stages and are presented in Table No.2.12.

The Indian Standard specifications regarding various properties of such permitted steel are listed in Table 2.13 a, b, c, d. In case of strands there are

Table 2.12 Permissible Stresses in Prestressing Steel

Stages	IS 1343 - 1980 [1]	ACI 318 [2]	BS 8100 - 1985 [3]	FIP 1984 [5]	ENV 1992 [6]
Initial prestress or prestress at transfer limitations	Initial prestress should not exceed 80% of ultimate strength of tendons (*fpu*) = $[\frac{Characteristic\ strength}{Load\ Factor}]$ just behind anchorages	Tendon jacking force should not exceed 94% of vield strength (0.2% proof stress level) and not exceed 80% of Specified tensile strength. ***Immediately after transfer fee*** It should not exceed 82% of yield strength and 74% of tensile strength whichever is smaller. For post tensioning tendons at anchorages and couplers (stress should not exceed 70% of tensile strength) immediately behind the anchorage	The jacking force should not normally exceed 75% of characteristic strength of tendon and may be increased to 80% provided additional consideration is given to safety. ***At transfer*** the initial prestress should not normally exceed 70% of characteristic strength and in no case should it exceed 75% (even if additional consideration is given to safety)	The initial prestress in tendons should not exceed the following values (a) at the moment of tensioning 80% of 0.2% proof stress (or) 90% of 0.1% proof stress (b) at transfer of force 75% of 0.2% proof stress (or) 8.5% of 0.1 proof stress	The transferred force shall not exceed 80% of 0.2% proof stress whichever is less fr pretensioned products and the force at transfer shall not exceed 75% of 0.2% proof stress and 85% of 0.1% proof stress which ever is smaller
Final prestress after accounting for losses and live load effects	Not less than 45% cf characteristic strength f_p of steel	......	Not greater than 60% pf characteristic strength of steel	.....	......

Note: at transfer or after transfer - means after elastic shortening losses

Properties of High Tensile Steel

2.13. a. HTS wires (cold drawn stress relieved)

Standard	Nomial Diameter mm	Tolerance on Diameter of Element Wire	Tensile Strength, Min	min % of elongation on a G.L. of 200 mm	Radius of Jaws (For Reverse Bend Test)	0.2% Proof Load (N)
	mm	mm	N/mm²	Min	mm	
IS 1785 - Part I (1987)	2.50	+ 0.025	2010	2.5	7.50	min of 85% of tensile strength
	3.00	+ 0.04	1865	2.5	10.0	
	4.00	+ 0.05	1715	3.0	12.5	
	5.00	+ 0.05	1570	4.0	15.0	
	7.00	+ 0.05	1470 *	4.0	20.0	
	8.00	+ 0.05	1375 *	4.0	25.0	

2.13. b. HTS wires - as drawn

Standard	Nomial Diameter mm	Tolerance on Diameter of Element Wire	Tensile Strength, Min	min % of elongation on a G.L. of 200 mm	Radius of Jaws (For Reverse Bend Test)	0.2% Proof Load (N)
	mm	mm	N/mm²	Min	mm	
IS 1785 - Part II (1983)	3.25	+ 0.02	1765	2.5	10.0	min of 85% of tensile strength
	4.00	+ 0.03	1715	3.0	12.5	
	5.00	+ 0.03	1570	4.0	15.0	

2.13. c. HTS Wires Indented

Standard	Nomial Diameter mm	Tolerance on Diameter of Element Wire	Tensile Strength, Min	min % of elongation on a G.L. of 200 mm	Radius of Jaws (For Reverse Bend Test)	0.2% Proof Load (N)
	mm	mm	N/mm^2	Min	mm	
IS6003 (1987)	5.00 4.00 3.00	3.1 2.0 1.5	1570 1715 1865	4.00 3.00 2.50	15.0 12.5 10.0	min of 85% of tensile strength

2.13. d. Normal H.T.S. Strand (stress relieved)

Standard	Designation	Nominal Diameter of Element Wire	Tolerance on Diameter of Element Wire	Nominal Cross-sectional area of strand	Norminal mass of strand	Breaking load	0.2percent Proof Load
		mm	mm	mm^2	Kg/m	N	N
---	2-ply 2mm	2.0	+ 0.03	6.28	0.0493	12 750	10 840
	2-ply 3mm	3.0	+ 0.03	14.14	0.111	25 500	21 670
	3-ply 3mm	3.0	+ 0.03	21.21	0.166	38 250	32 460
IS 6006 - 1983	6.3mm 7-ply	6.3	+ 0.4	25.1	0.195	44 480	37 810
Class I	7.9 mm 7-ply	7.9	+ 0.4	37.4	0.295	68 950	58 600
	9.5 mm 7-ply	9.5	+ 0.4	51.6	0.408	93 410	79 400
	11.1 mm 7-ply	11.1	+ 0.4	70.3	0.555	124 540	105 860
	12.7 mm 7-ply	12.7	+ 0.4	92.9	0.730	164 580	139 900
	15.2 mm 7-ply	15.2	+ 0.4	138.7	1.094	226 860	192 830
IS 6006 - 1982	9.5 mm 7-ply	9.5	+ 0.66	54.8	0.435	102 310	86 960
					- 0.15		
Class II	11.1 mm 7-ply	11.1	+ 0.66	74.2	0.585	137 890	117 210
			- 0.15				
	12.5 mm 7-ply	12.7	+ 0.66	98.7	0.775	183 710	156 150
1.102 261 440	222 230						
			-- 0.15				

% Elongaton min on G.L. 600 (mm) = 3.5%; Pitch of the strand (lay length) between 12D & 16D when D is the nominal diameter of the strand.

2.13. e. Low Relaxatioon H.T.S Strand (stress relieved)

Standard	Nominal dia of strand	Tolerance	Nominal Area of strand	Nominal Mass of strand	Breaking Strength of strand	Breaking Strength of strand	0.2%Proof Load (90% of Breaking Strength	0.2% Proof Load (90% of Breaking Strength
IS 14628-1995	mm	mm	mm^2	kg/km	kN	kgs	kN	kgs
Class I	9.5	± 0.4	51.6	405	89.0	9078	80.1	8170
	11.1	± 0.4	69.7	548	120.1	12250	108.1	11026
	12.7	± 0.4	92.9	730	160.1	16330	144.1	14698
	15.2	± 0.4	139.4	1094	240.2	24500	216.2	22052
Class II	9.5	+0.66 - 0.15	54.8	432	102.3	10434	92.1	9394
	11.1	+0.66 - 0.15	74.2	582	137.9	14065	124.1	12658
	12.7	+0.66 - 0.15	98.7	775	183.7	18737	165.3	16860
	15.2	0.66 0.15	140.0	1102	260.7	26592	234.6	23929

% Elongation min on G.L. 600 (mm) = 3.5%; Pitch of the strand (lay length) between 12D & 16D when D is the nominal diameter of the strand.

two classifications namely Class I type and Class II type which have different properties especially with respect to strength and percentage elongation.

2.3.4 Inelastic Properties of Steel

Creep and relaxation are the two inelastic properties, which have to be considered in assessing the quality of steel. Creep is defined as the change in deformation of the material with time under a constant stress. It is affected by the temperature, humidity, applied stress level and also the manufacturing process. For sufficiently low stresses, creep effect is negligible. Such limiting levels of stress is called technical creep limit. ***Technical creep limit*** is defined as a constant stress [13] which brings an additional 3 percent strain (creep strain) corresponding elastic strains due to applied load within a period starting from 6 hours to 1000 hours at a constant temperature of 20°C. For prestressed concrete works, creep effect in steel is indirectly considered in the form of relaxation of force, which is described later. For heat-treated and stress relieved wires, the technical creep limit is very high and hence creep loss is low. This value will also be proportional to 0.01 percent proof stress value for any steel. If 0.01 percent proof stress is high, the technical creep limit also will be high. Hence steel is also judged by 0.01 percent proof stress for its creep properties.

Relaxation is defined as the change in stress in steel under a constant strain (elongation). This is directly connected to creep. If the strain due to an applied stress is maintained constant in a steel element, the applied stress slowly reduces with time due to creep effects. This is termed as relaxation of force in steel. This phenomenon, occurs in prestressed concrete works since the length of the prestressed steel is kept constant. Relaxation is primarily a function of applied stress and time. But it is affected by variation in temperature, humidity and applied stress level. The rate of loss of force due to relaxation is increasing with higher applied stresses. For low applied stress levels, say upto technical creep limit, there will be only very little (negligible) relaxation of force with time.

IS 1343–1980 assumes the technical creep limit at 0.5 times the characteristic strength of steel. Hence for applied stress level higher than 0.5 f_p, only this code recommends the loss in stress in steel due to relaxation. Definite values have been given for different level of applied address in IS 1343–1980 as shown in Table 2.14.

Table 2.14 Relaxation Losses for Prestressing Steel at 1000 H at 27°C

Initial Stress	Relaxation Loss (N/mm^2)
$0.5f_p$	0
$0.6f_p$	35
$0.7f_p$	70
$0.8f_p$	90

NOTE : f_p is the characteristic strength of prestressing steel.

For *qualifying the material for acceptance* for use in prestressed concrete works, IS code define the relaxation test to be conducted at controlled temperature of 21° C to 24° C. They also specify that the loss should not be greater than 3.50 percent of applied stress, and 5 percent of applied stress, after 100 hours and 1000 hours respectively if the specimen is stressed to a level equal to 0.7 times its ultimate stress (nominal). Loss in stress due to relaxation abnormally increases if the applied stress goes beyond 0.7 f_p. This is one of the reasons that most of the statutory regulations limit the level of applied stress steel for prestressing. Now-a-days low relaxation steel is produced and it is accepted/preferred for use in important structures. IS 14268–1995 [20] gives the stipulation relating to low relaxation steel specially for strands.

BS 8110 states that loss in balance due to relexation in prestressed concrete structures can be evaluated by multiplying the loss predicted in 1000 hrs relaxation tests conducted at the prestressed level by a factor as detailed below.

Pretensioned product	Applied Stress Level
1.2 for Class II type of steel 1.5 for Class I type of steel	Should correspond to stress level in steel just after the time of application
Pretensioned product	**Applied Stress Level**
1.5 for Class II type of steel 2.0 for Class I type of steel	Should correspond to stress level in steel just after transfer of force to concrete

FIP 1984 recommends the loss due to relaxation to be considered in proportion to applied stress level as follows:

For low relaxation steel	3% to 10%	For applied stress level of	0.6 times to 0.8 times	of proof stress
For normal relaxation steel	6% to 12.5%	For applied stress level of	0.6 times to 0.8 times	of proof stress

Normally the applicd stress level is about 0.6 times the 0.2% proof stress level when the material suffers sustained loading and hence 3% can be assumed for low relaxation steel and 6% can be assumed for normal relaxation steel.

The loss of stress due to relaxation can be reduced by initially overstressing for a few minutes and then decreasing the stress to the required level. A correlation between the relaxation at 100 hours and at 1000 hours has been evaluated after a vast series of experimentation at I.I.T Madras [21].

2.3.5 Fatigue properties of steel

Fatigue strength of steel is ano her important property of steel, which is to be considered in the prestressed concrete works. In dynamically loaded

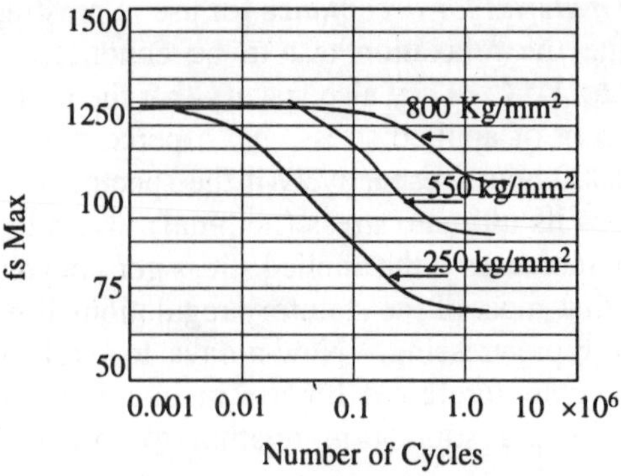

Fig 2.11 Typical Strength Cycle Diagrams For P. S. Steel (Wohlers's Diagram)

structures, prestressing steel is subjected to varying tension with time. It can be generally said that often repeated loading leads to a decrease in the capacity of material. It also depends on the frequency and amplitude of variation in stress level due to repeated loading. This property of steel is defined by 'Wohlers curves' and 'Smith's diagram [13] as explained below.

Several tests have been carried out on steel wires/rods with same lower limit on stress $f_{s\ min}$ but with different amplitudes of stress (namely different upper limit on stress $f_{s\ max}$). The number of cycles that the specimens have withstood before failure, have been recorded. From the series of experiments, it has also been noted that if a specimen crosses 2 million cycles with a given lower and upper limit of stress, it does not fail at all then after, for infinite cycles. Hence 2 million cycles has been identified as the number of cycles that the specimen has to pass, for a given range and lower limit of stress, if the specimen has to be qualified for fatigue conditions. Fig 2.11 shows a graph between the upper limit stress and the number of cycles

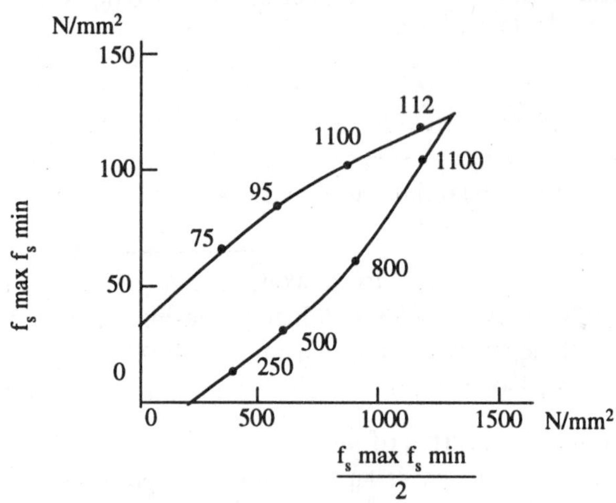

Fig 2.11Smith's Diagram

that the specimens have withstood without failure, for a given lower limit stress. It is seen from the figure that the upper limit decreases if the specimen has to cross-higher number of cycles. All graphs become asymptotic at 2 million cycles Hence 2 million cycles have been fixed as the fatigue limit for testing. These graphs are called 'Wohlers Curves'. The upper limit stress, which can be applied without failure, is called the fatigue strength for a particular lower limit. From the fatigue strength so obtained for different lower stress limits, a fatigue diagram is plotted, as shown in Fig 2.12. This is called Smith's diagram. The limit $f_{s\ max}$, $f_{s\ min}$ are plotted in ordinates and the abscissa represents $(f_{s\ max} + f_{s\ min})/2$. There will be a stress amplitude, for which at given maximum and minimum stress respectively no fatigue failure is expected. This amplitude reduces for increasing lower limit till the lower and upper limit meet and become one and the same namely the static strength.

It is well known that in prestressed concrete works, the ***actual stress variation*** due to service loads is very small (100 N/mm^2). Hence for fatigue strength, the reduction of upper limit of stress in steel from its 0.2% proof stress will be very small in case of prestressed concrete structures. There is very little danger of fatigue failure in prestressed concrete elements. However fatigue tests have to be carried out both for prestressing steel and anchorages before they are actually used in structures. ***The fatigue behaviour will be critical near the anchorages for prestressing steel because of local bending possibilities. Indian Standards do not give any specifications and acceptance criteria for prestressing steels based on fatigue properties. Hence as on date, BS 4447 is being adopted for such specification and acceptance criteria for use of prestressing steel in important prestressed concrete structures in India.***

2.4 Corrosion of Prestressing Steel

Corrosion of prestressing steel in concrete is an electro chemical process. The electro-chemical potential is generated (a) due to differences in the concentration of dissolved ions such as chloride and oxygen in the vicinity of steel and (b) due to 2 dissimilar metals embedded in concrete such as high tensile steel and sheathing metal. As a result, one of the metals becomes anode and other cathode and the electro chemical process starts with moisture in the concrete forming an electrolyte. Even in the same material (steel) because of dissimilar cells there is potential difference generated. One portion becomes the anode and other portion becomes the cathode. The water/moisture in the concrete forms the electrolyte and the corrosion process starts. By corrosion, there is deposition of iron oxide in the form of rust at the anode and depletion of iron ions at cathode. This deposition of iron oxide slowly increases the total volume can be as high as 600% of the original volume leading to pressure on surrounding concrete [22]. This leads to cracking of concrete. Hence corrosion cracks are mostly parallel to reinforcement. It must be noted that the anodic reaction involving ionization of metallic ion cannot progress unless flow of electrons from the cathode to anode is maintained. Further for corrosion process to start, an electrolyte and free surfaces of anode and cathode are necessary.

In steel normally there is always a thin ion oxide film available on the surface, which is impermeable. Till this film is existent, electrolyte process cannot start. This passive oxide film is broken either by the acidic environment such as existence of free chloride ions or due to reduction of alkalinity of concrete due to carbonization.

The carbon-dioxide present in atmosphere combines with hydrated concrete and neutralizes the alkaline nature of the concrete. This process is called carbonization. This leads to a reduction of pH value of concrete from around 12 to 9. Under these conditions it is easy to break the passive film on steel reinforcement. Further free chloride ion present in the concrete surrounding the reinforcement may also react with alkaline solution to form hydrochloric acid, which would break the protective layer on steel. Hence chloride attack and carbonization of the concrete, are specific causes, which would break the passive film on steel and leading to corrosion.

Free Hydrogen in the form of nascent ions present in concrete or steel may lead to formation of hydrochloric acid and further lead to cutting of the passive films causing corrosion.

2.4.1. Hydrogen embrittlement

When normal HYSD bars or high tensile steel gets in contact with acid, nacent hydrogen at atomic stage is generated. That penetrates into the steel surface and makes the steel brittle. On application of stress, due to want of ductility, the material fails suddenly without giving sufficient warning. Even small amount of Hydrogen can cause considerable damage to the tensile behaviour of especially high tensile steel wires. The brittle nature of the steel caused by Hydrogen is called Hydrogen embrittlement. Even rain water may become acidic if it passes through atmosphere carrying acid vapour and such rain water settling on high tensile steel may cause hydrogen embrittlement. Hence, high tensile steel should be protected from surface impurities of acidic nature even when it is kept in stock. Hence while at stock and during transport, protection against such contact has to be ensured. This is normally achieved by covering the steel with bituminous crape paper or some kind of quoting of alkaline nature.

It is clear from the above discussions that the alkaline environment of the concrete around the steel, prestressed tendons or any reinforcing steel should be maintained to prevent carbonization, chloride attack, formation of acidic substances which may break the passive layer on steel, and hydrogen embrittlement. It is expected that good dense concrete of sufficient thickness provided as cover for the reinforcement/prestressing steel will ensure safety against the above. The cover concrete must have sufficient strength and it should not get cracked due to tensile stresses caused by non structural effects like shrinkage, temperature etc. The minimum cover for prestressing tendons varies depending on various aspects such as exposure conditions and type of concrete, the grade of cement used, water cement ratio, maximum aggregate size etc.

British Standards 8110 of 1985 [5] suggests cover varying from 20 mm for mild exposure conditions to 50 mm for extreme exposure conditions. This has to be increased for fire resistance safety against fire hazards.

Indian Standard codes [1,7,10] also define the cover requirements based on exposure conditions and it varies from 25 mm to 50 mm for normal structures and from 40 mm to 75 mm for special structures like bridges.

2.5. Modern Materials

2.5.1 Light Weight aggregate for prestressed concrete

The purpose of using prestressed concrete for structural elements is to use the material fully and economically and to get a slender and less heavy structure. This can be further achieved by using lightweight aggregate for concrete without much loss in structural capabilities in compression for concrete. This idea has been gaining importance since 1955.

With the present day knowledge of concrete mix design using light weight concrete, it is possible to get a compressive strength of anywhere between 30 to 50 MPa by choice of proper aggregates, good quality cement and a mix design. Foam slag, lytag, aglite are a few lightweight aggregates, which can be used in making concrete. Data giving the properties of lightweight aggregates available in United States is given by Shideler [23]. As the time dependent properties of such concrete namely creep and shrinkage etc do play a role, more information on such lightweight aggregates are required. Best et. al [24] and Jones et. al [25] have discussed these aspects as early as in 1957. The use of lightweight concrete for prestressing works has not picked up e in India. The modulus of elasticity of such concrete will be of thc order of 60 % of that of normal concrete and the unit weight will be of the order of 60 to 70% of the normal concrete. For more understanding of ulilisation of lightweight concrete in prestressed concrete structures, reference can be made to Gervick [26]. The applicability, the advantages and the limitations have to be properly understood before the use of lightweight concrete for prestressed concrete works.

2.5.2 Fiber glass/carbon tendons for prestressing elements

As advancement in prestressing materials fiberglass, manufactured by drawing fluid glass into fine filaments, can be used as the prestressing element replacing steel. This can be made into parallel chords, twisted strands, and parallel fibre embeded in plastic. The ultimate strength of 7000 MPa can be achieved in fiberglass tendons. The other aspects such as ductility, fatigue properties and time dependent properties like relaxations, should be carefully understood before use of such materials. Carbon fibres are also finding its place as prestressing element in modern days. The reader is referred to specialized literature for use of such materials in prestressed concrete works.

References

1. IS 1343 - (1980). Indian Standard **"Code of Practice for Prestressed Concrete"**, Bureau of Indian Standards, New Delhi.
2. ACI 318 - (1998). **"Building Code requirements for Structural Concrete"**, American Concrete Institute, Formington Halls, HI 48333, USA.
3. BS 8110 - (1985). **"British Standard for "Structural use of Concrete"**, British Standards Institution, London WIA 2BS.
4. ENV 1992 - 1,2 (1991). **Europen Standard. Design of Concrete Structures Part I - General rules and rules for Buildings.** European committee for Standardisation - B - 1050 Brussels.
5. FIP - (1984). **Recommendations for Practical design of Reinforced and prestressed concrete structures, based on CEB - FIP model code** (MC - 78), Thomas Telford Ltd, London 1984.
6. IS 10262 - (1982). Indian Standard Guidelines for **"Concrete Mix Design"**. Bureau of Indian Standards, New Delhi.
7. IRC - 21 - (1987) " **Standard specifications and Code of practice of Road Bridges" - Section III, Cement Concrete (plain and reinforced) Second revision,** The Indian Roads Congress, New Delhi.
8. MEHTA P.K., MONTEIRO P.J.M., (1997) **"Concrete", "Microstructure, Properties, and Materials" Indian Concrete Institute,** Chennai.
9. RAO P.S., RAJAGOPALAN. N., ARAVINDAN P.K., ACHYUTHA. H. (1975) **"Development of Prestressed Concrete Monoblock sleepers for Indian Railways",** Structural Engineering Laboratory, Indian Institute of Technology, Madras (unpublished).
10. IS 456: (2000) Indian Standard **"Code of Practice for Plain and Reinforced Concrete",** Bureau of Indian Standards, New Delhi.
11. LEONHARDT. F. & MONNIG. E., (1973), **"Vorlesungen "ueber Massivbau",** Erster Teil, Springer Verlag, Berlin.
12. RôS . M.U. SARRASIN. A., (1950), **Die Material - technischen Grundlagen und Probleme des Eisenbetons in Hinblick auf die Zukunftige Gestaltung der Stahlbeton - Bauweise.** EMPA Bericht Nr. 162 Zurich.
13. LEONHARDT. F, (1964) **"Prestressed Concrete Design and Construction",** Wilhelm Ernst and Sohn, Berlin,.
14. **National Building Code of India** (1970), Bureau of Indian Standards, New Delhi.
15. **"Precast segmental box girder bridge manual",** Prestressed Concrete Institute, Chicago, Illinois, U.S.A.
16. TROST. H (1967) **"Implications of the superposition principle in creep and relaxation problems for concrete and prestressed concrete"** Betons Stahlbetonbau 62(10) PP 230-238 & 62(11) PP 261-169, (in German).

17. IS 1785 (Part II) - (1983). Indian Standard **"Specification for Hard drawn (stress relieved) steel wire for prestressed concrete"**, Bureau of Indian Standards, New Delhi.
18. IS 6003 - (1983). Indian Standard **"Specification for indented wire for prestressed concrete"** Fourth revision, Bureau of Indian Standards, New Delhi.
19. IS 6006 - (1983). Indian Standard **"Specification for uncoated stress relieved strandsfor prestressed concrete"**. First revision, Bureau of Indian Standards, New Delhi.
20. IS 14268 - (1995). Indian Standard **"Uncoated stress relieved low relaxation seven-ply strand for prestressed concrete specification"**, Bureau of Indian Standards, New Delhi.
21. RAJAGOPALAN. N., etal (1979), **"Relaxation properties of prestressing steel"**, Proceedings of All India seminar on prestressed concrete structures, Thiagarajar College of Engineering, Madurai.
22. CRANE. A.P., (1983), **"Corrosion of Reinforcement in Concrete Construction"**, Ellis Horwood Limited, Chichester, West Sussex, P.O. 19, IEB, England.
23. SHIDELER. J.J, (1957), **"Lightweight Aggregate Concrete for Structural Use"**, J. Am. Conc. Inst., October 1957, (Proc., Vol 54), PP 299 -328.
24. Best.C.H., et.al, (1957) **"Creep of Lightweight Concrete"**, Magazine of Concrete Research, November 1959, PP 129-134.
25. JONES. T.R., et.al, (1957), **"Properties of Lighweight Related to Prestressing"**, Proceedings World Conference on Prestressed Concrete, San Francisco,.
26. GERWICK. B.C, (1971), **"Construction of Prestressed Concrete Structures"**, Wiley Inter Science, New York.

3. Prestressing Systems and Devices

3.1 General

Transferring prestress either in pretensioned or in post tensioned elements has to be done by systems and devices which have the facility of pulling the prestressing elements to the required force and anchoring them in the pulled or elongated state so that the prestressing force is transferred to the structural elements. Pulling is against a prestressing bed in case of pretensioning system or against the proposed prestressed concrete element itself, in the post tensioning system. Pulling and anchoring can takeplace from either end of the prestressing element or from one end only keeping the farther end fixed in position. In such a case, the farther end is called the dead end and the fixity provided there, is called dead end anchorage. The dead end anchorage can be by embedment or by external device. This chapter is devoted for discussing about the various systems and devices used for pulling anchoring and transferring the forces. Pulling is done usually with jacks. Jacks, nowadays, are hydralically operated. They have the facility of pulling the prestressing element directly or pulling the anchor plate to which the prestressing elements are already anchored. This is done by pumping oil and ram movement. On release of the pressure of oil, the hold on the prestressing element or the anchor plate by the jack is released and the prestressed element gets anchored at the same place or the anchor plate is fixed in the same place where the jack left it. For each patented system, the jacks are designed separately to suit the system. Various anchorages, have been developed by different organisations and patented. But in general they can be grouped in to two types based on their structural action namely anchorages by embedment and anchorages through external components or devices.

3.2 Anchorage by Embedments

Embedment types of anchorages are normally used for dead end anchorages and for pretensioned products. Here the transmission of force on to concrete is by bond and can be fully achieved only after a distance from the free end of the prestressing element namely steel. This length is called transmission length and explained in greater details in a later chapter on transmission and anchorage zone. To make this length of transmission a minimum, several methods are thought of. A few methods adopted in practice are by curving of the tendons,by wobbling the tendons(wires) or by increasing the friction and bond characteristics of tendon on its surface. These methods lead to eventual problems for which remedial measures are to be adopted.

3.2.1. Curving of tendons

While the transmission of force could be accommodated in a small distance, by curving the tendons, there are limitations on the curvature that could be given to steel. It is dependent of the material property of steel and the extra stresses induced in confined concrete, due to curvature forces transmitted by the steel component. Hence the strength of concrete has to be very high. Since steel as a material has defined stress strain properties, the maximum curvature it can take before the material reaches a stage of limiting stress (yield or 0.2 percent proof stress) is also defined. The method of curving of tendon is used only in dead end anchorages. In case of such anchorages, as the tendon is pulled, there is the tendancy for the tendon is to open out from the curved ends. This is prevented by concrete around the tendon and the tendon exerts a radial force on concrete. This radial force is a direct function of the force in the tendon, which is decreasing towards the dead end. Hence to have a uniform pressure on concrete, the curvature of the tendon is increased in the same proportion as the decrease in force in tendon. The natural solution for this to have a logarithmic spiral as shown in Fig.3.1. governed by the expression $R_x = Re^{-\mu\alpha_x}$ where R is the radius at beginning, μ = coefficient of friction between steel and concrete and α_x = change in the direction of point x.

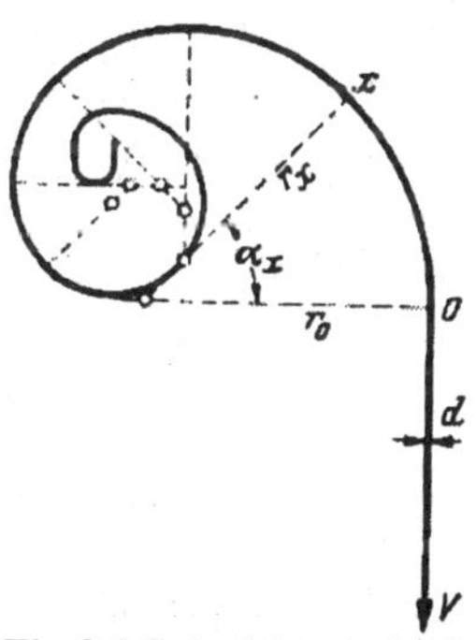

Fig 3.1 Spiral Hook Anchorage

Thus spiral anchorage system is patented by a French Company called Held and Franke Bau A.G.

3.2.2 Wobbling and surface characteristics

By providing undulations, wobbling, and increasing the surface roughness by indentions or ribs, the bond between steel and concrete increases. This decreases the embedment length. Two types of anchorages where wobbling is used are shown are shown in the Fig 3.2.

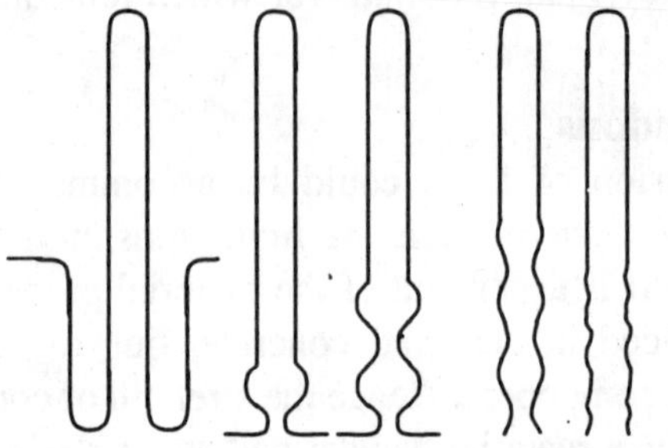

Fig 3.2 Wire Anchorage by Wobbling of Tendon

But in these types, tension is developed in concrete due to reversal of curvature in tendon. Leonhardt and Baur (1) have suggested a remedial measure to overcome the tensile stresses by providing steel around the curved tendons which would take care of any eventual tension and also increase the compressive strength of concrete due to confined nature. This system is called 'Leoba (Fig 3.3) anchorage used for dead end anchorages. Similarly other anchorage are also developed by many professionals, and one other is shown in Fig 3.4.

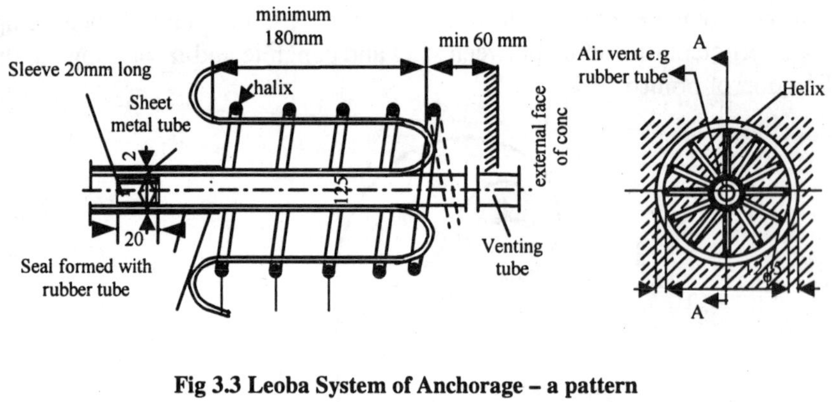

Fig 3.3 Leoba System of Anchorage – a pattern

By changing the surface characteristics of the material (steel), bond between concrete and steel can be increased through better adhesion, shear, and friction between steel and concrete. Adhesion is improved due to the fineness of cement paste and the paste fills the surface pores of steel. Surface roughness of steel increases the friction between concrete and steel. Undulation, such as ribs, increases the shear between concrete and steel.

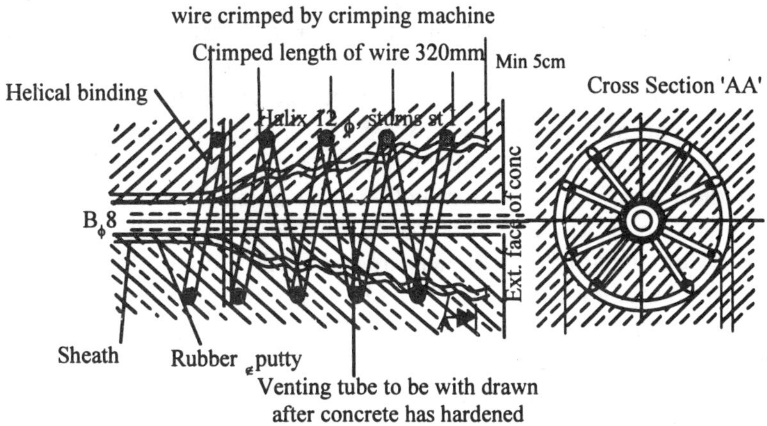

Fig 3.4 Helically Bound Fan type Anchorage

These prevent the separation between the mate rial and hence considered to have increased the bond. Increased bond by any one of the above, reduces the transmission length. The reason for smaller diameter or bars of wires being preferred to larger diameter bars is that, the smaller diameter bars give larger surface area of contact and hence better bond and smaller transmission length. Also the ribbed bars such as TOR steel have got smaller transmission length than smooth bars. Surface roughness can also be improved by identing the wires on their surfaces. All these aspects play a major role in pre-tensioned concrete products as the force is transferred only by means of bond in that system. In pretensioned products, the bond is improved further by lateral force developed by the wire itself. This was explained by Hoyer (2). As the wire or rod gets tensioned, its lateral dimensions are reduced to some extent. Then on release of force, they try to get back to their original diameter, which causes a swelling of rods near the free and zones. This swelling effect causes radial forces in concrete to an extent of $80N/mm^2$ which in turn give rise to frictional forces increasing the bond characteristics of the bars. This effect is called Hoyer effect. Transmission length is reduced because of this effect. More details on bond length, transmission length, Hoyer effect and anchorages by bond etc., are dealt in greater detail in chapter on Transmission and Anchorage Zone.

3.3 Anchorage by External Components

3.3.1 Parallel screw threads and nuts

One of the earliest methods of transferring prestressing force by an external system is by means of a bolt and nut arrangement. If the tendons are provided with threads at the ends, nuts can be used and tightened against the concrete, transferring a force on concrete. But while cutting the threads, on the tendon, the core section is reduced and that is not desirable. So an important improvement is later thought and the threads are got by hot upsetting the tendon ends, so that the core section of threaded zone remains equivalent to normal section of the tendon (rod). There is an American process patented by R.H. Dill in 1943 wherein screw thread is partially pressed and partially projecting beyond the normal diameter. The core section is larger than the cut screw threads area.

Dyckerhoff and Widman system in Germany called Dywidag (2) and Lee Mac call system in USA (3) are based on these principles. In case of Diwidag system, (Fig 3.5) the threads are obtained by cold rolling. Because of work hardening, the core of the thread has got almost the same capacity as the normal diameter portion. Flanged collar nuts are used to assure better distribution of force on all threads of the nut. The nuts are formed with a lip on the inside bearing face to relieve stress concentration.

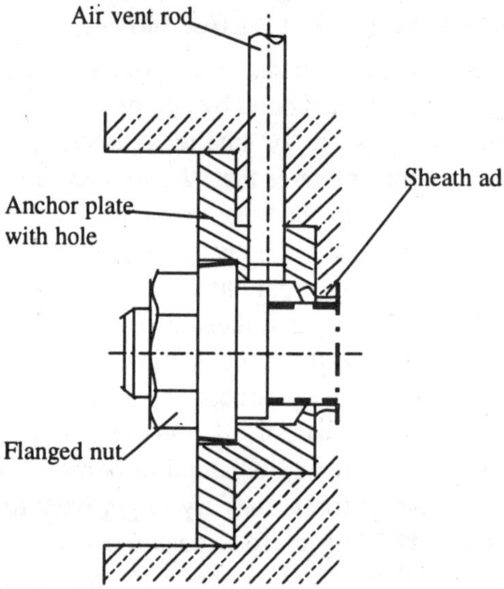

Fig 3.5 Dywidage Anchorage with Flanged nut nut (26mm bar)

The anchor plate has stepped recesses for the flanged nut and the anchor plate is also provided with grooves through which grout can be injected into tendon sheath. Because of stringent demand with regards to fatigue strength of the anchorage as per German Standards, Dywidag have further improved their system. To achieve the fatigue strength comparable to other types of anchorages, Dywdag brought in a tapered nut which would sit on the special anchor element, which is not a simple plate. In this type (Fig 3.6) the form of nut also resists the forces because of shell action. Arrangement for grouting is also provided in this system.

3.3.2 Anchorage with wedges

The prestressing elements (wire or a strand) are passed through or over a wedge piece which is in the form of a trancated cone and the wedge piece is

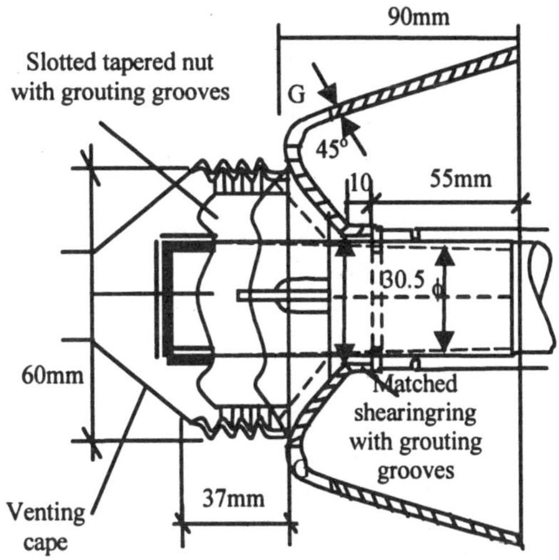

Fig 3.6 Dywidag Anchorage with Bell-Shaped Component

set into a cylindrical sleeve which has a similar hole to accommodate the wedge. The hole has a varying diameter from one face to other face of the sleeve component. When the prestressing element is pulled, it comes out of the sleeve through the wedge in which it is accomodated. But after applying the force, if the jack is released the steel element along the wedge tries to move into the sleeve. But the wedge can not move in and hence gets locked into the sleeve along with the prestressing element. The sleeve itself may either bear against the concrete or on a steel plate device embedded in concrete. Now the force is transferred from the tendons to wedges and sleeves and then to concrete by bearing. The wedges and sleeves are to be designed to resist the force transmitted by the tendons. There are two types of wedges. In one type, the wedges slip inside the sleeve and cause a resultant thrust on the sleeve and in the second type the wedges are rammed in by external forces. The first type of wedges are called the slip wedges and the second type of wedges are called the rammed wedges. The forces acting on these elements of wedges are discussed by Leonhart [1]. For patented systems, these aspects have been well taken care of by the manufacturers. Freyssinet system (4), Gifford ud-all system (5), Magnel Blaton system (6) etc., are certain standard types of wedge anchorages.

Freyssinet's System

The first anchorage of this system developed by Freyssinet has the anchor element consisting in two parts, a cylindrical piece with tapered hole (sleeve) and a cone (wedge). The cylindrical piece is a helically reinforced concrete unit cast with M60 concrete, and the tapered hole in the cylindrical piece is lined with closely wound helix of 2.5 mm ϕ steelwires of strength of 200 N/mm². This element is called the female cone. The idea of providing helical steel of capacity of 200 N/mm² is that it should be stronger than the wires pulled through the cone. The second part, namely the circular tapered wedge called the male cone is made of high strength concrete of M 100. A small steel tube is inserted inside the cone along its axis to provide facilities for grouting. The surface of this cone is grooved with 12 or 18 grooves depending on the number of wires to be stretched using the system. The width and depth of grooves are also defined by the diameter of the wire. Normally Freyssinet system is used for pulling 5 mm wires 12 to 18 Nos. The wires are taken over the male cone through the grooves and this unit slides into cylindrical unit (Female cone). As wires are stretched, stressed and released, male cone slips in to the sleeves. But this cone has to be thrust home with a force equal to one third of the intended prestressing force to

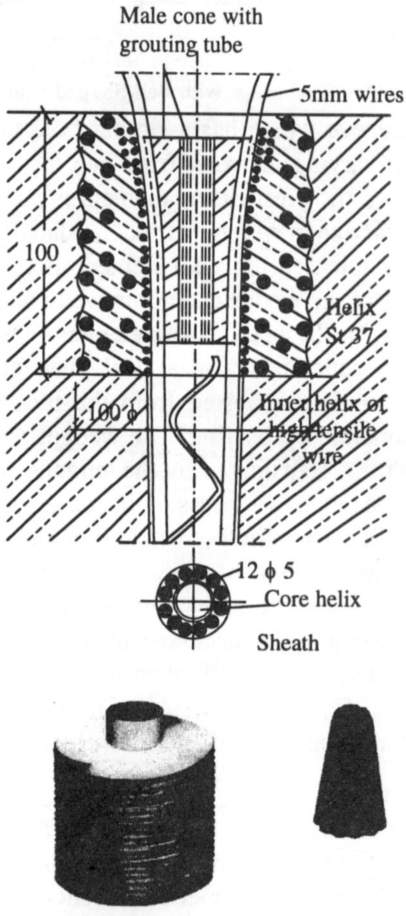

Fig 3.7 Freyyinet's 12 wire anchorage.

achieve full anchorage of the prestressing element. Even as the wires are released there will be minor slipping in and the average value of slip is 4 to 5 mm for cables with 12 wires of 5 mm and 6 to 8 mm for cables with 18 wires of 5 mm. The slip is taken into account while deciding on the elongation. 12 wires of 8 mm are also anchored by similar system. For high capacity of transfer of force central cone is designed as a heat-treated steel cone to take care of forces which comes on it. This system is shown in Fig 3.7.

As the demand of transferring higher prestressing forces inlarge industrial structures increased, this system is further developed and improved based on same principle. Strands are used instead of wires, as prestressing steel, which has a capacity of 180kN to 260kN per strand. They could be stressed up to 0.6 to 0.7 times their limiting capacity as per most of the national and international codes. Individual strands are pulled through slit wedges. (Three pieces or two pieces).

The slit wedges pass through tapered holes on a anchor plate. The number of holes in any one anchor plate may be 7, 12, 15, 19 as the case may be. The strands are passed through slit wedges, which are already placed, into the tapered holes of anchor plate. The anchor plate is pulled as one unit and all the strands are simultaneously pulled. The pulling jack has facility of

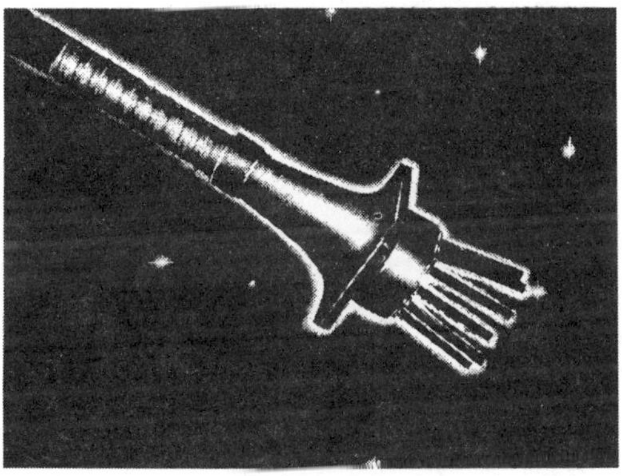

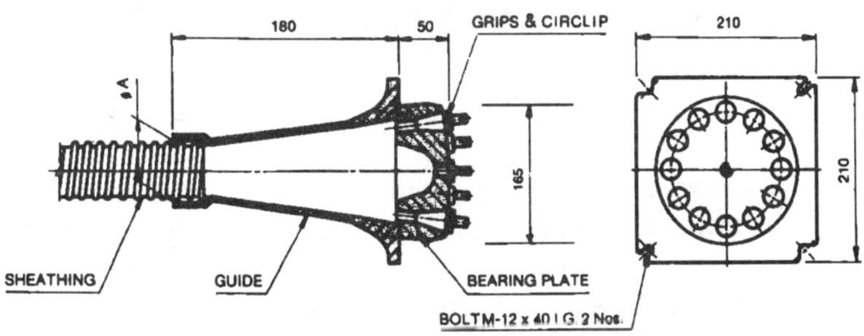

Fig 3.8 Freyssinet's System

anchoring and locking the strand to the jack itself. This is called lock anchorage. While the ram of the jack moves forward, the lock anchorage is moved backward pulling all the strands together, through the regular anchor plate and through the wedges. Once the required force is obtained (through required elongation), the strands are locked in the main anchor plate. When the jack pressure is released the lockanchorage moves to its original position. For this direction of movement, the strands are free of the lockanchorage stystem. By this process, the bulk pulling of the strand, is carried out. For more details, the patenting authorities have to be consulted. Bulk pulling of anchorage say 12nos of 7ply strand of 12.72mm dia is shown in Fig 3.8.

Magnel Blaton System

Another system, which follows the wedge type, is called Magnel Blaton system. In this system 8 wires are stretched and gripped in pairs by wedges. The anchor plate is provided with top and bottom wedge plates with tapered slots. For anchoring cables, number of such wedge plates are simply stacked. They are called sandwich plates. The wires are tensioned in pairs. The anchor plate and the wedges are of high strength steel to take care of the transfer forces. Finally it is the anchor plate which bears against concrete transmits the force to concrete. The pulling jack system is separately designed. The details could be seen in Fig 3.9.

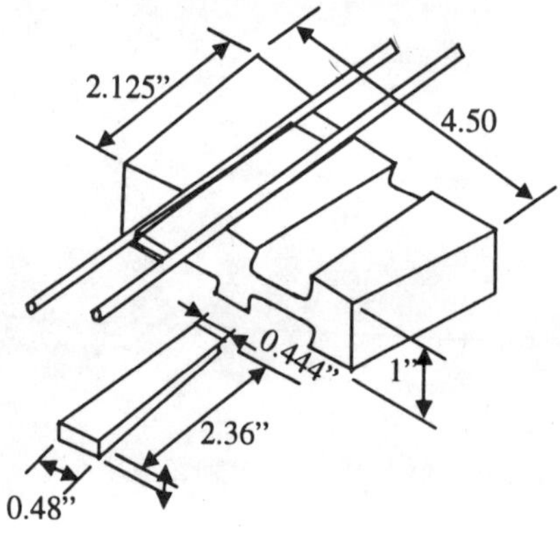

Fig 3.9 A Magnel Sand Wich Plate with Wedges

Gifford Udall System

Gifford Udall system is another wedge system wherein a small wedge is slit into 2 or 3 units and the cut pieces have teeth on the inner face. The individual wire or the tendon is gripped between the cut wedge pieces by means of the teeth. The wedge itself is housed into the sleeve. The force is transferred from wedge to sleeves is by wedge action. The wire or the tendon is kept fixed to the wedge by means of biting teeth. These are originally developed and used for single wires of sizes 2.6 mm to 7 mm. Number of such wire grips bear against a metal anchor plate which bears on concrete.

But this principle is used by more patented system for bulk pulling of wires / strands. The basic system is shown in Fig 3.10

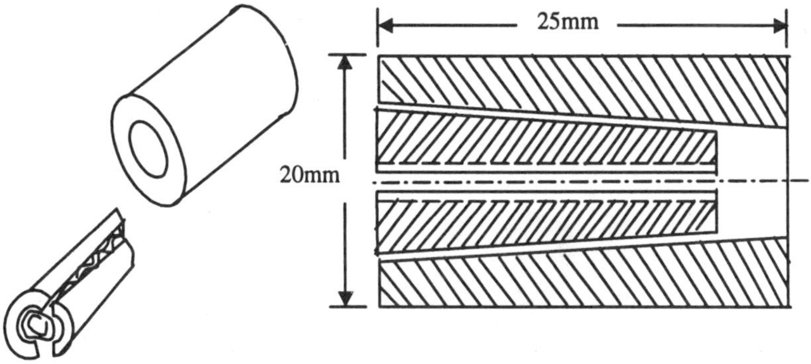

Fig 3.10 Gifford-Udall System of Wedge, Anchorage for One Wire or Strand

V.S.L. System (6)

This system follows the wedge type of anchorage. More number of wedges for individual strands are used and these wedges pass through tapered holes in an anchor plate.

The anchor unit is provided with internal and external screw threaded tubular attachment. The pulling device of the jack is attached to the inner screw thread. The annular nut, which secures the anchor in the tensioned position, is attached to the outer screw thread. The conical interior wedge has a bore with a screw threaded connection for grouting tube. V.S.L. tendons are made in 20 different sizes for prestressing forces 22 to 177 tones and hence it is more adoptable and versatile.

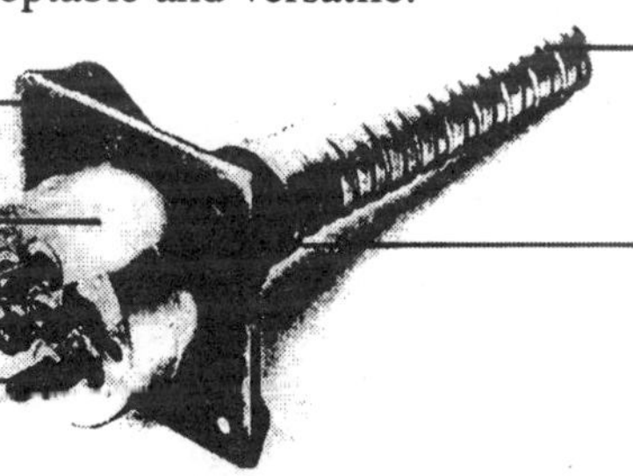

Fig 3.11 Stressing Anchorage VSL Type Screwed Thread 2 piece Head

3.3.3 Anchorage by enlarged ends (BBRV System)

If strain hardened steel is used as prestressing tendon s, the end of such tendons can be shaped into heads by cold or hot upsetting which would function similar to nut on a threaded bar. The head would bear against an anchor plate and transfer the force. One such system is developed by a group of swiss engineers Birkenmeyar, Brandestain, Ros, Vogt and called as BBRV system (3,4,7). The formation of the head be such that there is no stress concentration effect at the places where the flow of material changes direction. The length of barrel for forming a button should be 1.4d to 1.5 d where d is the diameter of the bar. Later the procedure for formation of the head was changed to a cold pressing approach. To start with BBRV system was developed for single wire pulling. Later it was developed for use for bulk pulling of wires Fig 3.12. While the formation of button head is done by the same cold prestressing approach, different types of jacks are used for pulling the anchor plate to which number of wires are anchored with button heads. BBRV systems have now introduced wedge system for bulk pulling.

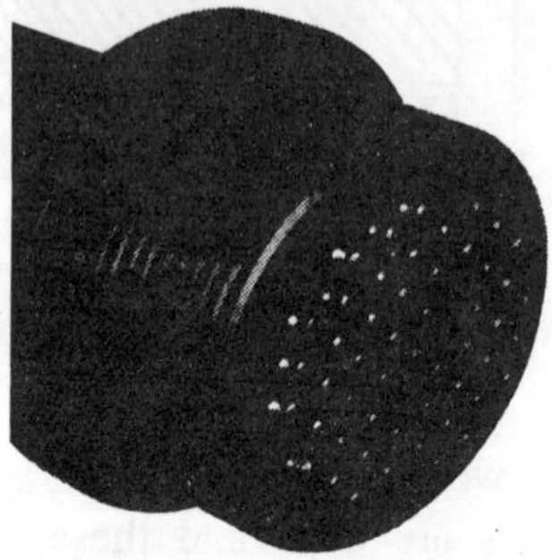

Fig 3.12 BBRV System

3.4 Couplers

With development in technology and application, prestressing is being used for many purposes in different types of structures, and at different construction stages. In some places prestressing has to be continued in the newly added portion and bring in continuously. When for this purposes all patented systems have developed special units called as couplers. These couplers have a tube unit, which could accommodate two anchorages one on either side of the tube. The coupler unit is attached to existing anchorage, which has been used as a live end anchorage at one end. At the other end of the coupler another anchorage unit without the prestressing tendons is attached. As and when the new continuity element is added, the prestressing element of the continuity unit is anchored to the free available anchoring unit of the couple and that acts at a dead end anchorage for the continuity unit. The continuity unit prestressing tendons can be prestressed from the farther end and anchored. Then the continuity is achieved between the new and old concrete units. There are other couplers for extending the tendons which have been already stressed and concreted by a prestressing system. A few of them are also shown in Fig 3.13 to 3.15.

3.5 Non-Conventional Prestressing

There are also methods of prestressing the concrete without special systems and devices and they are called non-conventional prestressing. Typical of them are Thermoelectrical prestressing and Chemical prestressing. In Thermo electrical prestressing steel wires are originally

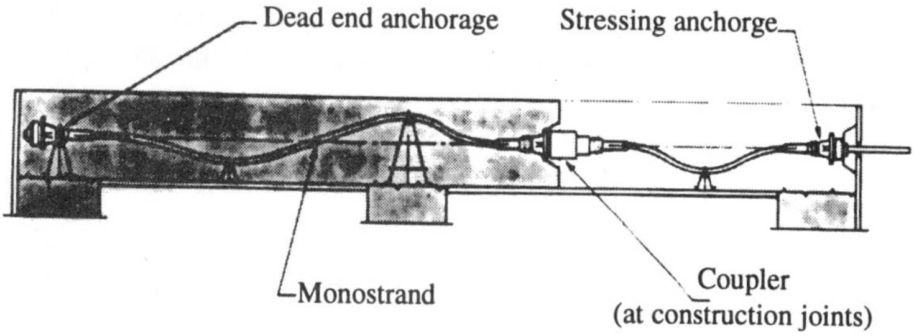

Fig 3.13 VSL Monostrand System Components

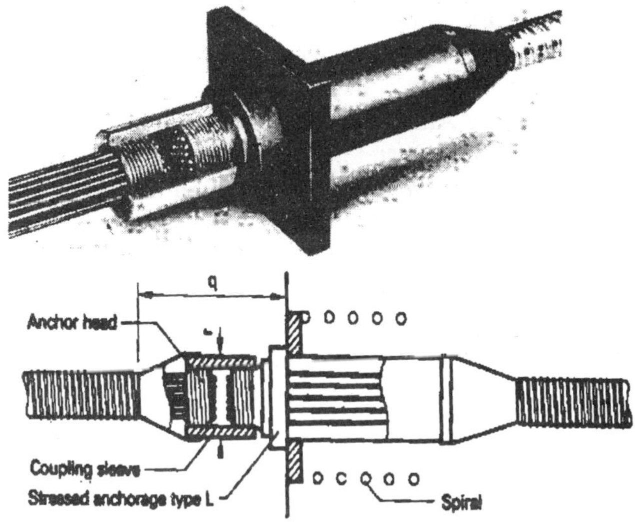

Fig 3.14 LFL Fixed type Coupler

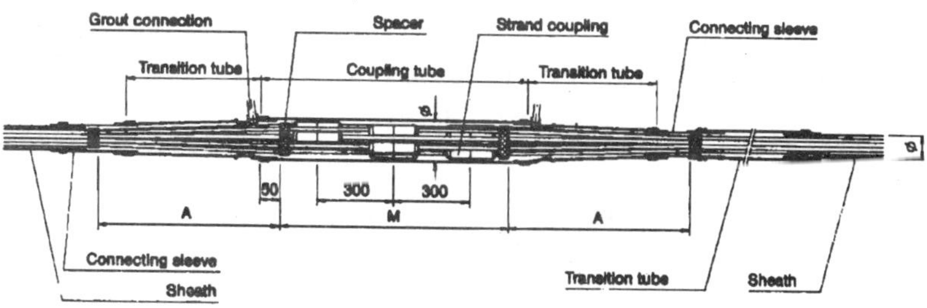

Fig 3.15 Strand Coupler Dywidag

elongated by applying current and heating them and then allowed to revert back to their original position by cooling.This process introduces compressive stress in concrete. The extension of the wires is done by temperature rise caused by a electrical currents and hence called Thermo electrical prestressing. By increasing the temperature of the bars to the level of 300°- 400°C an elongation is introduced up to 0.4% to 0.5% and the concreting is done under the elongated conditions of the wires. On cooling the prestressing wires, they try to shorten leading to a compressive stress on concrete. The cooling time is of the order of 12 to13 minutes and in this process a stress level of 500-600 N/mm^2 can be achieved on steel. This are more commonly used for pretensioned products. The temperature has to be properly controlled not to exceed 500°C lest the crystal structure of the material steel could get converted, leading to different properties. Because of the complications in prestressing and also of the non-assurance of the exact quantity of the prestress, this type of prestressing has not found wide range of acceptance from professionalists.

In chemical prestressing the expansion properties of cement are made use. The chemistry of the cement can be altered without offence to cementicious properties Alumina and Gypsum are normal additives which provide for expansion of concrete. This expansion is not allowed by high tensile steel and compressive stress is developed in concrete. The stresses introduced are of a very low value and probably may be sufficient in some products like pipes, shells etc. Chemical prestressing also has not found good acceptance with professionalists.

Hence this two types of processing are not dealt with any greater detail in this book. The readers are referred to special literature of Graduck (8), Lossier (9), Mikhailov (10), T.Y. Lin (11) and others.

References

1. LEONHARDT, F. (1964) **Prestressed Concrete Design and Construction,** 2nd edn, Wilhelm Ernst and Sohn, Berlin, Munich, p. 143.
2. HOYER E and E. FRIEDRICH. (1939), **"Beitrag zur frage der Hatspannung in Eisen betonbauteilen",** Beton and eisen (Berlin), vol.38, No.6, pp 107-110.
3. ABELES, P.W. (1958), **An Introduction to Prestressed Concrete Theory and Design,** Chapman and Hall, London, pp 3-6.
4. EVANS, R.H. and E.W. BENNETT (1958), **Prestressed Concrete Theory and Design,** John Wiley & Sons, tric., New York, 294p.
5. MAGNAL. G (1954), **Prestressed Concrete, Concrete** Publications Ltd., London, England, 3458.
6. **VSL Post tensioning system a publication of VSL Singapore Pvt Ltd.,** 151, Chin Swee road, Manhattan House Singapore 0316.
7. **BBR Post tensioning users Manual,** BBR (India) Ltd., Bangalore.
8. GRADUCK, I.I. (1968), **Prestressed Concrete,** Oxford and I.B.H., New Delhi, pp 13-16.
9. LOSSIER, H. (1946), **"Cements with controlled expansion and their application to prestressed concrete,"** The Structural Engineer, Vol.24, No.10, pp 505-534.
10. MIKHAILOV, V.V. (1957), **Self-stressed concrete, Moscow 1955, and New developments in self-stressed concrete,** Proceedings of the World Conference on Prestressed Concrete, San Francisco, pp 25.1-25.12.
11. LIN, T.Y. and A. KLEIN (1963), **"Chemical prestressing of concrete elements using expanding cements,"** A.C.I. Journal Proceedings, Vol.60, No.9, 1187-1218.

4. Analysis for Flexure
(Prestressed Concrete Elements)

4.1 General

A structure is analyzed for the effect of certain loads, which may act internally or externally. Prestressing also could be considered as load and it could be applied internally or externally. The structure could be in single plane (the planar structure) or in space (a space structure). The scope of this chapter is limited to analysis of planar structures. For such an analysis knowledge on the physical behaviour of the system under a given system of loading is necessary. For this purpose, in addition to the loading on the system, the following details are required i.e. (a) geometry of the structure (b) boundary conditions (c) cross-sectional properties (d) material properties especially constitutive relationship (stress-strain relationship) and (e) a failure criteria. The physical behaviour is first mathematically modeled and that mathematical model is considered for analysis. Certain assumptions have to be made both in the conversion of physical structure to the mathematical model and in the behaviour of materials. Accordingly, the behaviour of prestressed concrete elements in flexure is discussed followed by the basic assumptions made before actual analysis aspects are detailed.

4.2 Behaviour of Prestressed Concrete Elements in Flexure

When any structural concrete element is subjected to flexure, a portion of the element is in tension and another portion in compression. In reinforced concrete elements, the entire tensile force is resisted by reinforcements and compression by concrete. But in case of prestressed concrete element subjected to flexure, both compression and tension are resisted only by the concrete. The tension is resisted by initially applied compression in concrete, whereas the compression is resisted by its own inherent quality of higher strength in compression. This behaviour is valid upto the level of stage of decompression at the extreme tension fibre and the tensile stress in the concrete going upto cracking stress. Stage of decompression is defined as a stage when the tensile stress generated in the corresponding fibre, nullifies the existing precompression generated by prestress. But after cracking, the prestresssed concrete element behaves as if it is a reinforced concrete element and the tensile forces are being resisted only by steel components namely untensioned reinforcement if any and the high tensile steel. The complete behaviour of prestressed concrete element under flexure can be understood by studying the variation in stress in concrete and steel from zero load to failure load. The stress in steel is dependent on the fact whether the prestressing steel element is bonded to concrete element or unbonded. Generally, a structural element is designed for a ductile failure which means, that the design or the behaviour is controlled by the stresses in steel. Hence, here also, the behaviour of the prestressed concrete element is explained with reference to steel stress as discussed by LYN (1).

The stress in steel from zero load to failure load is shown in Fig 4.1. As soon as the structural element is prestressed, there is stress in steel due to applied prestressing force and there is no external load. This stress level is identified by point B on the diagram (Fig 4.1) and is defined as f_{spo}. This includes the loss in stress due to slip of anchorages and elastic shortening. When the formwork or shuttering is removed, the self-weight of the element starts acting and increases in the stress level in pretensioned steel due to dead load moment. The increase is seen in the diagram as movement of stress values from B to C. From the point of transfer of stress, the concrete goes through creep deformation and the steel goes through relaxation in stress. Both these cause a reduction in prestressing force as could be seen in the Chapter on Losses in Prestress. Hence there is a continuous reduction in the applied pretress for about 3 to 4 years. But the rate of reduction of stress is a fast decreasing function. Accordingly the loss in stress and the resultant stress level in steel could be predicted. When service load acts on the structure, the stress in steel may be assumed to be $f_{sp\infty}$ [(f_{sp0} – loss in stress due to creep, shrinkage and relaxation)]. This point is marked as point D in the figure. When the external load (service load) is applied, there is further increase in steel stress. In case of bonded tendon, because of strain compatibility between concrete and steel, this increase in steel stress is equal to modular ratio times the stress in concrete at the steel level and it is very small. In case of unbonded tendon this increase in strain and corresponding stress in steel is much lower as explained later in article 4.7.2. Upto the service load, the element is supposed to be in the stage of compression throughout. Even, if at all, a small amount of tension permitted under service load at extreme fibre, it would be below the cracking stress. Hence the increase in stress in steel would be very small when compared to the applied stress.

The applied stress in steel due to prestressing is normally of the order of 0.6 fpu, say around 1000 N/mm^2 in high tensile steel. The increase in stress in steel due to the dead load and live load, moment upto the service load would be of the order of modulus ratio times the concrete stress at steel level. It could be around 6 × 10 N/mm^2 assuming that 10 N/mm^2 is the compressive stress in concrete at the steel level due to prestressing which could be utilized to nullify the tension due to external loads and 6 as the modular ratio. This 60 N/mm² is negligible when compared to the applied pretension of 1000 N/mm². This increase in stress in steel is presented as D to E in the Fig 4.1. As the load is further increased, beyond the service load, at one load stage, the initial compressive stress in concrete at the extreme fibre level is totally lost and it moves to tensile zone. As the tensile stress at the extreme fibre reaches the cracking stress (modulus of rupture for the concrete) the concrete cracks. The corresponding load is called the cracking load (cracking moment). Till this load stage, the increase in stress level in steel is only marginal, following the earlier pattern. This is shown as D to E at service load and E to F at cracking load, which is more than service load in prestressed concrete structures. The service load may be less than the stage of decompression with residual compression in concrete or upto the stage of decompression when concrete stress at the bottom fibre is zero

At this stage (cracking load stage) because of cracking, the section modulus of the cross-section becomes smaller. The crack suddenly develops further and goes up to the axis of bending in case of prestressed concrete elements. Because of this sudden movement of the cracking up to the axis of bending, the tensile stresses resisted by concrete in the tensile zone (from the tip of crack to extreme fibre) is transferred on to steel at the same load. Since the area of cross-section of prestressing steel is small, the entire tensile force when transferred to smaller area in steel, causes a sudden jump in steel stress as could

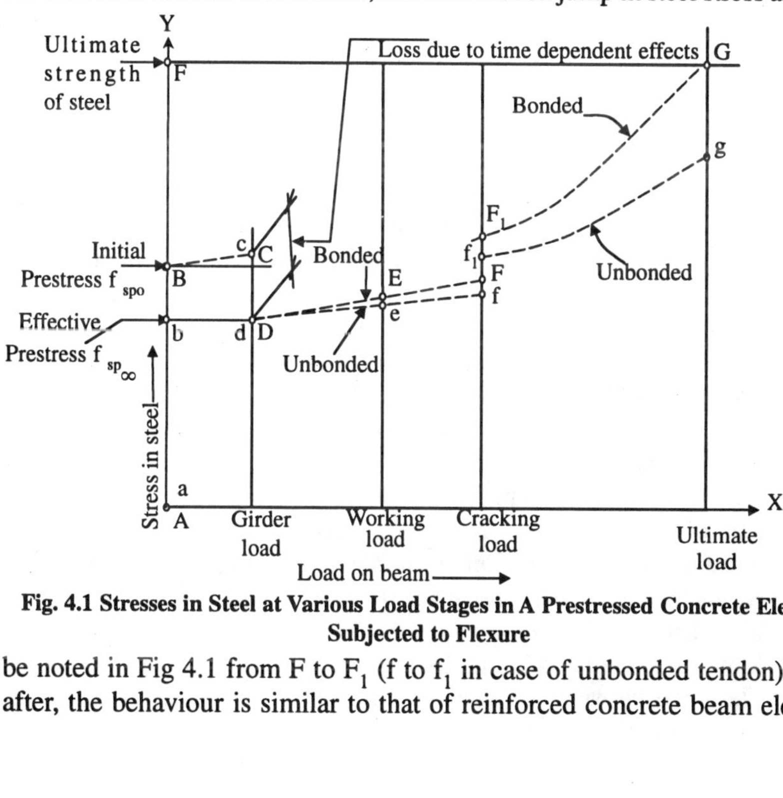

Fig. 4.1 Stresses in Steel at Various Load Stages in A Prestressed Concrete Element Subjected to Flexure

be noted in Fig 4.1 from F to F_1 (f to f_1 in case of unbonded tendon). Then after, the behaviour is similar to that of reinforced concrete beam element.

Any additional moment causes further increase in tensile force and that has to be taken only by steel.

Under these conditions the rate of increase of stress in steel is much larger than what was in the pre cracking stage. This could be seen from the slope of the curve in Fig 4.1, from F to G (f to g). At the point when the stress level in steel reaches its limiting value namely 0.2% proof stress, or (0.2% proof stress γ_m), the structural element is supposed to have reached its limiting moment (expected ultimate load). The above explanation holds good even for unbonded tendons, but the rate of increase of stress in steel will be lower as explained later in article 4.7.2. In Fig 4.1 the corresponding points are identified as a, b, c, d, e, f.

4.3 Basic Assumptions

It is well known that deformation is reality and stresses are only conjectures based on an assumed stress - strain relationship of material used in the structure. ***Material used is concrete and it is assumed to be homogeneous.*** The behavior of the prestressed concrete element upto service load alone is considered. ***Upto the level of deformation/stress corresponding to the service loads, concrete is assumed to behave linearly elastic.*** Prestressing force is transferred using steel elements/tendons. They may be bonded or unbonded. ***In case of bonded tendon, the bond between concrete and steel is assumed to be perfect and they act monolithically***. In case of unbonded tendon, the deformation compatibility between the full length of the tendon and the concrete element is assumed. When the prestressed element is subjected to flexure it is assumed, ***that the plane section remains plane throughout the behavioural stages and the curvature can be expressed as a function of the deflection as per "Navier's Hypothesis"***[2]. In the prestressed concrete element, steel is used as prestressing element and it is tensioned and anchored on the concrete. Hence there is an initial strain/stress in the material steel even before any external load starts acting on the concrete element. When such a prestressed concrete element is subjected to bending, the steel goes through additional strain due to introduced curvature leading to additional stress in the material. Depending on the conclusion that it is a bonded or unbonded tendon, the stress level in steel may vary. ***In any case, the increase in stress level in the prestressing steel of the prestressed concrete element due to bending caused by external loads, is assumed to be negligible.*** Summarizing, the assumptions are

(1) Navier's Hypothesis, namely plane section remaining plane throughout the behavioural at stages in flexure and that the curvature could be expressed as the second derivative of deflection with distance is valid.
(2) Material concrete is homogeneous.
(3) The bond between the prestressing steel and concrete is perfect in case of bonded tendons.

(4) The additional strain/stress in prestressing steel due to flexure caused by external loads is negligible when compared to applied stress due to pre-tensioning.

4.4 Notations

P = prestressing force in general (will be considered positive when it produces compression on concrete)

P^o = prestressing force on the jack or jacking force, or prestressing force applied on bulkheads or on self-straining frame in case of pretensioning

P_o = prestressing force transferred to the concrete element initially at the time (t_o).

P_∞ = prestressing force on the concrete element at any cross section at time t_∞.

e = eccentricity of the prestressing force

A = gross area of cross-section of concrete element.

A_n = net area of cross section of concrete

A_i = effective area of cross section of concrete

I = second moment of area or moment of inertia of gross concrete section

I_n = second moment of area of net section or moment inertia of net concrete section

I_i = second moment of area of effective section or moment of inertia effective concrete section

Z_t, Z_b = section modulus at the top and bottom fibre of the cross-section with gross sectional properties

y_t, y_b = distance of top and bottom fibres from the centroid of gross section

r = radius for gryration for gross sectional properties

(suffix n and i represent the corresponding property calculated for the concrete section considering net area and effective area respectively).

f_c = fibre stress in concrete

f_{ct} = fibre stress in concrete at top fibre

f_{cb} = fibre stress in concrete at bottom fibre

(the first suffix *'c'* represents material concrete, the second suffix *t'*, *'b'*, ….., represent the level of fibre such as top or bottom fibre. Any additional third subscript such as *p, D.L, M.L,* etc refers to the cause of the stress i.e. prestress, dead load or live load).

Normally, positive sign indicates that the stress is compressive in nature and negative sign indicates that the stress is tensile in nature.

$\bar{f}_c$ = permissible fibre stress in concrete
(the top horizontal stroke indicates the permissible stress values. This holds good for all stresses with suffixes also).

4.5 General Concept of Stresses

For explaining the concept of arriving at stresses, only gross sectional properties and total prestressing force are considered. Later it is shown that depending on the type of prestressing like pretensioning or post-tensioning, with bonded tendon or unbonded tendon, proper sectional properties have to be used to evaluate stresses.

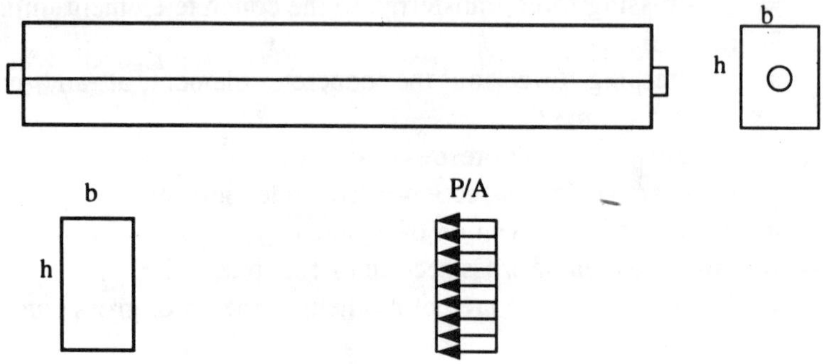

Fig.4.2 Concentric Prestressing & Stress

4.5.1 Concentric Prestress

When a concrete element is subjected to axial concentric force of P as shown in Fig 4.2 there is a uniform prestress applied on to concrete and the value is $\frac{P}{A}$ and it is compressive in nature.

4.5.2 Eccentric Prestress

Generally prestressing is meant to create initial stresses in concrete, which will be opposite in nature to that caused by externally applied loading. In a flexural element externally applied loads create varying stress throughout the depth of the cross section. In case of a beam, the externally applied load causes tension at bottom and compression at top. For countering such types of stresses, the prestressing force is applied with an eccentricity with respect to the center of gravity of cross section and the eccentricity is defined by a term *'e'*. Referring to Fig 4.2, where the prestressing force *'P'* is applied at an eccentricity 'e', the stresses developed at the top and the bottom fibre can be obtained using the relationships as detailed below.

$$f_{ctp} = \left[\frac{P}{A} - \frac{P}{I} \times e \times y_t\right] = \left[\frac{P}{A} - \frac{P \times e}{Z_t}\right] = \frac{P}{A}\left[1 - e \times \frac{y_t}{r^2}\right]$$

$$f_{cbp} = \left[\frac{P}{A} + \frac{P}{I} \times e \times y_b\right] = \left[\frac{P}{A} + \frac{P \times e}{Z_b}\right] = \frac{P}{A}\left[1 + e \times \frac{y_b}{r^2}\right]$$

4.5.2.1 Eccentric Prestress with dead load

As mentioned earlier in chapter 1, the effect of dead load of the element will be felt immediately after transfer of prestress. Hence the stresses on the element at top and bottom fibre have to be considered always with dead load when prestress is involved. They are

$$f_{ct1} = f_{ctp} + f_{ct\,D.L.} = \left[\frac{P}{A} - \frac{P}{I} \times e \times y_t + \frac{M_{D.L.}}{I} \times y_t\right]$$

$$= \left[\frac{P}{A} - \frac{P \times e}{Z_t} + \frac{M_{D.L.}}{Z_t}\right]$$

$$f_{cb1} = f_{cbp} + f_{cb\,D.L.} = \left[\frac{P}{A} + \frac{P}{I} \times e \times y_b + \frac{M_{D.L.}}{I} \times y_b\right]$$

$$= \left[\frac{P}{A} + \frac{P \times e}{Z_b} - \frac{M_{D.L}}{Z_b}\right]$$

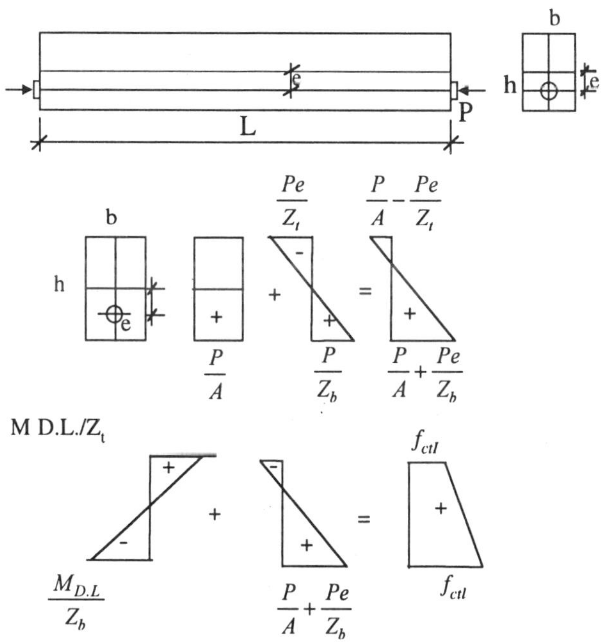

Fig 4.3. Eccentric Prestressing and Dead Load Effects Stress Distribution

4.5.2.2 Eccentric prestress with external loading

Consider a concrete beam element, which is eccentrically prestressed with a force of P at eccentricity 'e' towards the bottom fibre. If an external load on the beam is applied and it causes a moment 'M' at any cross-section, the stresses due to prestress and external moment can be superposed since it is assumed that the materials behaves in a linearly elastic way under such loads. Hence the stresses at the top and bottom fibres of beam due to the total moment of $M_t(M_{SL} + M_{LL})$ acting on the element are expressed as

$$f_{ctII} = f_{ctp} + f_{ct}(M_t) = \left[\frac{P}{A} - \frac{P}{I} \times e \times y_t\right] + \frac{M_t}{I} \times y_t$$

$$= \left[\frac{P}{A} - \frac{P}{I} \times e \times y_t\right] + \frac{M_{D.L.}}{I} \times y_t + \frac{M_{L.L}}{I} \times y$$

$$= \left[\frac{P}{A} - \frac{P \times e}{Z_t} + \frac{M_{D.L.}}{Z_t} + \frac{M_{L.L}}{Z_t}\right]$$

$$f_{cbII} = f_{ctb} + f_{cb}(M_t) = \left[\frac{P}{A} + \frac{P}{I} \times e \times y_b\right] - \frac{M_t}{I} \times y_b$$

$$= \left[\frac{P}{A} + \frac{P}{I} \times e \times y_b\right] - \frac{M_{D.L}}{I} \times y_b - \frac{M_{L.L}}{I} \times y_b$$

$$= \left[\frac{P}{A} + \frac{P \times e}{Z_b} - \frac{M_{D.L.}}{Z_b} - \frac{M_{L.L}}{Z_b}\right]$$

The pattern of stresses through the depth of cross section is shown in Fig.4.4.

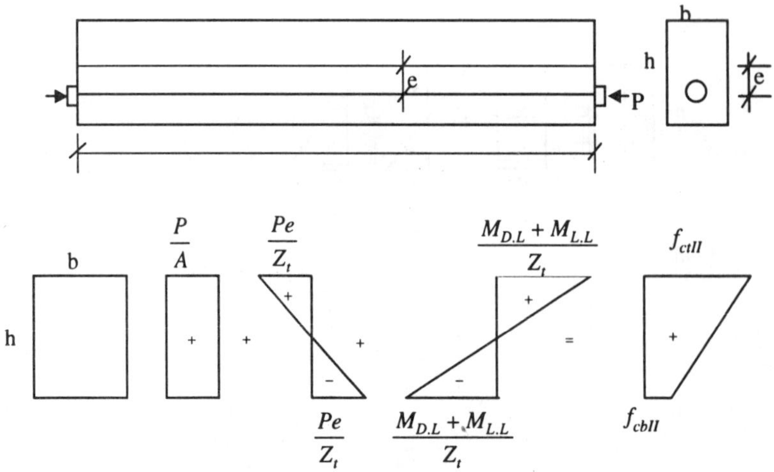

Fig.4.4 Eccentric Prestressing with Dead Load and Live Load Effects (Stress Distribution)

Example 4.1

A concrete beam of rectangular cross section of size 300*mm* × 500*mm* is prestressed with 750 kN at an eccentricity of 100 mm from the center of gravity of the gross section with 500mm^2 of prestressing steel stressed to 1500 N/mm^2. Analyze the section for stresses.

Solution

Properties of section

Cross section dimensions $= 300 \times 500\ mm$

Area of Cross section $= 15 \times 10^4\ mm^2$

Moment of Inertia of gross section $= \dfrac{1}{12} \times 300 \times 500^3$

$= 3125 \times 10^6\ mm^4$

Modulus of section at top $= 12.5 \times 10^6\ mm^3$

Modulus of section at bottom $= 12.5 \times 10^6\ mm^3$

Prestressing Force $= 750\ kN$

Eccentricity $= 100\ mm$

Stress in concrete at top fibre $= f_{cp}$

$$f_{cp_t} = \frac{750 \times 10^3}{15 \times 10^4} - \frac{750 \times 10^3 \times 100}{12.5 \times 10^6}$$

$= 5 - 6.0 = -1.0\ N/mm^2$ (tension)

Stress in concrete at bottom fibre = fcp

$$f_{cp_b} = \frac{750 \times 10^3}{15 \times 10^4} - \frac{750 \times 10^3 \times 100}{12.5 \times 10^6}$$

$= 5 + 6 = 11.0 N/mm^2$ (compression)

Example 4.2

The concrete beam given in Example 4.1 is covering a span of 7 *m*. It is subjected to live load of 20 *kN/metre*. Evaluate the stresses at top and bottom fibres at mid cross section due the effect of prestress, dead load and live load. Assume unit weight of concrete to be 24 kN/m^3.

Solution

Properties of section as given in Example 4.1

Dead load/self weight of the beam $= \dfrac{15 \times 10^4}{10^6} \times 24 kN/m = 3.6 kN/$

Bending moment due to dead load $= \dfrac{3.6 \times 7 \times 7}{8} kNm$

$= 22.05\ kNm$

Bending moment due to live load $= \dfrac{20 \times 7 \times 7}{8}$

$= 122.5\ kNm$

Stresses at top and bottom fibres due to dead load $= f_{ct\ D.L}, f_{cbD.L}$

$= \dfrac{22.05 \times 10^6}{12.5 \times 10^6} = 1.764\ N/mm^2$

(compression at top and tension at bottom)

Stresses at top and bottom fibres due to live load $= f_{ct\ L.L}, f_{cb} f_{L.L}$

$= \dfrac{122.5 \times 10^6}{12.5 \times 10^6} = 9.82 N/mm^2$

(compression at top and tension at bottom)

Stresses at top due to prestressing $= f_{ctp} = -1.0 N/mm^2$

(from example 4.1)

Stresses at bottom due to prestressing $= f_{ctp} = 11.0 N/mm^2$

(from example 4.1)

Extreme fibre stresses due to prestress + dead load (Stage I)

at top − 1.0 + 1.764 = +0.764 N/mm^2(compression)

at bottom 11.0 + 1.764 = 12.764 N/mm^2(compression)

Extreme fibre stresses due to prestress + dead load + live load (Stage II)

at top 1.0 + 1.764 + 9.82 = 10.564 $N\ mm^2$(compression)

at bottom 11.0 + 1.764 − 9.82 = 2.944 N/mm^2(compression)

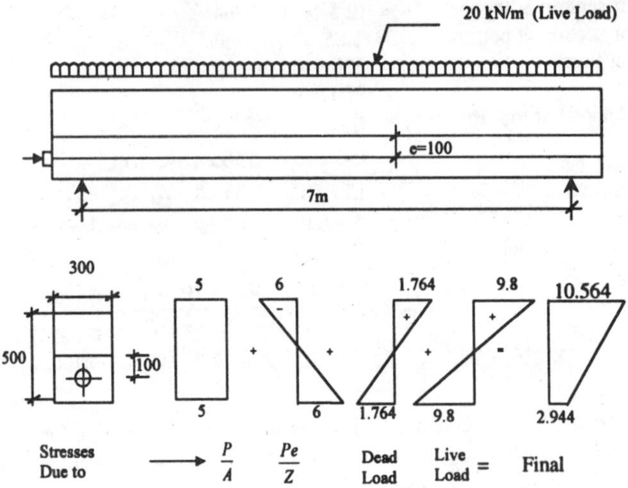

Fig.4.5 Stress Distribution

[Note: Both in Example 4.1 and 4.2 gross section properties are considered. Section properties to be considered depend on various aspects, which are discussed later. The prestressing force is taken as a given force. This also varies along the length of the cable depending on the losses, which are discussed later].

4.6 Prestressing Force Transferred by Pretensioning System

In case of pretensioning, the prestressing steel elements are pulled and anchored against bulkheads or against the self straining frames with a forceand the corresponding stress in steel is calculated as $\frac{P^0}{A_{sp}}$ where A_{sp} is the area of prestressing steel. This force is acting on the composite section at this stage since it is transferred from outside the section. The corresponding area of concrete can be expressed as

$$A_t = A + (m-1)A_{sp}$$

$$A_t = A_n + mA_{sp}$$

where m is modular ratio defined as $\frac{E_s}{E_c}$

When the steel element is cut and the force transferred to the concrete element, the concrete gets compressed and the elastic compressive strain in concrete is ε_{cp}. Prestressing steel now looses its strain to the same extent and the reduced strain in steel is defined as ε_{spo}. Hence the elastic compressive strain ε_{cp} can be espressed as a function of original steel strain $\varepsilon^0{}_{sp}$ and reduced strain ε_{spo}

$$\varepsilon_{cp} = \varepsilon^0{}_{sp} - \varepsilon_{spo}$$

Because of reduction in strain in prestressing steel, there is a change in the prestressing force on the element from P^0 to P_0 is transferred to the concrete from inside the section itself since the prestressing steel element is already cut off from bulk head or self straining frame.

The area of concrete which is resisting the prestressing force is only the net area A_n and the concrete strain ε_{cp} can be expressed in terms of reduced force P_0 and the net area of concrete.

$$\varepsilon_{cp} = \frac{P_0}{A_n \times E_c}$$

Strain in steel before transfer of force to concrete = $\varepsilon^0{}_{sp}$

$$\varepsilon^0{}_{sp} = \frac{P^0}{A_{sp} \times E_{sp}}$$

Strain in steel after the transfer of force to concrete = ε_{sp}

$$\varepsilon_{spo} = \frac{P_0}{A_{sp} \times E_{sp}}$$

Hence substituting for $\varepsilon^0{}_{sp}$ and ε_{spo}, ε_{cp} is expressed as

$$\varepsilon_{cp} = \frac{P^0}{A_{sp} \times E_{sp}} - \frac{P_0}{A_{sp}} \times E_{sp}$$

and it is rewritten as

$$\frac{P_0}{A_n \times E_c} = \frac{P^0}{A_{sp} \times E_{sp}} - \frac{P_0}{A_{sp} \times E_{sp}}$$

Substituting $\frac{E_{sp}}{E_c} = m$ and $A_i = A_n + mA_{sp}$

$$\frac{P_0}{E_c}\left[\frac{1}{A_n} + \frac{1}{mA_{sp}}\right] = \frac{P^0}{A_{sp} \times mE_c}$$

$$P_0 = \frac{P^0}{A_{sp}} \times \frac{1}{mE_c} \times \frac{A_n \times m \times A_{sp}}{A_n + m \times A_{sp}} \times E_c$$

$$A_i = A_n + mA_{sp}$$

$$\text{Hence } P_0 = \frac{P^0 \times A_n}{A_n + mA_{sp}}\ ;\ \frac{P_0}{A_n} = \frac{P^0}{A_i}$$

Stresses in steel and concrete

$$f_{cp} = \frac{P^0}{A_i} = \frac{P_0}{A_n}$$

$$f_{sp0} = f^0{}_{sp} + mf_{cp} = \frac{P^0}{A_{sp}} \times \frac{A_n}{A_i}$$

$$= f^0{}_{sp} \times \frac{A_n}{A_i}$$

Hence in pretensioning system, the stresses in concrete and steel can be calculated either using the force on the bulkhead (P^0) and effective sectional properties or using the force after transfer (P_0) and net sectional properties.

Similarly for eccentric tendons, it could be proved that

$$f_{cp} = \left[\frac{P_0}{A_n} \pm \frac{P_0 \times e_{np}}{I_n} \times y_n\right] \text{ or}$$

$$f_{cp} = \left[\frac{P^0}{A_i} \pm \frac{P^0 \times e_{ip}}{I_n} \times y_i\right]$$

$$f_{sp0} = \frac{P^0}{A_{sp}} + mf_{cp}$$

$$P_0 = f_{sp0} \times A_{sp}$$

Example 4.3

A pretensioned member of size 200 × 300 mm is prestressed with steel of 500 mm^2 at center of gravity of section, stressed to a level of 1050 N/mm^2. Assuming m = 6 compute the stresses in concrete and steel just after transfer.

Solution

Properties of section

A_n = 200 300 – 500 = 59500 mm^2

A_i = 200 300 + 500 × 5 = 62500 *mm*^{2}

$$f_{cp} = \frac{P^0}{A_i} = \frac{P^0}{A_n}$$

$$f_{cp} = \frac{1050}{62500} \times 500 = 8.4 \ N/mm^2$$

$$f_{sp} = \frac{P_0}{A_{sp}} + mf_{cp}$$

= 1050 – 8.4× 6 = 999.6 *N/ mm*^{2}

P_0 = 999.6× 500 = 499800 N = 499.8 *kN*

Approximate solution

Loss due to elastic shortening = $6 \times \dfrac{525000}{300 \times 200} = 52.5\ N/mm^2$

Steel stress after loss $= 1050 - 52.5 = 997.5\ N/mm^2$

Steel in concrete $= \dfrac{997.5 \times 500}{60000} = 8.3\ N/mm^2$

Example 4.4

A pretensioned member has a section 200× 300 *mm* and is 5*m* long. It is prestressed with 500 mm^2 of high tensile steel at 100 *mm* above the bottom fibre. The steel is stressed to a level of 1050 N/mm^2. Assuming *m*=6, compute stresses in concrete just after transfer

Solution

Properties of section (effective)

$A_{sp} = 500\ mm^2$

The shift of neutral axis below the geometrical center of gross section = y_0

Then $y_0 = \dfrac{5 \times 500 \times 50}{60000 + (5 \times 500)} = 2\ mm$

Effective moment of inertia $= I_i$

$$I_i = \frac{1}{12} \times 200 \times 300^3 + 60000 \times 2^2 \times 2500 \times 48^2$$

$$= 4500 \times 10^5 + 2.4 \times 10^5 + 57.6 \times 10^5$$

$$= 4560 \times 10^5\ mm^4$$

Stresses in Concrete due to prestress alone

Top fibre stress $= f_{ctp} = +\dfrac{500 \times 1050}{62500} - \dfrac{525000 \times 48 \times 152}{4560 \times 10^5}$

$= +8.4 - 8.4 \doteq 0$

Bottom fibre stress $= f_{cbp} = \dfrac{52500}{62500} + \dfrac{525000 \times 48 \times 148}{4560 \times 10^5}$

$= 8.4 + 8.19 = +16.59\ N/mm^2$ (compression)

Stress at steel level $= f_{csp} = 8.4 + \dfrac{525000 \times 48 \times 48}{45600 \times 10^5}$

$8.4 + 2.65 = +11.05\ N/mm^2$ (compression)

Due to self weight

Self weight of the beam $= \dfrac{60000 \times 1 \times 25000}{1000 \times 1000} = 1500\ N/m$

Self weight bending moment $= \dfrac{1500 \times 5 \times 5}{8} = 4687.5\ N/m$

Top fibre stress $= f_{ctD.L} = \dfrac{4687500}{Z_t} = \dfrac{4687500 \times 152}{4560 \times 10^4}$

$= +15.625\ N/mm^2$

Bottom fibre stress $= f_{ctD.L} = \dfrac{4687500 \times 148}{4560 \times 10^4}$

$= 15.214\ N/mm^2$

Stress at steel level $= f_{csD.L} = 4.935\ N/mm^2$

Final stresses in concrete

Top fibre stress $= f_{ctl} = 0 + 15.625 = 15.625\ N/mm^2$

Bottom fibre stress $= f_{cbl} = +16.59 - 15.214 = 1.376\ N/mm^2$

Steel level $= f_{csl} = +\ 11.05 - 4.935 = 6.115\ N/mm^2$

Final stress in steel $= f_{sp} = 1050 - 6 \times 6.115 = 1013.31\ N/mm^2$

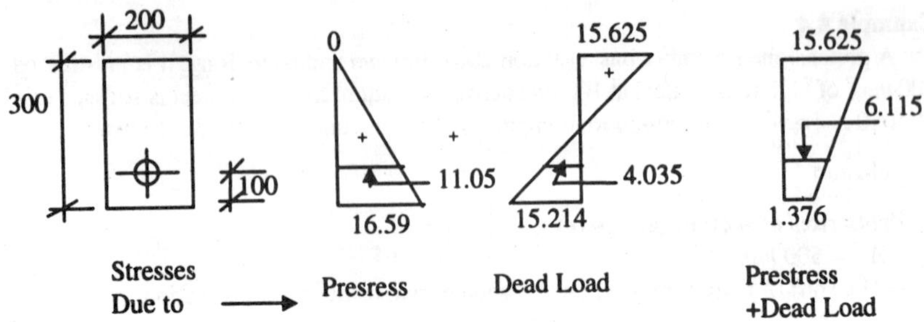

Fig 4.6 Stress Distribution

Approximate solution

Loss in prestress $= \dfrac{6 \times 52500}{60000}$

$= 52.5\ N/mm^2$

Steel stress $= 1050 - 52.5 = 997.5\ N/mm^2$

Extreme fibre stresses in concrete

$$= +\frac{997.5 \times 500}{60000} \pm \frac{997.5 \times 500 \times 150 \times 150}{45000 \times 10^4}$$

$= +\ 8.3 \pm 8.3$

$f_{ctp}, f_{cbp} = 0{,}16.6N/mm^2$ (at top and bottom respectively)

4.7 Prestressing Force Transferred by Post Tensioning System

In case of post-tensioning type of prestress, the concrete element is cast first with the provision for threading (introducing) the prestressing steel through the duct (holes) provided already. After the concrete element gains sufficient minimum strength to take care of the stresses caused by prestressing, the cable is threaded in the duct provided. The cable is stretched from either end by two jacks or it is stretched from only one end after keeping the other end anchored, with special type of anchorages. If it is pulled from both ends the anchorages at both ends are called live end anchorages. If it is pulled only from one end, the pulling end is supposed to have the live end anchorage and the other end is supposed to have the dead end anchorage.

The jacking force is referred as P^0, but when the same is transferred to the concrete, the concrete goes through the elastic shortening, thus reducing the elongation given during pretensioning. This reduces the prestressing force level from P^0 to P_0. Thisforce P_0 varies from section to section due to loss in force caused by friction. The tendon duct may be grouted or may be left ungrouted. If it is grouted, then a bond is created between the tendon and the concrete of the structural element through the grout. Such a tendon is called bonded tendon. Then for any structural action, compatibility of deformation between the prestressing tendon and the adjacent concrete fibres is assured. If the tendon cable is not grouted, elongation or deformation of the entire structure (concrete element) is taken as uniformly distributed deformation on the cable throughout the length, whereas the deformation need not be uniform on the main concrete element. Such a tendon is called an unbonded tendon. This means that, in case of beams with unbonded tendon, the total elongation of the beam due to the introduced bending moment/curvature, got by integrating the curvature from one end to another end, is reflected as the uniform strain on the cable. The same is a varying strain on the bonded tendon depending on the bending moment distribution, maximum being at the point of maximum bending moment.

4.7.1 Stresses in concrete elements with bonded tendons

In case of bonded tendons, prestressing at any cross-section is calculated as the force applied at the end of the beam minus loss due to friction and wobbing. The prestressing force also changes with respect to time due to time dependent losses caused by creep of concrete and relaxation of steel. The details about the losses are explained in the next chapter. The final prestressing force at the anchored end itself becomes less and it is referred as P_∞. The corresponding force at various cross-sections are also reduced by

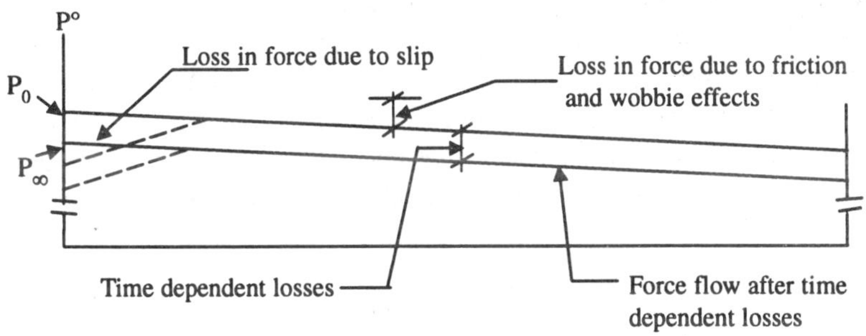

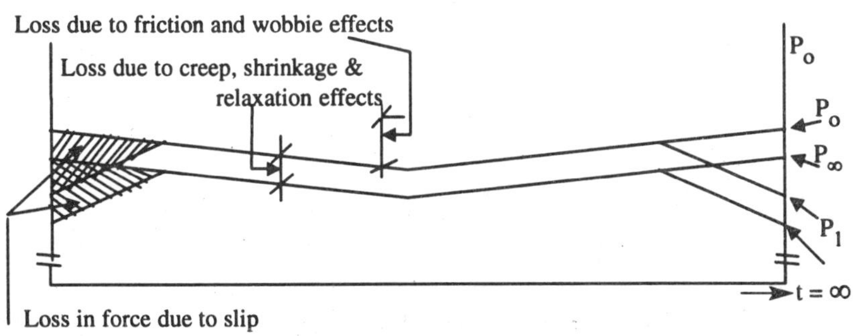

Fig 4.7 Typical Force Flow Diagrams for Beams Prestressed (a) From one end or (b) From Both Ends

the same amount equal to the loss in force due to creep, shrinkage and relaxation. It is necessary to draw the force flow diagram along the length of the beam. There is also a loss in force due to slip of anchorage at the pulling end and this affects the force flow for a definite length along the beam. A typical force flow diagram for a beam where prestress is applied at one end or both end are shown in Fig 4.7.

The stresses at any cross-section in the structural element subjected to prestressing by post tensioning method with the bonded tendon are calculated considering the effective section properties and the actual force at that cross-section (i.e. the applied force minus loss in force at that cross-section due to friction, elastic shortening, creep, shrinkage, relaxation etc). Hence the stresses at the top and the bottom fibre in the concrete element due to prestressing are evaluated as follows:

$$f_{ctp} = \frac{P_i}{A_i} \pm \frac{P_i \times e_i \times y_{it}}{I_i}$$

$$f_{cbp} = \frac{P_i}{A_i} \pm \frac{P_i \times e_i \times y_{ib}}{I_i}$$

where the subscript *i* refers to the effective section property.These stresses will be further reduced with time. When dead load, live load and any other load act on the structure, the stresses in concrete at salient levels due to those loads are algebrically added at the corresponding load stage. The resultant stress when dead load alone acts, is represented as f_{cI} (at Stage I). When all the loads are acting, the stresses in concrete are defined as f_{cII} (at Stage II). Accordingly, the fibre stresses at top and bottom fibre levels in concrete are expressed as. f_{cbI}, f_{cbII}, f_{ctI}, f_{ctII} Similarly the stresses in concrete at level of steel (prestress

$$f_{ctI} = f_{ctp} + f_{ctD.L.}$$

$$= \frac{P_0}{A_i} - \frac{P_0 \times e}{Z_{ti}} + \frac{M_{D.L.}}{Z_{ti}}$$

$$f_{btI} = f_{ctp} + f_{ctD.L.}$$

$$= \frac{P_0}{A_i} + \frac{P_0 \times e}{Z_{bi}} - \frac{M_{D.L.}}{Z_{bi}}$$

$$f_{ctII} = f_{ctp} + f_{ctD.L.} + f_{ctL.L}$$

$$= \frac{P_\infty}{A_i} - \frac{P_\infty \times e}{Z_{ti}} + \frac{M_{D.L.}}{Z_{ti}} + \frac{M_{L.L}}{Z_{ti}}$$

$$f_{cbII} = f_{cbp} + f_{cbD.L.} + f_{ctL.L}$$

$$= \frac{P_\infty}{A_i} + \frac{P_\infty \times e_i}{Z_{bi}} - \frac{M_{D.L.}}{Z_{bi}} - \frac{M_{L.L}}{Z_{bi}}$$

tendon) is represented as f_{csI}, f_{csII} in Stage I and Stage II respectively. The subscripts '*s*' refer to fibre stress at steel level. They are evaluated as follows:

[**Note** : eccentricity '*e*' is assumed positive when it is below the centroid of the section. P_0 and P_∞ are the effective prestressing force at the section under consideration at the time when stresses are evaluated with reference to that time at which the force is transferred. P_∞ includes losses due to time dependent effects].

The stresses in steel are evaluated as follows:

$f_{spI} = f_{spo}$ due to prestress (P_0) + f_{sp} due to dead load

= f_{spo} due to prestress + m times $f_{csD.L.}$

After the live load acts f_{spII} equals $f_{sp\infty}$ due to prestress P_∞ + m times $(f_{csD.L} + f_{csL.L})$

4.7.2 Stresses in concrete elements with unbonded tendons

In case of unbonded tendons, the prestressing force is transmitted at the ends where the cables are anchored. Any additional moment which comes on the structural element causes additional curvature on the element introducing stresses in concrete and steel at various sections. As far as the concrete is concerned, the additional moment caused by dead load, live load or super imposed loads introduces additional stresses and the same is evaluated using net section properties of the section (i.e. I_n, A_n, Y_n etc). But when steel is considered, the additional bending moments due to dead load, live load and any other load, introduce additional curvature and elongation over the length of the beam at the level of steel fibre. It is uniformly distributed for full length of the cable as uniform strain, unlike in case of bonded tendon where the strain in steel fibre at any cross section should be compatible with strain in concrete at that level, which is not uniform over the length of the beam. Hence in unbonded tendon, the increase in stress in steel due to loads inclusive of dead load is calculated based on the uniform increase in strain in steel for the full length to match the increased curvature. The details of calculation of stresses in concrete at various levels and the steel are shown below.

Stresses in concrete

due to prestress

bottom fibre $f_{cbp} = \frac{P_0}{A_n} + \frac{P_0 e_n}{I_n} y_{bn}$

top fibre $f_{ctp} = \frac{P_0}{A_n} + \frac{P_0 e_n}{I_n} y_{bn}$

due to dead load

bottom fibre $f_{cbD.L.} = -\frac{M_{D.L.}}{I_n} y_{bn}$ (tension)

top fibre $f_{ctD.L.} = +\frac{M_{D.L.}}{I_n} y_{tn}$ (compression) due to live load

bottom fibre $f_{cbL.L.} = -\frac{M_{L.L.}}{I_n} y_{bn}$ (tension)

top fibre $f_{ctD.L.} = +\frac{M_{L.L.}}{I_n} y_{tn}$ (compression)

At Stage I

$$f_{cbI} = f_{cbp} + f_{cbD.L.}$$
$$= \frac{P_0}{A_n} + \frac{P_0 e_n}{I_n} y_n - \frac{M_{D.L.}}{I_n} . y_{bn}$$

$$f_{ctI} = f_{ctp} + f_{ctD.L.}$$
$$= \frac{P_0}{A_n} - \frac{P_0 e_n}{I_n} y_n + \frac{M_{D.L.}}{I_n} . y_{tn}$$

At Stage II

$$f_{cbII} = f_{cbp} + f_{cbD.L.} + f_{cbL.L}$$
$$= \frac{P_0}{A_n} + \frac{P_0 e_n}{I_n} - \frac{(M_{D.L.} + M_{L.L.})}{I_n} . y_{bn}$$

$$f_{ctII} = f_{ctp} + f_{ctD.L.} + f_{ctL.L}$$
$$= \frac{P_0}{A_n} - \frac{P_0 e_n}{I_n} y_n + \frac{(M_{D.L.} + M_{L.L.})}{I_n} . y_{bn}$$

Stress in Steel

Initial stress due to Prestress $f_{sp_0} = \frac{P_0}{A_{sp}}$

due to Dead Load $f_{spD.L.}$

due to Live Load $f_{spL.L.}$

4.7.3 Stresses in steel due to dead load and live load ($f_{spD.L}, f_{spL.L}$)

The total elongation of steel over the length of the beam is evaluated as ν

$$\Delta = \int \epsilon \, dx = \int \left[\frac{M_{D.L} + M_{L.L}}{E_c} \right] \frac{Y_n}{I_n} \times dx$$

$$\frac{\Delta}{L} = \frac{1}{L} \int \left[\frac{M_{D.L} + M_{L.L}}{E_c} \right] \frac{Y_n}{I_n} \times dx$$

$$f_{spD.L, L.L} = E_s \frac{\Delta}{L} = \frac{E_s}{E_o} \times \frac{1}{L} \int \frac{(M_{D.L.} + M_{L.L})}{I_n} \times y_n dx$$
$$= \frac{m}{L} \int \frac{(M_{D.L.} + M_{L.L})}{I_n} \times y_n dx$$

If dead load and live load bending moment distribution along the length of the beam are assumed parabolic, the bending moment at any cross-section at a distance '*x*' from center cross section is expressed as function of maximum bending moment at the center M_0

$$M_x = M_0\left[1-\left(\frac{x}{l/2}\right)^2\right]$$

For such bending moment distribution (namely parabolic), the cable profile also will be parabolic. The equation of the profile is expressed as

$$y_x = y_0\left[1-\left(\frac{x}{l/2}\right)^2\right]$$

where y_0 is the maximum eccentricity, (the dip of parabola) and x is the distance measured from the point of maximum dip.

Then $\in_s \times \frac{\Delta}{L}$ may be rewritten as

$$\in_s \times \frac{\Delta}{L} = f_{sp} = \frac{m}{LI_n}\int_{-L/2}^{+L/2} M_0 \times Y_0\left[1-\left(\frac{x}{L/2}\right)^2\right]^2 dx$$

$$= \mathrm{m}\frac{M_0Y_0}{LI_n}\left[x - 2/3 \times \frac{x^3}{(L/2)^2} + \frac{x^5}{5} \times \frac{1}{(L/2)^4} + \ldots\ldots\right]_{-L/2}^{+L/2}$$

$= 8/15$ (corresponding additional stress in steel in case of bonded tendon)

Example 4.5

A post tensionsed bonded prestressed concrete beam has a cross-section of 300 × 600 mm. This beam spans distance of 10*m* and carries a uniformly distributed load of 16 kN per meter in addition to its own self-weight. A prestressing force of 1200*kN/m* is applied at a distance of 140*mm* from the soffit. Evaluate the stresses at the bottom fibre and top fibre at the mid span cross-section both at Stage I (just after prestressing) and at Stage II (service load stage). If 15% loss could be expected in prestressing force at time t=∞, evaluate the stresses at top and bottom fibre in Stage II considering the loss in prestressing force.

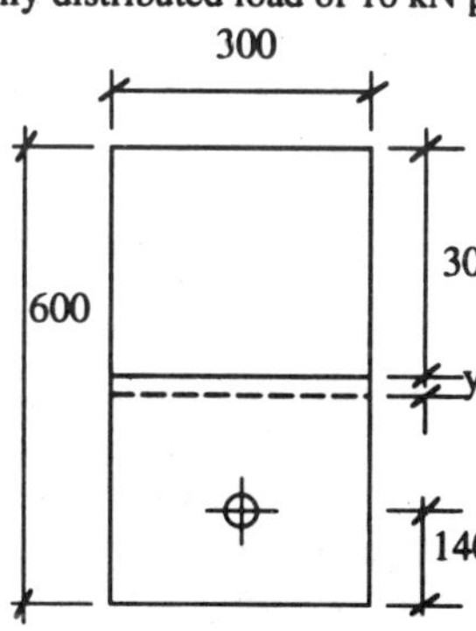

Fig 4.8

Solution

Properties of section

$A_g = 600 \times 300 = 18 \times 10^4 mm^2$

$I_g = \frac{1}{12} \times 300 \times 600^3 = 54 \times 10^8\ mm^4$

Assume y is shift distance of the effective center of area from the center of area of gross section

Taking moment above effective center of area

$180000y = 6000(160 - y)$

$$y = \frac{960000}{186000} = 5.16mm$$

Considering effective section

$$I_i = I_g + 600(160 - 5.16)^2 + 18 \times 10^4 \times 5.16^2$$
$$= 5400 \times 10^6 + 143.8 \times 10^8 + 479.26 \times 10^4$$
$$= 5548.59 \times 10^6\ mm^4$$

Stresses in Concrete (Due to prestress, initial stage)

$$\text{at bottom } = f_{cbp} = \frac{P_0}{A_i} + \frac{P_0 \times e_i}{Z_{bi}}$$

$$= \frac{1200 \times 10^3}{186000} + \frac{1200 \times 10^3 \times 154.84}{18.82 \times 10^6}$$

$$= 64.5 + 9.87$$
$$= 16.32\ N/mm^2 \text{ (Compression)}$$

$$\text{at top } = f_{ctp} = \frac{P_0}{A_i} - \frac{P_0 \times e_i}{Z_{ti}}$$

$$= \frac{1200 \times 10^3}{186000} - \frac{1200 \times 10^3 \times 154.84}{18.183 \times 10^6}$$

$$= 64.5 - 10.22$$
$$= -3.77\ N/mm^2 \text{ (tension)}$$

Due to dead load

$$\text{at bottom } = f_{ctD.L.} = \frac{54 \times 10^6}{18.82 \times 10^6} = -2.869\ N/mm^2 \text{ (tension)}$$

$$\text{at top } = f_{cbD.L.} = \frac{54 \times 10^6}{18.183 \times 10^6} = 2.969\ N/mm^2 \text{ (compression)}$$

Due to live load

$$\text{at bottom } = f_{cb\,L.L.} = \frac{200 \times 10^6}{18.82 \times 10^6} = 10.63\ N/mm^2 \text{ (tension)}$$

$$\text{at bottom } = f_{ct\,L.L.} = \frac{200 \times 10^6}{18.183 \times 10^6} = 11.00\ N/mm^2 \text{ (Compression)}$$

Stage I

Stresses due to prestress and dead load

$$f_{cbI} = 16.32 - 2.869 = 13.451\ N/mm^2 \text{(compression)}$$
$$f_{ctI} = -3.77 + 2.969 = -0.80\ N/mm^2 \text{ (tension)}$$

Stage II

Stresses due to prestress + dead load + live load

$$f_{cbII} = 16.32 - 2.869 - 10.63 = 2.821\ N/mm^2 \text{(compression)}$$
$$f_{ctII} = -3.77 + 2.969 = +11.00 = 10.199\ N/mm^2 \text{ (compression)}$$

If losses upto 15% is considered

f_{cbII} = 13.872 - 2.869 - 10.63 = 0.373 N/mm^2(compression)

f_{ctII} = -3.205 + 2.969 = +11.00 = 10.764 N/mm^2 (compression)

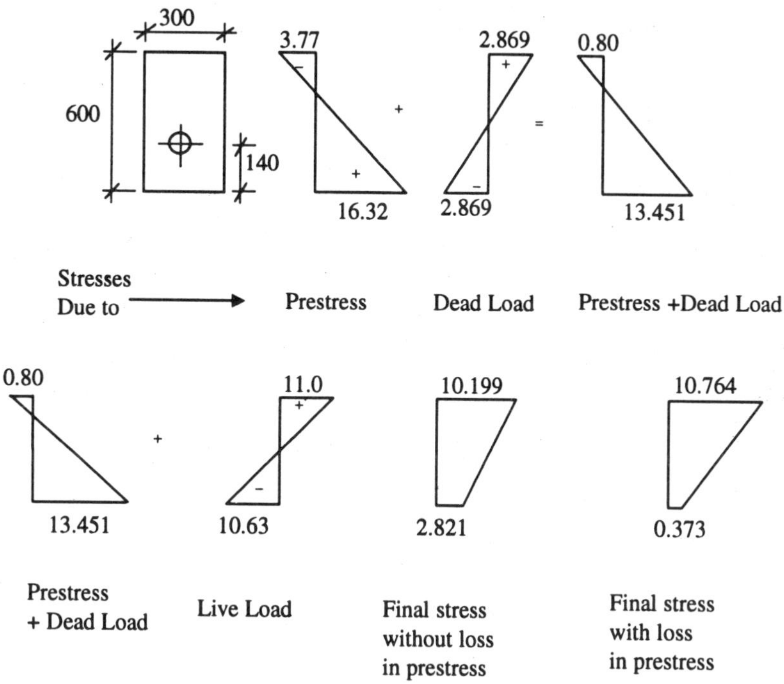

Fig 4.9 Pattern of Stress Distribution

Example 4.6

In the example 4.5, if the permissible stresses in concrete in compression and in tension are limited to as 16 N/mm^2 and 1.5 N/mm^2 respectively, determine the maximum uniformly distributed load that could be permitted on the beam for the same prestress and eccentricity.

Solution

Properties of section (effective)

$A_i = 186000\ mm^2$

$y_{ti} = 305.16\ mm$

$y_{bi} = 294.84\ mm$

$Z_{bi} = 18.82 \times 10^6\ mm^3$

$Z_{ti} = 18.183 \times 10^6\ mm^3$

Stage I Stresses have been calculated in Example 4.5

f_{cbp} = 16.32 N/mm^2 (compression)

f_{cbp} = -3.77 N/mm^2(tension)

$f_{cbD.L}$= -2.869 N/mm^2(tension)

$f_{ctD.L}$ = 2.969 N/mm^2(compression)

f_{cbI} = 13.451 N/mm^2(compression) within permissible limits

f_{ctI} = -0.80 N/mm^2(tension) within permissible limits

Stage II

Assuming no loss in the prestress, namely the liveload is applied even bofre any loss has occurred. (This is hypothetical) and assuming M_{LL} to be the permissible live load bending moment.

Top fibre stress $-0.80 + \dfrac{M_{\cdot L.L.}}{Z_t} \le 16$

Bottom fibre stress $13.451 + \dfrac{M_{\cdot L.L.}}{Z_b} \ge -1.5\ N/mm^2$

$$\frac{M_{\cdot L.L.}}{Z_t} = 16 + 0.80 = 16.80$$

$$M_{.L.L} = 16.80 \times 18.183 \times 10^6 = 305.4744 \times 10^6\ Nmm$$

$$\frac{M_{\cdot L.L.}}{Z_b} = 14.951$$

$$M_{.L.L} = 14.951 \times 18.82 \times 10^6 = 281.3778 \times 10^6\ Nmm$$

The limiting value of the live load moment which will satisfy the stress requirements $= 281.3778 \times 10^6\ Nmm$

The corresponding uniformly distributed load

$$= \frac{281.3778 \times 10^6 \times 8}{10 \times 10 \times 10^6}$$

$$= 22.51\ N/mm = 22.5\ kN/mR$$

If 15% loss in prestress is considered, the permissible Live Load Bending moment will get reduced. Assuming it to be M_{LL} stresses at service load condition can be evaluated as follows.

Bottom fibre stress $= 13.872 - 2.869 + \dfrac{M_{\cdot L.L.}}{Z_b} \ge -1.5\ N/mm^2$

Top fibre stress $= 13.872 - 2.869 + \dfrac{M_{\cdot L.L.}}{Z_t} \ge -16\ N/mm^2$

$$\frac{M_{L.L}}{Z_t} = 16 + 0.80 = 16.80$$

$$M_{.L.L} = 16.80 \times 18.183 \times 10^6 = 305.4744 \times 10^6\ Nmm$$

$$\frac{M_{L.L}}{Z_b} = 14.951$$

$$M_{.L.L} = 14.951 \times 18.82 \times 10^6 = 281.3778 \times 10^6\ Nmm$$

The limiting value of live load moment which will satisfy the stress requirement $= 235.30 \times 10^6\ Nmm$

The corresponding uniformly

distributed load $= \dfrac{235.30 \times 10^6 \times 8}{10 \times 10 \times 10^6} = 18.824\ kN/m$

Example 4.7

An unsymmetrical I Girder shown in Fig 4.10 is designed to carry a Live Load Bending moment of 648 *kNm* at mid span. It is prestressed with a force of 1600 *kN* at 150mm from soffit. Evaluate the stresses at transfer stage and at service load stage at top and bottom fibres. Assume 15% loss in prestress at service load stage. Evaluate also the limitation of position of cable if the permissible stresses in concrete and compression and tension are limited to $16 N/mm^2$ and $1.0\ N/mm^2$.

Solution

Properties of section

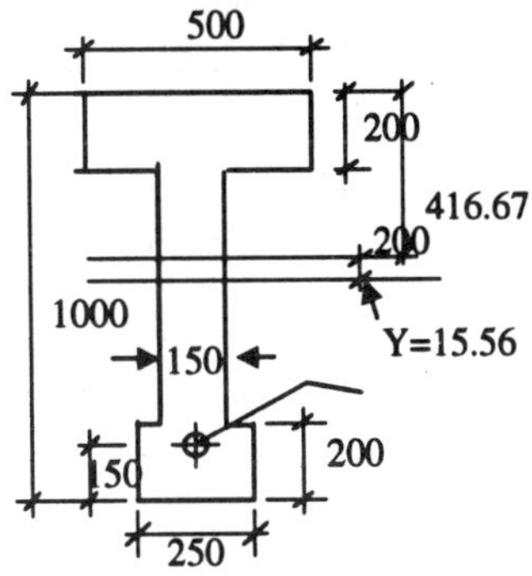

Fig 4.10

$Ag = 500 \times 200 + 600 \times 150 + 200 \times 250$
$= 240000\ mm^2$

$240000y_t = (1500 \times 200 \times 100) + (600 \times 150 \times 500)$
$+ (200 \times 250 \times 900)$

$y_t = 416.67\ mm$

$y_b = 1000 - 416.67 = 1583.33\ mm$

$$I_g = \frac{150 \times 200^3}{12} + (150 \times 200)(316.67)^3$$
$$+ \frac{150 \times 600^3}{12} + (150 \times 600)(83.33)$$
$$+ \frac{150 \times 200^3}{12} + (250 \times 200 \times 483.33)$$
$$= 2.5553 \times 10^{10}\ mm^4$$

Effective section properties

$A_i = 240000 + 5 \times 1800$
$= 249000 mm^2$

Let be $\bar{y}$ shift of neutral axis towards bottom for effective section

$240000\ \bar{y} = 9000\ (583.3 - 150 - \bar{y}\ )$

$240000\ \bar{y} = 3899970$

$\bar{y} = 15.66 mm$

$y_{bi} = 567.67 mm$

$y_{ti} = 432.33 mm$

$I = I_g + 9000(567.67\text{-}150)^2 + 240000 \times 15.66^2$
$= 2.553 \times 10^{10} +.157 \times 10^{10} +.00588 \times 10^{10}$
$= 2.7159 \times 10^{10}$

$$Z_b = \frac{2.7159 \times 10^{10}}{432.33} = 4.784 \times 10^7\ mm^3$$

$$Z_t = \frac{2.7159 \times 10^{10}}{432.33} = 6.2820 \times 10^7\ mm^3$$

Stresses in concrete

Due to prestress, initial stage

$$f_{cbp} = \frac{P_0}{A} + \frac{P_0 \times e_i}{Z_{bi}}$$
$$= \frac{1600 \times 10^3}{249000} + \frac{1600 \times 10^3 \times 417.67}{4.784 \times 10^7}$$
$$= 6.425 + 13.97$$
$$= 20.395 N/mm^2 \text{ (compression)}$$

$$f_{ctp} = \frac{P_0}{A} - \frac{P \times e_i}{Z_{ti}}$$
$$= 6.425 - \frac{1600 \times 10^3 \times 414.67}{6.2820}$$
$$= 6.425 - 10.64$$
$$= -4.215 N/mm^2 \text{ (tension)}$$

Due to Dead load

Dead Load Bending moment $= \dfrac{(0.24 \times 24) \times 18^2}{8} = 233.28 kNm$

$$f_{cbD.L.} = \frac{233.28 \times 10^6}{4.784 \times 10^7} = 4.876 N/mm^2$$

$$f_{ctD.L.} = \frac{233.28 \times 10^6}{6.2820 \times 10^7} = 3.71 N/mm^2$$

Stresses at Stage I, due to Prestress + Dead Load

fctI (top fibre) = -4.215+3.71=-0.505 *N/mm²* (tension)

fcbI (bottom fibre) = +20.395-4.876=15.519 *N.mm²* (compression)

Due to Live Load

Live Load Bending moment = 648 *kNm*

$$f_{cbL.L} = \frac{648 \times 10^6}{4.784 \times 10^7} = 13.545 N/mm^2 \quad \text{(tension)}$$

$$f_{ctL.L} = \frac{648 \times 10^6}{6.2820 \times 10^7} = 10.315 \quad \text{(compression)}$$

Stresses at Stage II

(top fibre) = 0.85 × (– 4.215) +3.71+10.315

= -3.583+3.71+10.315

= 10.445 (tension)

(bottom fibre) = 0.85 -4.876-13.545

= +17.34 - 4.876 - 13.545

= -1.081 (tension)

For deciding the maximum eccentricity, the Stage I, bottom fibre stress in compression, and top fibre stress in tension will be considered. If the maximum compressive stress permissible is 16*N/mm²* and the tensile stress is 1*N/mm²*, they have to be used to find the *'e'* value

Consider Stage I

$$f_{cbI} = \frac{P}{A} + \frac{P \times e}{Z_b} - \frac{M_{D.L.}}{Z_b}$$

$$= 6.425 + \frac{1600 \times 10^3 \times e}{4.784 \times 10^7} - 4.876 = 16 N/mm^2$$

$$e = 432.085 mm$$

$$f_{ctI} = \frac{P}{A} - \frac{P \times e}{Z_t} + \frac{M_{D.L.}}{Z_t}$$

$$= 6.425 - \frac{1600 \times 10^3 \times e}{6.2820} + 3.71 = -1.5 N/mm^2$$

$$e = 456.82 mm$$

Hence maximum eccentricity permitted = 432.085*mm*.

For deciding about the minimum eccentricity, the Stage II bottom fibre stress in tension and top fibre stress in compression will be considered

Stage II

$$f_{cbII} = \frac{P}{A} + \frac{P \times e}{Z_b} - \frac{M_{D.L.}}{Z_b} - \frac{M_{L.L}}{Z_b}$$

$$= (6.425).85 + \frac{0.85 \times P \times e}{Z_b} - 4.876 - 13.545 = -1.5N/mm^2$$

$$= 5.461 + \frac{1360 \times e \times 10^3}{4.784 \times 10^7} - 18.421 = -1.5N/mm^2$$

$$e = 403.125mm$$

$$f_{ctII} = \frac{P}{A} - \frac{P \times e}{Z_t} + \frac{M_{D.L.}}{Z_t} + \frac{M_{L.L}}{Z_t}$$

$$= 5.461 - \frac{1360 \times e \times 10^3}{6.2820 \times 10^7} + 3.71 + 10.315 = 16N/mm^2$$

e = 161.02mm,

The minimum 'e' required to be given 403.125mm.

The cable can be located in a zone with *e* value ranging from 403.125 to 432.085 to see that the stress limitation are satisfied. This is called cable zone.

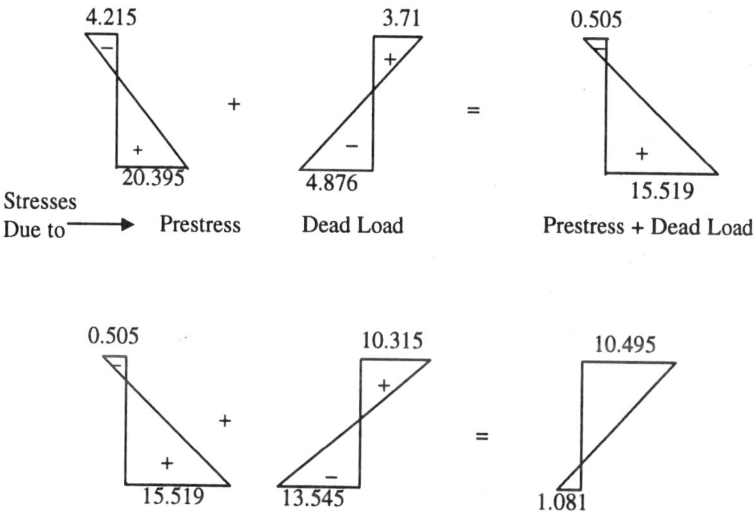

Fig. 4.11 a. Stress Distribution

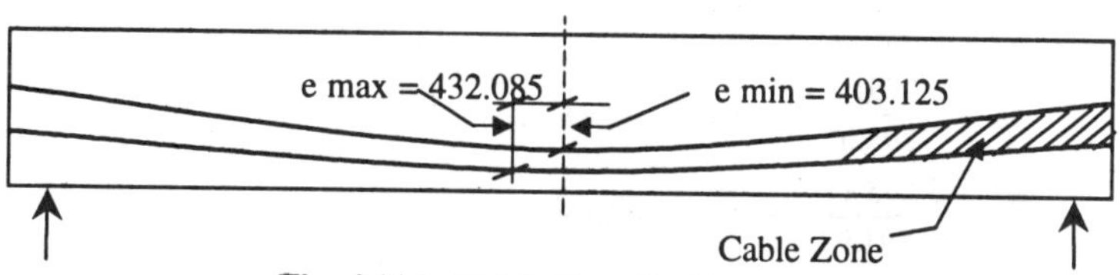

Fig. 4.11 b. Cable Zone for the Beam

Example 4.8

A prestressed concrete post-tensioned beam with unbonded tendons has a cross-section as shown in Fig 4.12. It is prestressed with 2 cables 6T 13 (6 number of 12.7 mm strand in one cable) with 2 ducts of 90 mm diameter placed at 100 mm from the soffit. The effective duct profile is parabolic and the effective center of anchorage is at the C.G. of the net cross section at the supports. The cables are prestressed so as to have a force of 1490*kN* initially (at mid span cross section) and the time dependent losses is 12% at time t = ∞. The beam spans 18.5 m. It carries a Live Load Bending Moment of 480 *kNm*. Evaluate the stresses in concrete at salient levels at Stage I and Stage II and also the steel stresses.

Solution

Properties of section (Ref Table 4.2 for relative properties of gross section)

$D_{ft} = D_{fb} = 200$

$\frac{D_{ft}}{D} = \frac{200}{1000} = 0.2$

$A = 0.48 \times 800 \times 1000$
$= 384000\ mm^2$

$I = 0.0510 \times 800 \times 1000^3$
$= 40.8 \times 10^9\ mm^4$

$y_b = 0.582 \times 1000 = 58.2$

$y_t = 0.418 \times 1000 = 41.80$

$\gamma^2 = 0.1086 \times 1000^2$

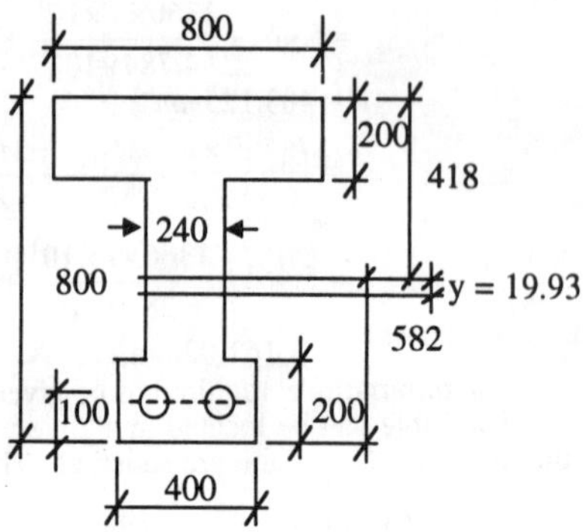

Fig 4.12

$$k_t = \frac{\gamma^2}{\gamma_b} = 0.186 \times 1000$$

$$k_b = \frac{\gamma^2}{\gamma_t} = 0.260 \times 1000$$

$P = 1490\ kN$ with 2 Nos of 6T 13 cables and 2 ducts 90 mm dia located at 100 mm from soffit.

Area of duct = 12717 mm^2

Net section properties

$A_n = 40.8 \times 10^9 - 12.717 \times 25.193 \times 10^7 + 1.525 \times 10^8$
$= 371.2 \times 10^3\ mm^2$

$384 \times 10^3\ \bar{y} = 12.717 \times 103\ (582 - 100 + \bar{y})$

$(384 \times 10^3 - 12.717 \times 10^3)\ \bar{y} = 12.717 \times 582$

$\bar{y} = 19.930\ mm$

$y_{bn} = 582 + 19.93 = 601.93$ mm

$y_{tn} = 418 - 19.93 = 398.07$ mm

$y_{sn} = 482 + 19.93 = 501.93$ mm

$= 40.8 \times 10^9 - 12.717 \times 25.193 \times 10^7 + 1{,}525 \times 10^8$

$= 40.8 \times 10^9 - 3.204 \times 10^9 + 1.5253 \times 10^8$

$= 37.75 \times 10^9\ mm^4$

$$Z_{bn} = \frac{37.75 \times 10^9}{601.93} = 6.272 \times 10^7 mm^3$$

$$Z_{tn} = \frac{37.75 \times 10^9}{398.07} = 9.483 \times 10^7 mm^3$$

$$Z_{sn} = \frac{37.75 \times 10^9}{501.93} = 7.52 \times 10^7 mm^3$$

Stresses in concrete

Stresses due to Prestress

$$f_{cbp} = \frac{P}{A} + \frac{P \times e_n}{Z_{bn}}$$

$$= \frac{1490 \times 10^3}{371.283 \times 10^3} + \frac{1490 \times 10^3 \times 501.93}{6.272 \times 10^7}$$

$= 4.013 + 11.924$
$= 15.937\ N/mm^2$(compression)

$$f_{ctp} = \frac{P}{A} - \frac{P \times e_n}{Z_{tn}}$$

$$= 4.013 - \frac{1490 \times 10^3 \times 501.93}{9.483 \times 10^7}$$

$= 4.013 - 7.886$
$= -3.873$ (tension)

Stresses due to Dead load

$$\text{Dead load bending moment} = \frac{0.3713 \times 24 \times 18.5^2}{8}$$
$$= 381.232 kNm$$

$$f_{cbD.L} = \frac{381.232 \times 10^6}{6.2728 \times 10^7} = 6.08\ N/mm^2 \text{ (tension)}$$

$$f_{ctD.L} = \frac{381.232 \times 10^6}{9.483 \times 10^7} = 4.02 N/mm^2 \text{ (compression)}$$

$$f_{csD.L} = \frac{381.232 \times 10^6}{7.52 \times 10^7} = 5.07 N/mm^2 \text{ (compression)}$$

Stresses in Stage I

$f_{ctI} = -3.873 + 4.02 = +0.147\ mm^2$ (compression)
$f_{cbI} = 15.937 - 6.08 = 9.857\ N/mm^2$ (compression)

Stresses due to Live Load

Live Load bending moment = 480 kNm

$$f_{cbL.L} = \frac{480.0 \times 10^6}{6.2728 \times 10^7} = 7.65 N/mm^2 \text{(tension)}$$

$$f_{ctL.L} = \frac{480.0 \times 10^6}{9.483 \times 10^7} = 5.06 N/mm^2 \text{(compression)}$$

$$f_{csL.L} = \frac{480.0 \times 10^6}{9.52 \times 10^7} = 6.38 N/mm^2 \text{ (tension)}$$

Stresses in Stage II

$f_{ctII} = [-3.783 \times 0.88] + 4.02 + 5.06$
$= 3.41 + 4.02 + 5.06 = 5.67\ N/mm^2$
$f_{ctII} = [15.937 \times 0.88] - 6.08 - 7.65$
$f_{ctII} = 14.02 - 13.73 = 0.29\ N/mm^2$

$$f_{sp_0} = \frac{1490 \times 10^3}{6 \times 98.7 \times 2} = 1242 N/mm^2$$

$$f_{sp\infty} = 1242 \times 0.88 = 1093 N/mm^2$$

$$f_{sp\infty} + f_{spD.L} + f_{spL.L} = 1093 + [6 \times 6.38 + 6 \times 5.07]\frac{8}{15}$$

$$= 1093 + (11.45 \times 6)\frac{8}{15} \quad = 1093 + 68.7 \times \frac{8}{15}$$

$$= 1093 + 36.64 = 1129.64 N/mm^2$$

Table 4.2.1 Constants for T-Sections

Section	$\frac{b_w}{h}$	$\frac{D_t}{h}$	A	Y_b	Y_t	I	r^2	K_t	K_b
1	0.1	0.1	0.19 bh	0.714 h	0.286 h	0.0179b h^3	0.0945 h^2	0.132h	0.333 h
2	0.1	0.2	0.28	0.756	0.244	0.0192	0.0688	0.0910	0.282
3	0.1	0.3	0.37	0.755	0.245	0.0193	0.0520	0.0689	0.212
4	0.1	0.4	0.46	0.735	0.265	0.0202	0.0439	0.0597	0.165
5	0.2	0.1	0.28	0.629	0.371	0.0283	0.1010	0.1610	0.272
6	0.2	0.2	0.36	0.678	0.322	0.0315	0.0875	0.1290	0.272
7	0.2	0.3	0.44	0.691	0.309	0.0319	0.0725	0.1050	0.234
8	0.2	0.4	0.52	0.684	0.316	0.0316	0.0616	0.0900	0.195
9	0.3	0.1	0.37	0.585	0.415	0.0365	0.0985	0.1690	0.237
10	0.3	0.2	0.44	0.626	0.374	0.0408	0.0928	0.1480	0.248
11	0.3	0.3	0.51	0.645	0.355	0.0417	0.0819	0.1270	0.231
12	0.3	0.4	0.58	0.645	0.355	0.0417	0.0720	0.1120	0.203
13	0.4	0.1	0.46	0.559	0.441	0.0440	0.0954	0.1710	0.216
14	0.4	0.2	0.52	0.592	0.408	0.0486	0.0935	0.1580	0.229
15	0.4	0.3	0.58	0.609	0.391	0.0499	0.0860	0.1410	0.220
16	0.4	0.4	0.64	0.612	0.388	0.0502	0.0785	0.1280	0.205
17	1.0	1.0	1.00	0.500	0.500	0.0833	0.0833	0.1670	0.167

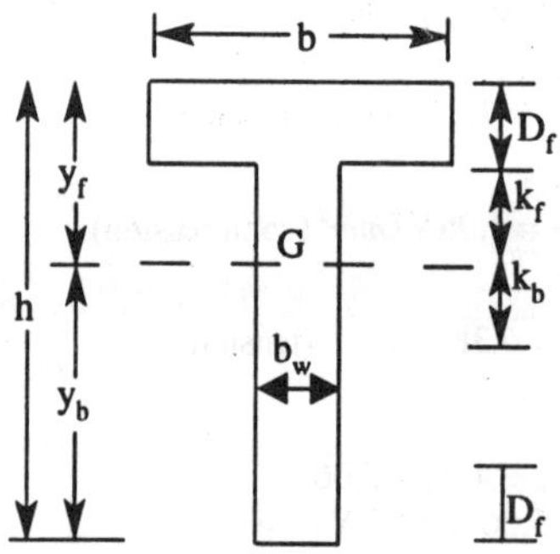

Table 4.2.2 Constants for Unsymmetrical I-Sections (Ratio of bottom to top flange width = 0.3)

Section	$\frac{b_w}{b}$	$\frac{D_t}{h}$	A	Y_b	Y_t	I	r^2	K_t	K_b
1	0.1	0.1	0.21 bh	0.650 h	0.350 h	0.0260 bh^3	0.1236 h^2	0.190 h	0.354 h
2	0.1	0.2	0.32	0.675	0.325	0.0345	0.1080	0.160	0.332
3	0.1	0.3	0.43	0.672	0.328	0.0387	0.0900	0.134	0.274
4	0.2	0.1	0.29	0.610	0.390	0.0316	0.1090	0.179	0.280
5	0.2	0.2	0.38	0.647	0.353	0.0378	0.0994	0.153	0.282
6	0.2	0.3	0.47	0.655	0.345	0.0402	0.0856	0.131	0.248

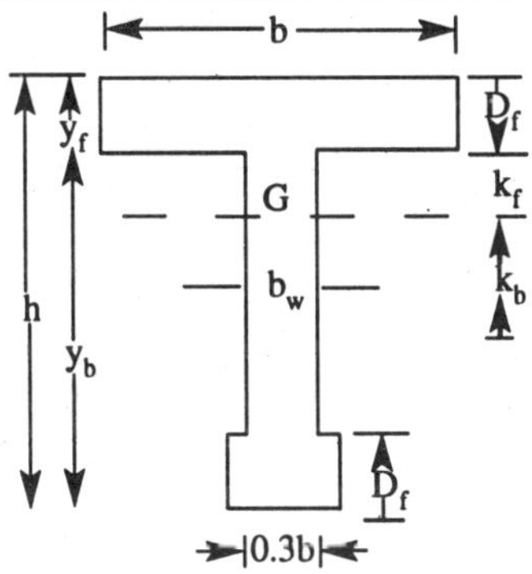

Table 4.2.3 Constants for Unsymmetrical I-Sections (Ratio of bottom to top flange width = 0.5)

Section	$\frac{b_w}{b}$	$\frac{D_t}{h}$	A	Y_b	Y_t	I	r^2	K_t	K_b
1	0.1	0.1	0.23 bh	0.597 h	0.403 h	0.0326 bh^3	0.1420 h^2	0.238 h	0.352 h
2	0.1	0.2	0.36	0.611	0.389	0.0464	0.1288	0.210	0.331
3	0.1	0.3	0.49	0.606	0.394	0.0535	0.1090	0.180	0.274
4	0.2	0.1	0.31	0.572	0.428	0.0373	0.1204	0.210	0.282
5	0.2	0.2	0.42	0.595	0.405	0.0488	0.1260	0.195	0.286
6	0.2	0.3	0.53	0.599	0.401	0.0540	0.1020	0.170	0.254
7	0.3	0.1	0.39	0.557	0.430	0.0443	0.1103	0.198	0.250
8	0.3	0.2	0.48	0.582	0.418	0.0510	0.1065	0.183	0.255
9	0.3	0.3	0.57	0.592	0.408	0.0553	0.0970	0.164	0.238

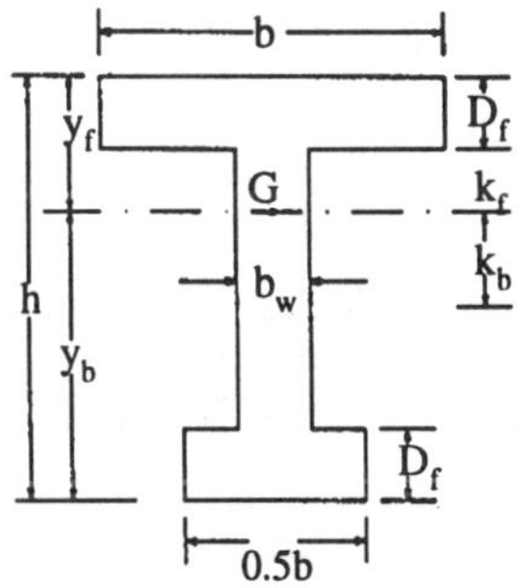

Table 4.2.4 Constants for Unsymmetrical I-Sections (Ratio of bottom to top flange width = 0.7)

Section	$\frac{b_w}{h}$	$\frac{D_t}{h}$	A	Y_b	Y_t	I	r^2	K_t	K_b
1	0.1	0.1	0.25 bh	0.554 h	0.446 h	0.0381 bh^3	0.1525 h^2	0.276 h	0.342 h
2	0.1	0.2	0.40	0.560	0.440	0.0560	0.1391	0.248	0.316
3	0.1	0.3	0.55	0.557	0.443	0.0651	0.1182	0.212	0.267
4	0.2	0.1	0.33	0.540	0.460	0.0425	0.1290	0.239	0.280
5	0.2	0.2	0.46	0.552	0.448	0.0578	0.1250	0.228	0.281
6	0.2	0.3	0.59	0.553	0.447	0.0657	0.1113	0.202	0.249
7	0.3	0.1	0.41	0.534	0.466	0.0467	0.1140	0.214	0.244
8	0.3	0.2	0.52	0.546	0.454	0.0598	0.1150	0.210	0.254
9	0.3	0.3	0.63	0.550	0.450	0.0663	0.1051	0.191	0.234

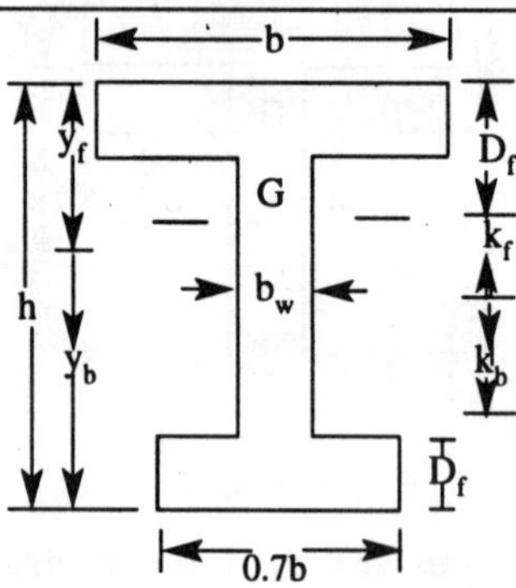

Table 4.2.5 Constants for Unsymmetrical I-Sections (Ratio of bottom to top flange width = 0.3)

Section	$\frac{b_w}{b}$	$\frac{D_t}{h}$	A	Y_b	Y_t	I	r^2	K_t	K_b
1	0.1	0.1	0.21 bh	0.350 h	0.650 h	0.0260 bh^3	0.1236 h^2	0.354 h	0.190 h
2	0.1	0.2	0.32	0.325	0.675	0.0345	0.1080	0.332	0.160
3	0.1	0.3	0.43	0.328	0.672	0.0387	0.0900	0.274	0.134
4	0.2	0.1	0.29	0.390	0.610	0.0316	0.1090	0.280	0.179
5	0.2	0.2	0.38	0.353	0.647	0.0378	0.0994	0.282	0.153
6	0.2	0.3	0.47	0.345	0.655	0.0402	0.0856	0.248	0.131

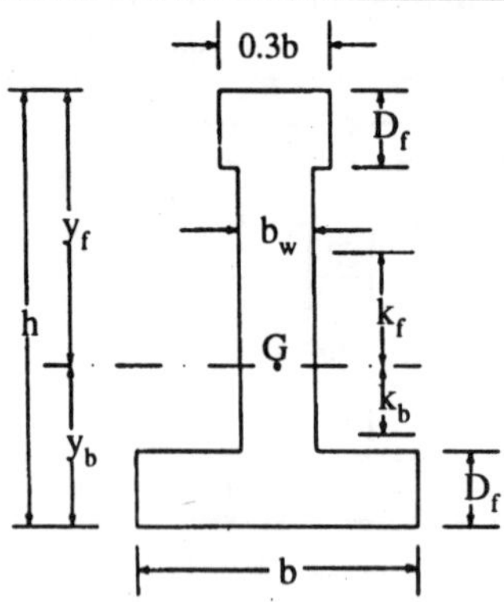

TABLE 4.2.6 Constants for symmetrical I And Box Sections

Section	$\frac{b_w}{b}$	$\frac{D_t}{h}$	A	Y_b	Y_t	I	r^2	K_t	K_b
1	0.1	0.1	0.28 bh	0.500 h	0.500 h	0.0449 bh³	0.160 h²	0.320 h	0.320 h
2	0.1	0.2	0.46	0.500	0.500	0.0671	0.146	0.292	0.292
3	0.1	0.3	0.64	0.500	0.500	0.0785	0.123	0.246	0.246
4	0.2	0.1	0.36	0.500	0.500	0.0492	0.137	0.274	0.274
5	0.2	0.2	0.52	0.500	0.500	0.0689	0.132	0.264	0.264
6	0.2	0.3	0.68	0.500	0.500	0.0791	0.117	0.234	0.234
7	0.3	0.1	0.44	0.500	0.500	0.0535	0.121	0.243	0.243
8	0.3	0.2	0.58	0.500	0.500	0.0707	0.122	0.244	0.244
9	0.3	0.3	0.72	0.500	0.500	0.0796	0.111	0.222	0.222
10	0.4	0.1	0.52	0.500	0.500	0.0577	0.111	0.222	0.222
11	0.4	0.2	0.64	0.500	0.500	0.0725	0.113	0.226	0.226
12	0.4	0.3	0.76	0.500	0.500	0.0801	0.105	0.211	0.211

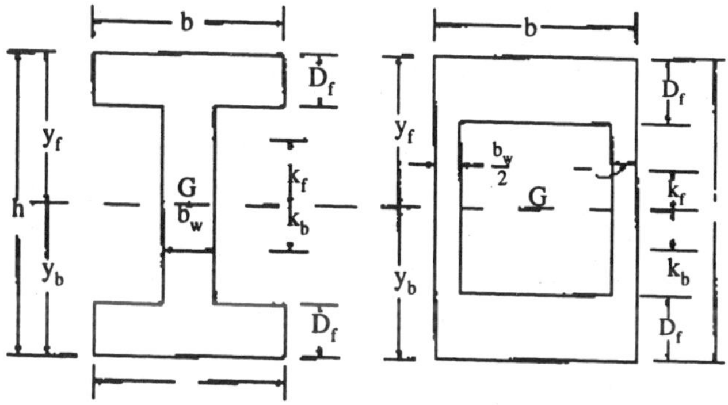

4.8 Resultant Compression Line (C.G.C line)

As indicated in Chapter 1, the prestressed concrete element may also be analyzed by ***"force concept"*** wherein, both causes and effects are to be considered as forces. The prestressing forces are felt along the alignment of prestressing tendon. It is identified as C.G.S. line. If there are no other forces acting on the structural element, the resultant compression will be collinear with the prestressing force line, so that the equilibrium is maintained by these 2 opposite forces. As soon as the dead load bending moment acts, which is instantaneous, (as and when the prestressing is applied), it has to be resisted by an internal couple. While the tensile force acts along the line of the prestressing tendon (C.G.S line) the resultant compression which is dictated by the internal distribution to stresses, is shifted so that the tensile force in the prestressing tendon and resultant compression concrete form the internal resisting couple. The distance between the resultant compression line (referred as C.G.C line) and C.G.S line forms the lever arm to resist external bending moment acting on the section. Hence as soon as the dead load bending moment starts acting there is redistribution to stress in cross-section so that the resultant compression shifts so as to create a lever arm for the resisting couple.

The total tensile force acting on the steel, [(i.e. prestressing force+ additional tensile force in steel due to curvature introduced by external moment (which is negligible)] is equilibrated by compressive stresses on concrete sections. The resultant of these compressive stresses namely the total compression 'C' acts at a distance 's_1', from the line of action of total tensile forces 'P' and 's_1' is evaluated as $\frac{M}{P}$ or $\frac{M}{C}$ where M is the total moment acting on the section and P is the force in prestressing tendon and C is the resultant compressive force. In stage I when dead load alone is acting the shift of the center of compression (the resultant compression line) is equal to $\frac{M_{D.L.}}{P_0}$. In stage II, when service load starts acting the prestressing force would have also been reduced due to losses and the resultant prestressing force will be P_∞. Ofcourse the additional force due to curvature given by the moment is neglected. Hence the distance between the C.G.C line (resultant compression line) and the C.G.S line (the position of prestressing cable) is equal to $\left(\frac{M_{total}}{P_\infty}\right)$. Hence the analysis of any section for the effects of prestressing and the external moments can be described by the position of C.G.S line and C.G.C line at any Stage. While the center of tension line will be the same as profile of prestressing cable, the center of compression line is moving depending on the external moment applied on the section. The locus of points of center of compression at various cross-sections is called the center of compression line for the structural element. If the center of compression line is known, the stresses across the section can be calculated using the total compression and the eccentricity of the compression line from the center of gravity effective/net cross-section.

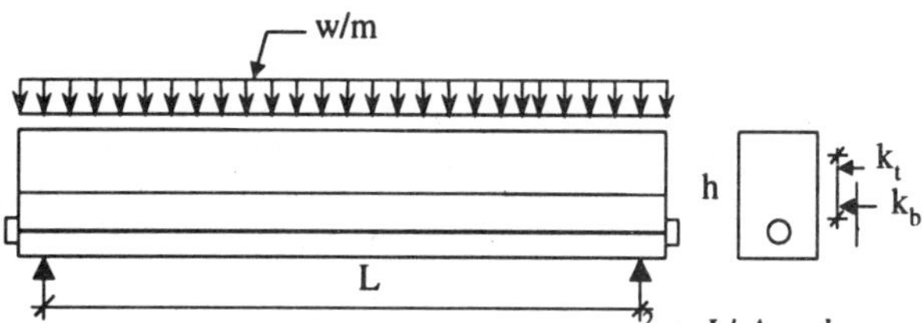

k_t = top kern point distance from C.G. $= \frac{r^2}{y^b} = \frac{I/A}{y^h} = \frac{d}{6}$

k_b = bottom kern point distance from C.G. $= \frac{r^2}{y_t} = \frac{I/A}{y_t} = \frac{d}{6}$

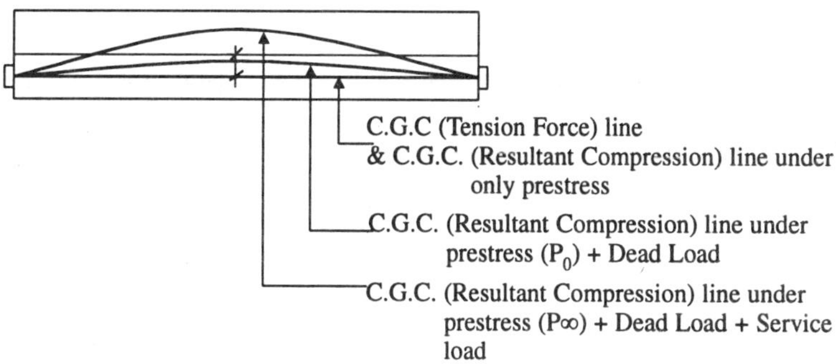

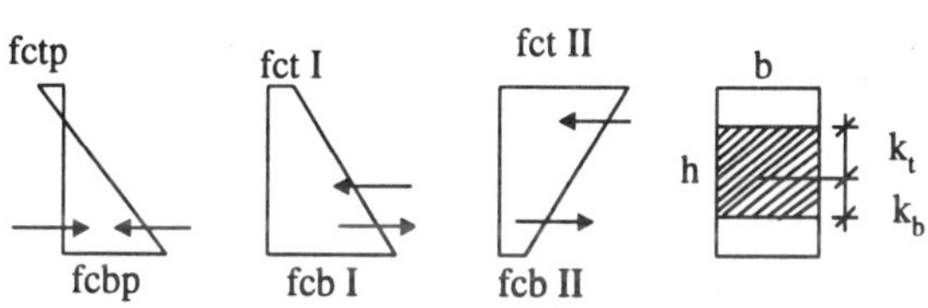

Note: If the C.G.C. line is above the shaded zone (kt - kb) bottom fibre will be in tension. If the C.G.C. line is below the shaded zone, the top fibre will be in tension.

Fig 4.13

In this approach the force flow of compression and tension are defined and the distance between these two, multiplied by the actual force gives the resisting moment. In order that tension does not occur at any point of time, either at the bottom fibre or at the top fibre both in Stage I and II, the force flow of the compression line should be within the kern point distances at the bottom and at the top. If it is going beyond the kern point distance at the top,tension will be generated at the bottom fibre and if the force flow of the compression line lies below the bottom kern point, tension will be generated at the top fibre.

Hence, once the force flow of compression line and the position of the kern points of the given cross-section are known, it can easily be noted whether the structural element is subjected to tension at all and the amount of tension generated, if at all, is dependent on the distance of the force flow line away from the kern points either at the bottom or at the top.

Example 4.9

A prestressed concrete member has a cross-section 600 × 300 mm and it is 12 m long. It carries a live load of 12k*N/m* in addition to its self-weight. It is prestressed with 2000 mm^2 high tensile steel located at 175 mm from soffit. The cable profile is straight for full length of the beam and it is stressed to a level of 800 N/mm^2. It is bonded to concrete. Determine the location of Thrust line in the beam and plot its position at end, quarter span/ and mid span sections. Assume $E_s/E_c = 6.0$

Solution

Properties of section

$A = 600 \times 300 = 180000$

$A_{sp} = 2000 mm^2$

$A_i = 180000 + 5 \times 2000 = 190000$

y_0 is the shift of centroid of effective section from its position as a gross section

$19000\, y_0 = 5 \times 2000 \times 125$

$y_0 = 6.5 mm$

$y_{bi} = 300 - 6.5 = 293.5 mm$

$y_t = 300 + 6.5 = 306.5 mm$

$$I_t = \frac{300 \times 600^3}{12} + 600 \times 300 \times 6.5^2 + 5 \times 2000 \times 118.5^2$$

$= 540\ 10^7 + .76\ 10^7 + 14.04 \times 10^7$

$$Z_{bi} = \frac{554.854 \times 10^4}{293.5}$$

$= 1890.47 \times 10^4\ mm^3$

$P_0 = 2000 \times 800 = 1600\ kN$

$e = 125 - 6.5 = 118.5 mm$

Stresses in concrete due to prestressing force

$$f_{ctp} = \frac{P}{A} - \frac{P \times e}{Z_t}$$

$$= \frac{1600 \times 10^3}{190 \times 10^3} - \frac{1600 \times 10^3 \times 118.5}{1810.3 \times 10^4}$$

$= 8.424 - 1.48$

$= -\ 2.056\ N/mm^2$

$$f_{cbp} = \frac{P}{A} + \frac{P \times e}{Z_b}$$

$$= 8.424 + \frac{1600 \times 10^3 \times 118.5}{1890.47 \times 10^4}$$

$= 18.454 N/mm^2$ (compression)

Bending moments

Self-weight $= 24 \times 0.18 = 4.32 kN/m$

Bending moment at mid span due to self-weight $= \dfrac{4.32 \times 12 \times 12}{8} = 77.76\ kN/m$

Bending moment at quarter span due to self-weight $= \dfrac{4.32 \times 12 \times 12}{8} \times \dfrac{3}{4.0} = 58.32 kNm$

Bending moment at mid span due to Live Load $= \dfrac{12 \times 12 \times 12}{8} = 216 kNm$

Bending moment at quarter span due to Live Load $= 12 \times 12 \times 12 \times \dfrac{3}{4.0} = 162 kNm$

Total bending moment at mid span $= 216 + 77.76 = 293.76$

Total bending moment at quarter span $= 162 + 58.32 = 220.32$

Shift of center of compression from C.G of steel in Stage I atmid span $= \dfrac{77.76 \times 10^3 \times 10^3}{1600 \times 10^3} = 48.6 mm$

Shift of center of compression from C.G. of steel in Stage I at quarter span $= \dfrac{58.32 \times 10^3 \times 10^3}{1600 \times 10^3} = 36.45 mm$

Eccentricity of pressure line from C.G. at mid span $= 118.5 - 48.6 = 69.9$ *mm*

Eccentricity of pressure line from C.G. at quarter span $= 118.5 - 36.45 = 82.05$

Shift of Center of Compression from C.G. of steel in Stage II at mid span $= \dfrac{293.76 \times 10^3 \times 10^3}{1600 \times 0.85 \times 10^3} = 216.0$mm

Shift of Center of Compression from C.G. of steel in Stage II in quarter span $= \dfrac{220.32 \times 10^3 \times 10^3}{1360 \times 10^3} = 162 mm$

Eccentricity of pressure line from C.G. mid span $= 216.0 - 118.5 = 97.5$ upwards

Eccentricity of pressure line from C.G at quarter span $= 162 - 118.5 = 43.5$ upwards

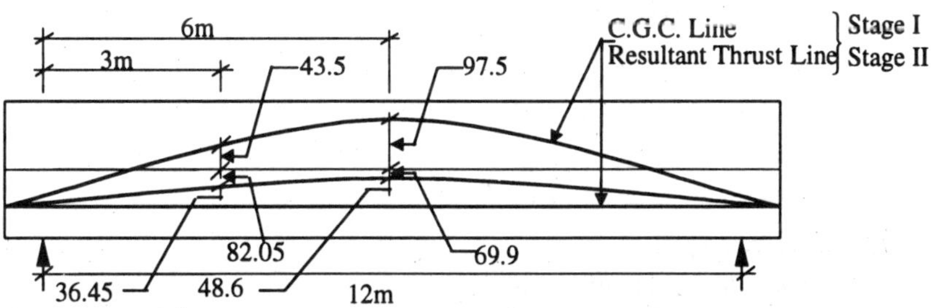

Fig 4.14 Center of Compression Profile At Stage I And Stage II

Example 4.10

Evaluate the stresses in Stage I and Stage II in case of Example 4.9 using Force Concept.

Solution

Properties of section and other details from example 4.9

A = 180000

A_{i0} = 190000

A_{sp} = 2000

$\frac{E_s}{E_c} = 6$

$y_0 = 6.5$

$y_{bi} = 293.5$

$y_{ti} = 30.65$

$I_i = 554854 \times 10^7\ mm^4$

$Z_{bi} = 1890.47 \times 10^4\ mm^3$

$Z_{ti} = 810.3 \times 10^4\ mm^3$

$P_0 = 1600kN$

$P\infty = 1360kN$

$e_i = 118.5\ mm$

Eccentricity of Resultant compression (C.G.C)
line from center of gravity at mid span at
Stage I & II = 69.9 downwards and 97.5 upwards respectively
Eccentricity of Resultant compression from
center of gravity at quarter span at
Stage I & II = 82.05 downwards and 43.5 *upwards* respectively

$P_0 = 1600\ kN$; $P_\infty = 1360\ kN$

Evaluation of stresses

Stresses at Mid cross section

$$f_{cbI} = \frac{1600 \times 10^3}{190 \times 10^3} + \frac{1600 \times 10^3 \times 69.9}{1890.47 \times 10^4}$$

$$= 8.42 - 5.92 = 14.34\ N/mm^2$$

$$f_{ctI} = 8.42 - \frac{1600 \times 10^3 \times 69.9}{1810.3 \times 10^4}$$

$$= 8.42 - 6.18 = 2.24 N/mm^2$$

$$f_{cbII} = \frac{1360 \times 10^3}{190 \times 10^3} - \frac{1360 \times 10^3 \times 97.5}{1890.47}$$

$$= 7.15 - 7.014 = 0.136 N/mm^2$$

$$f_{cbII} = 7.15 - \frac{1360 \times 10^3 \times 97.5}{1810.3 \times 10^4}$$

$$= 7.15 + 7.32 = 14.47\ N/mm^2$$

Stresses at Quarter Span

$$f_{cbI} = \frac{1600 \times 10^3}{180000} + \frac{1600 \times 10^3 \times 82.05}{1890.47 \times 10^4}$$

$$= 8.42 + 6.94 = 15.36 N/mm^2 \text{ (compression)}$$

$$f_{ctI} = \frac{1600 \times 10^3}{190000} - \frac{1600 \times 10^3 \times 69.9}{1810.3 \times 10^4}$$

$$= 8.42 - 7.25 = 1.17 N/mm^2 \text{ (compression)}$$

$$f_{cbII} = \frac{1360\times10^3}{190\times10^3} - \frac{1360\times10^3\times43.5}{1890.47\times10^4}$$

$$= 7.15 - 3.13 = 4.02 N/mm^2 \text{ (compression)}$$

$$f_{cbII} = 7.15 + \frac{1360\times10^3\times43.5}{1810.3\times10^4}$$

$$= 7.15 + 3.27 = 10.42\ N/mm^2 \text{ (compression)}$$

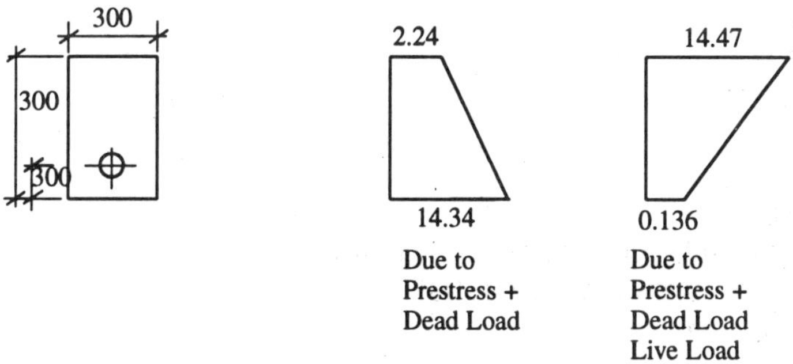

Fig 4.15 Stress Distbution At Mid Span

Example 4.11

Consider the Prestressed Concrete beam presented in Example 4.8 with bonded tendons. Draw the center of compression line for the same at Stage I and Stage II. Check for stresses at mid span both in Stage I and Stage II adopting the Force Method. Adopt m = 6.

Solution

Properties of Section

$I = 0.0510 \times 800 \times 1000$

$= 40.8 \times 10^9\ mm^4$

$D_{ft} = D_{fb} = 200$

$$\frac{D_f}{D} = \frac{200}{1000} = 0.2$$

$$\frac{b_w}{b_{ft}} = \frac{240}{800} = 0.3$$

$A = 0.48\times800\times1000$

$= 384000\ mm^2$

$y_b = 0.582 \times 1000$

$= 582\ mm$

$y_t = 0.418 \times 1000$

$= 418\ mm$

$\gamma^2 = 0.1086 \times 1000^2$

$= 1086 \times 100$

P = 2 Nos of 6T 13 cables with 2 ducts of 90 mm dia located at 100 mm from soffit.

$P_0 = 1490 kN$

Area of duct = 12717 mm^2

Effective section properties

At = $[384 \times 10^3 + 12 \times 98.72 \times 5]$

$= 384000 + 5923.2$

$= 389923.2\ mm^2$

$= 384 \times 10^3 \times \bar{y} = (5 \times 98.72 \times 12)(582 - 100 - \bar{y})$

$= (384\ \ 103 + 5923.2)\ \bar{y} = (5923.2 \times 482)$

$= 2854982.4$

$\bar{y} = 7.32mm$

$y_{bi} = 5827.32$

$y_{ti} = 418 + 7.32$

$= 425.32$

$y_{si} = 482 - 7.32$

$= 474.68$

$I_i = I_g + 384 \times 10^3 \times 7.32^2 + 5 \times 98.72 \times 12 \times 474.68^2$

$= 40.8 \times 10^9 + .0206 \times 10^9 + 1.335 \times 10^9$

$= 42.1566 \times 10^9$

$$Z_{bi} = \frac{42.1556 \times 10^9}{574.68} = 7.335 \times 10^7$$

$$Z_t = \frac{42.1556 \times 10^9}{425.32} = 9.912 \times 10^7$$

$$Z_{bi} = \frac{42.1556 \times 10^9}{474.68} = 8.88 \times 10^7$$

Bending Moments

Load Bending Moment at mid span $= \dfrac{0.3899 \times 24 \times 18.5^2}{8} = 400.33$kNm

Dead Load Bending Moment at quarter span $= 03899 \times 24 \times \dfrac{3}{3^2} \times 18.5^2 = 300.248$kNm

Live Load Bending Moment at mid span = 480 *kNM*

Live Load Bending Moment at quarter span = 360 *kNM*

P_0 =1490kN

$P_\infty = 0.88 \times 1490 = 1311.2kN$

Shift of Line of compression thrust from C.G of steel at mid span at Stage I = $\dfrac{400.33 \times 10^6}{1490 \times 10^3}$ = 268.68 *mm*

Shift of Line of compression from C.G. of steel at Quarter span at Stage I = $\dfrac{300.248 \times 10^6}{1490 \times 10^3}$ = 201.54 *mm*

Position of line of compression from C.G. of section at mid span in Stage I = 474.68 - 268.68 = 206.0 *mm* below of line of compression from

C.G. of section at quarter span in Stage I = 356.01 - 201.54 = 154.47mm

Shift of line of compression from C.G Steel at mid span at Stage II $= \dfrac{(480+400.33)\times 10^6}{0.88\times 1490\times 10^3}$ = 671.4mm

Shift of line of compression from C.G of steel at quarter span at Stage II $= \dfrac{(360+300.248)\times 10^6}{0.88\times 1490\times 10^3}$ = 503.54mm

Position of line of compression from C.G section at mid span in Stage II = 671.4 - 474.68 = 196.72 above

Position of line of compression from C.G section at quarter span in Stage II = 503.54 - 356.01 = 147.53 above

Stage II at top $= \dfrac{1311.2\times 10^3}{389923.2} + \dfrac{1311.2\times 196.72\times}{9.912\times 10^7}$

$= 3.362 + 2.602 = 5.962\ N/mm^2$

at bottom $= 3.362 - \dfrac{1311.2\times 196.72\times 10^3}{7.335\times 10^7}$

$= 3.362 - 3.515 = -0.155$ N/mm2

Stresses at mid section at Stage

I at bottom $= \dfrac{1490\times 10^3}{389923.2} + \dfrac{1490\times 10^3\times 206}{7.335\times 10^7}$

$= 3.82 + 4.185 = 8.005\ N/mm^2$

at top $= \dfrac{1490\times 10^3}{389923.2} - \dfrac{1490\times 10^3}{389923.2}$

$= 3.82 - 3.096 = 0.724 N/mm^2$

4.9 Load Balancing Concept

This concept is explained in Chapter 1 and is based on the fact that both causes and effects in the prestressed concrete structural element are expressed as loads. The loads for which any structural element is designed are primarily the self-weight of the element itself and externally imposed loads. The self-weight distribution is uniform and the distribution of externally imposed loads is given as data.

The prestressing force, and its location at different cross-section and the locus of the prestressing force flow along the length of the element defined of the profile of the prestressing tendon may be decided so as to see that it introduces an equivalent and opposite loads of that self-weight and externally imposed loads of structural element.

While profiling of the cable along the length of the element defines the pattern of load that would be generated by prestressing. The quantum of such load depends on the amount of prestress imparted. The types of tendon profile and the corresponding patterns of loads that are generated are given in Table 4.1. It should be noted that while representing the prestressing force

Table 4.1

No	Prestressing Tendon Profile	Equivalent Loading Pattern	Value of Load in Term of P & e
1.	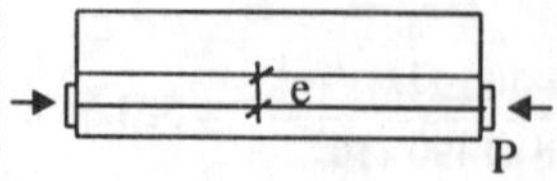	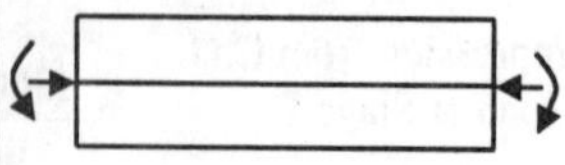	$M = Pe$
2.	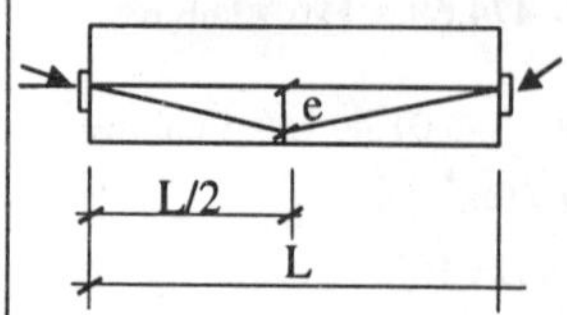	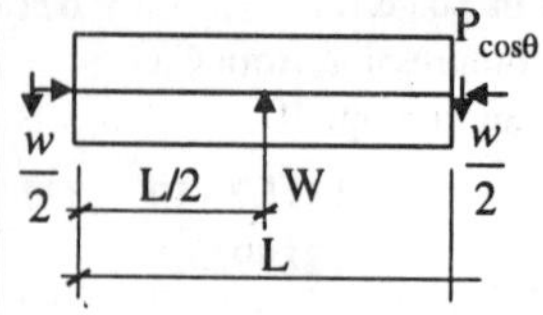	$W = \frac{4Pe}{L}$
3.	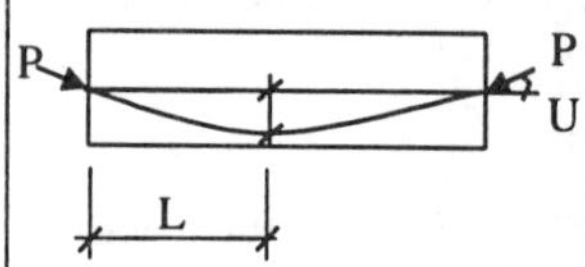	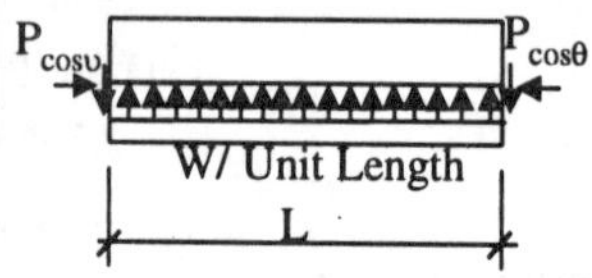	$W = \frac{8Pe}{L^2}$
4.	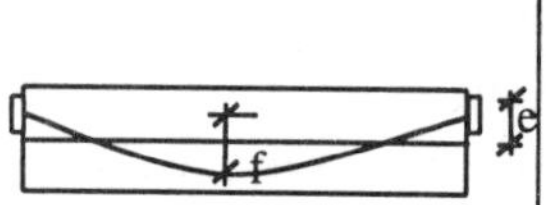	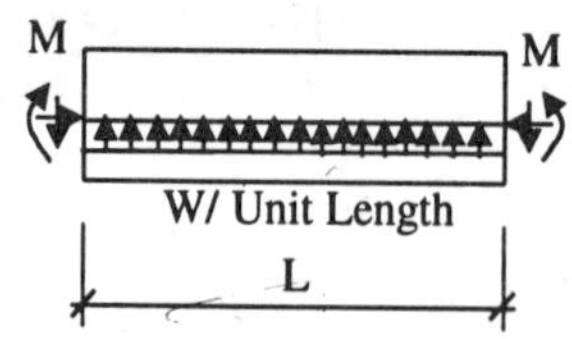	$W = \frac{8Pe}{L^2}$ $M = Pe$
5.	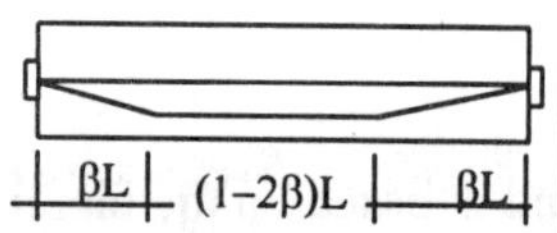	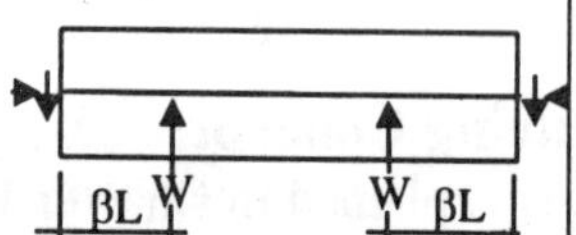	$W = \frac{Pe}{\beta L}$

along the length of the cable or replacing the prestressing tendon by an equivalent force pattern, one should not forget to introduce the end reaction at the anchored place in the equivalent force system. Without the end reaction, the structural element will not be in equilibrium, only with the equivalent force caused by prestressing. Hence in a simply supported prestressed beam, the prestressing cables are normally anchored at the ends where the beam is supported. The vertical component of the end reaction in the equivalent loading system does not affect the flexural stresses on the beam and the reaction forces is straight away taken by the support. The horizontal component of the end reaction leads to stresses, which have to be considered in the evaluation of final stresses. Any kink or sudden change in the slope of the cable profile leads to concentrated force on the structural element at that place. The method of approach for evaluation of the required prestressing force, profiling of the prestressing cable and the acceptable eccentricity are explained through examples.

Example 4.12

A rectangular beam of cross-section 350 × 700mm span 10 m and carries uniformly distributed load of 20*kN* the beam has to have a minimum cover distance of 150*mm* from soffit anywhere on the beam. Suggest a suitable cable profile and the prestressing force. Assume density of concrete to be $24kN/mm^2$. Assume gross properties of section for analysis.

Solution

Properties of section

Area of cross section $= 350 \times 700 = 245 \times 10^3 mm^2$

Moment of Inertia $= \dfrac{350 \times 700^3}{12} = 10004.16 \times 10^6 mm^4$

Z_t, Z_b $= \dfrac{10004.16}{350} \times 10^6$

$= 28.58 \times 10^6 mm^3$

Self weight of beam $= 350 \times 700 \times 10^{-6} \times 24 = 5.88kN/m$

$= 20kN/m$

Live Load on beam

Total Load $= 25.88\ kN/m$

Maximum possible eccentricity at

mid section $= \dfrac{700}{2} - 150 = 200mm$

If the end section should not have any moment there should not be any eccentricity at end. Since the bending moment profile due to loads is parabolic, the prestressing tendon profile is also made parabolic.

Maximum dip of parabola = 200mm

The *e.u.d.l* to be resisted = 25.88 *kN/m*

Prestressing force to be given in a Parabolic profile with maximum dip of 200mm at center = P

$$P = \frac{25.88 \times l^2}{8 \times 0.2}$$

$$= \frac{25.88 \times 100}{8 \times 0.2}$$

$$= 1617.5 \text{ kN}$$

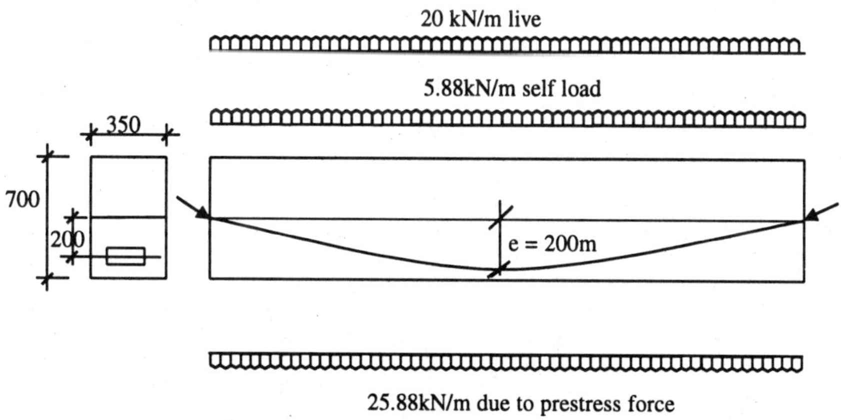

Fig 4.16 Equivalent Load By Load Balancing Method

It has been explained earlier while discussing about the resultant trust (C.G.C) line that the position of the CGC line defined the existence of tension or otherwise at extreme fibres. If the CGC line is above the top kern point the bottom fibre experiences tension and if the CGC line falls below the bottom kern point the top fibre experiences tension. The movement or shift of CGC line is decided by applied moment and prestressing force and expressed as *M/P*. If no tension is permitted at the top fibre in Stage I (Prestressed P_0 + *DL*). The C.G.C line can be located below the kern point only by an amount equal to M_{DL}/P_0. As soon as the dead load moment is realized (it occurs immediately after prestress) the CGC line moves up and takes the position at the place at or above the kern point so that no tension is generated at the top. These points are linked up in the load balancing approach to arrive at the required force and eccentricity at any cross-section (profile of the cable) to counteract external loads.

Example 4.12 and 4.13 explains the use of load balancing method in the analysis of prestressed concrete structural elements.

Example 4.13

In the Example 4.12, if end eccentricity is permitted and if no tension is to be generated at any point of time, what is the prestressing force that could be allowed in the beam.

Solution

Properties of Section

Area of cross section	$= 245 \times 103\ mm^2$
Moment of Inertia	$= 10004.16\ \ 106\ mm4$
Z_t, Z_b	$= 28.50 \times 106\ mm^3$
Self weight of the beam	$= 5.88 kN/m$
Live Load	$= 20\ kN/m$
Total Load`	$= 22.80$

Top kern point distance $= k_t = \dfrac{700}{6} = 116.66 mm$ from C.G.

Bottom kern point distance $= k_b = 116.66 mm$ from C.G.

Trial No.1

Assume that the cable is anchored at top kern point at support and has '*e*' = 200 at mid section. The parabolic profile will have a dip = 200 + 116.66 = 316.66

$$\text{The corresponding } P = \frac{25.88 \times l^2}{8 \times 316.66} = 1021.6 kN$$

Maximum possible eccentricity at mid span

$$\text{so that no tension is generated at top fibre} = \left[116.66 + \frac{M_{D.L.}}{P}\right]$$

$$= \left[116.66 + \frac{5.88 \times 100}{8 \times 1021.6} \times 10^3\right]$$

$$= 116.66 + 71.94 = 188.6 < 200 \text{ provided}$$

Trial No.2

Assume the cable is anchored at top kern point and has e = 190*mm* at mid section. The parabolic profile will have a dip of (116.66+190) 306.66 at mid span

$$\text{Corresponding } P = \frac{25.88 \times 100}{8 \times 306.66} = 1055.2kN$$

Maximum possible eccentricity at mid span

so that no tension is created at top fibre in Stage I = $\left[116.66 + \frac{M_{D.L.}}{P}\right]$

$$= 116.66 + \frac{5.88 \times 100}{8 \times 1055.2}$$

$$= 116.66 + 69.6 = 186.26 < 190 \text{ provided}$$

Trial No.3

Assuming the cable to have 'e' = 184 at mid span, the max dip of cable

$$= 116.66 + 184$$

$$= -300.66$$

$$\text{Corresponding } P = \frac{25.88 \times 100}{8 \times 300.66} = 1076kN$$

Maximum possible eccentricity permissible

$$\text{in Stage I} = \left[116.66 + \frac{5.88 \times 100}{8 \times 1076}\right]$$

$$= 116.66 + 68.3 = 184.96 > 184mm$$

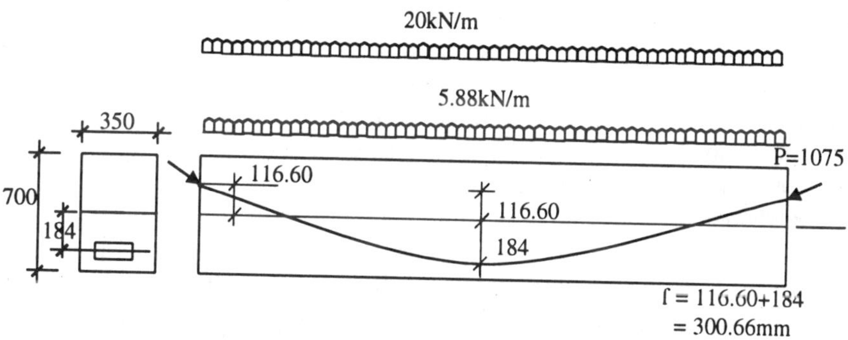

Fig 4.17 Maximum Dip Otimum Eccentricity And Cable Profile To Counter (20+5.88 kN/m)

References

1. LIN. T.Y., (1963), ***"Design of Prestressed Concrete Structures"***, John Wiley & Sons, Inc. New York.
2. PLAHN. J., (1971), "***Introduction to Prestressed Concrete"***, Lecture Notes, Department of Civil Engineering, Indian Institute of Technology 1971 (unpublished).
3. LEONHARDT. F., (1964) ***Prestressed Concrete Design and Construction***, Wilhelm Ernst and Sohn, Berlin,.
4. LIBBY.J.R., (1977) ***"Modern Prestressed Concrete"***, Van Norstrand Reinhold Company, Newyork.

5. Losses in Prestress

5.1 Introduction

The first attempt to produce prestressed concrete in the late 19th Century and early 20th Century was not successful since the prestressing force transferred on to concrete using normal steel, found to be lost totally after few months and these losses were later found to have been caused by the creep properties of concrete and relaxation properties of steel. Hence the ***loss in prestress due to time dependent material properties*** were identified as early as 1923 [1]. This led to the evolution of high tensile steel as mentioned in Chapter I.

In addition to the losses in prestress due to material properties such as of creep and relaxation, there are losses in prestress from the applied level of force due to various other causes. The initial prestress achieved by pulling the cable/prestressing steel by the jack system is identified as P^0. When the same is transferred on to the concrete, the concrete element elastically deforms and there is strain in concrete and that affects the applied elongation in steel, reducing the prestressing force to P_0. The loss in stress in steel due to elastic shortening of concrete happens ***predominantly in pretensioned elements.***

But in case of post-tensioned elements, the pulling of cables is always against the prestressing elements themselves. Hence pulling the cables and the elastic deformation of the concrete element happen simultaneously. If all the cables are pulled together simultaneously, the jack reads only the force, which includes the loss due to elastic shortening, and hence the jack force is already P_0. If in case of multiple stages of pulling or when the cables are pulled one after another, there is a reduction in the force in the first cable when the second cable is pulled and anchored and so on. ***Hence, in case of post tensioned elements the loss due to elastic shortening is nil only if there is only one stage of pulling and anchoring. There will be a definite loss in total prestress due to elastic shortening of concrete if pulling is done in number of stages.***

It may be noted that, in case of post tensioning, cables are anchored on the structural elements through external anchorage's and these anchorages, as seen in Chapter 2, are mostly wedge type or button type. In either case while anchoring the cables, there is a possibility of the slip of the cable with respect to anchorage, thereby reducing the already given elongation by prestress. This means there is ***loss in the prestressing force caused by slip of anchorages. This happens only in post tensioned elements***. Further in case of post tensioned concrete elements, the prestressing cables are pulled through the duct holes already provided in the element. During pulling, the elongations at the pulling ends are measured and these values are also considered for evaluating the force applied in the cable. These elongations could be due to straightening of the cables which have some wobble effect originally and the prestressing force read by the jack, may also include the force used up for straightening the wobble effect of the cables. Further the cable adheres to the duct surface if it were a single element or adheres to the adjacent elements if there are more than one prestressing element in the duct. In either case, during pulling, if a force P is applied at one end, the force P is not felt uniformly along the full length of the cable because of certain ***force loss in overcoming the frictional effects*** between the cable duct and the element themselves, or between the individual elements themselves. ***This happens only in post tensioned elements.***

After the prestressing force is transmitted to the element either by pretensioning approach or by post tensioning approach, the concrete is in compression and the steel is in tension (with constant elongation between the anchoring ends). For the same applied stress in concrete, the concrete deforms further with time due to creep of concrete. This creep deformation affects the effective elongation left over on the cable. Simultaneously the ends of prestressing steel are kept in the same elongated position for long duration. This causes a loss in its stress value due to relaxation of material. Hence, there is a reduction in force in steel. In addition to this, because of water evaporation caused by the temperature variation, there is shrinkage of material concrete leading to the loss in the total length of the element correspondingly also the total elongation in steel. Hence the creep and shrinkage of concrete, and the relaxation in steel cause reduction in applied force with respect to time. But the rate of reduction caused by these three effects is a decreasing function and becomes practically nil after five years. Hence the ***losses in prestress due to creep and shrinkage of concrete and relaxation of steel,*** which are time dependent, are calculated for a period of 5 years while arriving at the effective prestress on concrete at service load conditions. Summarizing, the losses in prestress due to various aspects mentioned above is given in the tabular form in the Table 5.1.

In general losses are categorized under 2 groups, namely losses in prestress immediately after transfer referred to as ***immediate losses*** and losses in prestress due to time dependent behaviour of materials concrete (creep, shrinkage) and steel (relaxation) referred as ***time dependent losses***.

Loss due to elastic deformation of concrete, loss due to friction between the materials and cables, loss due to wobbling effects and loss due to slip of anchorages come under immediate losses. Loss due to creep and shrinkage of concrete and loss due to relaxation of steel come under time dependent losses. They are discussed in detail below.

Table 5.1 Types of Losses in Prestress

No. Pretensioning		No. Post-tensioning
1. Elastic deformation of concrete	← Immediate→ losses	1. No loss due to elastic deformation if all the wires (cables) are simultaneously tensioned. If the wires are successively tensioned, in stages there will be loss of prestress due to elastic deformation of concrete. 2. Anchorage slip 3. Friction between cables and between cable duct.
2. Creep of concrete. 3. Shrinkage of concrete 4. Relaxation of stress in steel	←Time dependent→ losses	4. Creep of concrete 5. Shrinkage of concrete 6. Relaxation of stress in steel

5.2 Immediate Losses

5.2.1 Loss due to elastic deformation of concrete (elastic shortening)

This happens specially in case of pretensioning. In case of post tensioning there is a loss in prestress due to elastic shortening only when individual wires/cables are tensioned one after another and ***not simultaneously***. The loss due to elastic deformation depends on the average stress in concrete at the level of steel and the modular ratio $(\frac{E_s}{E_c})$. When a force of prestressing namely P^0 is applied at the bulkheads or in a self straining frame and then transferred to concrete, the concrete suffers a compressive stress at the level of steel defined as f_{csp} (fibre stress in concrete at steel level due to prestress). There is a corresponding compressive stress in steel, which could be evaluated as mf_{csp}, because of compatibility between steel and concrete, reducing in tensile stresses in steel called loss due to elastic shortening. Hence the loss in stress in steel due to elastic shortening of concrete is mf_{csp}. The stress in concrete f_{csp} can be calculated with P^0 and effective sectional properties or P_0 with net sectional properties as discussed in Article 4.5.6. Since P_0 is not yet known it is always preferable to take P^0 and effective sectional properties to evaluate the concrete stress at the level of steel. Hence the loss in prestress due to elastic shortening in pretensioned elements is evaluated as $mf_{csp} = m\left[\frac{P^0}{A_i} \pm \frac{P^0}{I_I} e_i^2\right]$. For approximate calculations, f_{csp} is also calculated using gross sectional properties as illustrated in Example 5.1.

In case of post tensioning, if all cables are tensioned together, there are no losses due to elastic shortening of concrete, since even while pulling, the concrete element gets compressed and at the time of transfer, the elastic shortening has completely occurred. If the cables are tensioned one after another, there is no loss of force due to elastic shortening when the first cable is tensioned. When the second cable is tensioned, it does not have any loss in force due to elastic shortening, but the already stressed and anchored first cable suffers a loss due to elastic shortening defined by *m* times the concrete stress at that fibre level (first cable level) due to stressing of second cable. The force transferred by the second cable on to the concrete (P_{02}), and the proper section properties at that point of time have to be considered for arriving at the stress in concrete at the first cable level. The loss in force in the first cable due to elastic shortening of concrete caused by stressing the 2nd cable $= \left[\frac{P_{02}}{A_n} + \frac{P_{02} e_{i2}}{I_n} y_{1n}\right]$ where the subscript 1 and 2 refers to the corresponding cables location and distances, *n* represents net section properties.

If the cables are not straight and are curved along the length of the element the stress on concrete at the level of the cable is not constant throughout the length of the element. In such a case the average stress in concrete at steel level has to be as illustrated in Example 5.3.

Example 5.1

A prestressed concrete sleeper produced by pretensioning method has a rectangular cross section of 300 mm × 250 mm (b × d). It is prestressed with 9 Nos. of straight 7 mm wires at 0.8 times the ultimate stress of 1570 N/mm^2. Estimate the percentage loss of stress due to elastic deformation of concrete. Consider $m = 6$.

Solution

Properties of gross section

$A = 300 \times 250 = 75000\ mm^2$

$I = \frac{300 \times 250}{12} = 390.625 \times 10^6\ mm^4$

$m = \left(\frac{E_s}{E_c}\right) = 6$

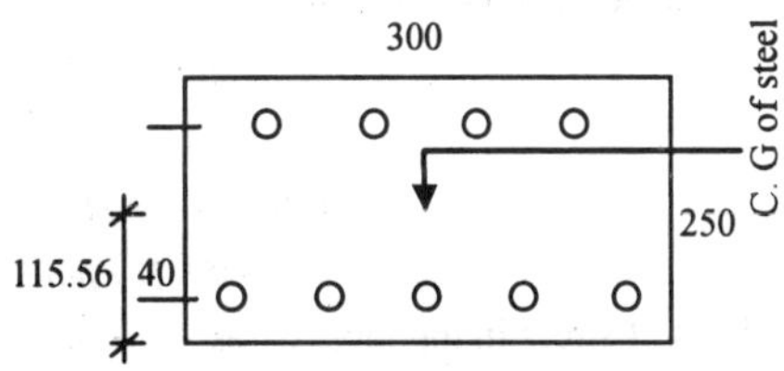

Fig. 5.1

Level of C.G. of steel from bottom fibre = $y = \left[\frac{4 \times 210 + 5 \times 40}{4 + 5}\right] = 115.56mm$

Effective Section Properties

$$A_i = 75000 + (6-1)\times 9 \times \pi \times \frac{7^2}{4}$$

$$= 76731 mm^2$$

Shift of Center of gravity of effective cross section be y_0 from Center of gravity of gross section towards bottom fibre

Taking moment of areas about the original Center of gravity position

$$76731 \times y_0 = \frac{\pi \times 7^2}{4} \times 5$$

$$y_0 = 0.0025 mm$$

The Center of Gravity of effective cross section is at 125.0025*mm* from top fibre and 124.9975*mm* from bottom fibre.

Moment of Inertia of effective cross section = I_i

$$I_i = 390.625\times 10^6 + 4\times\frac{\pi}{4}\times 5\times(125.0025-40)^2 + 4\times\frac{\pi}{4}\times 5\times(124.9975-40)^2$$

$$= 403.131 \times 10^6\ mm^4$$

Force details

Area of one 7*mm* wire = $\frac{\pi \times 7^2}{4} = 38.48 mm^2$

Initial stress in steel $= 0.8 \times 1570$

$= 1256\ N/mm^2$

Total applied force $= 9 \times 38.48 \times 1256$

$= 434.98 kN$

$$e = 124.9975 - 115.56 = 9.4375 mm$$

Stresses in Concrete

For calculating stresses in concrete, applied force without loss due to elastic shortening (P^0) and effective section properties are considered (Ref 4.5.6)

At the level of top wires $= \frac{434.98\times 10^3}{76731} - \frac{434.98\times 10^3 \times 9.4375\times(125.0025-40)}{403.131\times 10^6}$

$= 5.670 - 0.8655$

$= 4.8045\ N/mm^2$

At the level of bottom wires

$= \frac{434.98\times 10^3}{76731} - \frac{434.98\times 10^3 \times 9.4375\times(124.9975-40)}{403.131\times 10^6}$

$= 5.670 + 0.8655$

$= 6.5355\ N/mm^2$

Loss of stress in top level wires $= 6 \times 4.8045$

$= 28.827\ N/mm^2$

Loss of stress at bottom level wires $= 6 \times 6.5355$

$= 39.213\ N/mm^2$

Total force in top level wires = [1256 - 28.827] 4 × 49 × 0.785 = 188812.84

= [1256 - 39.213] 5 × 49 × 0.785 = 234018.56

Total force in Steel = 4228314.4 *N*

= 422.83 *kN*

Force transmitted to concrete before

losses due to elastic shortening = $1256 \times 9 \times 49 \times 0.785 = 434808.4$

= 434.8*kN*

Loss in force due to elastic shortening = 434.8 – 422.83

= 11.97 *kN*

Percentage loss in force due to elastic shortening

$$= \frac{11.97}{434.8} \times 100 = 2.753\%$$

Approximate Solution

Stresses in concrete

For approximate solution, it is sufficient to consider gross section properties, '*e*' with respect to gross section center of gravity = 9.44*mm*

Stress in concrete at the level of top wires

$$= \left[\frac{434.98 \times 10^3}{75000} - \frac{434.98 \times 10^3 \times 9.44}{390.625 \times 10^6} \times (125 - 40)\right]$$

= 5.80 – 0.893 = 4.907 N/mm^2

At the level of bottom wires $= \left[\frac{434.98 \times 10^3}{75000} - \frac{434.98 \times 10^3 \times 9.44}{390.625 \times 10^6} \times (125 - 40)\right]$

= 5.80 + 0.893 = 6.693 N/mm^2

Loss of stress in wires at top = 6×4.907 = 29.442 N/mm^2

Loss of stress in wires at bottom = $6 \times 6.693 = 40.158$ N/mm^2

Loss in forces in top wires = $29.442 \times 4 \times 38.48 = 45.32$ *kN*

Loss in forces in bottom wires = $40.150 \times 5 \times 38.48 = 7.726$ *kN*

Total = 12.258*kN*

Initially applied force = $1256 \times 9 \times 38.46 = 434.8$

% loss in force $= \frac{12.250}{434.8} = 2.819\%$

Example 5.2.

A prestressed concrete beam has a cross-section of 300 × 600 *mm* and it is post tensioned with 12 nos of 12.72 dia meter strand in 4 cable ducts. 3 strands form one cable. 40 *mm* duct is provided for each cable of 3 strands. Two cable ducts are located at a level of 100 *mm* from soffit and two more cable ducts at 200 *mm* from soffit. Each cable with 3 strands is prestressed at one stretch with a force of 3 × 128.6 *kN* which is 0.7 times its capacity. The bottom cables 1 & 2 are stressed first and then the second level cables 3 and 4 are stressed. The cables have straight profile. Evaluate the loss in force of prestress due to elastic shortening. Assume *m* = 6.

Solution

Since it is a post tensioned beam, there are duct holes and hence only net cross sectional properties are to be used for arriving at the stresses in concrete properties of net section.

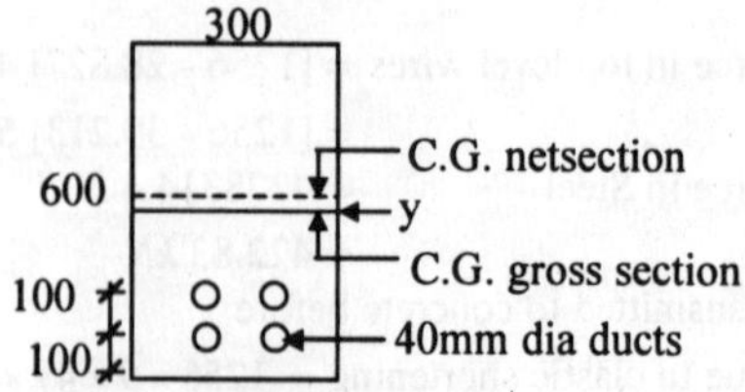

Fig. 5.2

Properties of net section

Area = 600 × 300 – 4 × 40 × 0.785

= 174976 = 17.497 × 10^4

Center of gravity for net section be shifted by 'y_0' from center of gravity of gross section

Taking moment of areas about the center of gravity of cross section

$$= 17.497 \times 10^4 \times y_0 = 2 \times 40 \times 0.785 \times (200 - y_0) + 2 \times 40 \times 40 \times 0.785(100 - y_0)$$

$$= 18 \times 10^4 \times y_0 = 3 \times 40 \times 0.785 \times 200$$

$$y_o = \frac{3 \times 40 \times 40 \times 0.785 \times 200}{18 \times 10^4}$$

$y_0 = 4.186\ mm = 4.2\ mm$

Distance of extreme top fibre from center of gravity of net section = 295.8*mm*

Distance of extreme bottom fibre from center of gravity of net section = 304.2

$$I_n = \left[300 \times \frac{600^3}{12}\right] + [600 \times 300](4.2)^2 - [40 \times 40 \times 0.785](104.2)^2 \times 2$$

$$-[40 \times 40 \times 0.785](204.2)^2 \times 2$$

$$= 5.4 \times 10^9 + 3.175 \times 10^6 - 2.2727 \times 10^7 - 10.4745 \times 10^7 = 5.276 \times 10^9$$

40*mm* duct is provided to carry 3 strands together as one cable

When cable 1 is stressed and anchored there is no loss due to elastic shortening in cable 1.

When cable 2 is stressed and anchored, there will be loss due to elastic shortening in cable 1, which is already stressed and anchored.

Stress in concrete at the level of cable 1.

$$f_{csp} = \frac{3 \times 128.6 \times 10^3}{174.97 \times 10^4} + \frac{3 \times 128.6 \times 10^3 \times 204.2 \times 204.2}{5.276 \times 10^9}$$

= 2.205 + 3.049 = 5.254 N/mm^2

Loss in stress in cable 1 due to elastic shortening

$$= mf_{csp} = 6 \times 5.254 = 31.524\ N/mm^2$$

Loss in stress in cable 2 = 0

When cable 3 is stressed and anchored there will be loss of stress in cable 1 and cable 2 due to elastic shortening

Stress in concrete at the level of cable 1 & 2.

$$f_{csp} = \frac{3 \times 128.6 \times 10^3}{17.497 \times 10^4} + \frac{3 \times 128.6 \times 10^3 \times 104.2 \times 104.2}{5.276 \times 10^9}$$

= 2.205 + 1.555

= 3.760 N/mm^2

Loss in stress in cable 1 $= mf_{csp} = 6 \times 3.760$

$= 22.560\ N/mm^2$

Loss in stress in cable 2 $= 22.650\ N/mm^2$

Loss in stress in cable 3 $= 0$

When cable 4 is stressed and anchored there will be loss in cable 1, 2 and 3.

Stress in concrete at the level of cables 1 and 2

$$f_{csp} = \frac{3 \times 128.6 \times 10^3}{17.497 \times 10^4} + \frac{3 \times 128.6 \times 10^3 \times 104.2 \times 104.2}{5.276 \times 10^9}$$

$= 3.760\ N/mm^2$

Stress in concrete at level of cable 3

$$f_{csp} = \frac{3 \times 128.6 \times 10^3}{17.497 \times 10^4} + \frac{3 \times 128.6 \times 10^3 \times 104.2 \times 104.2}{5.276 \times 10^9}$$

$= 2.205 + 0.794$

$= 2.999\ N/mm^2$

Loss of stress in cable 1 = 6 × 3.760 = 22.560 N/mm^2

Loss of stress in cable 2 = 6 × 3.760 = 22.560 N/mm^2

Loss of stress in cable 3 = 6 × 3 = 18 N/mm^2

Loss of stress in cable 4 = 0

Total losses due to elastic shortening in cable 1

$= 31.524 + 22.560 + 22.560$

$= 76.645\ N/mm^2$

Total losses due to elastic shortening in cable 2

$= 22.560 + 22.560$

$= 45.12\ N/mm^2$

Total losses due to elastic shortening in cable 3 = 18 N/mm^2

Total losses due to elastic shortening in cable 4 = 0

Total loss in force in cables

Cable 1 = 76.645 × 98.7 × 3 = 22.695

Cable 2 = 45.120 × 98.7 × 3 = 13.360

Cable 3 = 18 × 98.7 × 3 = 5.3298

41.3985*kN*

$$\%\ \text{Loss in force} = \frac{41.3848 \times 100}{128.6 \times 12} = 2.68\%$$

Example 5.3

Evaluate the losses in forces due to elastic shortening only in cables 1, 2, 3, 4 given in Example 5.2, if the cable alignment is straight for cable 1 and 2 at 100*mm* from soffit, and cable 3 and 4 have a parabolic profile with their position at 200*mm* from soffit at mid span to a position of 100*mm* from top at the end section. All other details are same as given in Example 5.2. The schedule of pulling operations is as given in Example 5.2.

Solution

Since it is a post tensioned beam, there are duct holes and hence net cross sectional properties only are to be considered for arriving at the stresses in concrete.

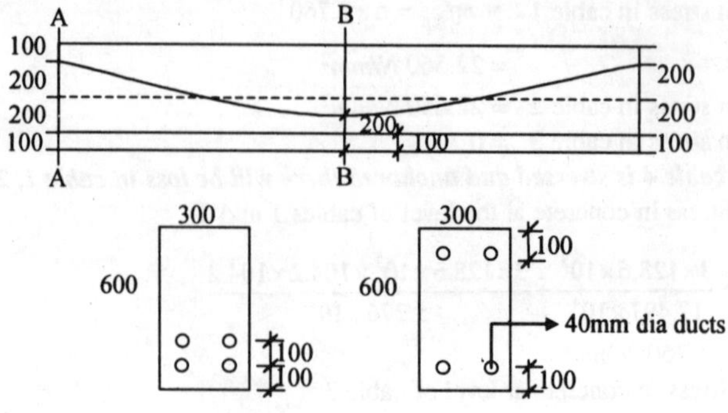

Mid cross section (B B)   End cross section (A A)

Fig. 5.3

Properties of net section

Area = 17.497 × 10^4 (From Example 5.2)

Distance of extreme top fibre from center of gravity of net section at mid cross section = 295.8*mm* (From example 5.1)

Distance of extreme bottom fibre from center of gravity of net section at mid cross section = 304.2*mm* (From example 5.2)

$I_n = 5.276 \times 10^9\ mm^4$ (From Example 5.2)

I_n for end section will be marginally different.

But it is assumed as $5.276 \times 10^6\ mm^4$

40*mm* duct is provided to carry 3 strands together as one cable

When cable 1 is stressed and anchored there is no loss due to elastic shortening in cable 1.

When cable 2 is stressed and anchored, there will be loss due to elastic shortening in cable 1, which is already stressed and anchored.

Stress in concrete at the level of cable 1.

$$f_{csp} = \frac{3\times 18.6\times 10^3}{174.97\times 10^4} + \frac{3\times 128.6\times 10^3\times 204.2\times 204.2}{5.276\times 10^9}$$

= 2.205 + 3.049 = 5.254 N/mm^2

Loss in stress in cable 1 due to

elastic shortenins g = mf_{csp} = 6 × 5.254 = 31.524 N/mm^2

Loss in stress in cable 2 = 0

When cable 3 is stressed and anchored there will be loss of stress in cable 1 and cable 2, which are already stressed and anchored

Stress in concrete at the level of cable 1 and 2 at mid section

$$f_{csp} = \frac{3\times 128.6\times 10^3}{17.497\times 10^4} + \frac{3\times 128.6\times 10^3\times 104.2\times 204.2}{5.276\times 10^9}$$

= 2.205 + 1.555 = 3.760 N/mm^2

Stress in concrete at level of cable 1 and 2 at support section

$$f_{csp} = \frac{3\times 128.6\times 10^3}{17.497\times 10^4} - \frac{3\times 128.6\times 10^3\times 195.8\times 204.2}{5.276\times 10^9}$$

= 2.205 – 2.923 = – 0.718 N/mm^2 (Tension)

Average stress in concrete over the length

$$= -0.718 + \tfrac{2}{3}(3.76 + 0.718)$$

$$= -0.718 + 2.985 = 2.267\ N/mm^2$$

Loss in stress in cable 1 & 2 = 6 × 2.267

= 13.602 N/mm^2

Loss in stress in cable 3 = 0

When cable 4 is stressed and there will be loss in stress in cable 1, 2, and 3

Stress in concrete at level of cable 1 and 2

at support = – 0.718 N/mm^2 (Tension)

at mid section = 3.760 N/mm^2

Average stress in concrete at the level of cable 1 & 2

$$= -0.718 + \tfrac{2}{3}(3.76 + 0.718)$$

$$= 2.267\ N/mm^2$$

Stress in level of cable 3 at support

$$= \frac{3\times 128.6\times 10^3}{17.497\times 10^4} - \frac{3\times 128.6\times 10^3\times 195.8\times 195.8}{5.276\times 10^9}$$

$$= 2.205 + 2.803$$

$$= 5.008\ N/mm^2$$

Stress in level of cable 3 at mid span

$$= \frac{3\times 128.6\times 10^3}{17.497\times 10^4} + \frac{3\times 128.6\times 10^3\times 104.2\times 104.2}{5.276\times 10^9}$$

$$= 2.205 + 0.794$$

$$= 2.999\ N/mm^2$$

$$\approx 3\ N/mm^2$$

Average stress at level of cable 3 = $(5.008 - 3)\tfrac{2}{3} + 3$

$= 1.339 + 3 = 4.339\ N/mm^2$

Loss in stress in cable 1 & 2 = 6 × 2.267 = 13.602 N/mm^2

Loss in stress in cable 3 = 6 × 4.339 = 26.034 N/mm^2

(This is the maximum. It could be less since it is unbonded parabolic tendon) (as discussed in chapter 4)

Total loss in stress in cable 1 = 31.524 + 13.602 + 13.602

= 58.728

Total loss in stress in cable 2 = 13.602 + 13.602

= 27.204

Total loss in stress in cable 3 = 26.034

Loss in force in cable 1 = 58.728 × 3 × 98.7 =17389.36

Loss in force in cable 2 = 27.204 × 3 × 98.7 = 8055.10

Loss in force in cable 3 = 26.034 × 3 × 98.7 = 7708.6

33.15306 kN

$$\%\ \text{Loss} = \frac{33.15306}{12\times 128.6}\times 100 = 2.148\%$$

5.2.2 Loss due to friction and curvature force

As explained earlier, losses in prestress due to friction and curvature forces happen only in case of post-tensioned members. If the cables are linear, there is only loss due to straightening of the cables called wobbling effect. If the cables are curved in profile, depending on the design requirements, there will be loss in stress while pulling, due to friction between the tendons and the ducts or the tendons themselves. Normally in case of post tensioned elements, cable profiles run as per the bending moment diagram since the prestressing is meant to counteract flexural stresses. In post tensioned elements the tendons are most probably placed in a curved profile.

The loss in prestress due to friction in such a case depends on the change of angle of the profile between the two points under consideration and the wobble effects are also normally represented as equivalent change in angle between the points under consideration. The friction co-efficient between the cable and the duct is defined as μ and the friction effect due to straightening otherwise called wobble coefficient is represented by k_f. (This k_f is defined as a co-efficient per unit length under normal conditions). Evaluation of loss due to friction in curved cables is explained as follows:

Any change in direction of prestressing force from the point of application due to curvature introduces radial forces because of curvature [Fig 5.4]. The curvature and the change in direction of tendon are interrelated.

If the radius of curvature is represented by *R*, the change in force in the tendon when it changes its direction by angle Δα is ΔP, and the total curvature force in the element within that angle is Δ*k*. Then Δ*k* is expressed as equal to $kR\Delta\alpha$. This is resolved into radial and tangential components as $k_nR\Delta\alpha$ and $k_tR\Delta\alpha$.

By equilibrium conditions, referring to Fig 5.4

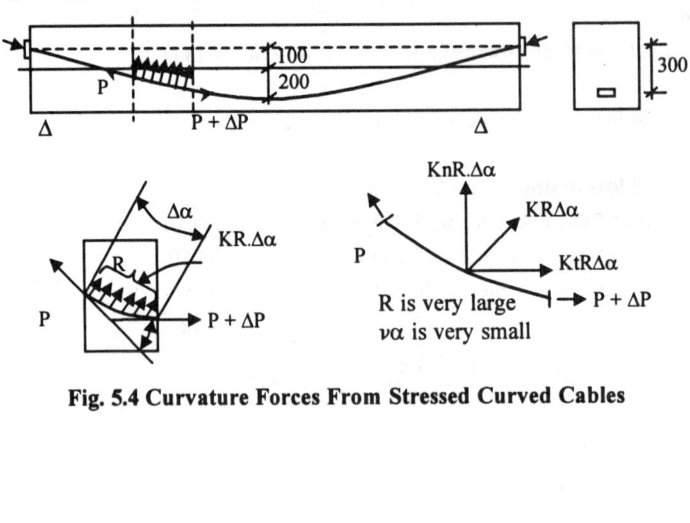

Fig. 5.4 Curvature Forces From Stressed Curved Cables

$$k_n R\Delta\alpha = 2P\sin\frac{\Delta\alpha}{2} + \Delta P\sin\frac{\Delta\alpha}{2} \quad \text{and}$$

$$k_t R\Delta\alpha = -\Delta P\cos\frac{\Delta\alpha}{2}$$

ΔP is small when compared to P, $\sin\frac{\alpha}{2} = \frac{\alpha}{2}$ where is very small, then

$$k_n = \frac{P}{R} \text{ and}$$

$$k_t = -\frac{1}{R}\frac{dP}{d\alpha}$$

Let P and $P + \Delta P$ be forces at the ends of the cable within a small distance, where the cable profile deviates through an angle $\Delta\alpha$. Then the force-generated normal to the axis of the beam k_n is equal $\frac{P}{R}$. The force along the axis of the beam k_t is equal to $-\frac{1}{R}\frac{dP}{d\alpha}$. While the normal force is resisted by self-weight of the beam and superposed loads, the tangential orce is resisted by the friction between the cable and the beam proper. This is also equal to the normal force multiplied by friction coefficient μ.

Hence equating these forces

$$-\frac{1}{R}\frac{dP}{d\alpha} = \mu\frac{P}{R}$$

$$\frac{dP}{d\alpha} + \mu P = 0$$

For the boundary conditions that the force at pulling end is P_0, the solution of the above equation is expressed as

$$P(\alpha) = P_0 e^{-\mu\alpha}$$

This expression shows that friction loss is effective only if there is a change in curvature. But experiments show that even in case of straight cables friction losses occur depending on the length of cable. So the straightening effect of cable (wobbling) is added separately for the loss.

Replacing $d\alpha = dx / R$

$$\frac{dP}{P} - \mu\frac{dx}{R} - k_f d_x = 0$$

where k_f is the wobbling coefficient [frictional force mobilized while straightening the wobbled cables] per unit length. The solution for the above differential equation is

$$P_{(x)} = P_0 e^{-(\mu x/R + k_f x)}$$

with $\alpha = x/R$, P at any distance x, from the boundary is expressed as

$$P_{(x)} = P_o e^{-(\mu\alpha + k_f.x)}$$

$$P_0 e^{-\mu\left(\alpha + \frac{k_f}{\mu}x\right)}$$

$$= P_0 e^{-(\alpha+\beta x)\mu}, \text{ where } \beta = \frac{k_f}{\mu}$$

The coefficient of friction μ and the wobbling coefficient k_f are varying, depending on the materials and grouping of cables. IRC 18 - 1985 [2] recommends these values for k and μ as given in Table 5.2.

These recommendations are based on the experimental work done by Cooley [3].

Beyond this, friction losses due to jamming, due to faulty arrangement of wires, due to pressure and due to wet conditions, are also to be considered during prestressing in the actual fieldwork. But these are either neglected or assumed to be 2% of the applied force in the field.

Table 5.2

Type of high Tensile steel	Type of duct or sheath	Values recommended to be used in design	
		k per metre	μ
Wire Cables	Bright metal	0.0091	0.25
	Galvanized	0.0046	0.2
	Lead coated	0.0046	0.18
	Unlined duct in concrete	0.0046	0.45
Uncoated stress Relieved strands	Bright metal	0.0046	0.25
	Galvanized	0.0030	0.20
	Lead coated	0.0030	0.18
	Unlined duct Concrete	0.0046	0.50

For simplicity in calculation

The formula $P = P_0 e^{-u(\alpha+\beta_x)}$ is modified as

$$P = P_0 e^{-\mu\gamma} \text{ where } \gamma = (\alpha + \beta x)$$

$$e^{-\mu\gamma} = 1 - \mu\gamma + \tfrac{1}{2}\mu^2\gamma^2 - \frac{1}{3.2}\mu^3\gamma^3 + \frac{1}{4.3.2}\mu^4\gamma^4$$

Neglecting higher order terms, for small values of γ

$$e^{-\mu\gamma} = 1 - \mu\gamma$$

$$= 1 - \mu\left(\alpha + \frac{k_f}{\mu}.x\right)$$

$$= \quad -\mu\alpha - k_f l$$

Hence $P = P_0(1 - \mu\alpha - k_f l)$

Sometimes the difference in prestressing force from pulling end to anchor end becomes too large. This is normally the case in case of long beams. In such cases pulling from both the ends is resorted to.

5.2.3 Methods of prestressing operations

The different methods of prestressing operations are as suggested by Leonhardt [4]

(*i*) Tensioning the cables from one end
(*ii*) Tensioning the cables from both ends
(*iii*) Tensioning with over stressing if the loss due to friction is negligible (5 to 10 percent of P)
(*iv*) Tensioning by overstressing and slackening repeatedly.

5.2.3.1 Tensioning from one end

Tensioning from one end is normally adopted, only when there is no facility of tensioning from both sides and in case of short beams. Tensioning from one end causes a heavy loss in force due to friction especially in long beams. Moreover there is no uniformity of prestress along the length.

5.2.3.2 Tensioning from both ends

This procedure is quite commonly adopted in practice. Instead of pulling the same cable (tendon) from both ends, alternate cables (tendons) are pulled from opposite sides, if more number of cables (tendons) are used for prestressing.

5.2.3.3 Tensioning by overstressing

This approach is adopted if the anticipated loss due to friction is not very much. The prestressing is increased by an amount, equal to the calculated loss in force at mid section in case both ends are pulled or (equal to the calculated loss in force at farther end, in case of pulling only from one end) This leads to higher prestressing forces at all other sections. It is to be noted that the increase in stress in the cable should not go beyond the limiting value. The percentage of force loss due to friction, which is overcome by this method, shall be less than 10 percent.

5.2.3.4 Overstressing and slackening

This is the most practical approach for long span structures. If the change in force ΔP is more than 10 percent of P, it is advisable to overstress by an amount such that this increase in stress level at pulling end is balanced (by the loss of force due to friction) at the farther end. Then such an increase in stress level at anchor end is released. The release of force (slackening) does not affect the prestress for the full length of the tendon (beam) because of the return friction associated with return movement of the cable. Referring to Fig 5.5

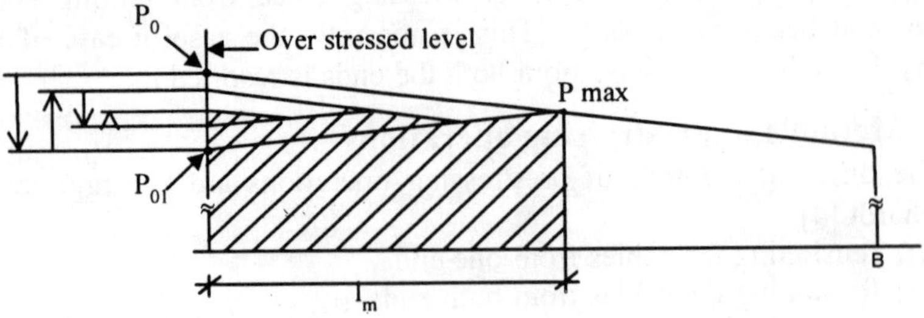

Fig. 5.5 Force Flow by Overstressing, Slakenking and Restressing

$P_{max} = P_0(1 - \mu' l_m)$ where μ' is the coefficient of friction in the return direction, l_m is the length upto which the effect of return friction is felt. If P_{01} is the force at the anchored end after slackening (releasing) and if the forward friction coefficient and the return friction coefficient are the same μ'.

$$P_{max} = \frac{P_0 + P_{01}}{2}$$

$$\frac{P_0}{2} + \frac{P_{01}}{2} - P_0 + \mu' l_m P_0 = 0$$

$$l_m = \frac{P_{01} - P_0}{2\mu' P_0}$$

By repeatedly doing this operation a more uniform tension force diagram along the length of cable can be obtained. The force flow is shown in Fig 5.5.

Example 5.4

A post-tensioned prestressed concrete beam has a parabolic cable profile with zero eccentricity at end and an eccentricity of 400 mm at the center, with respect to gross centroid of the section. It covers a span of 25*m* and a prestressing force of 400*kN* is applied from one end A. Determine the loss in prestressing due to friction at the farther end B and hence evaluate the force at B. $\mu = 0.30$, wobble effect = 0.25°/unit length of span.

Solution

$P_0 = 400\ kN$

Let α_1, α_2 be the slope of the cable profile at ends of the beams, and assuming tan α = α for small values of α

$$\alpha_1 = \alpha_2 = dy/dx = \frac{4r(1-2x)}{l^2} = \frac{4 \times 400}{25000 \times 25000} \times 25000$$

$$= 0.064 = 3° \ 40'$$

$$\alpha_1 = \alpha_2 = 3^0 40'$$

Change in angle of the cable profile from one end to other end = $\alpha_1 + \alpha_2$

$$= \alpha_1 + \alpha_2 = 7^0 20' = 0.1280^c$$

$$\beta = 0.25^0 / \text{unit length}$$

$$\gamma = (7.33 + 0.25 \times 25) = 13.58° = 0.237°$$

Then $P_B = P_0 e^{-mr}$ $= 400 \times e^{-0.3(0.237)}$

$= 400\ (1 - 0.071)$

$= 371.6\ kN$

Loss in prestress at the farther end (anchored or dead end) = 28.4 *kN*

Force Flow diagram is shown in Fig 5.6.

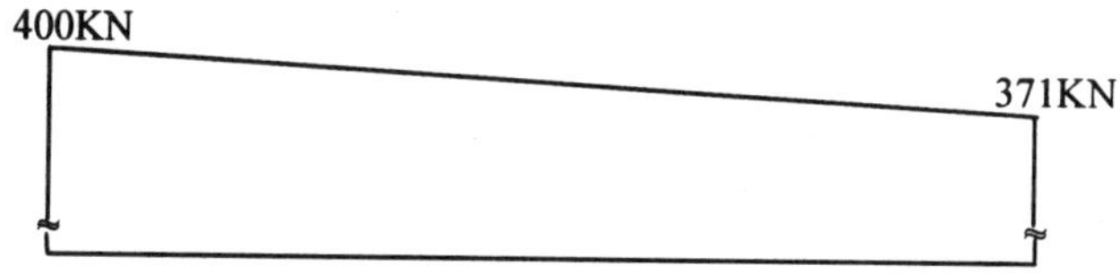

Fig. 5.6 Force Flow Diagram

Example 5.5

A two span continuous beam is having a cable profile as shown in Fig 5.7. The details of eccentricity at salient points and cable profile are given in the Figure. A prestressing force of 500 *kN* is applied at end A. Determine the effective prestress at point B and B'. Assume a μ value of 0.30 and β = 0.5°/ unit length of span.

Solution

Referring to Fig 5.7.

Slope of cable profile at the pulling end (A) be α_1

Expressing the equation of the cable profile as

$$y = \frac{4r}{l^2} \times (l - x)x$$

$$\frac{dy}{dx} = \frac{4r}{l^2} \times (1 - 2x)$$

$$\text{At } x = 0 \quad \frac{dy}{dx} = \frac{4r}{1} = \frac{4 \times 0.61}{20} = 0.122$$

$$\tan\ \alpha_1 = \alpha_1 = 0.122 \text{ radiams}$$

$$\alpha_1 = 6.95°$$

Slope of the cable profile at point B' will also be very near to α_1

The change in angle of profile from A to B" = 2 α_1 = 13.9°

At support point B the slope of the cable profile is zero (horizontal). The change in angle of the profile be α_2. Then referring to Fig 5.7.

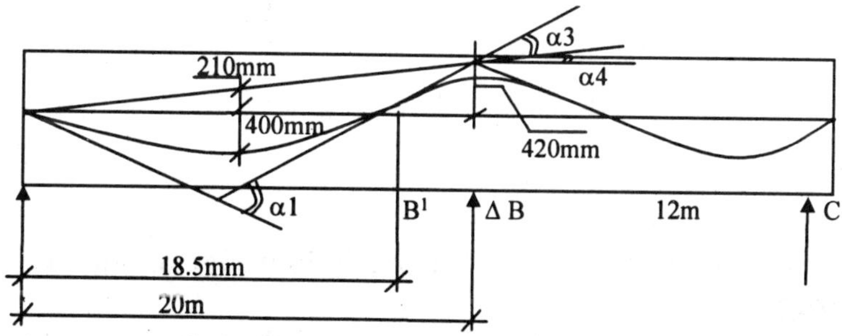

Fig. 5.7 Continuous Beam with Cable Profile

Then $\alpha_2 = \alpha_3 + \alpha_4$

$$\tan\alpha_3 = \frac{4 \times 0.61}{20} = 0.122$$

$$\alpha_3 = 6.95^\circ$$

$$\tan\alpha_4 = \frac{0.42}{20} = 0.021$$

$$\alpha_4 = 1.20^\circ$$

$$\alpha_2 = \alpha_3 + \alpha_4$$

$$= 6.95 + 1.2^\circ$$

$$= 8.15^\circ$$

At Point B', the change in the angle of cable profile A to B' = 13.9^0 and γ at B' = $13.9^0 + 0.5\ (20 - 1.5) = 23.15^0 = 0.405^c$

At point B, Change of angle of profile from A to B = $13.9^0 + 8.15^0 = 22.05^0$

$$\gamma_B = 22.05 + 0.5 \times 20 = 3205^0 = 0.559^c$$

$$P_{B'} = 500e^{-0.30\times0.405} = 500 \times 0.886 = 443kN$$

$$P_{B'} = 500e^{-0.30\times0.559} = 432kN$$

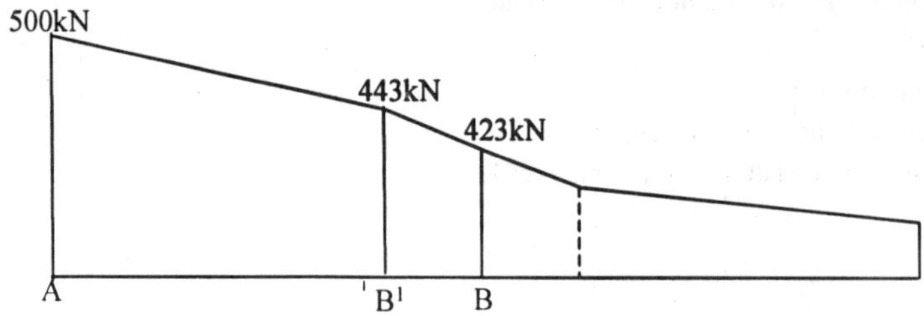

Fig. 5.8 Force FlowDiagram

5.2.4 Determination of extension of cables including friction effects

The force distribution along the length of tendon is drawn inclusive of the effect of friction loss over the length and the modulus of elasticity of material steel is known. The value of steel elongation is worked out as

$$\Delta l_z = \epsilon_{spx}\,.dx = \int_0^{l_z} \frac{\sigma_{sp.x}}{E_{sp}} dx = \frac{P_x.dx}{A_{sp}.E_{sp}}$$

Now $\int_0^{l_x} P_x\, dx$ is the area if tension force diagram. This divided by $A_{sp}\ E_{sp}$ gives Δl_z.

5.2.5 Loss of tensioning force due to slip in wedge or slip at anchorages

In all post-tensioning works, there is definitely certain slip when the tendon force is transferred from the jack to the anchoring ends. Because of this slip of wire through the anchorage's certain quantity of prestress is released. This amount of slip is generally known from the tests or from the patentors of the anchorage system. For example, the details of the slip given in certain anchorage systems are given below:

a) Freyssinet system
   (i) 12 Nos. of 5 mm dia = 4 mm
   (ii) 12 Nos. of 8 mm dia = 6 mm

(b) Magnel blaton system = 8 mm

(c) Dywidag system = 1 mm

The slips at anchorage is constant and is expressed as percentage of total elongation in the tendon. This depends on the length of the cable. Hence the percentage loss due to slip in long span construction is small whereas the same in short span construction is considerable. The loss in force due to slip

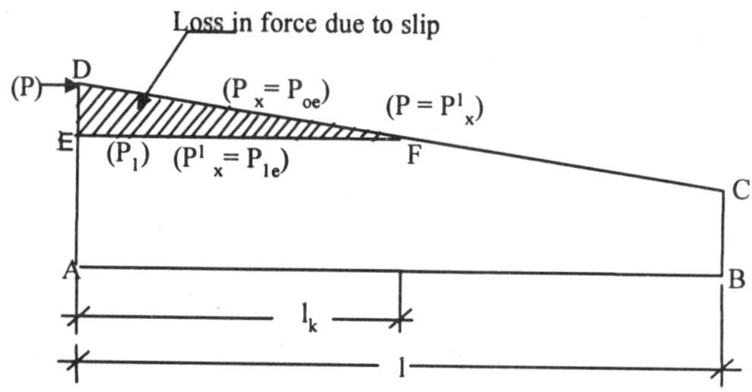

Fig. 5.9 Force Flow Diagram Including Loss Due To Slip In Anchorages

at anchorages also varies along the length of the beam due to return friction of the cable.

Accordingly the force flow diagram along the beam is affected. The force flow diagram for a beam with a loss in force due to friction and slip at anchorage is shown in Fig 5.9.

For calculating the total extension of the cable, the force flow diagram area ABCD in Fig 5.6 has to be considered. Due to slip in anchorage there is a fall in force at end A by an amount, ΔP. This reduction in force ΔP is not felt for full length of the beam. Because of the friction in return direction for the cable, ΔP value goes on decreasing and at point *F*, this value becomes zero and the force flow diagram after the slip of anchorages is given by ABCFE. To evaluate the distance of the section *F*, where the effect of slip in anchorages is not going to be felt, one should consider the force at any section (x) P_x, after the loss of force due to slip in anchorages. P_x can be calculated from *P* or from $P_1 = (P - \Delta P)$ as detailed below

$$P_x' = (P_0 - \Delta P)e^{+u'\gamma} \; ; \quad P_x = P_0 \Delta e^{-ur}$$

where μ' is coefficient of friction in return direction, μ is coefficient of friction in forward direction.

$\gamma = (\alpha + \beta l_x)$ where α is the change in the angle of profile of cable from end to the section at distance l_x from end and β the equivalent change in angle due to wobble effect/unit length of the beam.

Equating the expressions for P_x and P_x' at Section F.

$$P_0 e^{-\mu\gamma} = (P_0 - \Delta P)e^{+\mu\gamma}$$

$$P_0 - \Delta P = P_0 e^{-(\mu+\mu')\gamma}$$

$$\Delta P = P_0\left[1 - e^{-(\mu+\mu')\gamma}\right]$$

By simplified version $P_x = P_0\,(1 - \mu\gamma)$

We get $P_0 - P_0[1 - (\mu + \mu')\gamma] = \Delta P$

$$P_0[(\mu + \mu')\gamma] = \Delta P$$

$$P_0\left[1 + \frac{\mu'}{\mu}\right]\mu\gamma = \Delta P$$

$$= [P_0 - P_x]\left[1 + \frac{\mu'}{\mu}\right] = \Delta P$$

the force flow pattern, the force described by the area DEF can be calculated in terms of l_x [distance of section *F* from section *D*]. Dividing this force by A_{sp}. E_{sp}, the slip of the tendon Δl can be expressed in terms of l_x. If amount of slip say Δl is known, the distance through which it affects the force flow (l_k) can be evaluated and the correct force flow diagram can be drawn.

The details of calculations for evaluation of l_k are as follows

The force at pulling end = P_A

The force at the farther end, after loss ducts friction and wobble effect = P_B

The loss in force at the pulling end due to slip in anchorages = ΔP

The distance through which the force flow diagram is effected is due to loss caused by slip in anchorage along the length of beam $= l_k$

Coefficient of friction in forward and return direction be equal to μ and μ'

Then Force at any section $x = P_x$

$$P_x = P_A - \frac{P_A - P_B}{1}\, l_x$$

$$= P_{A-}P_A.\frac{l_x}{1} + P_B.\frac{l_x}{1}$$

$$= P_A\left[1 - \frac{l_x}{1}\right] + P_B.\frac{l_x}{1}$$

$$P_A - P_x = P_A.\frac{l_x}{1} - P_B.\frac{l_x}{1}$$

$$(P_A - P_z)\left(1 + \frac{\mu'}{\mu}\right) = \Delta P$$

$$\frac{l_x}{1}[P_A - Px]\left(1 + \frac{\mu'}{\mu}\right) = \Delta P$$

$$\frac{l_x}{1}[P_A - P_x]\left(1 + \frac{\mu'}{\mu}\right)\frac{l_k}{2}\frac{1}{A_{sp}}.\frac{1}{l_k}.\frac{1}{E_{sp}}.l_k = \Delta P$$

$$\frac{l_x^2}{21}\frac{[P_A - P_B]\left[1 + \frac{\mu}{\mu}\right]}{A_{sp}}.\frac{1}{E_{sp}} = \text{slip}$$

$$l_x^{\,2} = [\text{slip}]\,[E_{sp}] \times 21\,\frac{A_{sp}}{(P_A - P_B)\left(1 + \frac{\mu'}{\mu}\right)};$$ This distance (l_x) is equal to l_k

Having found l_x, $P_{(x)}$ can be calculated P can be evaluated.

The force flow diagram inclusive of the effect of slip at anchorages can be drawn. The effective force at any cross section can also be arrived at.

Example 5.6

In the Example 5.5, a slip of 4 *mm* occurs at the end A. Calculate the loss in prestressing force at A and also plot the force diagram. Area of steel is 600 mm^2. $\mu' = 1.5\mu$

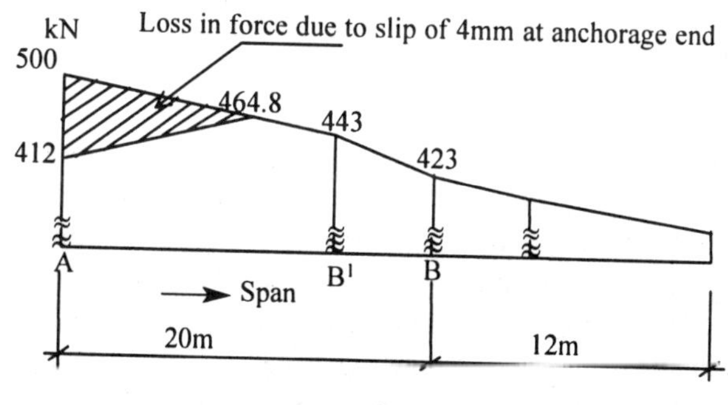

Fig. 5.10 Force Flow Diagram Inclusive of Slip Losses

Solution

The length through which the slip at anchorages will be felt as loss in force be l_k

$$l_k^2 = [0.004 \times 2.1 \times 10^{11} \times 2 \times 18.5] \times \left[\frac{6 \times 10^{-4}}{57 \times 2.5}\right] \times 10^3$$

$$= 130.86\ m^2$$

$$l_k = 11.43\ m$$

Force at distance $l_k = P_x$

$$P_x = 500 - \frac{(57)}{18.5} \times 11.43$$

$$= 500 - 3.520 = 464.8$$

The final force at pulling end after loss in force due to slip in anchorages

$$\Delta P = (500 - 464.8)\ (1 + 1.5) = 88.0$$

$$P - \Delta P = P_i = 500 - 88 = 412kN$$

5.3 Time Dependent Losses

5.3.1 Loss due to creep of concrete

The property creep of concrete has been discussed already in the earlier Chapter on Materials. This deformation of concrete is dependent on various aspects namely (a) concrete strength (b) strength of concrete at the time of transfer of load (c) applied stress level (d) humidity and atmospheric condition, (e) time at which creep deformation is measured. Such a creep deformation in concrete causes a loss in stress in steel (prestressing force). The exact calculation of creep strain is rather complex because of (a) continued variation of applied stress caused by creep itself, even at the anchored end and (b) variation of concrete stress at steel fibre level along the length of the beams. Any how an attempt can be made to reach the nearest value of creep deformation which is good enough for a reasonably correct design. The creep deformation (strain) is proportional to elastic strain at lower stress levels in concrete as already mentioned in Chapter 2. Since in most of the designs the applied stress level in concrete is around 0.33 times the cube strength and the proportionality of creep strain with elastic strain can be assumed to concrete. Hence most of the statutory code provisions always express creep strain as a factor times elastic strain and the factor is called as creep coefficient (φ). IS 1343 of 1980 also defines the creep strain as equal to creep coefficient × elastic strain. The creep coefficient varies from 2.2 to 1.1 depending only on the age concrete at the time of transfer of prestress. While calculating the creep strain, it should be kept in mind that the average of the stress coming on the structural element at the beginning and at the end of the period for which the creep strain is calculated. It is also to be noted that the stress at the starting time as well as at the end of the period must be the average over the length of the section. If there is a variation in stress in concrete due to the cable profile, the average has to be worked out based on the eccentricity at the various cross sections. The relevant extract of the IS code [5] is given under:

Loss of prestress due to creep of concrete

"The loss of prestress due to creep of concrete under load shall be determined for all the permanently applied loads including the prestress.

The creep loss due to live load stress, erection stress and other stresses of short duration may be ignored. The loss of prestress due to creep of concrete is obtained as the product of the modulus of elasticity of the prestressing steel and the ultimate creep strain of the concrete fibre integrated along the line of centre of gravity of the prestressing steel over its entire length.

The total creep strain during any specific period shall be assumed for all practical purposes, to be the creep strain due to sustained stress equal to the average of the stresses at the beginning and end of the period".

National Building Code adopts a different formula for calculating the creep strain which involves not only the age of the concrete at which the stress is transferred, but also the cube strength, relative humidity, etc, as detailed below:

$$\text{Creep strain} = 1.75\left[\frac{1.25-h}{h}\right]\sqrt{\frac{400}{F_{ck}}}\cdot\frac{F_{ck}}{F_r}\times 10^{-6}\,cm/kg \quad \text{per unit applied stress.}$$

The Example 5.7 solved below would give the details of the iterative procedure followed for arriving at creep deformation and the corresponding loss in stress in prestressing tendons.

Example.5.7

A post tensioned prestressed concrete beam for a span of 10m has a 'I' shaped cross section. The total height of the beam is 800 mm. The top flange is 300*mm* wide and 100*mm* thick and the bottom flange is 200*mm* wide and 200*mm* thick. The web portion is 500*mm* high with 100*mm* thickness. The beam is prestressed with an initial force of 628*kN* using 628 *sq mm* of prestressing steel. The cable duct area is 2500 mm^2. The cable is located at 100 *mm* from the soffit and has a straight profile for the full length. The force is transferred after the concrete has attained 100% of the designed strength of 50 N/mm^2. The beam is exposed to humid atmosphere. Evaluate the loss in stress in steel due to creep of concrete, using I.S. Code creep coefficient approach and also using National Building Code formula for creep.

The tendon is fully bonded. The value of $m = \dfrac{E_s}{E_c}$ is 6.

Solution

Since it is a post tensioned beam, grouted, the effect of creep is felt only on the effective cross-section

Properties of Section (effective)

$$A_i = 120000 + 628 \times (6-1) = 123140\ sqmm$$

The depth of centroid of the effective section from top tube

$$y_i = \frac{(300\times100\times50+500\times100\times350+200\times200\times700+3140\times700)}{123140} = 399.5mm$$

$$I_i = \frac{300\times100^3}{12} + 300\times100\times349.5^2 + 100\times\frac{500^3}{12} + 50000\times49.5^2 + 200\times\frac{200^3}{12} + 40000$$

$= 859902 \times 10^4\ mm^4$

Initial prestressing force $= 628kN$

Eccentricity $= 300.5mm$

Dead Load on the girder $= 2880\ N/m$

Bending moment $= \frac{2880\times10^2}{8} = 36000Nm\ = 36kNm$

Stresses in concrete & fibre stress at steel level due to self-weight bending moment

$$\frac{36\times10^6}{859902\times10^4}\times300.5 = 1.258N/mm^2 \text{ (compression)}$$

Fibre stress at steel level due to Presstress $\Big\} = \frac{628\times10^3}{123140} + \frac{628\times10^3\times300.5}{859902\times10^4}\times300.5$

$= 5.0998 + 6.594$

$= 11.6938\ N/mm^2$

Total stress in concrete at the steel level = 11.6938 - 1.258 = 10.436

Assume a fibre stress at steel level after creep loss = 9.324 N/mm^2

Average stress during creep of concrete

$$\frac{10.436+9.324}{2} = \frac{19.76}{2} = 9.88N/mm^2$$

Elastic strain at steel level in concrete $= \frac{9.88}{E_c}$

Creep strain $= 1.6\times\frac{9.88}{E_c}$

(Assumed the specimen to be prestressed at the age of 28 days)

Loss in stress in steel due to creep strain $= \left[+1.6\times\frac{9.88}{E_c}\right]\times E_s$

$= [+1.6\times9.88\times6] = 94.85N/mm^2$

Stress in steel after creep loss = 905.15 N/mm^2

Stress in concrete fibre at steel level $= +[5.0998+6.594]\times\frac{905.15}{1000} - 1.258$

$= 10.584 - 1.258 = 9.326\ N/mm^2$

Assumed stress = 9.324 which is very nearly same as the stresses got. Hence average stress assumed in concrete is OK.

Accordingly the loss in stress in steel = 94.85 N/mm^2

Percentage of loss = 9.485%

Applying National Building Code approach (C.G.S. units)

$$\text{Creep deformation, C} = 1.75\left[\frac{1.25-0.50}{0.25}\right]\sqrt{\frac{40}{50}\times\frac{40}{50}}\times10^6$$

$= 4.70\times10^{-6}\ cm^2/kg$

Assuming average concrete stress of 96 kg / cm^2

Creep strain = $4.70 \times 10^{-6} \times 96 = 4.51 \times 10^{-4}$

Loss in prestress $= 4.51 \times 10^{-4} \times 2.1 \times 10^{6}$

$= 947.1\ kg / cm^2$

Stress in steel $= 9052.9$

Stress in concrete fibre at steel level $= 116.938 \times \dfrac{9052.9}{10000} - 12.58$

$= 105.86 - 12.58$

$= 93.28\ kg / cm^2$

Average stress $= \dfrac{(104.36 + 93.28)}{2}$

$= 98.82\ kg / cm^2 > 96\ kg / cm^2$ (assumed)

Assume an average compressive stress of 98.6 kg / cm^2

Creep strain $= 4.70 \times 10^{-6} \times 98.6 = 463.42 \times 10^{-6}$

Loss in stress in steel $= 973.2\ kg / cm^2$

Stress in steel $= 9026.8\ kg / cm^2$

Stresses in concrete fibre $= 116.93 \times \dfrac{9026.80}{10000} - 12.58$

$= 105.55 - 12.58 = 92.97$

$= \dfrac{92.97 + 104.36}{2} = 98.66\ kg / cm^2$

This average stress in concrete is OK. Hence the loss in stress in steel = 973.2 kg / cm^2 and percentage loss is 9.73%.

5.3.2 Additional aspects of creep loss

It is known that creep strain can be calculated as $\epsilon_{cr} = \epsilon_{el}\,\varphi$ where φ is the creep coefficient. At lower levels of applied stress this is constant and dependent only on the age at which the force is transferred. Accordingly the loss in prestress due to creep has been calculated in Example 5.7. But at higher levels of applied stress it is not only dependent on the age of the concrete at which the prestress is transferred but also on the applied stress itself. If the applied stress is more, φ is no further a constant and it is exponentially increasing with applied prestress. The loss in prestresss due to the creep with time will be larger. This is already explained in Chapter 2. Hence for higher level of applied stress, φ varies with the applied stress.

In case of prestressed concrete sections, the applied stress goes on decreasing due to time dependent losses in prestress and hence the rate of increase of creep strain also goes on decreasing because of the reduction in applied stress. These aspects may have to be considered in places where such deformations play important role (say in case of cantilever construction where precamber calculations are important). These aspects were first considered by Dischinger [6] who expressed the variation of strain in concrete during a small period of time, Δt, as follows.

$$d\in_{cs}=\frac{d\sigma_{cps}}{\in_c}+\frac{\sigma_{cps}-d\sigma_{cps}}{\in_c}d\varphi_{cr}+\frac{\in_{s\infty}}{\varphi_\infty}d\varphi_{cr}$$

It could be seen that the variation in strain at a small period of time Δt is given by 3 components. The first term is the strain due to change in the applied stress. The second term refers to the creep strain due to the applied stress existent at the point of time and the third term refers to shrinkage strain corresponding to that period of time, which is expressed in terms of creep coefficient. This expression includes only the creep strain due to the revised applied stress at that point of time. The creep coefficient is not altered due to the new applied stress.

This expression is further modified by Trost [7]. He expressed creep strain at the end of period (t) as follows.

$$\in(t)=\frac{\sigma_{cpo}}{\in_c}(1+\varphi)+\frac{\sigma_{cpt}-\sigma_{cpo}}{\in_c}(1+\rho\varphi)+\frac{\in_{s\infty}}{\varphi_\infty}\varphi_{cr}$$

This expression takes care of the variation in creep factor due to variation of applied stress. The first term takes care of elastic and creep strain due to applied stress at the time of application. The second term is a correction factor due to variation of applied stresses, the corresponding elastic strain and creep strain. For that creep strain an additional factor 'ρ' is added to creep factor. This 'ρ' (which is less than one) takes care of reduced creep factor because of reduced applied stress. The third term is the effect of shrinkage. The value of 'ρ' is got from Nomograpgh available from basic studies on creep as could be seen from Table 5.3. These are finer aspects of creep strain calculations, which have to be used in case of camber calculations or match casting constructions. Those interested may have to refer to specialist literature as given in reference [8,9].

Table 5.3

φ_∞ / Age at time of loading Days	1	1, 5	2	2, 5	3	3, 5
5	0,745	0,788	0,816	0,858	0,884	0,904
14	0,786	0,819	0,849	0,874	0,896	0,913
28	0,812	0,839	0,864	0,884	0,905	0,920
90	0,857	0,873	0,888	0,903	0,917	0,928

5.3.3 Loss in prestress due to shrinkage

Shrinkage is a reduction in the volume of concrete with time, due to the gradual loss of moisture as explained in Chapter 2 resulting in the change in the volume of deformation. It is a function of many variables like the amount of cement in concrete, relative humidity, type of cement and water cement ratio. Rich mixes of concrete with very fine cement leads to greater shrinkage. The rate at which the shrinkage strain develops in concrete is very similar to that of creep and hence the shrinkage strain at any point of time is taken as the maximum shrinkage strain multiplied by the ratio $\left(\frac{\varphi_t}{\varphi_\infty}\right)$ where φ_t and φ_∞ are creep coefficients at time t and at time ∞. But in the regular design of prestressed concrete structures, an absolute value of shrinkage strain of 0003 for pretensioned product or $\frac{0002}{(\log_t + 2)}$ for post tensioned beams are considered. Here t is the number of days at which the shrinkage strain is calculated from the date on which the prestress is transferred on to the concrete. IS 1343 defines these values of shrinkage strain and also states that it can be increased by 50% under dry atmospheric conditions for post tensioned members. British Code or other European Codes give the shrinkage strain in greater detail as discussed in article 2.2.4.2.

5.3.4 Loss of prestress due to relaxation of steel

As mentioned earlier in Chapter 2 relaxation of stress in steel kept under constant strain is a time dependent property and is a function of the applied stress level, temperature, humidity etc. When the applied stress levels are very low, namely within the technical creep limit, there will not be relaxation of force with time even when the material is under constant strain. If the applied stress level becomes more than 0.5 times the ultimate capacity of the steel, there is a possibility of loss in stress for constant applied strain on the steel. Considering the provisions under various codes regarding the applied stress level in high tensile steel in prestressed concrete works, it is noted that the applied stress level can be anywhere between 0.5 times *fpu* to .8 times *fpu* during life cycle period of the prestressed concrete element. IS Code has specific relaxation loss in definite quantities in terms of N / mm^2 for the various applied stress levels is given in Table 2.14 of Chapter 2. Similar recommendations for evaluating losses due to relaxation steel given in the other National and International Codes are also presented in Article 2.3.4. Typical example for calculating all losses in a post-tensioned beam has been worked out in detail in Example 5.8 for the benefit of readers.

Example 5.8

A post-tensioned prestressed concrete beam has got geometrical properties as shown in Fig 5.11. It is prestressed with 3 cables. Two cables have 4 nos of 7 ply of 12.7 mm diameter and one cable has 7 nos of 7 ply strand at 12.72mm. The seven-strand cable has a parabolic profile with the ends at 850mm from soffit and at the middle section it is located at 320mm from soffit. The two other cables run straight at 120mm level from soffit for full length of the beam. The cable with 7 strands with parabolic profile is stressed first with a force of 900 kN followed by the 2 straight cables with a force of 514.4kN. While prestressing each tendon, a slip of 6 mm is expected. The forces are transferred after the concrete has attained its 28 days strength. Evaluate the losses in prestress in various cables due to (a) elastic deformation (b) friction and wobble effects and (c) slip at anchorages and (d) creep (e) shrinkage (f) relaxation and find out the forces in each cable during service load period. Assume a friction coefficient of 0.25 and wobble coefficient of .0035. For other details mentioned clauses in IS 1343 shall apply.

Solution

Properties of Section (Ref Fig 5.11)

$A_g = 47.25 \times 10^4\ mm^2$

$A_n = 44.89 \times 10^4\ mm^2$

$I_n = 4.374 \times 10^{10}\ mm^4$

$Y_b = 505.48\ mm$

$Y_t = 494.52\ mm$

$L = 15\ m$

$m = 6$

$O = 0.25$

$k = 0.0035$

$P_{01} = 900\ kN$ (For cable 1)

$P_{02,3} = 514.4\ kN$ (For cable 2 & 3)

Cable 1- 7 strands 98.7 mm^2 each

Cable 2 & 3 - 4 strands 98.7 mm^2 each

The bottom cables are straight, situated 120*mm* above the soffit of the beam. The top cable has a parabolic profile with its ends situated 850*mm* above the soffit and the cable is situated at 320*mm* above the soffit at the mid span.

The cross-section of the beam as well as the cable profiles are as shown in figures 5.11.

Different types of losses occurring in post-tensioned beams are

- Friction loss
- Loss due to elastic shortening
- Loss due to creep of concrete
- Loss due to shrinkage of concrete
- Loss due to relaxation of stress in steel

Friction Losses

The Cable 1 with 7 strands is jacked from end A whereas cables 2 & 3 with 4 strands each are jacked from end B.

During post tensioning of cables, there is a stress in the members due to friction as well as elastic shortening.

Let P_0 = prestressing force at jacking end

P_x = prestressing force at farther end

k = friction coefficient for wobble effect

α = cumulative angle in radians through which the tangent to the cable profile has turned between any two points under consideration.

For parabolic cable,

$$y = \frac{4e}{l^2} x(l - x)$$

$$\frac{dy}{dx} = \frac{4e}{l^2} x(l - 2x)$$

Slope at ends $(x - 0) = \frac{4e}{l}$

For cable 1

$$\text{Slope at end } = \frac{4 \times 53}{15 \times 100} = 0.14$$

$\alpha = 2 \times 0.14 = 0.28$ radians

For cable 2 & 3

Since they have a straight profile

$\alpha = 0$

$P_x = P_0 e^{-(\mu\alpha + kx)}$

For small values of $(\mu\alpha + kx)$

Loss of Force = $P_0\ (\mu\alpha + kx)$

Loss of Force in Cable 1 $= 900\ (0.25 \times 0.28 + 0.0035 \times 15)$

$= 110.25 kN$

Loss of Force in Cable 2 and Cable 3

$= 514.4\ (0.25 \times 0.28 + 0.0035 \times 15)$

$= 27 kN$

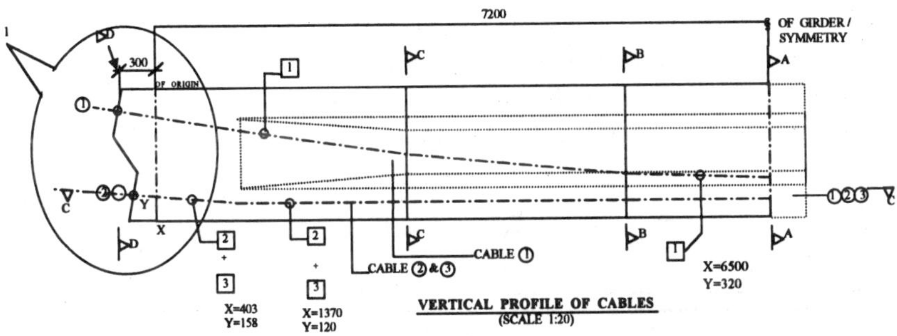

DISTANCE OF CABLE CENTRE LINE FROM SOFFIT OF GIRDER

DISTANCE FROM § OF ORIGIN X(mm)	-300	-130	000		1000	2000	3000	4000	5000	6000	7000	7200
CABLE ①	850		812		685	565	468	395	347	323	320	320
CABLE ②	-	200	190		126	120	120	120	120	120	120	120
CABLE ③	-	200	190		126	120	120	120	120	120	120	120

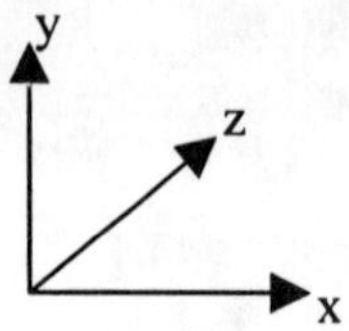

XY is the plane of the cross section bottom face centre at th from end face in the origin

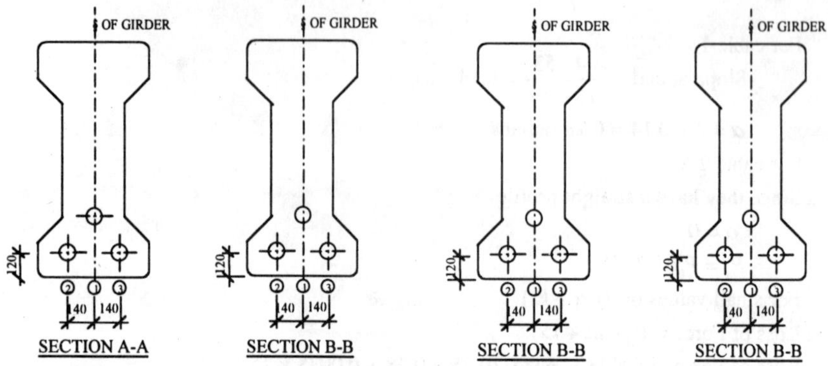

Fig. 5.11

Actually the loss of force in cables in post tensioned members due to friction as well as elastic shortening occurs simultaneously. But since principle of super position is valid, the force flow diagram considering friction losses is considered from which losses due to elastic shortening are evaluated. The force flow diagram for the 3 cable considering losses in prestress due to friction are shown in Fig 5.12 a (1,2,3).

Force in Cable 1 at jacking end (A) 900 *kN*; at farther end (B) 789.5 *kN*

Force in Cable 2 at jacking end (B) 514.4 *kN*; at farther end (A) 487.4 *kN*

Force in Cable 3 at jacking end (B) 514.4 *kN*; at farther end (B) 487.4 *kN*

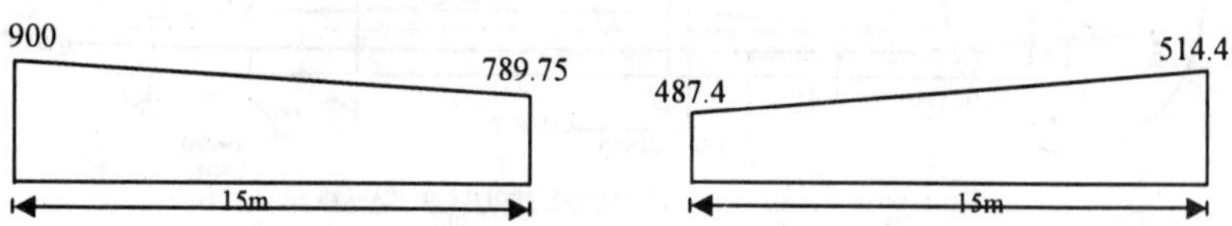

Fig. 5.12 a

Loss due to Elastic Shortening

When Cable 1 is stressed and anchored, there is no loss due to elastic shortening in Cable 1 itself.

When Cable 2 is stressed and anchored, there is a loss due to elastic shortening in Cable 1 which is already stressed and anchored.

Stress in concrete at level of Cable 1 due to stressing and anchoring of Cable 2.

At supports $f_{cs} = \dfrac{514.4 \times 10^3}{44.89 \times 10^4} + \dfrac{514.4 \times 10^3 \times 385.48 \times (-344.52)}{4.374 \times 10^{10}}$

$= -0.416\ N/mm^2$

At centre $f_{cm} = \dfrac{514.4 \times 10^3}{44.89 \times 10^4} + \dfrac{514.4 \times 10^3 \times 385.48 \times 185.48}{4.374 \times 10^{10}}$

$= 1.987\ N/mm^2$

Average stress in concrete at level of Cable 1 is

$= -0.416 + \frac{2}{3}\ (2.403) = 1.186\ N.mm^2$

Loss of stress in Cable 1 $= mf_c = 6 \times 1.186 = 7.116\ N/mm^2$

Loss of stress in Cable 2 $= 0$

When cable 3 is stressed and anchored, there is a loss of stress in Cable 1 & 2.

Stress in concrete at level of Cable 1

At support $= f_{c1s} = -\ 0.416\ N/mm^2$

At Centre $= f_{c1m} = 1.987\ N/mm^2$

Stress in concrete at level of Cable 2 (uniform for full length of the beam)

$$= f_c = \frac{514.4 \times 10^3}{44.89 \times 10^4} + \frac{514.4 \times 10^3 \times 385.48 \times 385.48}{4.374 \times 10^{10}} = 2.893\ N/mm^2$$

Loss of Stress in Cable 1 $= mf_c = 6 \times 1.186 = 7.116\ N/mm^2$

Loss of Stress in Cable 2 $= 6 \times 2.893 = 17.358\ N/mm^2$

Total loss of stress due to elastic shortening

In Cable 1 $= 7.116 + 7.116 = 14.232\ N/mm^2$

In Cable 2 $= 17.538\ N/mm^2$

Total loss of forces in cables due to elastic shortening

In Cable 1 $= 14.232 \times 7 \times 98.7 = 98.30\ N = 9.83\ kN$

In Cable 2 $= 17.358 \times 4 \times 98.7 = 6850\ N = 6.85\ kN$

In Cable 3 $= 0$

The Force flow diagram considering the elastic shortening losses is shown in Fig 5.12 b (1,2,3)

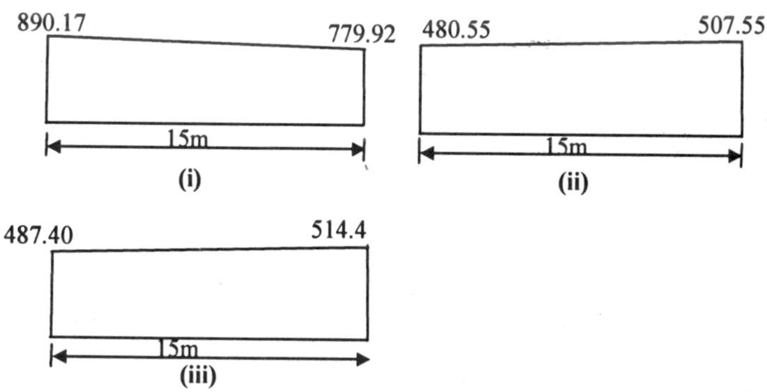

Fig. 5.12 b (i) (ii) (iii)

Force in Cable 1 at jacking end A = 900 – 9.83 = 890.17

Force in Cable 1 at farther end B = 789.75 – 9.83 = 779.92

Force in Cable 2 at jacking end B = 574.4 – 6.85 = 507.55

Force in Cable 2 at farther end A = 487.40 – 6.85 = 480.55

Force in Cable 3 at jacking end B = 514.40

Force in Cable 3 at farther end A = 487.4

Loss due to Anchorage Slip

The loss due to anchorage slip does not extend along the entire span, but is found to prevail for a certain distance which can be calculated using the given equation (Ref Art 5.25)

$\Delta = 6mm$

$$l^2 x = \Delta \in_{sp} (2l) \frac{A_{sp}}{(P_0 - P_x)\left(1 + \frac{\mu}{\mu}\right)}$$

here $\mu = 1.5\mu$

For Cable 1

$$= l_x = \sqrt{6 \times 1.95 \times 10^5 \times 2 \times 15 \times 10^3 \times \frac{7 \times 98.7}{110.25 \times 10^3 \times 2.5}}$$

$= 9.38m$

For Cable 2 & 3

$$= l_x = \sqrt{6 \times 1.95 \times 10^5 \times 15 \times 10^3 \times \frac{4 \times 98.7}{27 \times 10^3 \times 2.5}}$$

$= 14.238m$

Loss in Force at the Jacking end for Cable 1

$$\Delta P_1 = 2.5\left[6890.17 - \left(\frac{110.25 \times 5.62}{15}\right) - 779.92\right] = 172.36kN$$

Net force at jacking end A = 890.17 – 172.36 = 717.81

Loss in Force at the Jacking end for Cable 2

$\Delta P_2 = 64.476\ kN$

$\Delta P_3 = 64.47\ kN$

Net force at jacking end B for cable 2 = 507.550 – 64.476

= 443.074 kN

Net force at jacking end for cable 3 = 514.400 – 64.476

= 449.924 kN

The force flow diagram considering the losses due to anchorage slip, friction as well as elastic shortening is shown in Fig 5.12 c (1,2,3).

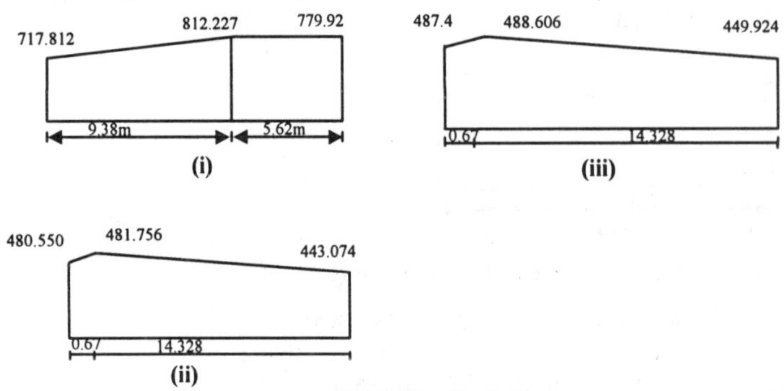

Fig. 5.12 c (1, 2, 3)

Stresses in Concrete at the levels of cables due to stressing of different cables are represented in tabular form in the table given below:

Stressed Cable	**Level of Stress**	**Stress at support section in Concrete** (N/mm^2)	**Stress at center section in Concrete** (N/mm^2)
Cable 1	Cable 1	4.44	0.69
	Cable 2	-0.728	3.476
	Cable 3	-0.728	3.476
Cable 2	Cable 1	-0.416	1.987
	Cable 2	2.893	2.893
	Cable 3	2.893	2.893
Cable 3	Cable 1	-0.416	1.987
	Cable 2	2.893	2.893
	Cable 3	2.893	2.893

Loss due to Creep of Concrete

IS 1343 Clause is considered for evaluating the creep loss. As the forces are transmitted at 28 days a creep coefficient of 1.6 is assumed. The stress in concrete at various levels are evaluated, based on the Force Flow Diagram and the average stress over the length of the beam is evaluated and shown in Table below.

Stressed Cable Stress	**Level of support** (l) **(Concrete)**	**Stress at left** (N/mm^2) **(Concrete)**	**Stress at Centre support (Concrete)**	**Stress at Right (Concrete)**	**Average Stress**	**Creep Loss In steel** ($N.mm^2$) ($1.6\in_{el}$) **m**
Cable 1	Cable 1	3.457	6.17	3.854	5.347	51.33
	Cable 2	-0.58	6.852	-0.631	4.366	41.91
	Cable 3	-0.58	6,852	-0.631	4.366	41.91
Cable 2	Cable 1	-0.389	5.549	-0.358	3.575	34.32
	Cable 2	2.702	6.366	2.492	5.11	49.056
	Cable 3	2.702	6.366	2.492	5.11	49.056
Cable 3	Cable 1	-0.394	5.576	-0.364	3.591	34.47
	Cable 2	2.742	6.405	2.531	5.149	49.43
	Cable 3	2.742	6.504	2.531	5.149	49.43

Total loss of stress in steel due to creep of concrete

In Cable 1 = 120.12 N/mm^2

In Cable 2 = 140.396 N/mm^2

In Cable 3 = 140.396 N/mm^2

Total loss of force due to creep of concrete is

In Cable 1 − 120.12 × 7 × 98.7 = 82.99 kN

In Cable 2 = 55.43 kN

In Cable 3 = 55.43 kN

The forces at end of cables considering losses due to friction, elastic shortening, slip and creep is as given below.

	END A	END B
Cable 1	717.81 - 82.99 = 643.82	779.92 - 82.99 = 696.93
Cable 2	480.550 - 55.430 = 425.12	443.074 - 55.430 = 388.634
Cable 3	487.400 - 55.430 = 431.970	449.824 - 55.430 = 394.394
Total force	1491.91	1479.958

% Loss in force at left support

$$= \left(\frac{1928.8 - 1491.91}{1928.8}\right) \times 100 = 22.65\,\%$$

% Loss in force at right support

$$= \left(\frac{1928.8 - 1479.958}{1928.8}\right) \times 100 = 23.28\,\%$$

Loss due to Shrinkage of Concrete

The shrinkage of concrete in prestressed members leads to shortening of tensioned wires and ehnce contributes to loss of stress.

For post tensioning $\in_{cs} = \dfrac{200 \times 10^{-6}}{\log_{10}(t+2)}$

Here $\in_{cs}$ = total residual shrinkage strain

t = age of concrete at transfer in days = 28 days

Loss of stress due to shrinkage $= \in_{cs} E_s$

$$= \frac{200 \times 10^{-6}}{\log_{10}(28+2)} \times 1.95 \times 10^{5}$$

$= 26.4\ N/mm^2$

Loss of force in Cable 1 = 26.4 × 7 × 98.7 = 18.24 kN

Loss of force in Cable 2 & 3 = 26.4 × 4 × 98.7 = 10.424 kN

Forces at end of cables after shrinkage losses are shown below

	END A	END B
Cable 1	634.820 - 18.24 = 616.580	696.93 - 18.24 = 678.690
Cable 2	425.12 - 10.424 = 414.696	388.634 - 10.424 = 378.210
Cable 3	431.970 - 10.424 = 421.546	394.394 - 10.424 = 383.970

Loss due to relaxation of stress in steel

Initially the stress applied to the cable is 0.7 f_{pu} but after the losses due to elastic shortening, friction and anchorage slip has occurred, the stress reduces to an average of 0.6 f_{pu}.

Relaxation loss corresponding to 0.6 f_{pu} stress = 35 N/mm^2

Loss of force in Cable 1 = 24.18 kN

Loss of force in Cable 2 = 13.82 kN

Force at end of cables after considering all the above losses is as given below

	END A	**END B**
Cable 1	616.58 - 24.18 = 592.40	678.690 - 24.180 = 654.51
Cable 2	414.696 - 13.82 = 400.876	378.210 - 13.82 = 364.390
Cable 3	421.546 - 13.82 = 407.726	383.970 - 13.82 = 370.150
Total force	1401.002	1358.050

% Loss in force at left support $= \left(\frac{1928.8 - 1401.002}{1928.8}\right) \times 100 = 27.37\,\%$

% Loss in force at right support $= \left(\frac{1928.8 - 1389.050}{1928.8}\right) \times 100 = 28.03\,\%$

The force flow pattern will be similar to what is shown in Fig 5.12 c. Hence the force flow diagram can be drawn with the above details. From the force flow diagram the prestressing force available at service load stage at any cross section can be evaluated.

References

1. EMPERGER. F., (1923) **"Concrete tubes for internal pressure", Concrete and Constructional Engineering,** Vol.18.
2. IRC - 18 - (1985) **"Design Criteria for Prestressed Concrete Road Bridges (Post - Tensioned Concrete),** Indian Roads Congress, New Delhi.
3. COOLEY, E.H., **"Friction in post-tensioned prestressing systems",** Research Report No.1, Cement and Concrete Association, Londaon, 1953.
4. LEONHARDT. F, (1964) **"Prestressed Concrete Design and Construction",** Wilhelm Ernst and Sohn, Berlin,.
5. IS 1343 - (1980). Indian Standard **"Code of Practice of Prestressed Concre te",** Sixth reprint, Bureau of Indian Standards, New Delhi.
6. DISCHINGER, F., (1937) **Untersuchugen Uber die Knicksicherheit, die elastichen Verformungen und das Kriechen des Betons bei Bogenbrucken.** Der Bauingenieur, Heft 18.
7. TROST. H (1967) **"Implications of the superposition principle in creep and relaxation problems for concrete and prestressed concrete"** Betons Stahlbetonbau 62(10) PP 230-238 & 62(11) PP 261-169, (in German).
8. **Precast Segmental Box Girder Bridge manual** (1975) Published by Prestressed Concrete Institute, Chicago, Illinois, USA.
9. **Concrete European due Beton/FIP** (1975). Bulletin d' information No.III, October.

6. Ultimate Strength in Flexure

6.1 General

The behaviour of prestressed concrete elements in flexure from zero load upto failure has been discussed in Article 4.2. Upto cracking of concrete in the tension zone, prestressing steel and untensioned steel, if any, act together with concrete and the entire effective cross section resist the rotation of the cross section, by means of stresses developed in concrete both in compression zone and tension one. Just after cracking, the tension resisted by concrete in that zone is transferred to the steel and the cracks propagate quickly upto the neutral axis. Thereafter the behaviour of the prestressed concrete element subjected to flexure is very similar to that of the reinforced concrete element and the theories for estimating the flexural capacities are one and the same.

The failure of prestressed concrete sections under flexure depends on various aspects such as percentage of pretensioned steel in the cross section and existence of untensioned steel, the bond between the tendons and the concrete, the maximum strain that the steel and concrete can take at the time of failure in tension and compression respectively, etc. Actual failure is always caused only by deformations, such as strains and rotations. But the designated failures are attributed to ***limiting stresses.*** These correspond to ***limiting strains***.

While deformations are realities, the corresponding levels of stresses are only conjectures, which are based on the assumed modulus of elasticity at various stages of loading got from the tests on the control specimens. The stress strain relationship of the materials in the prototype might be different from that obtained in control specimens. Since in actual practice, the concept of designated failure is related only to stresses, caused by load, bending moment etc, the stress - strain relationship has to be made use of, while arriving at the failure load or moment. The actual failure load or moment for the section subjected to flexure should be determined only on strain compatibility and assumed stress strain relationship.

In prestressed concrete sections, the initial strain in concrete and steel should be kept in mind while deriving the compatibility of strains at extreme fibres of the cross section for the given rotation caused by an applied moment. Just after cracking, the total tension is taken by the steel and the section rotates about the plane of zero stress. Hypothesis of the plane section remaining plain upto failure is assumed to be valid and hence for any additional load beyond the cracking load/causing moment on the cross section leads to additional tensile strain in steel and compressive strain in concrete till any one of them reaches the limiting values. Based on the stress concept, at time of failure, the steel reaches its limiting stress value and there is a corresponding strain to which it is subjected to, which is declared as the limiting strain. This strain ε_{spu} (corresponding to the limiting stress is steel as per Statutory Codes) is still not the strain ε_{sp} corresponding to failure stress fp. There is further additional capacity in terms of strain in steel (i.e. upto $\varepsilon_{sf} > \varepsilon_{sp}$) for the same failure stress level of f_p, as could be seen from Fig 6.1. Hence there is a facility of increase strain even after the steel reaching its limiting stress, because of the ductility of the material. This additional strain allows the cracks to open up and neutral axis to shift upwards. The reduction in the depth of compression zone increases the stresses in concrete at extreme fibre for the same amount of compressive force, thereby increasing the compressive strain leading to collapse. Hence even if the steel reaches its limiting stress, the final collapse happens only after the concrete reaching its limiting strain allowing for sufficient time before the collapse of the beam and there will be excessive deformation before such a collapse. This type of ductile failure happens in an under reinforced beam.

But if the concrete reaches its limiting strain (ε_{cu}) first, earlier to that of steel, the collapse becomes brittle and sudden and such a beam is declared as an over-reinforced beam.

Even earlier to the ductile failure of an under reinforced beam mentioned above, there is also the possibility of a sudden failure in beams subjected to flexure. This failure happens immediately after the cracking load. At the time of cracking of concrete in the tension zone, forces till then resisted by concrete are suddenly transferred to steel. If the area of steel provided is very small, the sudden increase in the force in steel may lead to a strain level in steel, which is beyond the actual failure strain level (ε_{sf}) of steel. In such a case, the steel fractures immediately causing sudden collapse.

In case of prestressed concrete beams, high tensile steel is used to transfer the prestress to concrete. Up till the cracking load is reached, it is the concrete, which takes the entire load. Only after the cracking, the prestressing steel also forms the structural element to resist the load. From this point onwards, any untensioned steel (intended for structural action or not intended to resist flexure but provided only to support stirrups) also participates in the structural action. This leads to higher capacity in the tension zone.

These 3 types of failures have been classified by Leonhardt (1). The methods of evaluating the flexural capacities of section before the failures occur

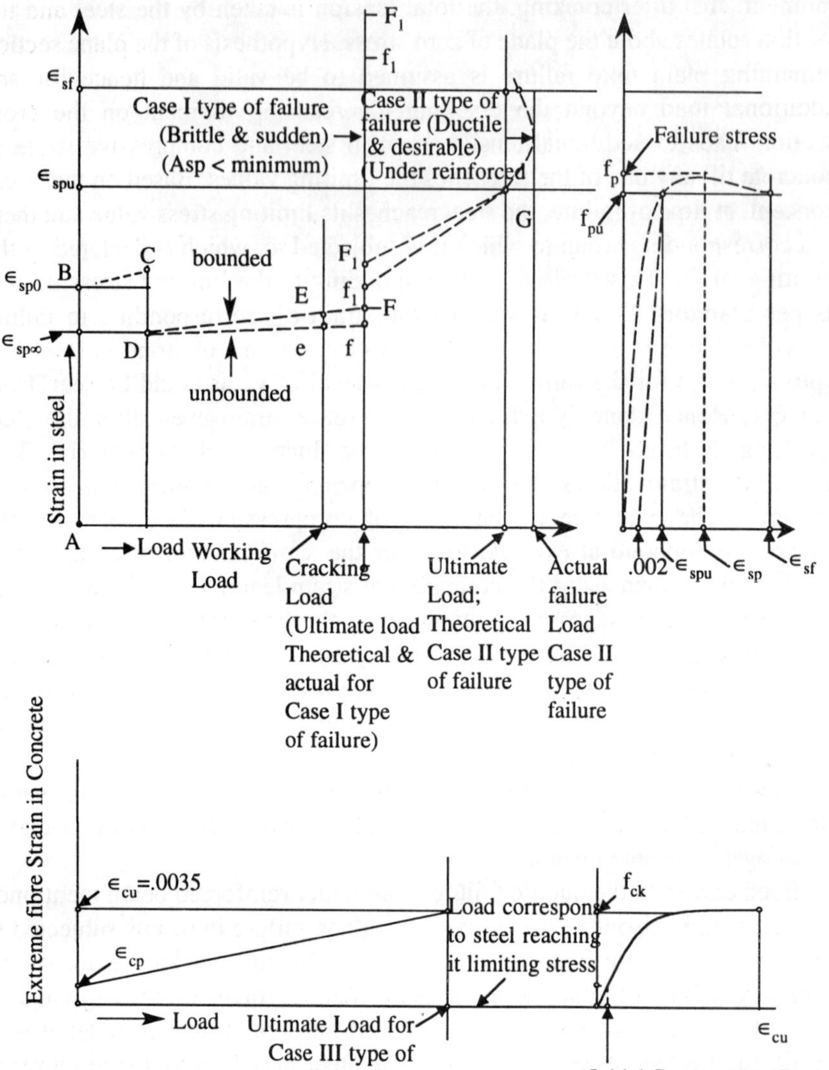

Fig.6.1. Strain in Steel at Various Load Stages in Prestressed Concrete Element Subjected to Flexure

and the methods of safety against any such failures before a predesignated load is reached, are discussed in this Chapter.

Since actual failure is controlled by materials reaching their limiting strain capacities, a figure showing the behaviour of prestressed concrete beams under increase in loads and corresponding increase in strains is presented in Fig 6.1.

6.2 Failure of Beam by Fracture of Steel

As mentioned earlier, after the stage of decompression in prestressed concrete elements, the concrete cracks as soon as the tensile capacity of the concrete is exceeded at the extreme fibre and the total tensile force carried by the concrete at that point of time suddenly shifts over to steel and also the crack develops faster up to the zone of compression. This sudden jump of force in steel is shown as F to F_1 in Fig 4.1. This increase in force causes additional strain in steel. This jump in strain in steel is shown as F to F1 in Fig 6.1. If the tensile strain exceeds the actual failure strain (ε_{sf}) in steel, ***(point F_1 goes beyond the level of ε_{sf})*** the steel may fracture and it leads to collapse of the beam. This type of failure is possible if the percentage of tensile reinforcement in beams is very low. To avoid this type of failure, various satutory codes provide for a minimum reinforcement of steel, defined as minimum percentage steel. If such a percentage of steel is provided, the total tensile force originally carried by the concrete before formation of cracks can be taken by this steel without even reaching to the strain [ε_{spu}] corresponding to limiting design failure stress f_{pu}. Such a minimum percentage of steel could be arrived at for concrete of different strengths with various levels of prestress as detailed in 6.2.1

Indian Standard Code IS 1343 - 1980 (2) prescribes minimum longitudinal reinforcement as 0.2% of the cross-sectional area in all case except in cases of pretensioned units of small sections. The minimum percentage of steel could be 0.15% if high tensile steel is used with the proviso, that with the percentage steel provided in terms of both tensioned and untensioned steel together should be sufficient so that when the concrete in the precompressed tensile zone cracks, steel is in a position to takeover the total force from concrete for reaching to its limiting strain. The other statutory codes like BS 8110 - 1985 (3) ACI 318 - 1998 (4), FIP - 1984 (5), ENV 1992 (6) provide for safety against such failure in different ways. BS 8110 (3) prescribes the minimum number of tendons such that the cracking of concrete precedes failure of steel by rupture. This condition is satisfied if the ultimate moment of resistance of the section is more than the cracking moment, i.e. the moment at the point of time, when the stress in concrete at extreme fibre crosses the limiting flexural tensile stress. For evaluating the stress at extreme fibre in concrete, the minimum prestressing force, which may be available after all losses, should be considered. ACI 318 - 1998 [4] specifies minimum area of bonded reinforcement as 0.004 times the area of concrete in the tension zone.

6.2.1 Minimum percentage of Steel

Minimum percentage of steel could be derived based on the philosophy mentioned in Article 6.2 above. The basis of the minimum percentage of steel recommended by IS Code is shown in the derivation below.

Total tensile force in steel before the steel reaches its ***limiting design stress*** $= A_{sp}.f_{pu}$

This force is reached in steel due to initial pretension, and additional forces which come on the same during external loading.

This could be expressed in the following form

$A_{sp} \cdot f_{pu} = P_{\infty} + C_T + mA_{sp} \cdot f_{cs}$

$A_{sp} \times 0.87 f_p = P_{\infty} + C_T + mA_{sp} \cdot f_{cs}$

where P_{∞} is the prestressing tension in steel after all losses

C_T = Tensile force taken by concrete, from stage of decompression upto the point of cracking, ie upto the point where the tensile stress in concrete at extreme fibre exceeds flexural tensile capacity.

f_{cs} = Stress in concrete at the level of steel fibre due to imposed loads

A_{sp} = Area of prestressing steel

By substituting the numerical values as applicable for a beam of given material properties like f_{ck}, f_p, the minimum area of prestressing steel (A_{sp}) required could be obtained so that the sudden transfer of tensile force from concrete to steel shall not lead to stress level in steel beyond acceptable design limiting stress (f_{pu}).

As per IS 1343 regulations, the flexural tensile capacity can be assumed as 0.7 $\sqrt{f_{ck}}$. Assuming the minimum acceptable value of f_{ck} for prestressed concrete works, namely 30 N/mm^2 . f_{cr} = 3.83 N/mm^2, assuming, a minimum acceptable strength for steel as 1500 N/mm^2. f_p = 1500 N/mm^2 and f_{pu} = 1500 × 0.87 N/mm^2, the concrete subjected to tension from the stage of decompression to stage of cracking at extreme fibre can be approximately assumed as 0.4 A_c, where A_c is the gross area of concrete section.

The maximum stress permitted in steel during service load as per IS 1343 is 0.6 f_p. Modular ratio for concrete used in prestressing works can be assumed to be 6. Rewriting the expression for total tension in steel, just at the time of cracking of concrete at extreme fibre.

$$0.87 \times f_p.A_{sp} = 0.6 f_p.A_{sp} + 0.4 A_c \left(\frac{f_{cr}}{2}\right) + 6 A_{sp}.f_{cr}$$

$$0.27 f_p.A_{sp} = 0.4 \times A_c \times \frac{3.83}{2} + 6 \times A_{sp} \times 3.83$$

$$\frac{A_{sp}}{A_c} = \frac{0.766}{382.02} = 0.002$$

% of steel = 0.2 %

If steel of higher f_p value is used (say 2000N/mm^2) the above expression leads to 0.15%.

6.3 Failure of Under Reinforced Beams

In prestressed concrete beams subjected to flexure, when the stress in the extreme fibre in tension zone crosses the limiting flexural capacity of concrete, cracks are developed in the tension zone and they propagate to the level of zero stress immediately. The tensile forces, till then resisted by concrete, are taken over by steel. If the reinforcement (High tensile steel and untensioned steel if any) is more than the minimum prescribed, the steel reaches a strain level which is still within designated failure capacity ($\in_{spu}$). As the load/external bending moment further increases, the additional tension is taken by the steel and the strain in steel increases and cracks start widening. Strain being proportional to the distance from the plane of zero stress is valid only if the strain is measured over a larger length covering the crackwidth in that zone. The strain in steel can go upto $\in_{spu}$. When the strain level reaches that value, the beam is said to have reached its designated failure load P_u or failure moment (M_u). But the stress level is only $f_{pu} = \dfrac{f_p}{\gamma_m} = 0.87\ f_p$. When there is further increase load/moment there is an increase in strain and stress level goes upto f_p. This increase in load is not considered in the design. Even after the steel reaches the stress level of f_p, the load will be sustained by the beam without collapse. But, for no additional load/moment, the strain in steel goes on increasing widening the crack. The tip of the crack moves upward reducing the compression zone. Since the total tension and total compression have to be equal, the compression block reduces in depth, and the compressive stresses at different levels increase and the compressive strain at the extreme fibres increases. When it reaches $\in_{cus}$ (the failure strain) concrete crushes and the beam collapses. This is a progressive collapse. From the times the steel reaching a stress limit of f_{pu} (corresponding to failure load P_u or moment M_u) till the actual collapse of the beam (when concrete reaches a strain level of $\in_{cu}$) large deformations take the place on the beam with widening of cracks, giving a warning before failure. Hence this type of failure is preferred in the design (Ref Fig 6.1 and 6.2 b).

If the beam has got some inadvertent/indeterminacy due to end conditions, there will be redistribution of the forces due to rotation of the cross section forming a hinge. That would lead to further increase in the capacity.

6.4 Failure of Over Reinforced Beams

In case of beam cross section, provided with large reinforcement (with high tensile steel and untensioned steel), for any applied load/moment, the tensile stress in steel is less than a corresponding under reinforced beam though the total tensile force is the same in both the beams. As the load further increases, the increase in moment is resisted by increased force in steel in tension zone and in concrete in compression zone. Because of large area of steel in Tension Zone, the stress level and the strain level increase in steel are not much. Because of higher area of tensile steel, the depth of compression block is more. Hence the rate of increase of compressive strain at the extreme fibre is high even for small increase in rotation due to applied moment. Accordingly, the extreme fibre in compression in concrete reaches the limiting strain in concrete before the steel reaches anywhere near the strain corresponding to the designated failure stress f_{pu} (strain being ε_{spu}) and the beam collapses without any warning. Such a beam is called over reinforced beam and the failure of over reinforced beam is primarily due to concrete reaching its limiting strain before steel and the failure is sudden (Ref Fig 6.1 b and 6.2 c).

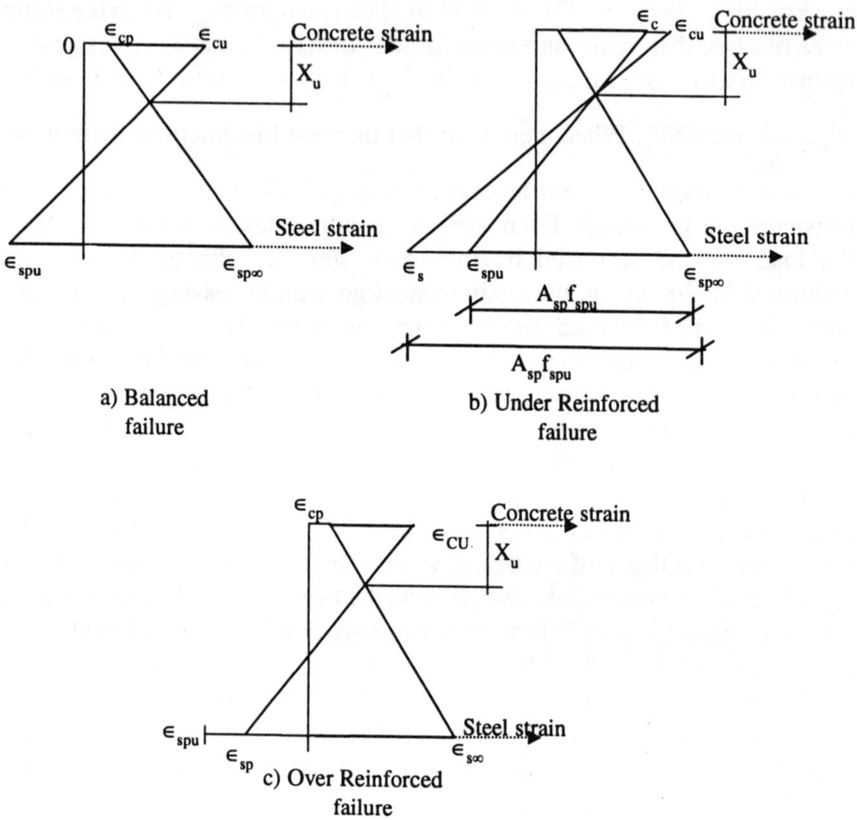

Fig. 6.2

6.5 Balanced Beam

If the beam is provided with sufficient reinforcement, so that the steel and concrete reach that desginated failure strain. (ε_{spu} & ε_{cu}) respectively, at the same time, the beam is called a balanced beam and the failure of such a beam is called balanced failure (Ref Fig 6.1 and 6.2 a).

6.6 Design Failure Loads and Moments

In the design concept of limit state of collapse, a safety factor for materials is expected to be provided. Hence the limiting strain, as far the design is considered (ε_{spu}) corresponding to f_{pu}, is smaller than the strain in

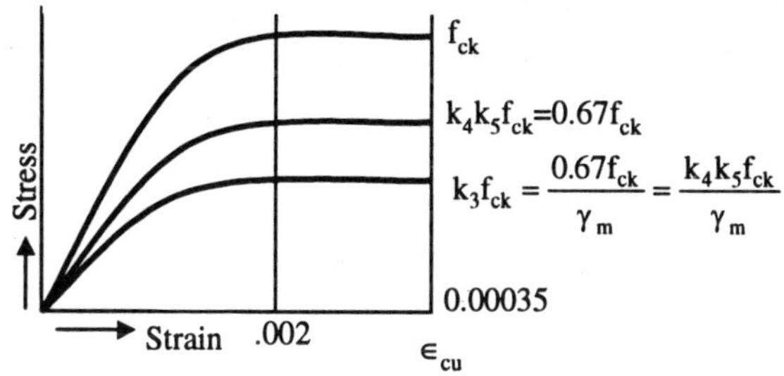

Fig.6.3.a.Stress Strain Curve for Concrete

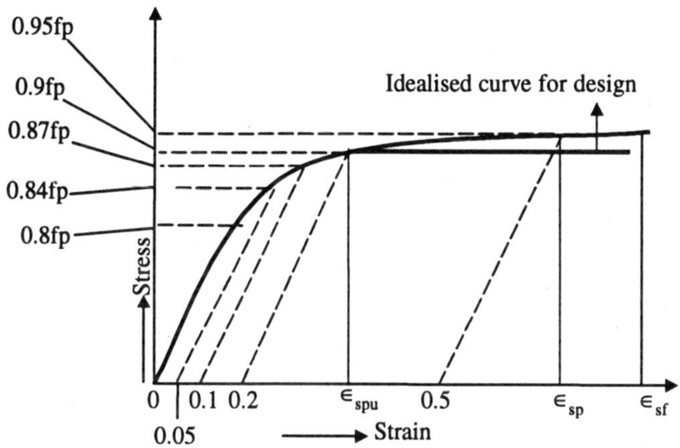

Fig.6.3.b. Stress Strain Curve for (Stress Relieved), Strands and Bars as per ISI343-1980

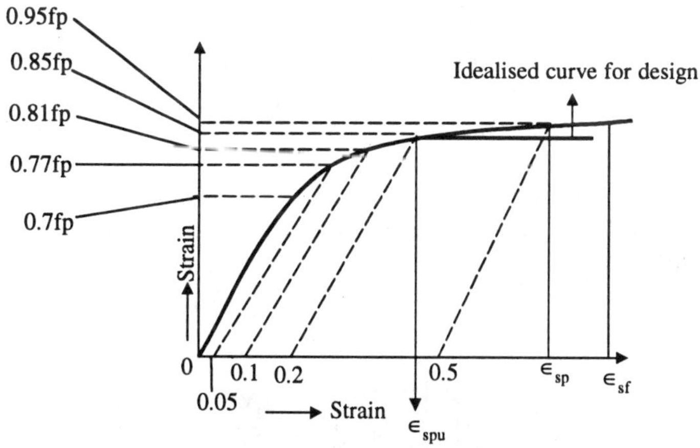

Fig.6.3.c. Stress Strain Curve for Wires (as-drawn) as per ISI343-1980

steel (ε_{sp}) corresponding to its material capacity of f_p. (It may be noted that the actual failure strain (ε_{sf}) will be much larger because of the ductility of the material). But in concrete the actual failure strain of ε_{cu} is assumed to occur even at the factored failure stress level, which is less than the actual failure stress f_{ck}. These aspects are seen in the stress strain curve for concrete and steel showing also the factored limiting stresses and the corresponding strains (Fig 6.3).

6.7 Evaluation of Failure Load/Failure Moment

As mentioned earlier, failure load/moment can be evaluated based on strain compatibility and assumed stress strain relationship in all cases of failure. In case of failure by fracture of steel or in case of under reinforced beam, the strain in steel is assumed to reach its limiting value as per design (ε_{spu}) corresponding to limiting design failure stress i.e. $f_{pu} = \left[\frac{f_p}{\gamma_m}\right]$ first and in case of over reinforced beam, the concrete reaches its limiting strain at design failure stress namely $\left[\frac{0.85 \times 0.8 \times f_{ck}}{\gamma_m}\right]$ and the strain at failure is assumed to be ϵ_{cu}. In under reinforced beams also, at the time of actual collapse the concrete reaches its limiting design failure stress, though this is caused by large scale elongation of steel for no increase in load after the steel reaches a stress value of f_{pu}. Hence the failure strain ϵ_{cu} in concrete is definitely to happen at the time of failure and the design failure stress in concrete should be reached. This point is incorporated in writing down the equilibrium equation at the time of failure. The basic assumptions made in formulating the strain compatibility expressions and the positions of the axis (Neutral axis) separating compression zone and tension zone and the explanations relating to these assumptions are detailed out first before actual expressions for evaluations of Failure loads and moments are written.

6.7.1 Basic Assumptions

(i) The distribution of strain across a section is linear

(ii) Tensile resistance of concrete is neglected

(iii) The strain in the reinforcement is the same as the strain in the concrete at the same level

(iv) At the time of failure the stress and strain of concrete are not mutually proportional and the shape of the concrete stress distribution follows the factored stress strain curve for concrete known as "Stress Block". This can be defined by three coefficients k_1, k_2 and k_3 (Fig 6.4).

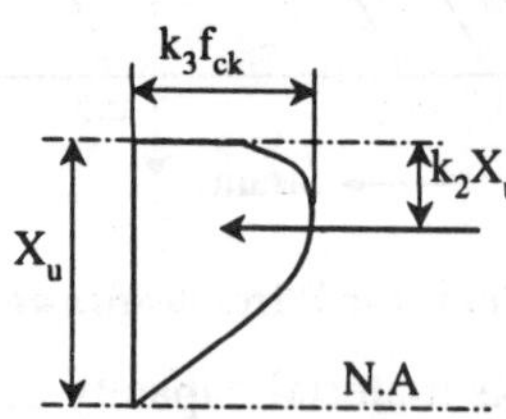

Fig 6. 4. Stress Distribution Concrete in Compression Zone at Failure

k_1 = ratio of average to maximum compressive stress permitted at the time of designated failure

k_2 = ratio of distance of center of compression from the extreme compressive fibre to the depth of neutral axis

k_3 = ratio of maximum compressive stress permitted at the time of designated failure to cube strength

(v) The concrete strain in compression at the time of failure has a specific value.

6.7.2 Explanations

(a) The assumption that the strain distribution is linear across the section is really valid for a prestressed concrete section only upto cracking. The strain in the concrete on the tension side of the neutra axis will vary considerably, at any given level, due to cracking. If, however the extension per unit length of a gauge length including several cracks is measured, it is found that this "apparent" tensile strain varies linearly with distance from the neutral axis.

(b) The second assumption can be considered to be reasonably correct. Any tension zone, which does exist at the ultimate strength, is very small indeed in a normally proportioned beam and its lever arm is also small. It is therefore reasonable to neglect any contribution by concrete in tension to be ultimate moment of resistance of a section.

(c) The assumption, of compatibility of strain between steel and concrete at the same level, just as in the case of assumption (a) is valid only when the gauge length covering several crack spacings is also considered. Thereby the "concrete strain" is to be evaluated from the actual elongation of the concrete plus the widths of cracks concerned within that gauge length. With proper detailing, crack spacings and crack widths are fairly small and hence the assumption holds good also at relatively high steel stresses, even when relatively small lengths of the beam are considered.

(d) Extensive research work was carried out regarding the compressive stress distribution in concrete at or near ultimate load. The relation between the compressive stresses and strains in the concrete of a given section depends on (i) the rate of application of loads (ii) the duration of loading (iii) the nature of the loading (iv) the quality of the concrete (v) the position of the neutral axis at the ultimate state or, in other words, the percentage of reinforcement and (vi) the geometrical shape of the section in question. The designer's task would become too complicated to take into account all the possible variations of the parameters listed above. Hence, the calculations of the ultimate flexural strength of sections, the properties of the stress block etc., are based on the experimental investigations by Hognested et al (7) Riisch (8) and others. Their research works have led to specific values for k_1, k_2, k_3, which take care of all variables for a safe design (Ref Fig 6.4 & 6.5).

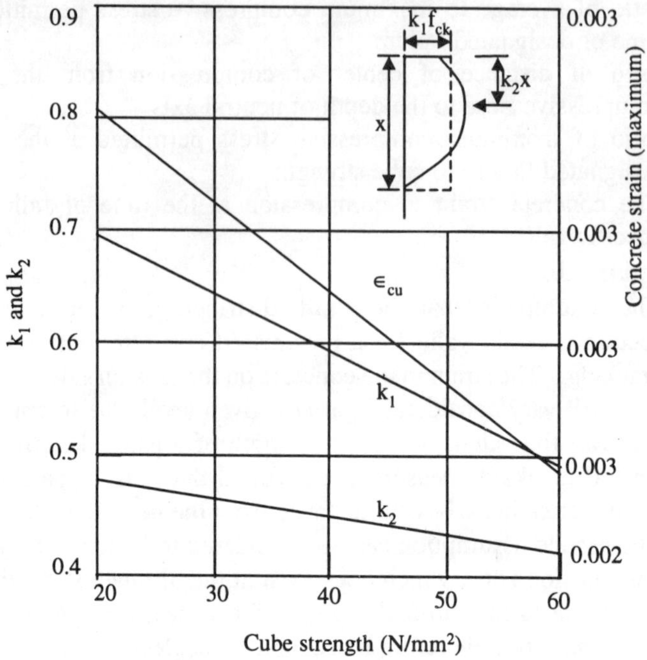

Fig 6.5. Characteristics of Hognestad et al's Stress Block

(e) The value of the maximum strain in concrete may vary considerably for loads applied for long periods or for special shapes of sections , It also depends on the strength of concrete as mentioned in the Chapter on Materials. Based on the tests conducted on reinforced concrete members a safe limiting strain of 0.0035 is assumed for design purposes for concrete of strength upto 60 Mpa in Indian Standard Codes.

6.7.3 Equilibrium conditions and strain compatibility at the time of failure

At the time of failure, equilibrium of forces can be written as total compression in concrete in the compression zone (C) equal to total tension in steel in tension zone (T)

$C = k_3 f_{ck} b x_u k_1$

$T = A_{sp} f_{pu}$ or $A_{sp} f_{sp}$(if it is an over reinforced beam and $f_{sp} < f_{pu} < f_p$)

$k_3 f_{ck}$ is the factored designated failure stress in concrete and at the time the failure strain in concrete is ϵ_{cu}. This factor k_3 encompasses the material safety factor desired to be provided by the various statutory authorities under limit state of collapse, the ratio of cylinder strength to cube strength, and a reduction factor in capacity due to sustained load conditions. f_{ck} is normally determined for a given concrete under a given rate of loading and a sustained loading for long duration.

f_{pu} is the limiting stress in steel (designated failure stress) which has the material safety factor inbuilt and it corresponds to a strain ε_{spu} which is designated failure strain and not corresponding to f_p.

It may be noted that at the stress level $k_3\ f_{ck}$ in concrete, the maximum concrete strain ε_{cu} (failure strain) is assumed to have been reached whereas in steel at the steel level fpu, the strain in steel is only ε_{spu} and not ε_{sp} and not to talk of ε_{sf}. This can be seen in the stress strain curve assumed for material concrete and steel for arriving at the failure capacity of the section under flexure.

The force which causes the failure, compressive force because extreme fibre in compression reaches its limiting strain or tensile force because steel reaches its limiting design strain first, will be used for evaluating the failure moment (T × L.A) or (C × LA).

It is known that in prestressed concrete beams, there is an initial strain and stress level in the steel and concrete respectively even before application of loads for which the beam is designed. There is an initial strain in steel i.e. $\varepsilon_{sp\infty}$ [corresponding to $f_{p\infty}$ after all time dependent losses in prestressing steel]. There is an initial strain in concrete at the extreme compression fibre (ε_{cp}). This strain can be compressive or an acceptable tensile strain within the limit of cracking.

As the load increases, the tensile strain in steel increases along with the tensile strain in concrete at that fibre since they are bonded. As the beam cracks in the tension zone, the tensile strain in steel increases at higher rate and compressive strain in extreme compression fibre also increases with increase in applied load. The increase in strain in steel and concrete goes on till either of them reaches it limiting strain ε_{spu} (corresponding to f_{pu}) or ε_{cu}. Hence the movement of strain, (corresponding to strain range) in steel is from $\varepsilon_{sp\infty}$ to ε_{spu}, whereas the same in concrete is ε_{cp} to ε_{cu}.

Depending on the type of beam (over reinforced, under reinforced or balanced), the depth of compression block can be arrived at using (a) strain compatibility conditions (b) equilibrium conditions for under reinforced beam and (c) a combination of both for over reinforced beam. Once the depth of compressive block or the depth of neutral axis is known, the position of centre of compression and lever arm between the centre of compression and centre of tension are also known. The same are used for evaluating the moment carrying capacity (failure moment) of a cross section.

6.7.4 Under reinforced prestressed concrete beams

In case of under reinforced prestressed concrete beams, the steel reaches its limiting strain $\in_{spu}$ corresponding to limiting stress of $f_{pu} = \dfrac{f_p}{\gamma_m}$ as per Limit State design concepts. The concrete is not reaching its limiting strain of $\in_{cu}$ at top fibre. At this point of time equilibrium of forces is maintained. But the concrete is not reaching its $\in_{cu}$. Now the beam does not fail. The steel elongates and strain increases both in steel (in tension) and concrete (in compression) (Ref Fig 6.2 b). When the concrete reaches a strain of $\in_{cu}$ the beam fails. But equilibrium of forces is not disturbed. Hence equilibrium of forces will define the depth of compression block at failure. The equilibrium equation at failure can be written as

$$A_{sp} \cdot f_{pu} = K_3 f_{ck}\, b x_u (k_1)$$

$x_u = \dfrac{A_{sp} f_{pu}}{k_1 k_3 f_{ck} b}$. With this depth of stress block, and the compressive strain at the extreme compression fibre ($\in_{cu}$), the strain in the steel should be such that it is more than the strain ($\in_{spu}$) corresponding the limiting stress of f_{pu}. For this check, the strain was originally available in steel due to pretensioning should also be taken into consideration i.e

$$\frac{(\in_{cu} - \in_{cp})}{x_u} = \frac{\text{Additional strain in steel due to loads}}{d - xu}$$

The additional strain in steel due to loading plus the original strain due to pretensioning ($\in_{sp\infty}$) should be more than $\in_{spu}$.

The moment of resisting capacity of the section is calculated

$$M_u = A_{sp} \cdot f_{pu}\, [d - k_2 x_u]$$

6.7.5 Balanced section failure

In case of prestressed concrete beam sections which has balanced percentage of reinforcement, at the time of failure the strain in steel and concrete should have simultaneously reached and the equilibrium of forces should also be achieved at that point of time. (Ref Fig 6.2 a)

Since the depth of compression block or depth of neutral axis is decided based on the strain compatibility

$$\frac{x_u}{d} = \frac{\in_{cu} \pm \in_{cp}}{\in_{cu} \pm \in_{cp} + \in_{spu} - \in_{sp\infty}},$$

this should match with x_u calculated with equilibrium equation namely $A_{sp} f_{pu} = k_1 k_3 f_{ck} b x_u$

The moment of resistance of the cross section is evaluated as $A_{sp} f_{pu} (d - k_2 x_u)$

6.7.6 Over reinforced beam failure

In case of over reinforced beam, the concrete at the extreme fibre in the compression fibre reaches its failure strain ε_{cu} earlier than steel in tension zone, reaching a strain ε_{spu} corresponding to the designated factored failure stress of f_{pu} (Ref Fig 6.2 c). But the actual strain in steel and the corresponding steel is not known and it has to be iteratively evaluated as follows.

First the equation of equilibrium of forces at the time of failure is written down assuming the steel to have reached the strain ε_{spu} and the stress of f_{pu} and x_u is solved

$$A_{sp}f_{pu} = k_1 k_3 f_{ck} b.x_u$$

$$x_u = \frac{A_{sp}f_{pu}}{k_1 k_3 b f_{ck}}$$

For this first value of x_u and the failure strain value of ε_{cu} at the top fibre, the strain level in steel is evaluated and for this strain level the corresponding stress level will not be f_{pu} and hence the actual stress level is noted from the assumed stress strain curve. With this stress level, the equilibrium equation is solved once again and x_u is evaluated. The operation is repeated till convergence is got and a value of x_u , which would satisfy the equilibrium conditions and compatibility of strains, is arrived at.

Considering I.S. Code provisions, from the shape of stress block at this time of failure, the depth of centre of compression from top fibre is arrived at (0.42 x_u) and the lever arm for resisting the external moment is evaluated. The failure moment capacity of the beam is then got using the expression

$$M_u = A_{sp} f_s (d\text{-}0.42x_u)$$

where f_s is the stress level in steel at the time when concrete in compression zone reaches a strain level of ε_{cu} and the designated factored failure stress at top fibre.

Since the failure of an over reinforced beam is brittle and sudden and without warning, various national codes and statutory authorities define stipulations so that the flexural element is not led to such a failure. This could be done by defining the upper limit on percentage of reinforcement. Any additional reinforcement, if available, or if provided, is not taken into consideration for evaluating the failure capacity. The upper limit of reinforcement is the percentage of reinforcement for balanced failure. Unlike in reinforced concrete beams where the strain compatibility at the time of failure is dependent only on the limiting strain in concrete and steel, in prestressed concrete beam, the strain compatibility at the time of failure (strains caused by external moment before the materials reaching their limiting values) includes the strains already available in concrete and steel at zero load which is a function of applied prestress. Further the limiting strains in steel at the time of failure stress are also dependent on the capacity of steel which is very much variant and it is a function of diameter, type of reinforcement etc (wire, rod, strand). Hence the percentage of reinforcement for balanced failure and depth of neutral axis for balanced failure are not uniquely defined in prestressed concrete beams. Various codes define clauses to avoid a brittle failure due to over reinforced section. IS 1343 – 1980 defines clauses to avoid brittle failure (over reinforced type) by limiting the total tensile force in steel for various percentage of reinforcement as given in Table 6.2 later in this chapter.

It could be seen from the Table 6.2, that when the ratio of tensile force capacity in the section to the compression force capacity $\left[\dfrac{A_{sp}f_p}{bd\,f_{ck}}\right]$ is increasing (as the beam goes towards over reinforced section, the designated f_{pu}(permitted for design)) is reduced by a factor. For eg. for a $\left[\dfrac{A_{sp}f_p}{bd\,f_{ck}}\right]$ value of 0.4,f_{pu} is reduced by a factor 0.75 of its full value. Hence the usable tensile force is kept less than in the case of a balanced beam tensile force capacity and the same could be proved. Because of this safety clause only, the moment of resistance is calculated using total tensile force multiplied by Lever arm though compression zone reaches its limiting capacity earlier.

BS 8110 – 1985 (2) also suggests similar procedure to avoid brittle failure for prestressed concrete beams. The limitations on f_{pu} is given in Table 6.4 under Article 6.8.2. ACI 318 - 1998 (3) also gives a reduction factor for the limiting stress value in steel for evaluating the moment carrying capacity, as could be seen later in Article 6.8.3.

In a rigorous calculation, using iterative approach as explained in article 6.7.3, the actual load/moment carrying capacity calculated would be higher than those calculated using code formulations.

Example 6.1

Evaluate true ultimate moment resisting capacity of a section 300 × 600 *mm* with 1000 *sqmm* of high tensile steel at 150 *mm* from bottom. Prestressing force of 1000 *kN* is transmitted to the cross section after losses. Strength of concrete is 40*N/mm*² and steel has a capacity of f_p = 1380 N/mm^2 and f_{pu} = 1200 N/mm^2. The stress strain properties of concrete and steel may be assumed as given in I.S. Code (1343 - 1980). $E_c = 5000\sqrt{f_{ck}}$. $E_s = 2\times10^5\,N/mm^2$.

Solution

True moment carrying capacity can be evaluated using strain compatibility approach for finding out the position of neutral axis at the time of failure and after deciding whether the beam is under reinforced, or over reinforced.

First, the depth of neutral axis is evaluated considering the beam to be under reinforced. If under reinforced, both concrete and steel must have reached the limiting capacity at the time of failure.

Writing the equation of equilibrium of forces namely

C = T the depth of neutral axis is evaluated

T = 1000 × 1200 = 12 × 105 N

C = $0.36 f_{ck} b x_u$

= $0.36 \times 40 \times 300 \times x_u$

Equating total tension to total compression

$0.36 \times 40 \times 300 \times x_u = 12 \times 10^5$

$$x_u = \frac{12\times10^5}{36\times40\times300} = 277.7\ mm$$

To validate the assumption that at the time of failure, both concrete and steel reach their designated limiting strain, the strain in steel is calculated with the failure strain in concrete at the top fibre and the depth of neutral axis is evaluated.

Initial stress in concrete at top $= \dfrac{1000 \times 10^3}{18 \times 10^4} - \dfrac{1000 \times 10^3 \times 150 \times 6}{300 \times 600^2}$

$= +5.55 - 8.33$

$= -2.78\ N/mm^2$ (Tensile)

Strain in concrete fibre at top $= \dfrac{-2.78}{3.162 \times 10^4}$

$= 0.0000879 \approx 0.000088$

Strain in steel due to prestress $= \dfrac{1000000}{2.0 \times 10^5 \times 1000} = 0.00500$

Assuming that concrete has reached its limiting strain of 0.0035 at the time of failure, the depth of neutral axis at 277.7*mm* and the compatibility of strain, the strain in steel due to applied moment is evaluated.

Strain in steel due to

applied moment $= \dfrac{(0.0035 + 0.0000889)}{277.7}(450 - 277.3)$

$= 0.0023$

Total strain steel at that time $= 0.005 + 0.0023$

$= 0.0073$

Strain at the time steel reaching $f_{pu} = 1200\ N/mm^2$

(Stress relieved steel)

$= \left[\dfrac{1200}{2 \times 10^5} + 0.001\right]$

$= 0.0070 \prec 0.0073$

Hence it can be assumed the steel has reached the design stress limit of $0.87 f_p = f_{pu}$ and the beam is under reinforced.

The ultimate moment carrying $= A_{sp} f_{spu}(d - 0.42 \times u)$

$= 1000 \times 1200(450 - 0.42 \times 277.3)$

$= 4000 \times 10^5\ Nmm$

$= 400\ kNm$

Example 6.2

Evaulate the True moment resisting capacity of the beam cross section given in Example 6.1 if the area of steel provided is 1110mm^2 and other details are same as given in the Example 6.1. $f_p = 1400 N/mm^2$, $f_{pu} = 0.85\ f_p = 1200 N/mm^2$ and the steel used is "*as drawn wires*".

Solution

Assuming the section to be under reinforced, and using the equations of equilibrium at the time of failure

$C = T$

$T = 1110 \times 1200 = 1332000$

$C = 0.36 \times 40 \times 300 \times x_u$

$x_u = 308.33$

To validate the assumption that the steel has reached its limiting strain when concrete at extreme compression face has reached the failure strain, the strain in steel for the depth of neutral axis of 308.33 is evaluated

Initial compressive strain in concrete at top

$$= \frac{1000\times10^3}{18\times10^4} - \frac{1000\times10^3\times150\times6}{300\times600^2}$$

$= -2.78$ *N/mm*^{2} (tensile)

Strain in top concrete fibre $= -\dfrac{2.78}{3.162\times10^4} = 0.0000879 \approx 0.000088$

Strain in steel due to prestress $= \dfrac{1000000}{2\times10^5\times1110} = 0.0045$

Assuming concrete has reached the limiting strain of 0.0035 at the time of failure, depth of neutral axis of 308.33 and compatibility of strain, the strain in steel due to applied moment

$$= \frac{0.003588(450-308.33)}{308.33}$$

$= 0.001648 \approx 0.0165$

Total strain in steel at that time $= 0.0045 + 0.00165$

$= 0.00615$

Strain at the time steel reaching f_{pu} = 1200*N/mm*^{2}

$$= \left[\frac{1200}{2\times10^5} + 0.002\right]$$

$= 0.0070 \succ 0.00615$

Hence steel has not reached the limiting stress of $0.87 f_p = f_{pu}$

The beam is a over reinforced beam. Hence, in equation of equilibrium of forces, the stress cannot be assumed as 1200 *N/mm*^{2}.

Assuming a stress level of 0.77 *fp* in steel at the time of failure

= 1062.6*N/mm*^{2}

Strain corresponding to the stress level of 1062.6 *N/mm*^{2} (From Fig.6.3.c)

$$= \left[\frac{1062.6}{2\times10^6} + 0.0005\right]$$

= [0.005313 + 0.0005]

= 0.005813

Strain due to applied moment = [0.005813 + 0.0045]

= 0.001313

If the strain in the extreme fibre is 0.0035, the depth of neutral axis is x_u

$= 0.003588(450 - x_u) = 0.001313x_u$

$= 1.6146 - 0.003588\ 0.001313\ x_u$

$$x_u = \frac{1.6146}{0.004901} = 329.14$$

C = 0.36 × 40 × 300 × 329.14 = 142188.5*N*

T = 1110 × 1380 × 0.77 = 1179486.0*N*

The equilibrium of forces is not satisfied. Hence as a next trial

Assume a stress level of $0.81 f_p$ = 1117.8 *N/m*^{2} in steel.

Strain at stress level of 1117.8*N/mm*^{2} $= \left[\dfrac{1117.8}{2\times10^5}\times0.001\right] = 0.0066$

Strain in steel due to applied moment= 0.0066 – 0.0045

= 0.0021

If strain in extreme fibre is 0.0035 the depth of neutral axis x_u

$$\frac{0.003588}{x_u} = \frac{0.0021}{450 - x_u}$$

$$1.6146 = (0.003588 + 0.0021)\, x_u = 0.0056\, x_u$$

$$x_u = \frac{1.6146}{0.0056} = 288.3$$

Checking equilibrium of forces

$$C = 0.36 \times 40 \times 300 \times 28.32 = 12455.50$$

$$T = 1110 \times 1117.8 = 1240758$$

There is a marginal difference between forces T and C. This could be removed by a small change in strain in steel and depth of neutral axis. That change will not make much difference in total forces. Hence taking the total force as the average of C and T

i.e. $\frac{1245580 + 1240758}{2} = 1243169$

the moment of resistance $= 1243619\ (450 - 0.42 \times 288.32)$

$= 1243619 \times 328.9$

$= 409026289.10$

$= 409.02\ kNm$

Example 6.3

Evaluate the True moment carrying capacity from fundamentals using strain comparability approach for a post-tensioned prestressed concrete beam section with bonded tendon as shown in Fig. 18 nos of 7mm stress relieved wires are used for prestressing. They are prestressed to a stress of 1000N/mm^2 and their ultimate strength is 1400N/mm^2. The strength of concrete is 40N/mm^2. Use I.S. provisions for material properties of concrete and steel.

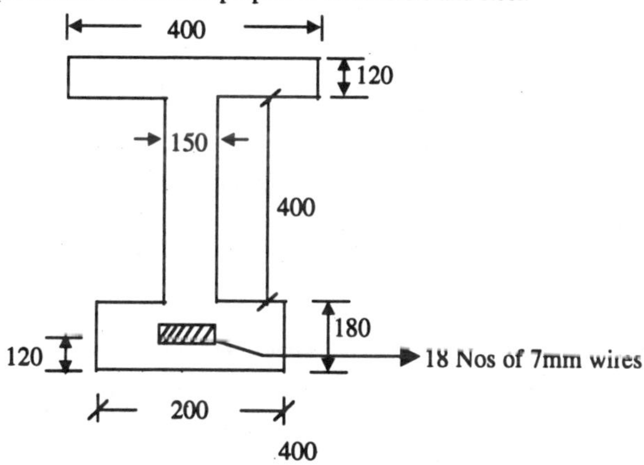

Solution

Section properties (Gross)

$A_g = 144000 mm^2$

$I_g = 8.69 \times 10^9 mm^4; y_b = 292.23 mm^2; y_t = 407.77 mm^2$

$$Z_1 = \frac{I_{xx}}{\bar{y}_1} = \frac{8.69 \times 10^9}{407.77} = 21.3 \times 10^6 mm^3$$

$$Z_1 = \frac{I_{xx}}{\bar{y}_b} = \frac{8.69 \times 10^9}{407.77} = 29.73 \times 10^6 mm^3$$

Area of prestress A_{sp} = 18 no of 7mm wires = 692.72mm^2

Tensile force at failure = $0.87 \times 1400 \times 692.72$

$= 843.73 \times 10^3$N

Prestressing force = $692.72 \times 1000 = 692.72 \times 10^3$N

Assuming the neutral axis is falls in the web and uniform stress in the flange area of 0.446 f_{ck} ie the strain at the bottom of the flange is >0.002.

Total Compressive force at the time of failure =

$$C = 0.36 f_{ck}\, b_w\, x_u + 0.446 f_{ck}(b_f - b_w)D_f$$

$$= 2160\, x_u + 13380$$

Equating total tension and compression at the time of failure and solving for x_u

$$2160\, x_u + 13380 = 843.73 \times 10^3$$

$$x_u = 384.42\text{mm}$$

Effective depth d = 700 – 120 = 580mm

$$M_u = 0.36\frac{x_u}{1}(1 - 0.42\frac{x_u}{1}) f_{ck} b_w d^2 0.45 f_{ck}(b_f - b_w)D_f(d - \frac{D_f}{2})$$

$$= 0.36 \times \frac{384.42}{580}(1 - 0.42 \times \frac{384.42}{580})40 \times 150 \times (580)^2$$

$$+ 0.45 \times 40(400 - 150)120(580 - \frac{120}{2})$$

$= 628.33 \times 10^6$ *Nmm*

$= 628.33$ *kNm*

Strain at the bottom of the flange has to be checked to validate the assumption that it is more than 0.0002

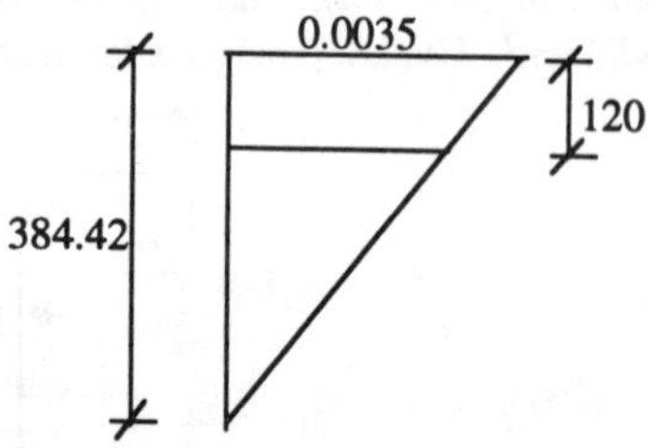

Strain at the bottom of the flange

$$= \frac{0.0035}{384.42}(384.42 - 120)$$

$$= 0.0024 \succ 0.002$$

Hence the above assumption is O.K.

The assumption that steel has reached to its failure capacity has to be checked

Initial compression of top $f_{ctp} = \dfrac{692.72 \times 10^3}{1000} - \dfrac{692.72 \times 10^3 \times 230}{19 \times 10^6}$

$= 4.81 - 8.385$

$= -3.57$ *N/mm2 (Tensile)*

Corresponding strain $= \dfrac{3.57}{3.6 \times 10^4} = 0.000099$ *(Tensile)*

Steel strain due to Prestress (P_∞) $= \dfrac{1000}{2 \times 10^5} = 0.005$

Strain in concrete due to external moment = 0.0035 + 0.000099

= 0.003599

Strain in steel due to applied moment for a depth of neutral axis.of 384.42mm

$$= \frac{0.003599}{384.42}(580 - 384.42)$$

$$= 0.001831$$

Total strain in steel $= 0.005 + 0.001831 = 0.00683$

Failure strain in steel $= \left[\dfrac{0.87 f_y}{E_s} + 0.001\right]$

$$= \frac{0.87 \times 1400}{2 \times 10^5} + 0.001$$

$$= 0.00709 > 0.00683$$

Hence the steel has not yielded and the calculated true moment M_u= 628.33kNm is not correct.

Assume the steel stress of $0.86f_p$

Strain for 0.87 f_p = 0.00709: Strain for 0.84 f_p = 0.00638

Strain Corresponding to a stress level of $0.86f_y = 0.0691$

Writing Equilibrium equation again and solving for x_u

$2160x_u + 13380 = 0.86 \times 1400 \times 692.72$

Neutral axis. depth x_u = 379.93mm

Strain in concrete due to external moment = 0.0035 + 0.00099

= 0.003599

Strain in steel due to applied moment for a depth of Neutral Axis of = 379.93mm

$$= \frac{0.003599}{379.93}(580 - 379.93)$$

Total strain in steel = 0.005 + 0.00190 = .0069

Assumed strain = 0.00691

Variance from assumed stress of 0.00691 is negligible small. Hence the force of $0.86f_p$ can be assumed to be O.K

True moment $M_u = 0.36 \times \frac{379.93}{580}(1 - 0.42 \times \frac{379.93}{580})40 \times 150 \times 580^2$

$$+0.45 \times 40(400 - 150)120(580 - \frac{120}{2})$$

$= .625.82 \times 10^6 Nmm$

$= .625.82 kNm$

Checking the strain at the bottom of a flange to validate the assumption.

Strain at the bottom of the flange

$$= \frac{0.0035}{379.93}(379.93 - 120)$$

$= 0.00239 \succ 0.002$

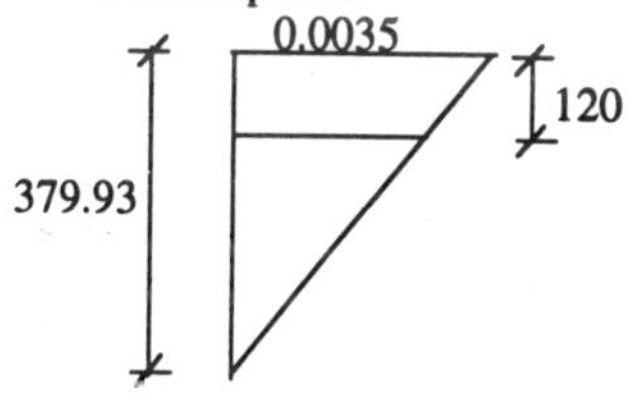

Hence O.K.

Moment resisting capacity of hc section = 625.82kNm

6.8 Code Provisions

While there are no differences in philosophy in evaluating the failure moment/load capacity of prestressed concrete section, different international and national codes provide different stress strain curves for concrete and steel, and correspondingly different failure strains for steel and concrete. So also the corresponding stress levels are variant from code to code. Equivalent area of compression blocks are also defined. Hence, the factors $k_1, k_2, k_3, k_4, k_5, k_6$ are different. By proper substitution it is possible to get the failure moment carrying capacity as per any defined code. The designated stress level in steel at the time of failure is different, based on the fact whether the section is over reinforced or under reinforced, whether compression steel is existent or not etc. The details in various code provisions are explained in Article 6.8.1, 6.8.2 and 6.8.3.

6.8.1 Indian standard code provisions [1343 - 1980]

Indian Standard Code IS 1343 - 1980 assumes a parabolic - rectangular stress strain curve for concrete with maximum concrete strain at failure to be ϵ_{cu} equal to 0.0035. The stress strain relationship is parabolic upto strain level of 0.0020 for a stress level of f_{ck} and then a strength line upto a strain of 0.0035 with increase in strain for no increase in stress and the values of k_1, k_2, k_3 in the stress block are defined as $k_1 = 0.8$, $k_2 = 0.42$, $k_3 = 0.447$, $k_4 = 0.85$, $k_5 = 0.8$ for the values as given in Fig 6.3.

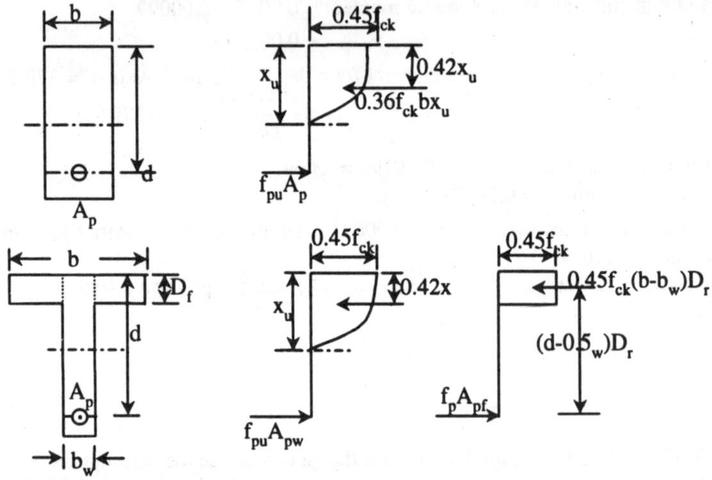

Fig 6.6. Moment of Resistance of Rectangular Sections & Flanged Section ($X_u > D_f$) (IS:1343-1980)

Hence the equilibrium ozf forces at a cross section at the time of failure can be written as $0.36 f_{ck} b_{xu} = A_{sp}\ 0.87 f_p$ for rectangular beam and flanged beams where the neutral axis falls within the flange and $0.36 f_{ck} b_{xu} + 0.447 f_{ck}(b_f - b_w)D_f$ where neutral axis falls outside the flange

Where x_u is depth of neutral axis

b = width of beam in rectangular beams

b = width of top flange in flanged beam

D_f = thickness of top flange

f_{ck} = cube strength of concrete

After evaluating proper x_u, the moment of resistance is calculated with proper values for f_{pu}

The Moment of resistance for rectangular beams and flanged beams where the neutral axis falls within the flange, is expressed as

$$M_u = A_{sp} \cdot f_{pu}(d - 042x_u)$$

where

M_u = ultimate moment of resistance of the beam section

f_{pu} = tensile stress developed in tendons at failure stage = f_p/γ_m

f_p = characteristic tensile strength of prestressing steel

f_{pe} = effective prestress in tendons after losses

A_p = Area of prestressing tendons

d = effective depth

x_u = depth of neutral axis

The value of f_{pu} is normalised with a factor $k \leq 1$ to take care of over reinforced beams where the steel may not reach to f_{pu} at the time of failure. f_{pu} is hence expressed as a function of $\left[\frac{A_p f_p}{bdf_{ck}}\right]$. Higher the percentage of steel (moving towards over reinforced sections) lower will be the k value as found in Table 6.2.

Table 6.2 Conditions at the Ultimate Limit State for Rectangular Beams with Pre-tensioned Tendons or with Post-tensioned Tendons having effective Bond

$\frac{A_p f_p}{bdf_{ck}}$	Stress in Tendon as a Proportion of the Design Strength		Ratio of the Depth of Neutral Axis to that of the Centroidt of the Tendon in the Tendon Zone Strength	
	Pretensioning	Post - tensioning with effective bond	Pretensioning	Post - tensioning with effective bond
(1)	(2)	(3)	(4)	(5)
0.025	1.0	1.0	0.054	0.054
0.05	1.0	1.0	0.109	0.109
0.10	1.0	1.0	0.217	0.217
0.15	1.0	1.0	0.326	0.316
0.20	1.0	0.95	0.435	0.414
0.25	1.0	0.90	0.542	0.488
0.30	1.0	0.85	0.655	5.88
0.40	0.9	0.75	0.783	0.653

In case of flanged beams where the neutral axis falls outside the flange, the moment of resistance is offered by the flange portion and the web portion. It is assumed that the flange portion has a uniform compressive stress of $0.447 f_{ck}$ throughout the depth (i.e. the strain is beyond 0.002). The total steel area is apportioned based on equivalent force to resist the compression in the flange and that in the web. Hence A_{sp} is apportioned as A_{sp1} and A_{sp2}. The compression generated in the flange portion of the beam at the time of failure namely $0.447 f_{ck}(b_f - b_w)D_f$ is resisted by equivalent tensile force $A_{sp1} \times 0.87 f_p$. The compression generated in the web portion of the beam at the time of failure, namely $0.36 f_{ck} b x_u$ is resisted by equivalent tensile force $A_{sp2} \times 0.87 f_p$. Hence A_{sp1} is directly evaluated as

$$\left[\frac{0.447 f_{ck}(b_f - b_w)D_f}{0.87 f_p}\right]$$

$$A_{sp2} = A_{sp} - A_{sp1}$$

For this A_{sp2}, the effective or factored value of f_{pu} is arrived at from Table 6.2 using $\left[\frac{A_{sp2} fp}{f_{ck} bd}\right]$. The moment of resistance is calculated as M_u $= f_{pu} \cdot A_{sp2}(d - 0.42x_u) + 0.447 f_{ck}(b_f - b_w) D_f(d - 0.5D_f)$

Table 6.3 Cconditions at the Ultimate Limit State for Post Tensioned Rectangular Beams having Unbonded Tendons

$\frac{A_p f_p}{bdf_{ck}}$	Stress in Tendon as the a Proportion of Effective Prestress f^{pu}/f_p For Values of of $\frac{l}{d}$ $\left[\frac{Effective\ Span}{Effective\ Depth}\right]$			Ratio of the Depth of Neutral Axis to that of the Centroid of the Tendon in the Tendon Zone x_u/d For Values of $\frac{l}{d}$ $\left[\frac{Effective\ Span}{Effective\ Depth}\right]$		
	30	20	10	30	20	10
(1)	(2)	(3)	(4)	(5)	(6)	(7)
0.025	1.23	1.34	1.45	0.10	0.10	0.10
0.05	1.21	1.32	1.45	0.16	0.16	0.18
0.10	1.18	1.26	1.45	0.30	0.32	0.36
0.15	1.14	1.20	1.36	0.44	0.46	0.52
0.20	1.11	1.16	1.27	0.56	0.58	0.64

It may be noted that IS 1343 1980 does not provide for the facility of any untensioned steel, if available in tension zone, resisting additional moment before failure. But the untensioned reinforcement, if any, will definitely add to the moment resisting capacity. That is ignored. Hence the resisting moment value got in such cases is much on conservative side. Similarly, any reinforcement in the compression zone is also not taken cognizance. There will be an increase of capacity in compression block. That will give increased moment resisting capacity, which is also ignored.

6.8.2. British Standard Code Provision [BS 8110 - 1985]

In British Standard Code (BS 8110 - 1985) the following assumptions are made while evaluating the ultimate moment of resistance of a section. Ofcourse most of the assumptions are the same or on similar lines as given in Indian Standard Code provisions.

(a) The strain distribution in the concrete in compression is derived from the assumption that plance sections remain plane.

(b) The design stresses in the concrete in compression are derived either from the stress-strain curve given in Figure 6.8 with $\gamma_m = 1.5$ or taken as $0.45 f_{cu}$ for a depth (from the compression face) equal to 0.9 times the depth of the compression zone.
Note. In both cases the strain at the outermost compression fibre is taken as 0.00035 Ref Fig 6.7.

(c) The tensile strength of concrete is ignored.

(d) The strains are bonded prestressing tendons and in any additional reinforcement, whether in tension or compression, are derived from the assumption that plane sections remain plane and the design stresses are derived from the appropriate stress-strain curves. The stress-strain curves for prestressing tendons are given in Figure 6.6 and those for reinforcement in Fig 6.8 (An alternative approach for obtaining the stress in the tendons is also given in Table 6.4).

(e) The design stress is unbonded prestressing tendons f_{pb} is limited to the values got by expression detailed below unless a higher value can be justified by a more rigorous analysis or on the basis of tests.

(f) The stress block at the time of failure is shown in Fig 6.9.

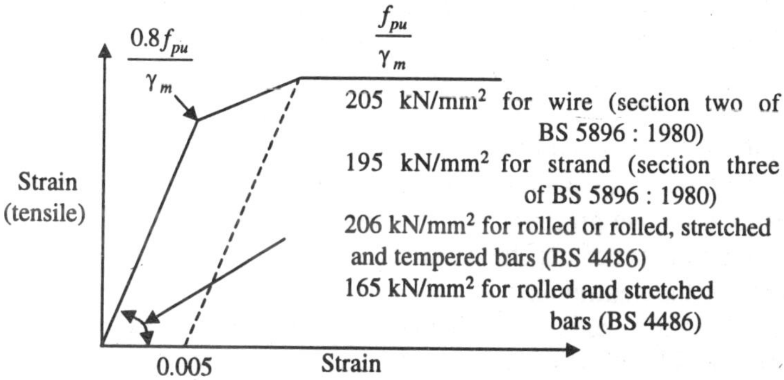

Note. f_{pu} is in N/mm²

Fig. 6.7. Short Term Design Stress-Strain Curve for Prestressing Tendons (BS 8110:1985)

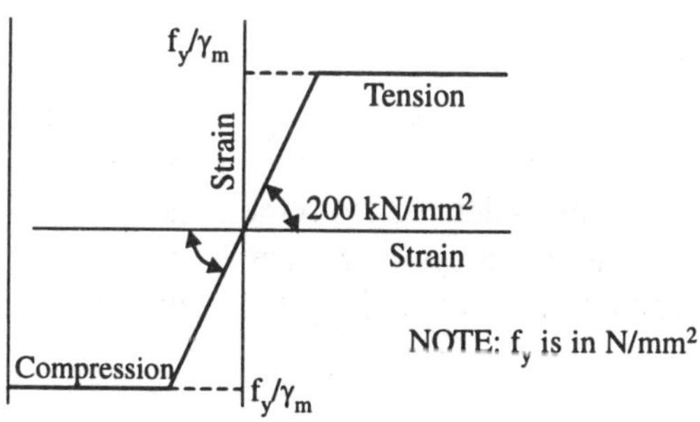

Fig. 6.8. Short Term Design Stress-Strain Curve for Reinforcement

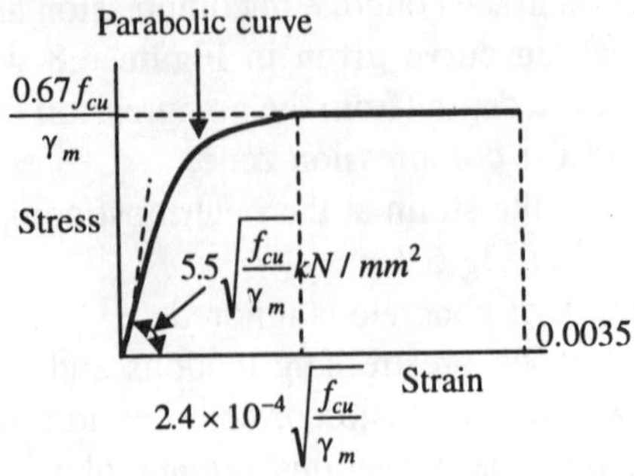

Note 0.67 takes account of the relation between the cube strength and the bending strength in a flexural member. It is simply a coefficient and not a partial safety factor.
Note 2. f_{cu} is in N/mm²

Fig. 6.9. Short Term Design Stress-Strain Curve for Normal-Weight Concrete (BS 8110:1985)

The notations adopted are

A_{ps} area of the prestressing tendons in the tension zone
A_s area of reinforcement
b width of effective width of the section or flange in the compression zone
d effective depth to the centroid of the steel area A_{ps}
b_ω width of web in case of flanged sections
D_f thickness of flange in case of flanged sections
d_n depth of the centroid of the compression zone
f_{pb} design tensile stress in tendons at the time of failure
f_{pe} design effective prestress in the tendons after all losses
M_u design moment of resistance of the section
x depth of the neutral axis

First depth of neutral axis is evaluated assuming that steel reaches its limiting value and then actual f_{pb} is evaluated as per code provisions. The moment of resistance is then calculated on the basis that the section is under reinforced so that the steel reaches its limiting value.

The resistance moment of a beam containing bonded or unbonded tendons, all of which are located in the tension zone, may be obtained from the following equation:

$M_u = f_{pb} A_{ps} (d - d_n)$. For a rectangular beam or a flanged beam in which the flange thickness is not less than $0.9x$, d_n, may be taken as $0.45x$.

For bonded tendons, values of f_{pb} and x may be obtained from Table 6.4. These values have been derived from the assumptions given earlier.

For unbonded tendons, values of f_{pb} and x may be obtained from the expressions given below. The value of f_{pb} should not be taken as greater than $0.7 f_{pu}$.

$$f_{pb} = f_{pe} + \frac{7000}{l/d}\left(1 - 1.7\frac{f_{pu}A_{ps}}{f_{cu}bd}\right)$$

$$x = 2.47\left[\left(\frac{f_{pu}A_{ps}}{f_{cu}bd}\right)\left(\frac{f_{pb}}{f_{pu}}\right)d\right]$$

Expression for f_{pb} has been derived by taking the length of the zone of inelasticity within the concrete as 10x. The length l should normally be taken as the length of the tendons between end anchorages. This length may be reduced in the case of continuous multi-span members when an analysis is carried out to determine the minimum number of zones of inelasticity associated with each arrangement of design load

For flanged sections where the neutral axis falls outside the flange the ultimate moment resisting capacity is got by adding the flexural capacity of web and of flange portions having stress blocks shown in Fig 6.9. The ultimate moment resisting capacity is expressed as

$$M_u = f_{pb} A_{sp2}(d - 0.45x) + 0.450 f_{ck}(b - b_w)(d - 0.5D_f)$$

where $A_{sp2} = A_{sp} - A_{sp1}$

$$A_{sp1} = 0.450 f_{ck}(b - b_w)\left(\frac{D_f}{f_{pu}}\right)$$

Table 6.4 Conditions at the ultimate limit state for rectangular beams with pre-tensioned tendons or post-tensioned tendons having effective bond

$\frac{f_{pu}A_{ps}}{f_{cu}bd}$	Desing stress in tendons as a proportion of the design strength, $f_{pb} / 0.87 f_{pu}$			Ratio of depth of neutral axis to that of the centroid of the tendons in the tension zone, x/d		
	f_{pe} / f_{pu} =			f_{pe} / f_{pu} =		
	0.6	**0.5**	**0.4**	**0.6**	**0.5**	**0.4**
0.05	1.0	1.0	1.0	0.11	0.11	0.11
0.10	1.0	1.0	1.0	0.22	0.22	0.22
0.15	0.99	0.97	0.95	0.32	0.32	0.31
0.20	0.92	0.90	0.88	0.40	0.39	0.38
0.25	0.88	0.86	0.84	0.48	0.47	0.46
0.30	0.85	0.83	0.80	0.55	0.54	0.52
0.35	0.83	0.80	0.76	0.63	0.60	0.58
0.40	0.81	0.77	0.72	0.70	0.67	0.62
0.45	0.79	0.74	0.68	0.77	0.72	0.66
0.50	0.77	0.71	0.64	0.83	0.77	0.69

Allowance for additional reinforcement in the tension zone. In the absence of a rigorous analysis, the area of reinforcement A_s may be replaced by an equivalent area of prestressing tendons $A_s f_y/f_{pu}$. It may be noted that the effect of reinforcement in compression zone is not included in formulation of expressions for evaluating the moment carrying capacity of the section.

6.8.3 American Code Provisions [ACI 318 - 1998]

While adopting ACI 318-1998 specifications for arriving at the flexural strength of Prestressed Concrete sections ***it may be noted that the units***

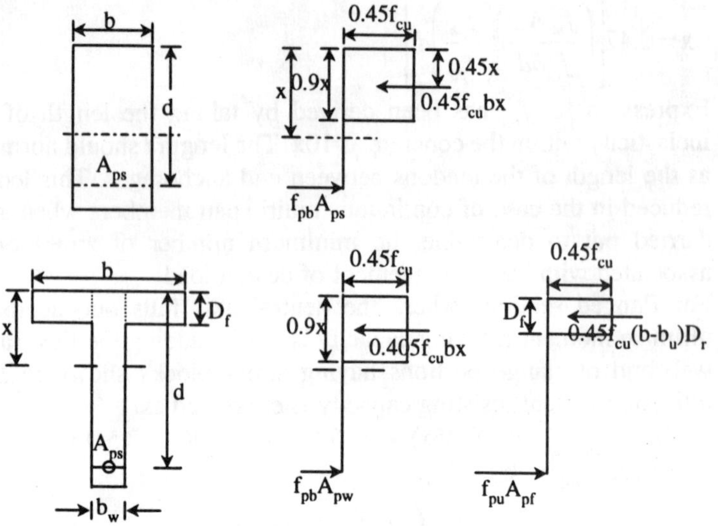

Fig. 6.10. Moment of Resistance of Rectangular Sections & Flanged Sections ($X>D_f$) (BS:8110 - 1985)

adopted as FPS Units and the following notations may be understood.

A_{ps} = area of prestressing steel in tension zone
A_s = area of untensioned steel in tension zone
A'_s = area of untensioned steel in compression zone
b = width of compression zone
d = effective depth of untensioned steel
d_p = effective depth of prestressed steel
f'_c = Specified compressive strength (cylinder)
f'_{ci} = Compressive strength of concrete at the time of transfer
f_{ps} = Stress in prestressing steel at limit state of failure corresponding to f_{pu} in IS Code provisions
f_{pu} = Specified tensile strength of prestressing steel corresponding to f_p in IS Code provision
f_{py} = Specified yield strength of prestressing steel (0.002 strain level)
f_{se} = effective stress in prestressing steel at P_∞
f_y = Specified yield strength of untensioned steel
h = overall thickness of the member

$$\rho = \frac{A_s}{bd}$$

$$\rho' = \frac{A'_s}{bd}$$

$$\rho_p = \frac{A_{ps}}{bd}$$

$$\omega = \frac{\rho f_u}{f_c'}$$

$$\omega' = \frac{\rho' f_u}{f_c'}$$

$$\omega_p = \frac{f_p f_{ps}}{f_c'}$$

ω, ω', ω_p are reinforcement indices and they may be computed only with web width in case of flanged sections.

$$\beta_1 = \left\{0.85 - \frac{(f_c' - 4000).05}{1000}\right\}$$

but not less than 0.65

γ_p = a factor for type of prestressing tendon

$= 0.55$ for f_{py}/f_{pu} not less than 0.8

$= 0.40$ for f_{py}/f_{ou} not less than 0.85

$= 0.28$ for f_{py}/f_{ou} not less than 0.90

Unlike IS Code provisions in which the limiting tensile stress at the time of failure (f_{pu}) is considered as $\frac{f_p}{\gamma_m}$ in case of under reinforced section, and $f_p/\gamma_m \times$(*factor*) for over reinforced section, in ACI Code provisions the limiting tensile stress at time of failure, designated as f_{ps} is expressed as a function of concrete strength, area of untensioned steel, if any, area of steel in compression zone, if any, and whether the section is over reinforced or under reinforced. The evaluation of f_{ps} is described separately for sections with bonded and unbonded tendons.

(a) For members with bonded prestressing tendons

$$f_{ps} = f_{pu}\left\{1 - \frac{\gamma_p}{\beta_1}\left[\rho_p \frac{f_{pu}}{f_c'} + \frac{d}{d_p}(\omega - \omega')\right]\right\}$$

If any compression reinforcement is taken into account when calculating f_{ps} as above, then the term shall be taken not less than 0.17 and d' shall be no greater than 0.15 d_p.

(b) For members with unbonded prestressing tendons and with a span-to-depth ratio of 35 or less:

$$f_{ps} = f_{se} + 10{,}000 + \frac{f_c'}{100\rho_p}$$

but f_{ps} calculated as above shall neither be taken greater than f_{py}, nor (f_{se} + 30,000)

(c) For members with unbonded prestressing tendons and with a span-to-depth ratio of greater than 35:

$$f_{ps} = f_{se} + 10{,}000 + \frac{f_c'}{300\rho_p}$$

but f_{ps} calculated as above shall neither be taken greater than f_{py}, nor (f_{se} + 30,000)

Reinforcement indices are limited to avoid the over reinforced type of failure.

i.e ω_p $\left[\omega_p + \frac{d}{d_p}(\omega - \omega')\right]$ or $\left[\omega_{pw} + \frac{d}{d_p}(\omega_w - \omega_w')\right]$ should not exceed 0.36 β_1.

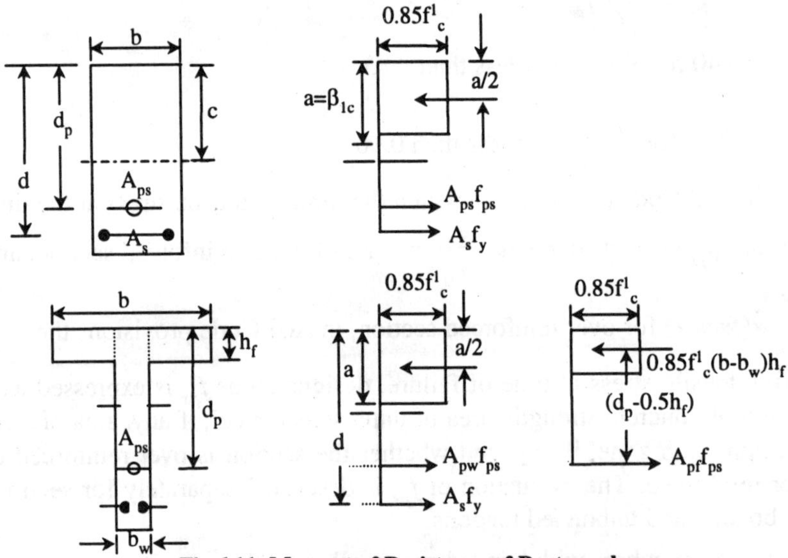

Fig.6.11. Moment of Resistance of Rectangular Sections & Flanged Sections (ACI:318 - 1998)

For evaluating the moment of resistance a rectangular stress block is assumed as shown in Fig 6.10 as per Whitney, C.S. [9], Jones [10] and Mattock et.al[11] and the depth of neutral axis is evaluated and depending on the fact whether the beam sis rectangular and flanged beam with depth of compression block within the flange or it is a flanged beam where the depth of compression block falls outside the flange depth. Accordingly f_{ps} is evaluated and the moment of resistance arrived at using proper formula.

(a) For rectangular beams, and for Flanged beams where depth of compression block 'a' is within the flange depth

$$M_u = \left[A_{ps}f_{ps}(d_p - \frac{a}{2}) + A_s f_y\left(d - \frac{a}{2}\right)\right]$$

$$a = \beta_1 \text{ c}$$

$$a = \left[\frac{A_{ps}f_{ps} + A_s f_y}{0.85 f_c^{'}.b}\right]$$

(b) For flanged beam where the depth of compression block is more than the depth of the compression flange

$$M_u = \left[A_{p2}f_{ps}\left(d_p - \frac{a}{2}\right) + A_s f_y (d - d_p) + 0.85 f_c^{'}(b - b_w)h_f(d_p - 0.5h_f)\right]$$

where $A_{sp2} = \left[A_{ps} - A_s\left(\frac{f_y}{f_{ps}}\right) - \frac{0.85 f_c^{'}(b_f - b_w)h_f}{f_{ps}}\right]$

$$a = \frac{A_{ps2} \cdot f_{ps}}{0.85 f_c^{'}.b}$$

In case of unbonded tendon where the span to depth ratio is less than 35 the stress in tendon at failure is computed using an expression

$$f_{ps} = f_{se} + 10{,}000 + \frac{f_c^{'}}{100\rho_p}$$

f_{ps} is limited to f_{py} or f_{se} +60000

If the span to depth ratio is more than 35

$$f_{ps} = f_{se} + 10{,}000 + \frac{f_c^{'}}{300\rho_p}$$

and f_{ps} is limited to f_{py} or f_{se} + 30000

In case of over reinforced beams, indicated by the reinforcement index greater than 0.36 β_1, if the depth of compression block is within the flange, or if it is only a rectangular beam

$M_u = f'_c\, bd^2\, p(0.36\beta_1 - 0.08\beta_1{}^2)$

and the depth of compression block is larger than the depth of flange

$M_u = [f'_c\, b_w\, d^2\, p(0.36\beta_1 - 0.08\beta_1{}^2) + 0.85\, f'_c\, (b_f - b_w)h_f\, (d_p - 0.5f_y)]$

The designated moment resistance capacity of the section is declared as φM_u when φ is the capacity reduction factor.

Strength reduction factor φ shall be defined as follows:

Tension controlled sections 0.90

Compression-controlled sections:

(a) Members with spiral reinforcement 0.75

(b) Other reinforced members 0.70

For sections in which the net tensile strain in the extreme tension steel at nominal strength is between the limits for compression-controlled and tension-controlled sections, φ shall be linearly increased from that for compression-controlled sections to 0.90 as the net tensile strain in the extreme tension steel at nominal strength increases from the compression controlled strain limit to 0.005. These capacity reduction factor takes care of material safety factor for concrete and time dependant effects on concrete.

It may be noted that the evaluation of moment of resistance looks complicated but it is more comprehensive. But the philosophy remains the same in all the codes.

ACI 318 - 1998 permits a maximum compressive strain of 0.003 only in concrete at the time of failure with the maximum stress level in concrete as $0.85\ f'_c$ (f'_c is the specified cylinder strength in *psi*) to a depth equal to β_i times the depth of neutral axis (x_u) from the fibre of maximum compressive strain. For evaluating the depth of neutral axis for an under reinforced or balanced section, the failure strain in steel may be taken as the strain at the specified yield strength of material. The additional strain in steel after reaching stage of decompression at steel level is limited to 0.005. The procedure for evaluating the capacity of the section is illustrated with examples.

The procedure given in IS 1343–1980 and BS 8110–1985 is very similar and the one given in ACI 318–1998 is different and comprehensive and hence more involved for calculations.

Example 6.4

A pretensioned prestressed concrete beam has a rectangular section of 200 × 350 mm. The High tensile steel has its centre of area at 280*mm* from top of beam 8*mm* H.T. wires are proposed to be used. It has a f_p value of 1500 N/mm^2. Determine the following details based on IS 1343 – 1980 Code provision.

a) the minimum area of steel to avoid failure by fracture of steel

b) the maximum area of steel which would ensure a ductile failure (under reinforced beam)

c) flexural strength corresponding to the case (b) above

Solution

Section details

b = 200*mm*

D = 350*mm*

d = 280*mm*

f_{sp} = 1500N/mm^2

f_{spu} = 1305 N/mm^2

a) Minimum area of steel required to avoid brittle failure by fracture of steel

= 0.15% of 200 × 200 as per clause 18.6 3.3

= 84 mm^2

But this clause precludes pretensioned products

Referring to Table 6.2 the minimum $\dfrac{A_{sp}}{bdf_{ck}}$ for a pretensined prestressed beam is 0.025

Using this value, A_{sp} can be evaluated as

$$A_{sp} = \frac{0.025 \times 200 \times 280 \times 40}{1500} = 37.33mm^2$$

Hence a minimum steel of $37.33mm^2$ is to be provided

b) If the section is under reinforced at the time of failure, f_{pu} is to be taken on $0.87 f_p$

i.e. $\frac{f_{pu}}{0.87 f_p} = 1.0$

This is the case till the value of $\frac{A_{sp} \cdot f_{sp}}{bdf_{ck}} = 0.15$ for pretensioned products (From Table 6.2)

$$\frac{A_{sp} \cdot f_{sp}}{bdf_{ck}} = 0.15$$

$$A_p = \frac{0.15 \times 200 \times 280 \times 40}{1500} = 224mm^2$$

Hence the maximum area of steel which would ensure a ductile failure (under reinforced beam failure) is $224mm^2$.

c) Flexural strength of cross section with

$A_{sp} = 224mm^2$

$M_u = 1305 \times 224(280 - 0.42x_u)$

Equating the total compression to total tension at the time of failure

$0.36 \times 40 \times 200 \times x_u = 224.0 \times 1305$

$$x_u = \frac{224.0 \times 1305}{0.36 \times 40 \times 200}$$

$x_u = 101.5mm$

$M_u = 1305 \times 224(280 - 0.42 - 101.5)$

$= 1305 \times 224 \times 237.37$

$= 69387998.4Nmm$

$= 69.388\ kNm$

Example 6.5

An unsymmetrical I section has a top flange width of 1000 and thickness of 150 mm, a web thickness of 200 mm, and the effective depth of cross section of 900mm. It is prestressed with 48 numbers of High tensile steel wires of 8*mm* diameter, which has ultimate tensile strength of $1500N/mm^2$. The effective centre of steel is at 100*mm* from the soffit of the beam. The steel is prestressed to a level of $900N/mm^2$ after losscs.

Solution

Section details

$b_f = 1000mm$

$d_f = 150mm$

$d = 900mm$

$b_w = 200mm$

$f_{spu} = 1305\ N/mm^2$

Concrete Strength $f_{ck} = 40N/mm^2$

Steel strength $f_p = 1500N/mm^2$

$f_{pu} = 1305N/mm^2$

$f_{p\infty} = 900N/mm^2$

A_{sp} $48 \times \frac{\pi}{4} \times 64 = 2411.52mm^2$

IS 1343 1980 limits the failure to be initiated by steel and accordingly the limits the steel stress at failure which varies from f_{pu} to 0.75 f_{pu}. Assuming the neutral axis to fall within the flange and writing the equilibrium equation at failure

$$0.36 \times 40 \times 1000 \times x_u = 2411.52 \times 1305$$

$$x_u = \frac{2411.52 \times 1305}{0.36 \times 40 \times 1000} = 218.54$$

Hence the neutral axis falls outside the flange

Rewriting the equilibrium equations

$$0.36 \times 40 \times 100 \times x_u + 0.447 \times 40 \times 800 \times 150$$

$$= 2411.52 \times 1305$$

$$0.36 \times 40 \times 200 \times x_u = 3147033.6 - 2145600$$

$$x_u = 347.72$$

The steel of $2009.6 mm^2$ is to be apportioned for resisting compressive force mobilized in the flange and that in the web portion as A_{sp1} and A_{sp2}.

$$A_{SP} = \frac{0.447 \times 40 \times 150 \times 800}{1305.0}$$

$$= 1644.14\ mm^2$$

$$A_{sp2} = 2411.52 - 1644.14$$

$$= 767.38 mm^2$$

$$\frac{A_{sp2}}{f_{ck} b_w d} = \frac{767.38 \times 1500}{40 \times 200 \times 900} = 0.1598$$

For this value of 0.1598 (reinforcement ratio)

$$f_{pu} = 0.87 f_p \times 0.99$$

Hence with the factored value of f_{pu}, the moment carrying capacity of the section is evaluated.

$$M_u = 0.87 f_p \times 0.99 \times 767.38(900 - 0.42 \times 347.72)$$

$$+ 0.447 \times 40 \times 800 \times 150(900 - \left(900 - \frac{150}{2}\right)$$

$$= 1291.95 \times 767.38\ (753.96)$$

$$+ 0.447 \times 40 \times 800 \times 150 \times 825$$

$$= 7.475 \times 10^8 + 17.7 \times 10^8$$

$$= 25.176 \times 10^8\ N/mm$$

$$= 2517.6 kNm$$

Example 6.6

A post tensioned unbonded beam of rectangular section has 500*mm* width and 800*mm* effective depth. It is prestressed with 64 Nos of 8*mm* dia High Tensile steel in 4 cables. The cables are unbonded. The steel is stressed to a level of 800 N/mm^2 after all losses. The ultimate tensile strength of steel is 1400 N/mm^2. Evaluate the moment of resistance of the beam as per IS Specifications. Assume a concrete strength of 40N/mm2. This is proposed to be used for a span of 16 *m*..

Solution

Section details

$b = 500 mm$

$d = 800 mm$

$f_{ck} = 40 N/mm^2$

$f_{pe} = 800 N/mm^2$

$f_{pu} = 1400 \times 0.87\ N/mm^2$

$$l/d = \frac{16}{0.8} = 20$$

$$A_p = 50.24 \times 64$$
$$= 3215.36\ mm^2$$

Reinforcement ratio = $\dfrac{f_{pe} A_p}{b.df_{ck}}$

$$\frac{800 \times 3215.36}{500 \times 800 \times 40} = 0.1608$$

From Table 6.3

$$\frac{f_{pu}}{f_{pe}} = 1.19$$

$$\frac{x}{d} = 0.484$$

$$f_{pu} = 1.19 \times 800 = 952\ N/mm^2$$
$$x_u = 0.484 \times 800 = 387.2$$
$$M_u = 3215.36 \times 952(800 - 0.42 \times 387.2)$$
$$= 3215.36 \times 952 \times 637.376$$
$$= 1951022417\ Nmm$$
$$= 1951\ kNm$$

Example 6.7

A prestressed concrete beam has an unsymmetrical I section with a top flange width of 1200 mm and a thickness of 180 *mm*. The web thickness is 200 mm. It has a total depth of 1400 mm. The centre of prestressing steel is at 1200 *mm*. Area of High tensile steel provided is 3200mm^2. The ultimate tensile strength of steel is 1500 N/mm^2. The steel is prestressed to 900 N/mm^2 after losses. Evaluate the moment carrying capacity of the section as per BS 8110 – 1985. Assume the tendons as bonded.

Solution

The notations given are as per BS 8110 - 1985 (Ref Article 6.8.2)

Section details

$b_f = 1200mm$

$D_f = 1800\ mm$

$b_w = 200\ mm$

$d - 1200\ mm$

$f_{pu} = 1500\ N/mm^2$

Assume the neutral axis to fall within the flange

Writing the equations of equilibrium at the time of failure

$0.36 \times 40 \times 1200 \times x_u = 3200 \times 1500 \times 0.87$

$x_u = 241.664\ mm \succ 180mm$

Hence the assumption is wrong. The neutral axis falls within the web and outside the flange

Apportion the steel area to resist compression in flange portion and web portion as Asp_1 and Asp_2

$$A_{sp1} = 0.45 \frac{f_{ck}(b - bw)D_f}{f_{pb}}$$

$$= \frac{0.45 \times 40(1000)180}{1305} = 2482.758mm^2$$

A_{sp2} = 3200 – 2482.758
= 717.242 mm^2

$$\frac{A_{sp2}.f_{pu}}{f_{cu}bw.d} = 0.112$$

From Table 6.4 the design stress in tendon for $\frac{f_{pe}}{f_{pu}}$ of 0.6 for a value of 0.112, the design failure stress. $f_{pb} = 1.0 \times 0.87 f_{pu}$

$$\frac{x}{d} = 0.22$$

$x = 0.22 \times 1200 = 264\ mm$

$M_u = fp_b \times A_{ps2}\ (d - 0.45x) + 0.45 \times 40\ (1200 - 200) \times 180 \times (1200 - 0.5 \times 180)$
$= 1305 \times 717.242\ (1200 - 0.45 \times 264) + 0.45 \times 40 \times 1000 \times 180 \times 1110$
$= 936000.81 \times 1081.2 + 35.96 \times 10^8$
$= 10.12 \times 10^8 + 35.96 \times 10^8$
$= 46.08 \times 10^8\ Nmm$
$= 4608\ kNm$

Example 6.8

A post tensioned T beam has a flange width of 1000*mm*, flange thickness of 250*mm*, web thickness of 300*mm*, and an effective depth of 1000*mm* with an effective cover of 200*mm*, for prestressing steel. It is prestressed with 7 ply strands having a total area of 3000 mm^2. The ultimate strength of the steel can be assumed as 1750 N/mm^2. The strands are prestressed to a level of 1050 N/mm^2 after losses. In addition 6 bars of 22*mm* dia of Fe 415 steel are provided at 100*mm* from the soffit to take care of surface cracking, stirrup binding etc. Estimate the moment carrying capacity of the section as per BS 8110 - 1985. Assume the tendons to be bonded and concrete strength of 40 N/mm^2.

Solution

The notations used are as per BS 8110 - 1985

Section details	
	A_{ps} = 3000mm^2
b = 1200mm	f_y = 415N/mm^2
D_f = 220mm	f_{pc} = 1000N/mm^2
b_w = 300mm	f_{pb} = to be evaluated
d_p = 1000mm	A_s = 2280 mm^2
	f_{ck} = 40 N/mm^2

Untensioned steel area is replaced by an equivalent area of prestressed steel which is given by

$$\frac{A_s.f_y}{f_{pu}} = \left(\frac{2280 \times 415}{1750}\right) = 541mm^2$$

Area of prestressing steel (equivalent) = 3541 mm^2

Assuming that neutral axis falls within the flange and equating the total compression and total tension at the time of failure

$3541 \times f_{pb} = b.f_{ck} \times 0.45 \times 0.9x$

Assuming $f_{pb} = 0.87 f_{pu}$

$$x = \frac{3541 \times 0.87 \times 1750}{1200 \times 40 \times 0.45 \times 0.9} = 277.32$$

Neutral axis falls outside the flange.

Hence the equilibrium equation is rewritten again assuming $f_{pb} = 0.87 f_{pu}$

$3541 \times f_{pb} = 0.45 f_{ck} (b - b_{\omega})D_f + 0.45 f_{ck} \times 300 \times 0.9x$

$3541 \times 0.87 \times 1750 - 0.45 \times 40(1200 - 300)220$

$= 0.45 \times 40 \times 300 \times 0.9x$

$x = (5391172.5 - 3564000)/ 4860$

$= 375.96$

Neutral axis falls outside the flange

Apportioning the steel area to resist the compressive force from flange and web portion

A_{spl} is the steel area equivalent to resist the compressive force from flange

$$A_{SP1} = \frac{0.45 \times 40 \times 900 \times 220}{0.87 \times 1750}$$

$= 2340.88\ mm^2$

A_{sp2} is the steel area equivalent to resist the compressive force from Web portion

$A_{sp2} = 3541 - 2340.88$

$= 1200.12\ mm^3$

$$\frac{A_{sp2} \cdot f_{pu}}{f_{ck} \cdot b_{\omega} \cdot d} = \frac{1200.12 \times 1750}{40 \times 300 \times 1000} = 0.1750$$

$$\frac{f_{pe}}{f_{pu}} = \frac{1050}{1750} = 0.6$$

$$f_{pb} = \frac{0.99 + 0.92}{2}(f_{pu} 0.87)$$

$= 0.955 \times 0.85 f_{pu}$

$= 0.831 f_{pu}$

The position of neutral axis has been arrived with assumed value of f_{pb}=0.87f_{pu}, where as f_{pb} 0.831 f_{pu} now.

It has to be iteratively corrected in case of flanged beams

Assuming $f_{pb} = 0.845 f_{pu}$ and evaluating x_u from equilibrium equation

$3541 \times 0.845 \times 1750 = 0.45 \times 40 \times 900 \times 2200 + 0.45 \times 40 \times 0.9 \times 300x_u$

$x_u = (5236253.75 - 3564000) / 4860$

$x_u = 344.08\ mm$

$$A_{SP1} = \frac{0.45 \times 40 \times 900 \times 220}{0.845 \times 1750}$$

$= 2410\ mm^2$

$A_{sp2} = 3541 - 2410 = 1131\ mm^2$

$$\frac{A_{sp2} \cdot f_{pu}}{f_{ck} \cdot b_{\omega} \cdot d} = \frac{1131 \times 1750}{40 \times 300 \times 1000} = 0.165$$

For this value $f_{pb} = 0.97 \times 0.87 f_{pu}$

$= 0.844 f_{pu}$

Our assumption of f_{pb} is OK.

Moment of Resistance of the Section

$= f_{pb} \cdot A_{sp2}(d\text{-}0.45x) + 0,45 f_{xk}(b - b_{\omega})(d - 0.5D_f)D_f$

$= 0.845$

$= 1672466.25 \times 845.164$

$= 1413508266 + 3171960000$

$= 4585468266\ kNm$

$= 4585.47\ kNm$

Example 6.9

Evaluate the moment resisting capacity of the post-tensioned beam section given in Example 6.8 as per BS 8110-1985 with following modifications. This may be considered as unbonded and the beam be considered without the untensioned reinforcement. The flange thickness is only 200mm.

Solution

The notations adopted are as per BS 8110 - 1985.

Section details

$b = 1200$ mm	$f_{pc} = 1000$N/mm^2
$D_f = 200$ mm	$f_{pb} = ?$
$b_w = 300$ mm	$f_{pu} = 1750$N/mm^2
$d_p = 1000$ mm	$f_{ck} = 40$N/mm^2
$A_{ps} = 3000$mm^2	

Assume that the neutral axis falls within the flange and equating the total compression and total tension at the time of failure

$3000 \times f_{pb} = b.f_{ck} \times 0.45 \times 0.9x$

Assuming $f_{pb} = 0.87\, f_{pu}$

$$x = \frac{3000 \times 0.87 \times 1750}{1200 \times 40 \times 0.45 \times 0.9} = 234.95mm$$

Neutral axis falls outside the flange and hence the equilibrium equation is rewritten.

Again assuming $f_{pb} = 0.87 f_{pu}$

$3000.f_{pu} = 0.45\, f_{ck}\,(b - b_{\omega})\, D_f + 0.45\, f_{ck} \times 300 \times 0.9x$

$3000 \times 0.87 \times 1750 = 0.45 \times 40 \times 900 \times 200 + 0.45 \times 40 \times 300 \times 0.9x$

$x = (4567500 - 3240000)/4860$

$= 273.15$mm

Neutral axis falls outside the flange

The reinforcement (H.T. Steel) should be apportioned to resist the compression from flange and the Web portion A_{spl} is the steel area equivalent to resist the compressive force from flange

$$A_{sp1} = \frac{0.45 \times 40 \times 900 \times 200}{0.87 \times 1750} = 2128.08mm^2$$

$A_{sp2} = 3000 - 2128.08 = 871.92mm^2$

$$f_{pb} = f_{pe} + \frac{7000}{bd}\left[1 - 1.17\frac{f_{pu}.A_{ps}}{f_{cu}.bd}\right]$$

$$= 1000 + \frac{7000}{35}\left[1 - 1.17\frac{1750 \times 871.92}{40 \times 300 \times 1000}\right]$$

$= 1170.24N/mm^2 \nless 0.7 \times 1750$

$$x = 2.47\left[\left(\frac{f_{pu}.A_{ps}}{f_{cu}.bd}\right)\left(\frac{f_{pb}}{f_{pu}}\right)d\right]$$

$$= 2.47\left[\frac{1750 \times 871.92}{40 \times 300 \times 1000}\left(\frac{1170.24}{1750.0}\right).1000\right]$$

$= 2.47\,[0.127155 \times 0.6688 \times 1000]$

$= 210.05mm$

$$M_u = f_{pb} \times A_{ps2}(d - 0.45x) + 0.45 f_{ck}(b - b_\omega)(d - 0.5D_f)D_f$$
$$= 1170.24 \times 871.92(1000 - 94.52) + 0.45 \times 40(900)(1000 - 100) \times 200$$
$$= 923911643.7 + 2916000000$$
$$= 383991164.4$$
$$= 3839.1164 kNm$$

Example 6.10

Evaluate the Moment carrying capacity of the section described in Example 6.5 as per ACI Code provisions.

Solution

The notations are as per ACI 318 1998 (Ref Article 6.8.3)

Section details

$b_f = 1000mm = 40in$

$d_f = 150mm = 6in$

$d = 900mm = 36in$

$b_w = 200mm = 8in$

Concrete strength $= f_{ck} = 40N/mm^2$

Corresponding $f'_c = 4765.55\ psi$

$f_{pu} = 1500N/mm^2$

$= 210240\ psi$

f_{ps} = to be evaluated

$f_{py} = 0.87 f_{pu}$

$A_{ps} = 2411.52mm^2 = 3.8584mm^2$

Untensioned steel area $A_s = 0$

Area of steel in compression zone $A'_s = 0$

Evaluation of M_u depends on the factor whether the neutral axis depth 'a' falls within the flange depth or outside

Hence equating the tensile and compressive forces at the time of failure

$$3.8584 \times f_{ps} = 0.85 \times 40 \times 4765.55 \times a$$

$$a = \frac{3.8584 f_{ps}}{0.87 \times 40 \times 4765.55}$$

Assuming $f_{ps} = 0.87 f_{pu}$ to start with

$$a = \frac{3.8584 \times 210240 \times 0.87}{0.87 \times 40 \times 4765.55}$$
$$= 4.355'' \prec 6''$$

Hence the neutral axis falls within the flange

Then

$$M_u = \left[A_{ps} \cdot f_{ps}\left(d_p - a/2\right) + A_s f_y\left(d - a/2\right)\right]$$

$$f_{ps} = f_{pu}\left\{1 - \frac{\gamma_p}{\beta_1}\left[\rho_p \frac{f_{pu}}{f'_c} + \frac{d}{d_p}(\omega - \omega')\right]\right\}$$

$$\beta_1 = \left[0.85 - \left(\frac{f_c' - 4000}{1000}\right)0.05\right]$$

but not less than 0.65

$\beta_1 = 0.812$; $0.36\,\beta_1 = 0.2923$

$$\gamma_p = 0.55 \text{ for } \frac{f_{py}}{f_{pu}} \not< 0.8$$

$$= 0.40 \text{ for } \frac{f_{py}}{f_{pu}} \not< 0.85$$

$$= 0.28 \text{ for } \frac{f_{py}}{f_{pu}} \not< 0.9$$

Here f_{py} is assumed as $0.87 f_{pu}$. Hence

$\gamma_p = 0.40$

$$\rho_p = \frac{A_{ps}}{bd} = \frac{3.8584}{40 \times 36} = 0.00268$$

$$\omega_p = \frac{0.00268 f_{ps}}{f_c'}$$

ω and ω' are not existent since there is no untensioned steel or compression steel

Hence $f_{ps} = 210240\left[1 - \dfrac{0.4}{0.812} \times 0.00268 \times \dfrac{210240}{4765.55}\right]$

$= 210240 \times 0.942$

$= 198046\ p_{si}$

$$\omega_p = \frac{0.00268 f_{ps}}{f_c'} = \frac{0.00268 \times 198046}{4765.55} = 0.1114 \prec 0.36\beta_1$$

Hence it is under reinforced beam

Now rewriting the equilibrium expression at the time of failure

$3.8584 \times 198046 = 0.85 \times 40 \times 4765.55 \times a$

$a = 4.716'' < 6''$

Hence M_u can be written as

$$A_{ps} \cdot f_{ps}\left(d_p - a/2\right)$$

$$= 3.8584 \times 198046\left(36 - \frac{4.716}{2}\right)$$

$= 3.8584 \times 198046 \times 33.642$

$= 25707220.97\,lb$ in

$= 2921.27\ kNm$

Applying a capacity reduction factor of 0.9 as the beam is under reinforced and the failure is tension controlled.

$M_u = 2921.27 \times 0.9 = 2629.1476\,kNm$

M_u as per IS Code provisions $= 2593.3\,kNm$

Example 6.11

A prestressed concrete beam of size 400 × 800*mm* is prestressed with 1000*mm*² of High tensile steel placed at an effective depth of 650*mm*. The steel is unbonded and has a failure stress f_{pb} = 1000*N/mm*^{2} . Concrete strength of 40*N/mm*^{2} may be assumed. Estimate the moment carrying capacity of the section as per ACI Specification. Assume a span to depth ratio $\prec 35$.

Solution

The Notations adopted are as per ACI Specifications

Section details

b = 400*mm* = 16*in*

d_p = 700*mm* = 28*in*

f'_c = 40 × 0.85*N/mm*^{2} = 4765.55*psi*

f_{pu} = 1600*N/mm*^{2} = 224261*psi*

f_{se} = 1000*N/mm*^{2} = 140163*psi*

A_{ps} = 1.6*sqcm* = 1000*mm*^{2}

Evaluation of depth of neutral axis

$$1.6 \times f_{ps} = 0.85 \times 16 \times 4765.5a$$

$$a = \frac{1.6 f_{ps}}{0.85 \times 16 \times 4765.5}$$

Assuming $f_{ps} = \left(f_{se} + 10000 + \frac{f'_c}{100\rho} \right)$

$$= 140163 + 10000 + \frac{4765.55}{100\rho}$$

$$\rho = \frac{1.6}{16 \times 28} = 0.00357143$$

$$f_{ps} = 140163 + 10000 + 13343.53$$

$$= 163506.53\ Psi$$

$$a = \frac{1.6 \times 163506.53}{0.85 \times 16 \times 4765.55} = 4.036''$$

$$\beta_1 = 0.85 - \left(\frac{4765.55 - 4000}{1000} \right) 0.05$$

$$= 0.812$$

$$= 0.36 \times 0.812 = 0.29232$$

$$\omega_p = \frac{\rho_p f_{ps}}{f'_c} = \frac{0.0035713 \times 163506.53}{4765.55} = 0.1225$$

$$\omega_p = \frac{\rho_p f_{ps}}{f'_c} = \frac{0.0035713 \times 163506.53}{4765.55} = 0.1225$$

$$\omega_p \prec 0.36\beta_1$$

Hence it is under reinforced beam

$$M_u = A_{ps} \cdot f_{ps} \left(d_p - a/2 \right)$$

$$= 1.6 \times 163506.53 (28 - \frac{4.036}{2})$$

$$= 6797162.66\ P_{si}$$

$$= 169929066.5\ Nm$$

$$= 169.92 kNm$$

Example 6.12

An unsymmetrical I section with a top flange of 400 × 100*mm* has an effective depth of 500*mm* and a total depth of 600*mm*. It is prestressed with 24 Nos of 3 × 3 ply strand (21.21 mm^2/strand) to a level of 800N/mm^2. The ultimate strength of strand is 1600N/mm^2 . The tendons may be considered as unbonded. Evaluate the moment resisting capacity of section as per ACI provisions. The span to depth ratio can be assumed to be less than 35 and the strength of concrete as 40N/mm^2.

Solution

The notations adopted are as per ACI Specifications

$b_f = 400 = 16in$

$h_f = 4''$

$f_{pu} = 1600N/mm^2 = 224261\ psi$

$$f_{ps} = f_{se} + 10{,}000 + \frac{f_c'}{100\rho_p}$$

$f'_c = (40 \times 0.85)N/mm^2$

$= 4765.55\ Psi$

$f_{se} = 800N/mm^2 = 112130.5\ psi$

$A_s = 0$

$A'_s = 0$

$$\beta_1 = (0.85 - \frac{765.55}{1000} \times 0.05$$

$$= 0.8118 \approx 0.81120$$

$$f_{ps} = f_{se} + 10000 + \frac{f_c'}{100\rho_p}$$

$$= 112130.5 + 10000 + \frac{4765.5}{100\rho_p}$$

$$\rho_p = \frac{A_{ps}}{bd} . \frac{36 \times 2121}{400 \times 500} = 0.0038178$$

$$f_{ps} = 112130.5 + 10{,}000 + \frac{4765.5}{0.38178} = 134612.8\ psi$$

$$a = \frac{A_{ps} \cdot f_{ps}}{0.85 f_c' b} = \frac{36 \times 21.21 \times 134612.80}{100 \times 625 \times 0.85 \times 4765.5 \times 16}$$

$$= 2.54'' \prec 4''$$

$$f_{pu} = \frac{36 \times 21.21 \times 134612.8}{100 \times 6.25}\left(\frac{500}{10 \times 25} - a/2\right)$$

$$= 350.4\ kNm$$

References

1. LEONHARDT. F, (1964) "**Prestressed Concrete Design and Construction**". Wilhelm Ernst and Sohn, Berlin.
2. IS 1343 - (1980). Indian Standard "**Code of Practice for Prestressed Concrete**". Bureau of Indian Standards. New Delhi.
3. BS 8110 - (1985). "**British Standard for Structural use of Concrete**". British Standards Institution. London WIA 2BS.

4. ACI 318 - (1998). **"Building Code requirements for Structural Concrete"**. American Concrete Institute. Formington Halls, HI48333, USA.
5. FIP - (1984). **Recommendations for Practical design of Reinforced and prestressed concrete structures, based on CEB - FIP model code** (MC - 78). Thomas Telford Ltd, London 1984.
6. ENV 1992 - 1.2 (1991). **Europen Standard, Design of Concrete Structures Part I - General rules and rules for Buildings.** European committee for Standardisation - B - 1050 Brussels.
7. HOGNESTAD, E., N.W. HANSON, and D. Mc HENRY (1955), **Concrete Stress Distribution in Ultimate Strength Design,** "Proceedings of the American Concrete Institute", Vol.52, No.4, Dec. pp 455 - 80.
8. RUSCH, H., (1960) **Research towards a General Flexural Theory for Structural Concrete,** "Proceedings of American Concrete Institute", Vol.57, No. 1, July, 1960, pp 1 - 28.
9. WHITNEY, C.S. (1942), **"Plastic theory in reinforced concrete design"**, Transaction of ASCE, pp.251 -282, Discussion - pp. 282 - 326.
10. JONES, L.L. (1986), **"Ultimate Load Analysis of Reinforced and Prestressed Concrete Structures"**, Chatto and Windus, London, pp 23 - 57.
11. MATTOCK, A.H., L.B. KRIZ. and E.HONGNESTAD (1961), **Rectangular Concrete Stress Distribution in Ultimate Strength Design,** "Proceedings of the American Concrete Institute", Vol, 57, No. 8, Feb. pp 875 - 928.

7. Philosophy of Design

7.1 Introduction

The purpose of design of a structure is fulfilled if the clients requirements are fully met. Normally, the clients requirements are that the structure shall not collapse under any type of loading, and it shall also have certain reserve strength to meet with any eventuality. When the first formal design was made by Prof. Morsch, the materials used were evaluated for their capacities namely failure stresses. ***A factor of safety*** was assumed for this failure stress to arrive at the working stress. This working stress was used to design the structure. The design was considered safe, if the stress in the material under the given loading did not exceed the factored stress allowable for the material namely working stress. At that time the concept of strain being the main cause for failure was not adequately understood. Hence, if the structure was designed for a load P and the corresponding stress level in the material did not cross the factored stress called working stress, it was assumed that the maximum load that the structure could carry, would be the design load multiplied by the factor of safety adopted for stresses. This assumes that the linear relationship adopted between the applied stress and the failure stress could be extended between the applied load and failure load. This method of approach is called the ***working stress approach.***

After the First World War, a necessity for more accurately predicting the ultimate load was felt. Experimental evidences proved that the ratio between ultimate load and working load was not the same as the ratio of failure stress to working stress. This lead to the development of the ultimate load method, wherein the design was made for the load which was a factor times more than the working load. That factor was called the ***load factor***. It indicated the reserve strength available in the structure. The approach gave the confidence to the designers that the structure could carry the working load safely. But the stress level in the material at working load conditions was not known. Hence the corresponding strain levels, deformation levels etc. in the structure at the working load could not be ascertained. But the psychological feelings of clients using the structure would be satisfied only if the deformation on the structure during the service load would be small. This necessitated that a) the design should be made with a reserve strength before its collapse b) it should be made sure that deformation under working load would be acceptable for the client and c) the deformational effects such as cracking should not also cause environmental distress to the structure such as corrosion.

This exercise finally ended up in a new philosophy of design called ***limit state design*** and first incorporated in the national code of the then U.S.S.R. Since, behaviour at the failure load levels and at service load levels were coupled, more research was carried out later. This lead to the present ***limit state concepts of design*** and the same has been universally accepted now. Hence, many countries have adopted limit state concept for the design of structures.

7.2 Concepts of Limit States and Partial Safety Factors

Two major limit states are defined in various statutory codes. They are ***limit state of collapse and limit state of serviceability***. Under the limit state of collapse, different types of collapses are identified, such as, collapse due to flexure, collapse due to shear, collapse due to torsion and a combination of them. The collapse may be due to the structural causes or due to stability problems.

The limit states of serviceability define the level up to which, the structure can be loaded so that the usage of the structure for the purpose for which its intended is served. Normally in concrete structures deflections and cracking under service load are considered important. If the structure is cracked badly or deformed beyond a limit, it is declared as unserviceable. Hence, limit states of serviceability are defined with respect to limits up to which cracking can be tolerated and the deflections are permitted.

Generally, if any structure is meant for certain system of loads it is designed for a higher load to ensure certain reserve capacity before failure. The higher design loads are called ***factored loads*** and defined by a ***partial safety factor for loads***. Similarly, even in case of materials, they are not allowed to reach their limiting capacity in terms of stresses or strains even under the increased (factored) loads. There also, a certain reserve strength is made available. This is defined by a ***partial safety factor for material capacities***. The partial safety factors are meant to take care of unforeseen loadings and material inadequacies. Hence, the design capacity will be more than required upper limiting capacity, for a given section.

7.3 Design Loads and Stresses

Any structure is designed for a particular load, which is more than the defined characteristic load that the structure is expected to carry, to provide for reserve strength for the structure. These design loads for various limit states are obtained from the characteristic loads multiplied by ***partial safety factors*** greater than one. Any force which is going to produce a favorable effect, on the structure while resisting the external loads, that favorable effect will be considered with a partial safety factor less than one. Prestress is an example for such a force causing favorable effect to resist the external load. Tables 7.1 to 7.4 give the partial safety factors adopted in various national codes for individual load and combination of loads.

Similarly the strength of material is not also used fully to resist the design load. If f_{ck} the characteristic strength of material defined by the relevant specification, the strength of the material that could be used in the design is expressed as f_{ck} / γ_m where γ_m is called the ***partial safety factor for strength***. If the material used is a factory produced one, like

Table 7.1 Values of Partial Safety Factor γ_f Loads as per IS 1343 - 1980 (1)

<table>
<tr><th>Load Combination</th><th colspan="3">Limit State ofCollapse</th><th colspan="3">Limit States of Serviceability</th></tr>
<tr><td>(1)</td><td>DL
(2)</td><td>LL
(3)</td><td>WL
(4)</td><td>DL
(5)</td><td>LL
(6)</td><td>WL
(7)</td></tr>
<tr><td>DL+LL</td><td colspan="2">1.5</td><td>-</td><td>1.0</td><td>1.0</td><td>-</td></tr>
<tr><td>DL+WL</td><td>1.5 OR 0.9
(See Note 2)</td><td>-</td><td>1.5</td><td>1.0</td><td>-</td><td>1.0</td></tr>
<tr><td>DL+LL+WL</td><td colspan="3">1.2</td><td>1.0</td><td>0.8</td><td>0.8</td></tr>
</table>

NOTE 1 - DL is the dead load, LL is the live load and WL is the wind load.

NOTE 2 - This value of 0.9 is to be considered when stability against overturning or stress reversal is critical.

NOTE 3 - While considering earthquake effects, substitute EL for WL.

NOTE 4 - For the limit states of serviceability, the values of gf given in this table are applicable for short-term effects. While assessing the long-term effects due to creep, the dead load and that part of the live load likely to be permanent may only be considered.

Table 7.2Partial Safety Factor γ_l as per ACI 318 - 1998 (2)

Load Combination	Limit State of Collapse					
	DL	LL	WL	EL	HL	TL
DL+LL	1.4	1.7	-	-	-	-
DL+LL+WL	1.05	1.28	1.28	-	-	-
DL+LL+EL	1.05	1.28	-	1.40	-	-
DL+EL	0.9	-	-	1.43	-	-
DL+LL+HL	1.4	1.7	-	-	1.7	-
DL+LL+TL	1.05	1.05	-	-	-	1.28
DL+WL	0.9	-	1.3	-	-	-
DL+TL	1.4	-	-	-	-	1.4

DL = Dead Load, LL = Live Load, WL= Earth quake Load, HL = Earth Pressure Load, TL = Different settlement, creep, shrinkage of temperature loads

steel, the partial safety factor will be small where as the same for man made material, like concrete it is more. While Indian, European and British codes follow the partial safety factor for materials, American code follows a factor called as strength reduction factor.

Table 7.3 Partial Safety Factor for Load γ_l as per BS 8110 :1985 (3)						
Load combination	**Load type**					
	Dead		**Imposed**		**Earth and water Pressure**	**Wind**
	Adverse	Beneficial	Adverse	Beneficial		
1. Dead and imposed (and earth and water pressure)	1.4	1.0	1.6	0	1.4	-
2. Dead and wind (and earth and water pressure)	1.4	1.0	-	-	1.4	1.4
3. Dead and wind and imposed (and earth and water pressure)	1.2 1.4	1.2	1.2	1.2	1.2	1.2

Table 7.4 Partial Safety Factors for Loads (γ_f) (FIP - 1984) (4)		
Type of loads	**Limit state of Collapse**	**Limit state of serviceability**
Permanent loads	1.35	1.00
Variable loads	1.50	1.00

The partial safety factors for material strength, (γ_m) as recommended by different national codes are listed in tables 7.5 and 7.6.

Table 7.5 Partial Safety Factor for Material (γ_m)		
Code	**Materials**	**Partial Safety Factor (γ_m)**
IS:1343 - 1980 and FIP - 1984	Reinforcement	1.15
	Concrete	1.50
BS:8110-1985	Reinforcement	1.15
	Concrete	
	(a) Flexure or axial load	1.50
	(b) Shear strength without Shear reinforcement	1.25
	(c) Bond Strength	1.4
	(d) Others (e.g., bearing Stress)	1.5

Table 7.6 Strength Reduction Factors (φ) (AC1 318 : 1998) (2)	
Structural element	**Factor(φ)**
Beam or slab (Flexure)	0.9
Column with ties	0.7
Columns with spirals	0.75
Colums carrying very small axial loads	0.7 to 0.9 or 0.75 to 0.9
Beam : shear or torsion	0.85

The characteristic load itself is arrived at based on a semi probabilistic approach. The probability of any load coming on the structure higher than the characteristic load during the life time of the structure is first defined as percentage. Accordingly the characteristic load is calculated from the average (mean) load occurring on the structure. If that probability is assumed as 5 % as in Indian Standard codes, the characteristic load will be equal to the mean load plus 1.64 times the standard deviation of the various loads which have been used to arrive at the mean load. In absence of any systematic data regarding the loads coming on the structure, the values suggested in national codes on loading such as I.S. 875 of 1987 may be treated as characteristic loads.

Similarly the characteristic strength is also arrived based of semi probabilistic approach. It is assumed to be less than the average strength based on tests on a minimum of 40 specimens. It is based on the percentage of probability, that the strength of the material used in the actual structure will be less than the mean. If it is assumed as 5 % as given in the Indian Standard codes the characteristic strength is arrived at as the mean strength minus 1.64 times, the standard deviation of various values of strength which have been used to arrive the mean strength.

The design is made it is made for a given characteristic load using materials of given characteristic strength.

7.4 Limit State of Collapse

Collapse of a structure may happen due to different structural actions. In planar structure, it can be due to axial forces, moment in the plane of structure (M) and moment normal to the plane of the structure namely torsion (T). Variation of moment along the length of the member is defined as shear in structural mechanics. If the variation of moment is very steep it may also lead to collapse called collapse due to shear (V). The combination of these may also lead to collapse. In addition, specially in prestressed concrete structures, collapse due to bursting or crushing of the end anchorages zones in post tensioned elements, collapse due to interface shear between prestressed concrete elements and cast in-situ reinforced concrete elements cannot be ruled out.

Any collapse can occur only due to excessive deformation namely elongation/rotation/twist/separation etc. defined by strains in a differential element. ***The strains are realities and stresses are only conjectures***. The stress corresponding to a strain, which can not be tolerated, is called the limiting stress or failure stress or strength of the material. For example, in flexure, concrete fails when the compressive strain reaches .0035 as per I.S. code provisions. The corresponding stress is called the characteristic strength for the concrete (f_{ck}). But this stress is reached even when the strain is only 0.002. Though the stress f_{ck} is reached at a strain level of 0.002, the material is not considered to have failed. Only when the strain reaches 0.0035, the material is supposed to have failed. But, under axial compression, the failure strain is only .002 and the stress level reached is fck. Hence, the concept of strain on the collapse is to be considered as basic. The corresponding stress level is to be considered as an indirect acceptance level for collapse.

Similarly in steel, the strain, which, when released leaves back a residual value of 0.002, is called failure strain and the correspondence stress level in steel is called failure stress or 0.2 percent proof stress.

Though designs are made based on limiting stress values at the time of collapse, it should be made sure that such stresses are corresponding to or less than limiting strains (deformations) relating to the type of stresses.

The strains and stresses corresponding to limit states of collapse have to be taken from the material constitutive relationship namely stress-strain relationship. In case, the collapse is due to a combination of structural actions like moment and torsion, designs have to be made considering the interaction, either following a failure criteria like von-Mieses, Drugger and Prager criteria, etc., or interaction equation or equivalent individual structural actions, (for example, equivalent total moment, and total shear.) National and International codes present design rules for provision of material like concrete, and steel in terms of individual equivalent structural actions like equivalent moment etc., But the interaction expressions and failure criteria have been made use of in arriving at design rules.

The design for limit state of collapse due to any one of the possible causes like flexure, shear and torsion are independently dealt with in chapters 6,8 and 9 and hence the procedure for evaluating the capacity of the structure under limit state of collapse is not discussed in this chapter. For arriving at the limit state of collapse the load combination and the partial safety factor both for load and materials have been given in tables 7.1 to 7.6. The adequacy of structure against the limit state of collapse could be checked up as shown in examples in chapters 6,8 and 9.

7.5 Limit State of Serviceability

Serviceability of a structure is generally defined as the clients' acceptance for continuous use of the structure for its lifetime under given loading condition without fear or any inhabitation. In reinforced concrete or prestressed concrete structures, cracking and deflection under service loads define the confident use of the structure at a given point of time. The vulnerability of structure for corrosion prone damages or detoriation of strength of materials leading to cracking and deformation define the long-term confidence level. These aspects are called durability. Hence, cracking, deflection and durability are the aspects, which fix up the limit states of serviceability.

In reinforced concrete and prestressed concrete structure, the serviceability limit states are indirectly defined and controlled by providing maximum acceptable span to depth ratio, incorporating certain specification of spacing of reinforcement, and minimum percentage reinforcement. Depending on the type of structure, and its importance, deflection and crack widths may have to be calculated and checked against the permissible values. This is particularly important in prestressed concrete structures. Prestressing reduces the deflection. Creep reduces prestressing and hence increases deflection. Creep of concrete itself introduces additional strains which further increases the deflection. This affects the effective prestressing force. Hence, long term deflection in prestressed concrete is affected by many phenomenon and should not be under estimated. Specialist's literature has to be consulted for arriving at exact deflection.

Accepting the common understanding of serviceability as a control deflection, crack width, and crack spacing, only those points are discussed in this chapter. Generally prestressed concrete structures are grouped into 3 types as difined earlier in chapter 1. IS1343-1980 does not permit Type 3 structures in Indian Constructions. Type 1 and Type 2 structures do not allow cracking during service loads and hence procedure for calculation of crack width and crack spacing has limited use for Indian Industry. Accordingly evaluation of deflection in prestressed concrete elements is covered first for Type 1 & Type 2 members, in this chapter. Later evaluation of deflection, crack width and crack spacing for Type 3 structures are discussed.

Deflection of the prestressed elements is dependent on applied prestress, loads, inclusive of self weight, and the time at which the deflection is measured in addition to the properties of materials like concrete and steel and cross sectional properties of the section itself. For a structural element of known material properties and loading, deflections are defined by E and I, identified by flexural rigidity of the section. They can be evaluated as short term and long term deflection.

7.5.1 Short-term deflection of uncracked members (Type1&Type 2)

The short-term or instantaneous deflection of structural components can be arrived at using the principles of theory of bending and structural mechanics. There are many simplified methods for evaluation of deflection. One of the most commonly adopted methods is by using Mohr's moment area theorems for simply supported beams. In this method the bending moment diagram due to a given system of loading in simply supported beam divided by flexural regidity (M/EI diagram) is considered as the loading diagram on the same span. The area of this loading diagram between two cross sections give the change of slope of the elastic curve of the original beam between these two cross sections and the moment of this loading between two cross sections about a vertical gives the intercepts of the tangents on the elastic curve at these two cross sections on the vertical about which moment is taken.

This could be used to evaluate the deflection as illustrated in the examples in cases of prestresses. In case of a simply supported beam having straight tendons with a prestressing force P, and constant eccentricity 'e' throughout the length of the beam, the maximum deflection (δ_{max}) occurs at the centre of the beam and can be evaluated as shown in Fig 7.1 From figure, δ_{max} = Moment of M/EI diagram about the vertical through the support section.

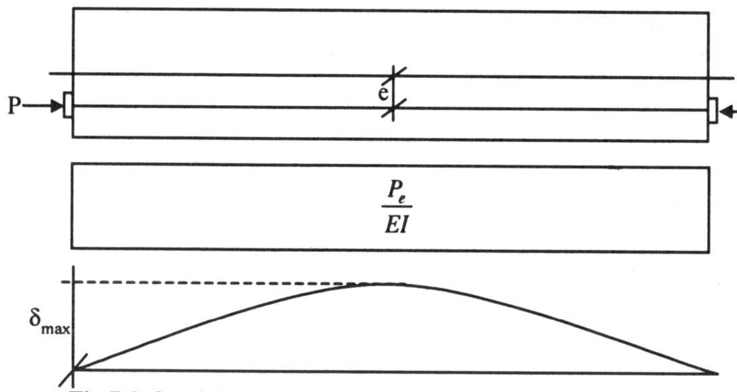

Fig.7.1. Straight Tendons with Prestressing Force P, and Constant Eccentricity 'e' throughout the Length of the Beam

$= (P.e/E.I) \times L/2 \times L/4$

$= P\,e\,L^2/8EI \uparrow$

In case of simply supported beam having parabolic tendons with eccentricity 'e'at the centre and zero at support, the maximum deflection (δ_{max}) occurs at the centre of the beam.

δ_{max} = Moment of M/EI diagram about the vertical through the support. (Fig 7.2)

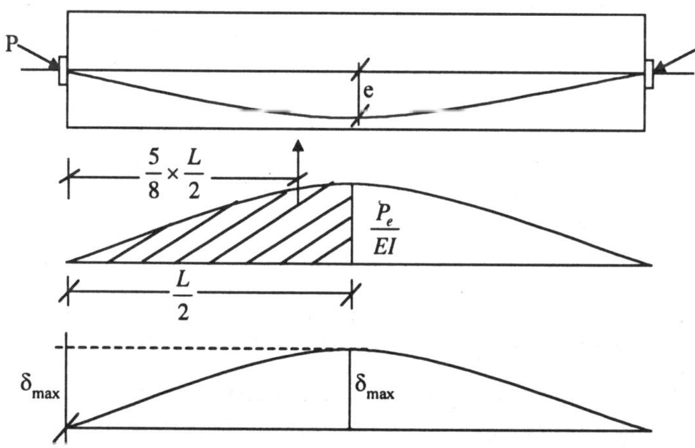

Fig.7.2 Parabolic Tendons with Eccentricity 'e' at the Centre and Zero at Support

$= Pe/EI \times 2/3\ L/2 \times 5/8\ L/2$
$= (5/48)\ PeL^2/EI \uparrow$

In case of simply supported beams having parabolic tendons with eccentricity e_2 upwards at the ends and eccentricity e_1 downwards at the centre, the maximum deflection dmax occurs at the centre of the span.

δ_{max} = Moment of the area of M/EI deagram about the vertical through the support line (Fig 7.3).

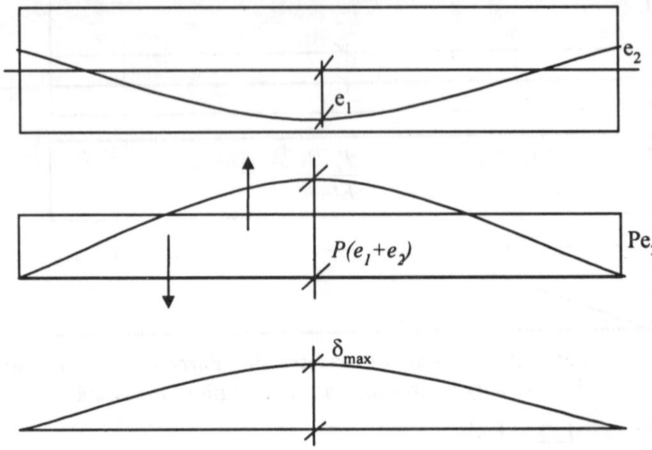

Fig 7.3 Parabolic Tendons with Eccentricity e_2 Upwards at the Ends and Eccentricity e_1

$= +(5/48)\ (PL^2/EI)\ (e_1 + e_2)\uparrow - Pe_2\ L^2/8EI \downarrow$
$= (PL^2/48\ EI)\ (5e_1 - e_2) \uparrow$

Similarly standard cases of sloping tendons with constant slopes, tendons with various continuous profiles or combinations can be considered and deflection evaluated.

At the time of transfer of prestress, the beam hogs up due to the effect of prestressing generally and immediately the self weight comes into effect. Hence, the deflection even in the first stage namely just after the transfer of prestress, has to be considered as a combination of the deflection due to prestress and that due to self weight. The self weight load being a uniformly distributed load, the deflection due to this can be computed as $\delta_g = \frac{5}{384} \cdot \frac{wl^4}{EI}$ where w is the weight of the beam per unit length.

Further the deflection due to super imposed loads or the construction loads can be evaluated using any one of the recognized elastic approaches.

7.5.2 Long term deflection of uncracked members

The final deflection and deflection at various stages should be compared with the permissible values and the checked for serviceability conditions as laid down in the statutory documents.

The deflections of beam may vary due to change in force caused by time dependent properties of materials, concrete and steel. Due to creep of concrete and relaxation of steel, there will be loss in prestress and hence the deflection due to prestress will change. Simultaneously due to creep effect, for the same force there will be additional deflection. But creep effects occur only due to stresses caused by sustained loads. Hence these aspects have to be considered and the long-term deflection calculated.

Creep causes strains in the elements which is reflected as a change in curvature and finally as a change in the deflection of the element. Number of authors like Corley, Sozen and Siess (5), Busemann (6) etc., have discussed this aspect in great detail. In the recent past, the variation of the creep coefficient due the variation of applied stress level caused by losses in prestressing, have also been included in the evaluation of deformation due to creep by H Trost (7). This has been discussed in the chapter of materials and also the chapter on losses in prestressing briefly. These sophistication's in calculation of deflections are required or taken care of only in special structures. ACI committee report (8) has presented a simplified procedure for deformation (deflection) due to prestressing inclusive of the creep effects.

If the initial deflection due to sustained transverse loads is δ_{is} the final deflection is calculated as $\delta_{is}(1 + \phi)$. If the initial deflection due to prestress is δ_{ip}, the final deflection after creep is

$$= \delta_{ip}\left[\frac{P_i - L_p}{P_i}\right] + \left[\frac{P_i + P_i - L_p}{2}\right]\phi\delta_{ip}$$

$$= \delta_{ip}\left(1 - \frac{L_p}{P_i}\right) + \left(1 - \frac{L_p}{2p_i}\right)\phi\delta_{ip}$$

$$= \delta_{ip}\left[\left(1 - \frac{L_p}{P_i}\right) + \left(1 - \frac{L_p}{2P_i}\right)\phi\right]$$

Hence the total deflection after including time dependant effects can be represented by the expressions.

$$\delta_f = \delta_{is}(1 + \phi) + \delta_{ip}\left[\left(1 - \frac{L_p}{P_i}\right) + \left(1 - \frac{L_p}{2P_i}\right)\phi\right]$$

Example 7.1

A post tensioned prestressed concrete with rectangular cross section 300mm × 400mm is prestressed with 1000mm^2 (2 cables of 500 mm^2 each) of high tensile steel wires. The steel is stressed to a level of 1500 N/mm^2. The beam is to span 8m. Calculate the initial deflection of the beam due to prestress and dead load, and a live of 20 kN/m. if

a) the cable profile is straight with a constant eccentricity 100 mm

b) the cable profile is parabolic with dip of 100 mm at mid section and zero eccentricity at the ends.

Assume E_{sp} = 2.1 × 105 N/mm².& E_c = 3.6 × 104 N/mm². Assume a average friction loss of 5 percent for the mid section of beam in case of parabolic profile. The cable is stretched alternatively from ends.

Solution

Section Properties

Area = 400 × 300 = 120000 mm²

Uncracked section moment of Inertia = $300 \times 400^3/12 = 16 \times 10^8$ mm⁴

Self weight = $12 \times 10^4 \times 10^{-6} \times 24$

= 2.88 kN/m

Prestressing force = P_0=1000 × 1500N

=15×10^5N = 1500 KN

Downward deflection due to dead load (Self weight)

$$= \frac{5}{384} x \frac{(2.88x10^3 x8)x8000^3}{3.6x10^4 x16x10^8}$$

= 2.66mm ↓

Upward deflection due to prestress $= \frac{P.e.L^2}{8EI}$

$$= \frac{1500 \times 100 \times 8000^2 \times 10^3}{8 \times 3.6 \times 10^4 \times 16 \times 10^8}$$

=20.83mm↑

Downward deflection due to live Load

$$= \frac{5}{384} \times \frac{20000 \times (8000)^3 \times 8}{3.6 \times 10^4 \times 16 \times 10^8}$$

= 18.5mm ↓

Final deflection = 2.66↓ – 20.83↑ + 18.5↓

= 21.16 – 20.83 = 0.33↓mm

If the cable has a parabolic profile

Deflection due to prestress $= \frac{5}{48} \times \frac{PeL^2}{EI}$

= 5/6 × 20.83 = 17.36↑ mm

Since the cable has a parobolic profile, loss due to friction can not be avoided.

5 percent loss at mid section is assumed. Loss in Deflection due to loss in prestress assuming linear variation of loss from 0% at end to 5% at the mid span.

$$= \frac{5}{100} \times 1500 \times 100 \times 8000/2 \times 2/3 \times 8000/EI$$

$$= \frac{5 \times 15 \times 100 \times 8000 \times 8000 \times 1000}{3 \times 3.6 \times 10^4 \times 16 \times 10^8}$$

$$= \frac{25 \times 64 \times 10^1}{3.6 \times 16 \times 10^{12}} = 2.78\text{mm}$$

Deflection due to prestress (upward) = 17.36 – 2.78 = 14.58mm↑

Downward deflection due to self weight and

Super imposed Load = 2.66 + 18.5 = 21.66↓

Net down ward deflection =7.08 mm↓

Example 7.2

A prestressed Concrete beam with cross section 150mm × 500mm is prestressed with a parabolic cable carrying a force of 400 kN. The cable has a dip of 100mm at centre and 60mm eccentricity at the support. It spans a distance of 12m and carries a super imposed load 2kN/m. Calculate the short-term deflection at mid span. E_{sp} = 3.6 × 10^4 N/mm^2. Creep coefficient =2.0. Loss in prestress may be assumed as 20 percent after 6 months. Estimate the long-term deflection, if the final prestressing force is 400kN and the initial prestress is 1.2 times the final prestress.

Solution

Initial prestress = 1.2 × 400 = 480 kN

Moment of Inertia of the cross section I = 1/12 × 150 × 400^3

= 8 × $10^8$$mm^4$

E_{sp} = 3.6 × 10^4N/mm^2

e = 100mm

ϕ = 2.0

Dead Load = 1440 N/m

Total = 3440 N/m

Loss in prestress = 0.2 P_I

Short term deflection

a) Due to prestress (Neglecting Friction Loss)

$$= \frac{5}{48} \times \frac{480 \times 10_3 \times 100 \times 12000 \times 12000}{3.6 \times 10^4 \times 8x10^8}$$

$$= \frac{5}{48} \times \frac{480 \times 12 \times 10^{11} \times 12}{3.6 \times 8 \times 10^{12}}$$

$$= \frac{5 \times 12 \times 12}{3.6 \times 8} = \frac{15 \times 12}{7.2} = 25mm \uparrow$$

b) Due to Dead Load

$$\frac{5}{384} \times \frac{1440 \times 12 \times (12000)^3}{3.6 \times 8 \times 10^{12}}$$

$$= \frac{5}{384} \times \frac{1440 \times 12 \times 12^3 \times 10^9}{3.6 \times 8 \times 10^2} = \frac{5 \times 18 \times 18}{120}$$

= 90 × 18/120 = 13.5mm ↓

c) Due to Live Load = 18.75 mm↓

Long Term Deflection

Due to Prestrressed + Dead load.

$$= 13.5(1+2) - 25\left[1 - \frac{0.2P_i}{P_i} + \left(1 - \frac{0.2P_i}{2P_i}\right)2\right]$$

= 13.5 × 3 – 25 (1 – 0.2+1.8)

= 13.5 × 3 – 25 × 2.6

= 40.5 – 65

= 24.5↑ mm

Since live load is not a sustained load, creep effect can not be included for deflection due to live load. Hence deflection due to live load is taken as 18.75mm.

Final deflection = (24.5 – 18.75)↑ = 5.75↑ mm

7.5.3 Short-term deflection of cracked members (Type 3)

In general, deflection is caused by rotation of cross sections and is evaluated by integrating the same over a length and these rotations are due to applied moments on the cross sections. Hence it is desirable to have an understanding of the moment rotation characteristics of cross sections for arriving at the deflection. While for uncracked sections the moment rotation characteristics could be defined by a straight line, for cracked sections it is bound to be non-linear depending on the development of cracks. The non-linear moment curvature relationship can be idealsied into number of linear units. Experimental evidences show that moment curvature diagram of a already cracked beam can be represented in four straight lines defined by 4 points namely (a) the moment causing decompression (M_c) of concrete at the bottom most fibre, the moment causing cracking (M_{cr}) at which tensile stress at the bottom is about equal to flexural tensile strength of concrete, moment (M_y) at the time of yielding of steel, and moment (M_u) leading to the limit state of collapse. For serviceability consideration, it is sufficient to consider the moment curvature relationship up to the point where cracking is permitted to limited crack width. A typical moment curvature relationship for a prestressed concrete element is shown in Fig.7.4. The curve up to point e is only necessary and is also adequate to calculate the deflection of cracked members under serviceability limit state. Since cracking is permitted only in Type 3 structures this is applicable only for Type 3 structures.

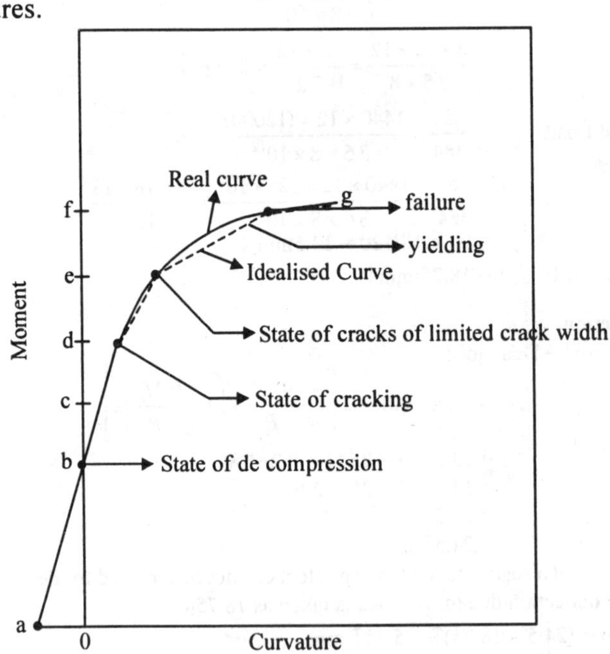

Fig.7.4 Typical Moment Curvature Relation of a Prestressed Beam.

European Concrete Committee (9) has recommended that deflection of a cracked concrete member can be calculated using the bilinear portion of the moment curvature relationship and deflection is expressed in the form $\delta = \left[\frac{ML^2}{E_c I}\right]k$.

Where M = the maximum moment on the element.

E_c = modulus of elasticity of concrete.

I = the equivalent moment of inertia of transformed section.

L = the span, and

k = a constant depending on the loading and support conditions.

k value for various loading & bounding curvature are shown in Table 7.7.

Table 7.7 Values of k for various bending moment diagrams

Loading	Bending moment diagram	k
	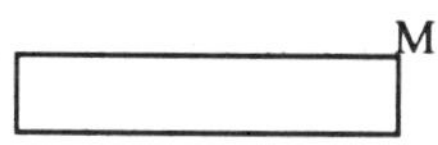	0.125
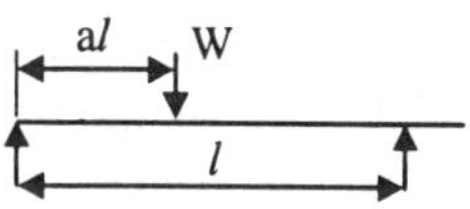	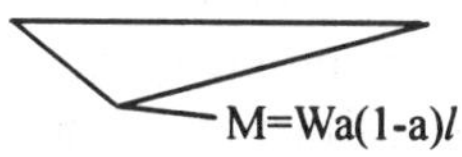	$\frac{3-4a^2}{48(1-a)}$ $if\ a = \frac{1}{2}, k = \frac{1}{12}$
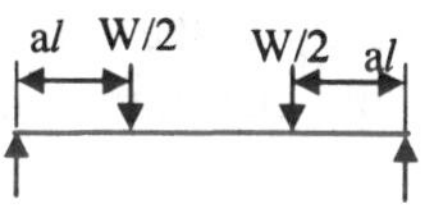	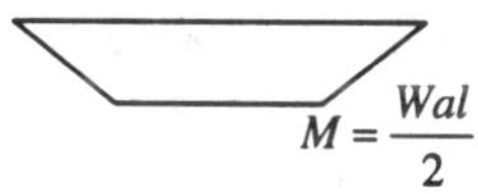	$0.125 - \frac{a^2}{6}$
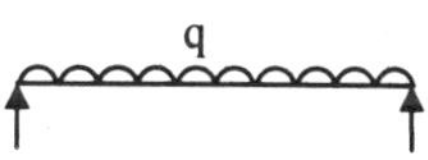	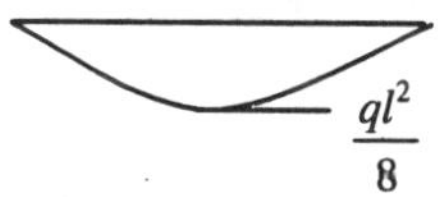	0.104
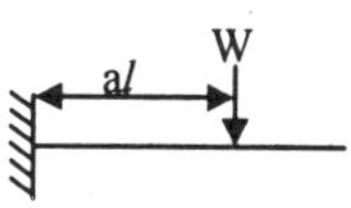	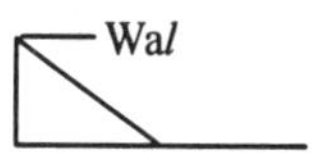	$\frac{a(3-a)}{6}$ if a = 1, k =0.333
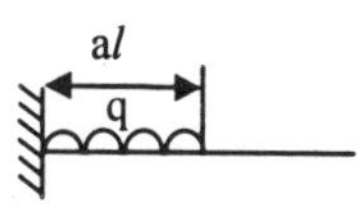	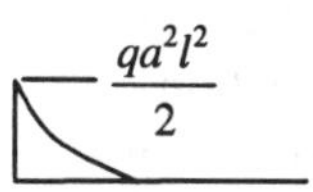	$\frac{a(4-a)}{12}$ if a = *l*, k = 0.25

If the deflection is caused by a load, which during application from zero to the concerned load, has caused variation in E_c and I, for each zone, proper moment, and proper E_c and I_c values are to be substituted and the deflection values added. Because bilinear moment curvature relationship has been universally accepted for deflection calculation, the deflection δ_{max} can be expressed as a sum of two terms as detailed below.

$$\delta_{\max} = kL^2\left[\frac{M_{cr}}{E_c I_g} + \frac{M - M_{cr}}{0.85 E_c I_{tr}}\right]$$

The second term gives the contribution to deflection by the moment which comes on the beam after the beam has cracked. Hence, proper value of moment of Inertia and modulus of elasticity have to be substituted. ACI 318-1998 (2) has recommended an expression for I_{tr} as a function of Moment at the time of cracking, flexural tensile capacity of concrete, and moments of inertia of gross and cracked sections.

$$I_{tr} = \left[\frac{M_{cr}}{M}\right] I_g + \left[1 - \left(\frac{M_{cr}}{M}\right)^3\right] I_{cr}$$

Where $M_{cr} = \dfrac{f_{cr}.I_g}{y_b}$ and $f_{cr} = [7.5\sqrt{f_c^i}$ plus initial precompression at the extreme bottom fibre in prestressed elements].

British code BS-8110-1985 (3) goes more into fundamentals to arrive at the curvature and then the deflection. The second term is represented by curvature at the maximum moment section. The curvature of the section could be expressed as a) a function of stresses in tension reinforcement, or b) or could be expressed by modified value of I_{cr} or c) could also be expressed by modified value M (beyond M_{cr}) keeping I_{cr} as the effective moment of Inertia. Accordingly for the second term, the curvature can be expressed as:

$$a) = \frac{1}{\gamma_b} = \frac{f_x}{(d-x)E_s} = \frac{f_c}{x.E_c}(or)$$

$$b) \frac{M.nett}{E_c I_{cr}}$$

Where M.nett = $M - \dfrac{1}{3} b \dfrac{(h-x)^3}{(d-x)} f_{cr}$

In these two formulations:

f_c = design service stress in concrete.
f_s = estimated stress in steel at service load.
d = effective depth of section.
h = depth of neutral axis.
h = total depth of beam.
f_{cr} = flextural tensile capacity of concrete.
b = width of tension zone of concrete.

Example 7.3

A prestressed concrete beam element has a symmetrical I section with maximum depth of 320 mm (area top and bottom flange of 160 mm width and 70mm thickness. The webs is 50 m thick) The properties of section are A=31400mm^2. I=3700 × 10^{5}mm^4 $Z_b = Z_t$ = 230 × 10^4 mm^3, self weight=0.76kN/m. It is prestressed with a force of 182 kN at an ecentricity 105 mm. Assume E_s/E_c=6, E_c=3.0 × 10^{4}N/mm^2 f_{cr}= 4N/mm^2. Evaluate the maximum deflection at the following stages.

a) Pressers + self weight of the beam.
b) Prestress + self weight + live load of 6 kN/m on a span of 6m
c) Cracking load
d) 1.5 times the working load.

Solution

Section Properties

A = 31400mm^2 P = 182000N
I = 3700 × 10^{5}mm^4 e = 105 mm
Z = 230 × 10^{4}mm^3 E_c = 3 × 10^{4}N/mm^2
L = 6000mm α = 6.0
$f_{cr} = 4N/mm^2$

Self weight the beam = 0.76kN/m = 0.76N/mm.

Deflection due to prestress ↑ $= \frac{P_e L^2}{8 E_c I_g} = \frac{18.2\times10^4\times10^5\times6000^2}{8\times3\times10^4\times3700\times10^5}$

$$\frac{18.2\times10^5\times36\times10^6}{8\times3\times37\times10^7} = \frac{18.2\times10.5\times36}{8\times3\times37} = 7.75mm\uparrow$$

Deflection due to self weight $= \frac{5}{384}\cdot\frac{wL^4}{E_c I_g}$

$$= \frac{5\times0.76\times6000^4}{384\times3.0\times10^4\times3700\times10^5}$$

$$= \frac{5\times0.76\times36\times36}{384\times3.0\times3.7}$$

= 1.155mm↓

Working load on the beam = (0.76 + 6.0) 6.76kN/m

Working moment = 6.76 × 6^2/8 kNm

= 30.42 kNm

Deflection due to working load $\frac{5}{384}\times\frac{6.76\times6000^4}{3.0\times10^4\times3700\times10^5} = 10.2mm$

Stress at bottom fibre due to Prestress and working load.

$$\frac{182000}{31400} + \frac{182000\times105}{230\times10^4} - \frac{30.42\times10^6}{230\times10^4} = 5.80 + 8.31 - 13.22$$

= 0.89(compression)*N/mm*^{2}

Cracking Load Corresponds to the load which causes a stress level of 4N/mm^2 tension at the bottom. The moment beyond the working moment which will cause cracking.

= (4.0 + 0.89) × 230 × 10^4
= 1124.7 × 10^{4}Nmm
= 11.25kNm.

Therefore, cracking moment = (11.25 + 30.42) = 41.67kNm

Cracking load = 41.67 × 8/6^2 = 9.26kN/m = 9.26N/mm

Deflection at the time of cracking moment

$$= \frac{5 \times 9.26 \times 6000^4}{384 \times 3 \times 10^4 \times 3700 \times 10^5} = 14.07 mm \downarrow$$

Properties of transformed section

Let 'x' be depth of neutral axis from top fibre and 'b' be width of cross section, 'd' is effective depth

$$\frac{b.x^2}{2} = \alpha_s A_s (d - x)$$

$$160.\frac{x^2}{2} = 6 \times 4 \times .785 \times 49(320 - x)$$

$$80x^2 = 923.16\,(320 - x)$$

$$80x^2 = 295411.2 - 923.16x$$

$$80x^2 + 923.16x - 295411.2 = 0$$

$$x^2 + 11.54x - 3692.64 = 0$$

$$x = 55.5\text{mm}$$

Second moment area of transformed section = I_t

$$= \left[\frac{M_{cr}}{M}\right]^3 I_g + \left[1 - \left(\frac{M_{cr}}{M}\right)^3\right] I_{cr}$$

$$I_{cr} = 160 \times \frac{(55.5)^2}{3} = 9.1175 \times 10^6 mm^4$$

$$I_g = 3700 \times 10^5\ mm^4$$

Deflection due to 1.5 times working load.

Load $= 1.5 \times 6.76 = 10.14$

Corresponding moment $= 10.14 \times 6^2/8 = 45.63$ kNm

Cracking moment $= 41.67$

$$\left[\frac{M_{cr}}{M}\right]^3 = \left(\frac{41.67}{45.63}\right)^3$$

$$I_{cr} = \left(\frac{41.67}{45.63}\right)^3 I_g + \left[1 - \left(\frac{41.67}{45.63}\right)^3\right] I_{cr}$$

$$= 28.3930 \times 10^7$$

Deflection $\frac{5}{48} L^2 \left[\frac{M_{cr}}{ECI_g} + \frac{M - M_{cr}}{0.85 + E_c.I_{tr}}\right]$

$= 14.07 + 2.05 = 16.12$mm

Deflection due to prestressing $= 7.75$ mm$\uparrow$

Due to prestressing +self wt. $= 7.75\uparrow - 1.155\downarrow = 6.595\uparrow$ mm

Due to superimposed load only $= 10.2\downarrow - 1.155\downarrow$

$= 9.045\downarrow$ mm

Due to working load $= 10.2$ mm$\downarrow$

Due to prestress + working load $= 10.2\downarrow - 7.75\uparrow = 2.45\downarrow$ mm

Due to prestress + cracking load (just before cracking) $= 14.07$mm$\downarrow - 7.75\uparrow$ mm

$= 6.32\downarrow$mm

due to prestress +1.5 working load $= 16.12\downarrow - 7.75\uparrow$

$= 8.37\downarrow$ mm

If long term deflection is called for, creep effect shall be used only for sustained loads namely prestress + Dead Load. The formula given earlier may be used.

7.5.4 Long term deflection of cracked members

As done in case of Long term deflection of uncracked members the simplest form of evaluation would be to multiply the short term deflection appropriately by a coefficient which would take care of creep, and shrinkage effects of concrete and relaxation properties of prestressed high tensile steel. While I.S. formulations do not permit cracked prestressed structures (Type 3) B.S.-8110-1985 (3) and ACI-318-1998 (2) provide simplified approaches. ACI-318-1998 recommends that the long term deflections can be arrived at by multiplying the short term deflection by a factor $\lambda = \left[\frac{\xi}{1+50p^1}\right]$ where $p' = A_s'/bd$ at the place of maximum bending moment, and A'_s – the area of compression zone reinforcement.

b - the width of element in compression zone.

d - the effective depth.

ξ -a factor dependent on time and the same can be assumed to be as follows:

5 years and more	2.0
1 year(12 months)	1.4
6 months	1.2
3 months	1.0

B.S 8110-1985 recommends the calculation of instantaneous curvature caused the permanent loads and the same to be subtracted from curvature caused by total load. This instaneous curvature due to permanent load alone should be modified by reducing the value of $E_{longterm}=E_c/1+\varphi$ where φ is the creep coefficient. Then the curvature due to shrinkage effects to be calculated separately as $\frac{1}{r_s} = \frac{E_{cs}\alpha_c S_e}{I_g}$

With this modification, the total curvatre has to be reevaluated and substituted in the expression for deflection. B.S 8110-1985(3) recommends that curvatures be may be considered as two terms a) instantaneous curvature of permanent loads and b) Instantaneous curvature due to other loads for long term deflections. The instantaneous curvature due to permanent load ($1/r_p$) should be modified for the reduced values of modulus of Elasticity of concrete namely ($Ec/1+\varphi$) and the curvature due to shrinkage ($1/r_s$) be added to it. This new value of curvature ($1/r_{et}$) shall be used alongwith instantaneous curvature caused by temporary load to evaluate the deflection using the formula in art 7.5.3 The curvature caused by shrinkage stresses is calculated using an expression.

$$1/r_s = \left[\frac{\in_{cs.}\,\alpha_{c.}S_c}{I_g}\right]$$

where $\in_{cs}$ = shrinkage strain
α = modulus ration E_s/E_{eff}
S_c = First moment of are of gross section.
I_g = Second moment of area.

In a simplified form, and as an approximate solution, the long-term deflection can be calculated using the short-term deflection and the formula adopted for uncracked elements for creep effects.

7.5.5 Limitation on deflection in various national and international codes of practice

It is the general practice in most of the codes to safeguard against excessive deflections under serviceability limit states, either indirectly by prescribing a minimum span to depth ratio for the member, or directly by specifying a maximum permissible deflection expressed as a fraction of the span.

The recommendations of the Indian standard code IS:1343-1980 (1) with regard to the limit state of deflection are as follows:

- The final deflection, due to all loads including the effects of temperature, creep and shrinkage measeured from the as-cast level of supports for all horizontal members should normally not exceed span/250.
- The deflection, including the effects of temperature, creep and shrinkage occuring after the erection of partitions and the applications of finishes, should not normally exceed span/350 or 20 mm, whichever is less.
- If finishes are to be aplied to the prestressed concrete members, the total upward deflection should not exceed span/300, unless uniformity of camber between adjacent units can be ensured.

The British code BS:8110-1985 (3) specifies a maximum deflection limit of span/250, beyond which the sag in a member will usually become noticeable. To prevent damage to non-structural elements, the code recommends that the deflection after the installation of finishes and partitions should not exceed the following values:

- Span/500 or 20 mm, whichever is less, for brittle materials.
- Span/350 or 20 mm, whichever is less, for non-brittle partitions or finishes.

FIP recommendation specify a maximum total deflection of span/200 to span/300 below the level of supports. The deflection that could occur after addition of partitions could be span/500 to span/1000. It suggests means of limiting deformations by

- Prestressing to a preferred degree.
- Use of high strength concrete with low water cement ratio.
- Careful casting of concrete (quality control).
- Removal of false works as late as possible.
- Proper dimensioning.
- Avoiding high span to depth ratio.
- By providing initial canber.

ACI 318-1998 specifies limitations to deflection as given in Table 7.8

Table 7.8 Maximum Permissible Deflections (ACI:318-1998)(2)

Type of member	Deflection to be considered	Deflection limitation
Flat roofs not supporting and not attached to non-structural element likely to be damaged by large deflections.	Immediate deflection due to the live load.	(Span/180)
Floors not supporting and not attached to non-structural elements likely to be damaged by large deflections.	Immediate deflection due to the live load.	(Span/360)
Roof or floor construction supporting or attached to non structural elements likely to be damaged by large deflections	That part of the total deflection which occurs after attachment of the non-structural elements (sum of the long time deflection due to all sustained loads and the immediate deflection due to any additional live load)	(Span/480)
Roof floor construction supporting or attached to non-structural elements not likely to be damaged by large deflections.		(Span/240)

7.5.6 Cracks and crack widths in type 3 elements

Cracking is primarily caused by incompatible strain levels. If the material capacity is less than the strain caused by flexure, there will be flexural cracks. If reinforcements are used along with concrete, the compatibility of strain between steel and concrete and the bond properties between then also play a role. Researches have established that crack width and cracks spacing in reinforced and prestessed concrete elements are functions of

- applied prestress.
- average strain in steel
- minimum concrete cover available for the tensile reinforcement
- overall depth of the member, and

- depth of that portion of the member under tension zone.

Three distinct schools of thought have emerged in the past for evaluation of crack width and crack spacing. They are

- bond-slip hypothesis
- no slip hypothesis, and
- crack stabilisation theory.

Expression have been arrived at for crack width and spacing with all these three philosophies and experimentally verified.

As for as prestressed concrete elements are concerned only type 3 elements are permitted to have crack width and they are limited to certain values depending on the exposure conditions of the structural elements for distress. Limitations of crack widths are defined differenly in different statutory authorities. Indian National codes I.S. 1343-1980 and British Code have provided for indirect control of crack width of values varying from 0.1mm to 0.2mm for various strength of concrete by defining hypothetical tensile stress on the elements at the extreme fibre level based on uncracked concrete cross sections. These values are given in Table (2.6). This is based on fictions tensile stress approach developed by Bebby (10) and the related formula for crack width is

$\omega = 1.75 f_t (h - x) E_c$

where, f_t is the fictions tensile in concrete, h is the overall depth of member, x is the depth of neutral axis and E_c modules of elasticity of concrete

British code B.S.8110 - 1988 however provides for more detailed procedures for calculation of crack width based on the work of Bebby (10) for reinforced concrete members. Hence it has to be judiciously applied for prestressed concrete members after the elements has reached the stage of decompression (ref Fig 7.4)

Crack width is evaluated as

$$\omega = \frac{3a_{cr} \in_m}{1 + 2\left[\dfrac{a_{cr} - c_{min}}{h - x}\right]}$$

where a_{cr} = the distance from the point of cracking to the surface of nearest longitudinal reinforcement.

$\in_m$ = the average strain in steel at the level of cracking allowing for tension stiffening on the concrete in the tensile zone.

c_{min} = minimum cover to tension steel.

h = over all depth of the member.

x = the neutral axis depth calculated as per cracked section assumption.

Using the value of x, $\in_m$ is evaluated as follows

$$\in_m = \in_1 - \frac{b_1(h-x)(a-x)}{3E_z A_s (d-x)}$$

where $\in_1$ is the strain at the level at which crack is considered

$$\in_1 = \frac{M.x_1}{0.5E_c I_c}$$

x^1 = distance compression force from the point at which crack is measured.

A_s = area of tension reinforcement.

b = width of section at the centroid of tension steel.

x_1 = is the distance measured from the neutral axis to the point at which is measured

m = is the moment at the section.

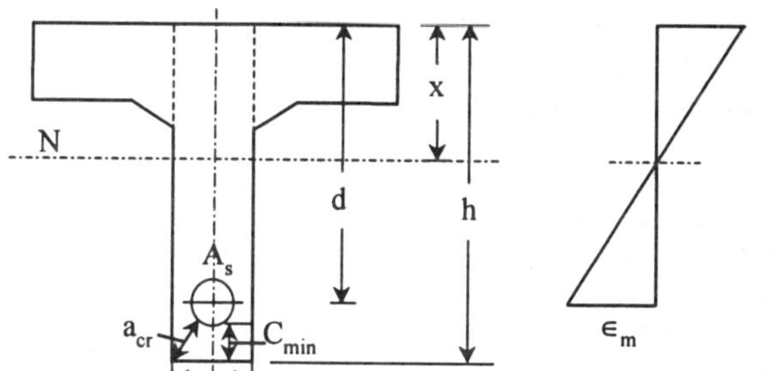

Cross Section Strain Diagram

FIP recommendations of 1984 (11) has suggested evaluation of crack width using bond step hypothesis and the same has included in the formula for crack width recommended in **CEB/FIP** model Code(12).

$$W_k = \left[1.5 + \frac{3.5c}{s}\right]\frac{\sigma_s}{E_s}\left[1 - \left(\frac{M_{cr}}{M}\right)^2\right]$$

where s is the spacing of bars.

c is cover for steel reinforcement.

E_s is modulus of elasticity of steel.

M_{cr} cracking moment.

M Moment under consideration.

Limiting values of crack width for prestressed concrete elements for different explosive conditions are given in different statutory codes are given in tables 7 9 & 7.10.

Table 7.9 Limitation on crack as per FIP - 1984 recommendation

Exposure	Limitations
Mild	w ≤ 0.2 mm.
Normal	w ≤ 0.1 mm.
Severe	No cracks.

Table 7.10 Limitation of crack width as perACI - Committee - 224 Recommedations

Exposure condition	Crack width mm (max)
Dry air protective member	0.41
Humidity, moist air	0.30
Chemical atmosphere	0.18
Sea Zone with salt web spray permissible	0.15
Water retaining structures	0.10

References

1. IS 1343 - (1980). Indian Standard **"Code of Practice for Prestressed Concrete"**, Bureau of Indian Standards, New Delhi.
2. ACI 318 - (1998). **"Building Code requirements for Structural Concrete"**, American Concrete Institute, Formington Halls, HI 48333, USA.
3. BS 8110 - (1985). **"British Standard for Structural use of Concrete"**, British Standards Institution, London WIA 2BS.
4. FIP - (1984). **Recommendations for Practical design of Reinforced and prestressed concrete structures, based on CEB - FIP model code (MC - 78),** Thomas Telford Ltd, London.
5. CORLEY, W.G., M.A.SOZEN., and C.P.SIESS, (1961) **"Time dependent deflections of prestressed concrete beams"**, Bulletin No.307, Highway Research Board, pp 1-25.
6. BUSEMANN, R., (1950) **Kriechberechnung von verbundtragern unter Bentzung von zwei Kriechfasern**, Der Bauingenieur, 25, Hesft 11, 1950, pp 418 - 420.
7. TROST. H (1967) **"Implications of the superposition principle in creep and relaxation problems for concrete and prestressed concrete"** Betons Stahlbetonbau 62(10) PP 230-238 & 62 (11) PP 261-169, (in German).
8. **ACI Committee** 435 Report, (1968) ACI Manual of concrete practice, part 2, American concrete institute Detroit.
9. **Recommendation for an international code of practice for reinforced concrete** (1964) American concrete Institute and Cement and Concrete Association London.
10. BEEBY, A.W., (1971) **"The prediction and control of flexural cracking in reinforced concrete members,"** Proceedings of the ACI Symposium on Cracking, Deflection and Ultimate Load of Concrete Slab systems, American Concrete Institute, Detroit.
11. **FIP Recommendation Practical design of reinforced & prestressed concrete structures** (1984) - Thomas Telford Ltd., London.
12. **CEB - FIP** (1970), **International recommendations for the design and construction of concrete structures** FIP sixth the congress, Prague.

8. Design for Flexure

8.1. Introduction

The Philosophy of design of any element has been explained in chapter 7 and this could be well understood by first studying the behaviour of that element under the same system of forces from zero load to failure. The behaviour of a prestressed concrete element under flexure from zero load to failure load has also been already explained in art. 4. 2 and 6.1 through the variation of stresses or strains in steel with load. These is based on actual stress strain relationship in steel. They are also represented in Fig 4.1. and 6.1. These graphs are linear in bits, say linear from B to C, a fall, linear from D to F, a jump, non linear from F_1 to G. For design purposes the strain hardening of steel and the corresponding increase in stresses are not considered and hence the graph is considered linear even from F_1 to G. These two approaches lead to stress based or deformation based designs. In both the methods, ***a ductile failure is preferred at the proposed limit State of collapse***. Hence failure of a reinforced or a prestressed concrete element is expected to be initiated by steel reaching its limiting stresses/strains first, before concrete reaches its limiting stress/strains since steel is more ductile than concrete. If the same graph is drawn for a reinforced concrete elements, and if it is assumed that the concrete does not have tensile strength, the concrete would be cracked even at zero load. Hence the steel takes the full tension even from zero load and the graph would start at the origin and the origin itself becomes point B. The graph is linear from B to G in one strech and the slope of this curve is E_s (modulus of elasticity of steel). This is depicted in Fig. 8.1(b). For purposes of comparision Fig.4.1. is repeated again as Fig.8.1a. In the design, based on stress criteria, the stresses in steel are monitored carefully, and the deformation will be observed and checked with respect to serviceability conditions. The service load in prestressed concrete elements is limited to pre cracking load, if it is Type1 or Type2 structures. It may be a little higher load even beyond cracking load for Type 3 type of structure with a permissable minimum crack width. Types 1,2,3 structural have been defined in I.S. 1343 specifications and are explained in chapter 1.

Failure load (Collapse load) is defined by the capacity of the section in flexure. As could be seen from Fig.8.1 (b), in a reinforced concrete element, if the ratio between service load and collapse load is known, for a given stress level at service load, the stress in steel at failure load (limit stage of collapse) can be evaluated, since the relationship between the load and stress is linear for full stretch starting from the origin. This is not true in case of prestressed concrete elements. It could be seen from

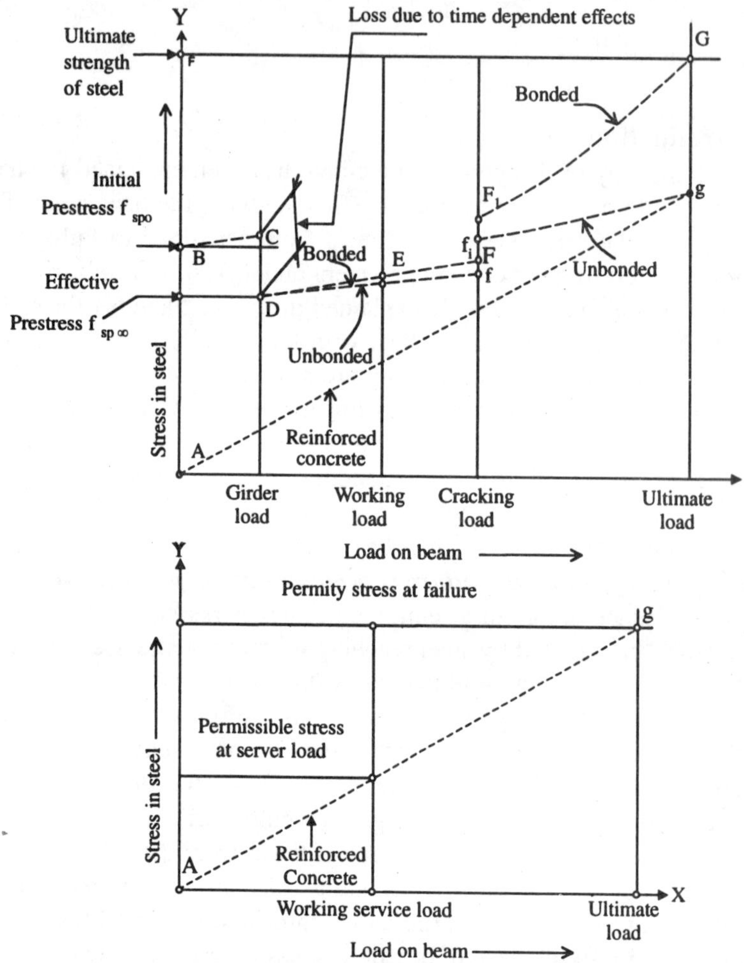

Fig. 8.1 Stresses in Steel at Various Load Stages in a Prestressed Concrete Element Subjected to Flexure

Fg. 8.1a that if the ratio between service load and the collapse load is known, the same cannot be used to evaluate the stress in steel at the failure stage (limit stage of collapse) for given stress level at the service load, since relationship between load Vs stress in steel graph is not linear for the full stretch and the graph does not also start at the origin. This is a major difference between the design of prestressed concrete element and the reinforced concrete element.

Hence if a reinforced concrete element design is done with limiting stress in steel at collapse and permissible stress in steel at service load, is known, the ratio of limits stress level at collapse load to stress level at service load could be used to evaluate the ultimate load carrying capacity of the section from the service load and vice versa. Hence design may be done either using limit state approach or service load (working stress) approach. But in prestressed concrete element design, if a design is made using service load and corresponding stress approach, it is necessary to calculate the ultimate load capacity of the section for given limiting stresses in materials and compare the same with required capacity of section.

Philosophy of design have discussed already in chapter 7 under two headings namely:

- Design based on allowable stress concept during service load .
- Design based on limit state concept or otherwise called limit state method

[The limiting state for the designs have also been broadly outlined as limit state of collapse and limit state of servicability and in either case, i or ii, a check has to be made for the other conditions.]

After the stage of decompression, and at limit state of collapse, the prestressed concrete section behaves only as reinforced concrete element. At the service load the behavior of prestressed concrete element is different from that of reinforced concrete element. To have a control on the design with respect to prestressing, flexural design is normally done based on allowable stresses at service loads and a check is made for the capacity of section (failure load) using the limiting stresses on the materials.

8.2 Flexural Design Using Allowable Stresses at Service Load

This approach should not be considered as working stress approach since the stresses are not evaluated from failure stress using a factor of safety. ***The stresses are based on serviceability of the member at the service load***. More over there will be check on failure capacity based on limit state approach. Hence it is called ***service load stress approach***.

In this approach, the stresses under service loads or any load from zero load stage to service load stage, control the design. Design of a prestressed concrete element consists of

- Selection and proportioning the concrete sectional dimension.
- Selection of proper values of prestressing forces to be applied at initial stage and the points of application of the forces called the eccentricities with respect to the centroid of the sections.

With the chosen section, prestressing forces and eccentricities, the stresses in the concrete elements at salient points should not exceed the permissible value under any combination of loads described in Table 7.1. and at any load stage from zero load to service stage. The concrete section properties are defined by the area provided and the distribution of areas. The distributions of areas are identified by the section modulus at top fibre and bottom fibre. The prestressing force is defined by two values namely P_i (initial) and P_∞ (final), related by a loss factor η. The location of prestress is defined by the term eccentricity 'e'. Hence the design variables are identified as A_c, Z_b, Z_t, P_i, η, e.

The important load stages are

- Initial load stage, i.e.,where the applied prestressing force is the maximum and the applied moment is minimum namely that due to self weight alone.
- Service load stage, i.e., Where the applied prestressing forces gets reduced due to various losses and hence a minimum and the external applied moment is maximum.

If the design satisfies the two extreme load stages mentioned above, it would be safe for any intermediate load stage. Similarly the stresses are maximum only at the extreme fibres (top and bottom) and the variation of stress through the depth may be assumed to be linear. With Navier's hypothesis for pure pending, the strain distribution through the depth of the section is linear. In service load stress approach the permissible stresses at service load are in a range with in which the stress strain relationship both for concrete and steel are linear. Hence the assumption of stress variation to be linear through the depth of the section is justifiable. Accordingly it is sufficient to check the stresses at extreme fibres against the limiting values. If that check is satified, it is justified to assume that the stresses at any other level through the depth of the section will be within limits.

Hence in the service load stress method, the design of the section, is governed by the four stress inequalities as expressed in equation 8.1.a. to 8.1.d. two just after prestressing (at transfer) and two at service load stages.

At transfer i.e., initial load stage:

Top fibre $\quad f_{ctp} + f_{ctdl} \geq \overline{f}_{cttr}$ .....(8.1a)

Bottom fibre: $\quad f_{cbp} + f_{cbdl} \leq \overline{f}_{cbtr}$ .....(8.1b)

Service load stage:

Top fibre: $\quad \eta f_{ctp} + f_{ctdl} + f_{ctll} \leq \overline{f}_{cts}$ .....(8.1c)

Bottom fibre: $\quad \eta f_{cbp} + f_{cbdl} + f_{cbll} \geq \overline{f}_{cbs}$ .....(8.1d)

Where $f_{cpt}, f_{cpb}, \eta f_{cpt}, \eta f_{cpb}$ are stresses at top (t) and bottom (b) fibres due to initial prestress and final prestress P_∞

$f_{ctdl}, f_{cbdl}, f_{cll}, f_{cbll}$ are stresses at top (t) and bottom (b) fibres due to Dead load (DL) or live Load (LL).

$\overline{f}_{cbtr}, \overline{f}_{cbs}$ are permissible stresses at top and both at transfer (tr) or service load (s) stages. Compressive stresses are assumed to be positive.

There are four stress inequalities to be satisfied with six unknowns to be solved to arrive at a satisfactory design. It is clear that there cannot be unique solution in such cases. Hence the number of unknowns have to be reduced and the problem solved by making some reasonable assumptions. Later the assumptions must be justified. If the assumptions are found to be wrong or unjustifiable, the design has to be altered. The first solution would give a better understanding for making more appropriate asumptions for the revised design. Hence the design is done in two stages namely. (a) preliminary design and (b) final design.

In the preliminary design, the cross section dimension, would be fixed up and in the final design, the required prestressing force, proper eccentricity and any other modification for the section to be made etc would be decided.

There are number of approaches for such designs namely i) Stress range approach, ii) LIN's approach, iii) Magnel's approach with dimension less factors etc., They are discussed one after other.

8.3 Stress Range Approach (1)

In this approach, the range of stresses through which the top or bottom fibre stresses can move from initial prestressing stage to service load stage is decided based on the permitted upper and lower limits of stresses on concrete at top and bottom fibre. Using the inequalities mentioned earlier, to be satisfied, the moduli of section required with respect to top fibre and bottom fibre are first evaluated. Using these values of moduli of section the required prestress at final stage and the permitted eccentricity for the prestress is decided. Afterwards proper dimensions of the section are chosen to achieve the moduli section.

8.3.1 Modulii section

Considering two critical combinations of prestress and moment already explained as important load stages in art 8.2. The permissible values of stresses in concrete at top and bottom fibres, expressions are derived from Z_b, & Z_t. In these expressions compressive stresses are considered positive and tensile stresses are negative.

a) At transfer:

top fibre:

$$\left[f_{ctp} + \frac{M_g}{Z_t}\right] \geq \bar{f}_{cttr} \qquad \text{..... (8.2a)}$$

bottom fibre:

$$\left[f_{cbp} + \frac{M_g}{Z_b}\right] \leq \bar{f}_{cbtr} \qquad \text{..... (8.2b)}$$

b) At service load:

top fibre: $$\left[\eta f_{ctp} + \frac{M_g}{Z_t} + \frac{M_a}{Z_t}\right] \le \bar{f}_{cts} \quad .....(8.2c)$$

bottom fibre: $$\left[\eta f_{cbp} + \frac{M_g}{Z_b} + \frac{M_a}{Z_b}\right] \ge \bar{f}_{cbs} \quad .... (8.2d)$$

From equation 8.2.a and 8.2.c

$$\frac{M_a + (1-\eta)M_g}{Z_t} \le (\bar{f}_{cbs} - \eta\bar{f}_{ctt}) \le \bar{f}_{ctrange} \quad .....(8.3a)$$

From equation 8.2c and 8.2d

$$\frac{M_a + (1-\eta)M_g}{Z_b} \le (\eta\bar{f}_{cbs} - \bar{f}_{cbs}) \le \bar{f}_{cbrange} \quad .... (8.3b)$$

Using 8.3a and 8.3b

$$Z_t \ge \frac{M_a + (1-\eta)M_g}{\bar{f}_{ctramge}} \quad .....(8.4a)$$

$$Z_t \ge \frac{M_a + (1-\eta)M_g}{\bar{f}_{ctramge}} \quad .... (8.4b)$$

η generally lies between : 0.8 to0.85 for post tensioned member 0.75 to 0.8 for pre tensioned member

8.3.2 Prestressing force

The section chosen may have the Z_t, Z_b values a little more than what is required as per expression 8.3.g. and 8.3.h. The prestressing force may also have an upper and a lower limit. The upper limit is governed by initial conditions (specially top fibre stress) and the lower limit by final or service load conditions (specially bottom fibre stress). Hence rearranging the equations 8.2.a and 8.2.b

$$f_{ctp} \ge \bar{f}_{ctr} - \frac{M_g}{Z_t} \quad .....(8.5a)$$

$$f_{cbp} \ge \frac{f_{cbs}}{\eta} + \frac{(M_g + M_a)}{\eta Z_b} \quad .....(8.5b)$$

Here actual Z_b, Z_t value have to be substituted.

The Values of f_{ctp} *and* f_{cbp} can be calculated as stresses due to prestressing force and its eccentricity. For eccentricity an initial value can be assumed providing for a accepted concrete cover from the effect.

$$f_{ctp} = \frac{P_o}{A_c} - \frac{P_o e}{Z_t}$$

Eliminating e from equation

$$P_o = A_i \left[\frac{f_{cpb} Z_b + f_{ctp} Z_t}{Z_t + Z_b} \right] \qquad .....(8.6a)$$

$$e = \left[\frac{Z_t Z_b (f_{cbp} - f_{ctp})}{A(f_{ctp} + f_{cbp})} \right] \qquad .....(8.7b)$$

This method of design is illustrated by example 8.1 later

8.4 LIN's Approach

For Lyn's approach (2) to be followed, an initial design arriving at the cross sectional dimensions of the concrete section has to be made.

8.4.1 Preliminary design for cross sectional properties.

The following empirical relations are made use for arriving at the preliminary dimensions of the cross sections (ref 3).

- Slenderness ratio (L/d ratio) may vary from 1/15 for 1/20 where L is the span and d is the total depth.
- Depth of section in meters may be expressed as a function of applied moment namely k $\sqrt{M}$. Where k may be assumed anywhere between 0.1 to 0.15 and M is the total bending moment to be resisted in tonnes meter.
- Live Load moment may be around 80% of total moment in short span structures and 30% to 40% in long span bridges and 60% to 70% in long span roof structures. Hence if the live load bending moment is known, the total bending moment can be evaluated.
- Failure moment Mu may be assumed as
  - M_u= 0.098 *fck* bd^2 to 0.170 *fck* bd^2 for post tensioned rectangular beams.
  - M_u= 0.101 *fck* bd^2 to 0.189 *fck* bd^2 for pretensioned rectangular beams.
  - M_u= 0.085 *fck* bfd^2 to 0.099 *fck* bfd^2 for post tensioned flanged beam.
  - M_u= 0.085 *fck* bfd^2 to 0.1028 *fck* bfd^2 for prestressed beams.

    Where *hf/d* = 0.2 and $b_w/b_f = 0.2$

    These expressions are based on assumption that the structure has a percentage of steel defined by the factor $\left[\frac{A_{sp}}{b_d t_{ck}}\right]$ anywhere between 0.15 to 0.4.

    Hence if *fck*, and '*d*' is known b_l , b_w, b_f etc., can be evaluated.
- I section is preferred for shortspan structures and 'T' section is preferred for long span structures.

8.4.2 Prestressing force and eccentricity

It is known that concrete is weak in tension and hence very little tension or no tension can be permitted in concrete at top or bottom at any load stage.

At service load stage normally the tension will be the resultant stress at the bottom and that should be limited to zero. The compressive stress at the top will be maximum and that should be limited to the permissible values $\bar{f}_{cts}$. The stress pattern will be triangular over the depth and the centre of compression will be at the top kern point. Assuming a rectangular section the top kern point would be at 0.66H from bottom fibre. The centre of tension will be at the level of steel, which could be obtained at a distance equal to cover from bottom soffit. The cover is normally assumed as 0.05H. Then the distance between the centre of compression and centre of tension is 0.61H. Then total tensile free or the compressed force is evaluated as total moment to be resisted at service load divided by the lever arm and expressed as $M_t/0.61H$. From this total prestressed force required, on the cross sections, the area of concrete that could resist that force is evaluated as the force/divided by average permissible compressive stress, Average compression stress is assumed as half the permissible stress $\bar{f}_{cts}$. Hence area of concrete required is evaluated as force divided by $0.5\,\bar{f}_{ets}$

Having known the area of concrete required, depth and width of section, the shape of the section are emperically decided and other section properties such as (y_{bg}, y_{tg}, I_g, Z_{bg}, Z_{tg}, k_{tg}, k_{bg}) are evaluated.

Flexural Design of prestressed concrete elements are grouped under three categories namely Type 1, Type 2, and Type 3 structures as already defined in chapter 1. But I.S. 1343 - 1980 (3) allows only Type 1 structures for constructions. Hence that design is dealt with first. Type 2 group of elements are also covered under prestressed concrete and hence their designs are discussed later. Type 3 group of elements are covered in the chapter on partial prestressing. I.S. 1343 does not allow partially prestressed concrete elements design for construction in India.

8.4.3 Design of Type 1 group of elements

In this design, no tension is permitted at any load stage up to service load stage in the elements. As mentioned earlier the design is normally done only for two load stages namely.

- Transfer Load Stage when prestress is just transferred. At this stage the section is subjected to maximum prestress (P_i) and minimum external moment (self weight moment M_g only)
- Service load stage, when the section is subjected the maximum external moment (M_t) caused by all loads inclusive of service load and minimum prestress (P_∞) after all losses.

The section properties like A_g, y_{bg}, y_{tg}, Z_{bg}, Z_{tg}, k_{bg}, k_{tg} for the cross section are known after the preliminary design. These values are used to arrive at the prestressing force and eccentricity. The loss factor η is also assumed.

Since ***no tension*** is permitted in concrete at any load stage, the distribution of stresses in concrete through the depth could at the most be triangular with zero stress at top just after transfer and zero at bottom at serivce load stage. The maximum compressive stress will be felt at bottom at transfer stage and at top at service load stage. The maximum compressive stresses shall be within permissible limits. At stage I (Transfer Stage) the externally applied moment will be only the self-weight moment (M_g). Just at the time of transfer of prestress, the total tensile force and the compressive forces are collinear. As soon as self weight moment (M_g) starts acting, the centre of compression shifts through a distance s_1 and the moment is resisted by the initially applied prestress (P_i) through the lever arm of s_i.

Hence $P_i\, s_i = M_g$ .....(8.8a)

At this point of time the centre of compression has to be at bottom kern point or above. Then only, the top fibre will not be subjected to tension. And the distance between the centre of compression and centre of tension (Location of steel area) is equal to s_i. Hence the centre of tension, namely the location of centre of area of steel should be at a distance equal to s_1 below the bottom kern point which is at a distance k_b from the centroid of the section. This distance of centre of tension namely the position of steel from the centroid of the section is normally referred as eccentricity 'e'.

$$e = s_1 + k_b = \frac{M_g}{P_i} + k_b \qquad .....(8.8b)$$

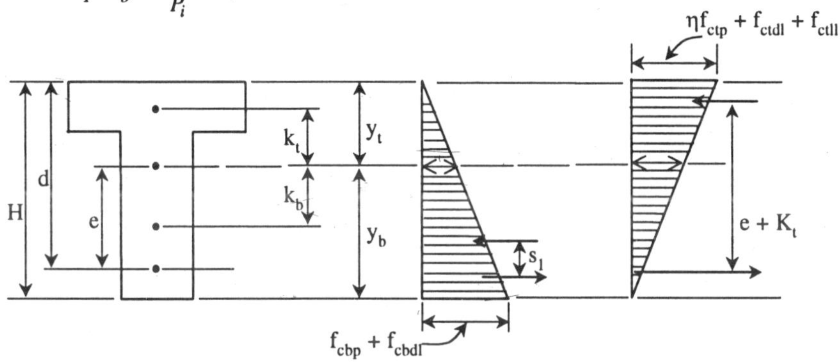

(a) Section Properties (b) Just after Transfer, C above bottom kern point (c) Under Working Load. C at top kern point

Fig 8.2a Stress Distribution No Tension in Concrete

With the approximately assumed value of P_i (from the preliminary design) eccentricity of 'e' can be evaluated. This value is to be checked against the available distance from the centroid of the section allowing for proper cover for outer most layer of prestressing steel.

As the beam is loaded further, and when the section is subjected to maximum service load, the section is subjected maximum bending moment caused by dead load, incidental load and service loads. And that time the applied prestress gets reduced to a minimum value (P_∞) because of losses in prestress.

Even at this stage, no tension should occur at any point in the section. Any tensile stress if at all, could occur only at the bottom fibre. If no tension is to occur at bottom fibre, the centre of compression should be at the top kern point of the section, namely at a distance kt above the centroid of section. The distance between the centre of tension (location of Centroid) of steel area) to the centre of compression will then be equal to $(e + k_t)$.

This is the lever arm for the forces to form a couple and resist the total moment (M_t) at the time of full service load. The total tensile force in steel will be equal to P_i plus additional tension, which would have come due to live load bending moment. Since concrete has not reached the stage of decompression at that time, steel is bonded to concrete, it will have only negligible additional tension beyond P_∞.

Even in case of type II structures where tension in concrete permitted upto the level of cracking, the additional strain in steel will be at the maximum equal to the tensile strain in concrete, since the section is still monolithic. The cracking strain in concrete, and the corresponding stress in steel also will not be very high. Hence the increase in tensile force in prestressing steel will not be much and could be neglected. Hence the total tensile force in steel may be assumed as P_∞ at service load stage.

M_t (Total Bending Moment) is resisted by a force P_∞ through a lever arm $(e + k_t)$. Hence the total tensile force in steel at service load stage can independently evaluated as

$$P_\infty = \frac{M_t}{(e + k_t)} \qquad \text{.....(8.8c)}$$

With this value of P_∞, with a given loss factor or with the variation of permissible stress in steel at transfer and at service load P_i can be calculated as

$$P_i = P_\infty / \eta \qquad \text{.....(8.8d)}$$

η = Loss factor or the ratio of permissible stress in steel at service load to that transfer load stage.

Hence having got the prestressing force and eccentricity, the earlier assumed values of the cross section properties, can be checked. The most important of those is the area of cross section. The condition that no tension is allowed to occur at transfer and at service load (at top and bottom respectively) have been made use of in arrriving at the eccentricity and pretressing force (Refer equation 8.4b and 8.4c). Hence the two other conditions limiting compressive stresses both at bottom and top during transfer stage and service load stage may be made use of for arriving at the minimum area of concrete required at both stages. The assumed cross section should have area more than those arrived using these conditions.

Writing the expression for limiting compressive stresses at the bottom fibre at transfer, and at top fibre at service load stage, area of concrete required can be evaluated

$$f_{cbtr} = \left[\frac{P_i}{A_g} + \frac{P_i.e_g}{Z_{bg}}\right] + \frac{M_g}{Z_{bg}} \geq \bar{f}_{cbtr} \qquad \text{.....(8.9a)}$$

$$\left[\frac{P_i}{A_g} + \frac{P_ie_g}{A_g.k_t} + \frac{M_g}{A_g.k_t}\right] \geq \bar{f}_{cbtr}$$

$$A_g = \frac{1}{\bar{f}_{cbtr}}\left[P_i + P_i \cdot \frac{e_g}{k_t} + \frac{M_g}{k_t}\right] \qquad \text{.....(8.9b)}$$

$$f_{cts} = \left[\frac{P_\infty}{A_g} + \frac{P_\infty.e_g}{Z_{tg}} + \frac{M_t}{Z_{tg}}\right] \geq \bar{f}_{cts} \qquad \text{.....(8.10a)}$$

$$\left[\frac{P_\alpha}{A_g} \quad \frac{P_\infty.e_g}{A_gK_b} + \frac{M_t}{A_gK_b}\right] \geq \bar{f}_{cts}$$

$$A_g = \frac{P_\alpha}{\bar{f}_{cts}}\left[P_\alpha + \frac{e_g}{k_b} + \frac{M_t}{k_b}\right] \qquad \text{.....(8.10b)}$$

Area of concrete should be more than the greater than the two arrived at using expression (8.9b, 8.10b). If the area provided is less than any one the areas, the section has to be redesigned and the procedure for arriving at eccentricity and prestressing force may be repeated.

(Note: if the area required at transfer load stage is only marginally inadequate, say up to 5% and the area required at service load stage is less than actually provided, the redesign can be avoided by increasing the strength of concrete, or postponing the transfer load stage, till concrete acquired sufficient strength, or by marginally increasing 'e' value or by temporary loading on the top.)

Once the design calculations as described are over the design parameters may be summed up as P_i, P_∞ and e.

The actual prestressing force that could be transmitted depends on the availability of type of tendons, and system of prestressing, Based on the market availability of steel and system, the actual prestressing force the type of steel and their locations may be decided. (Ref. Illustration examples). Then the eccentricity should be accordingly adjusted. After deciding about the type of steel, their location the nett and effective section properties may be calculated.

In case of post tensioned beams at transfer stage nett section properties may be used and at service load stage effective section properties may be used.

Finally the moment carrying capacity of the section may be calculated and compared with the factored required moment carrying capacity as per detail given in chapter 6. If the actual moment carrying capacity is less, untensioned reinforcement may be added in the Tension Zone and the moment carrying capacity increased to the required level.

8.4.4 Design of Type II group of elements

In this group of prestressed concrete elements, tensile stresses are permitted in concrete up to the point of cracking and the concrete under tension may also be considered for resisting the external moment. In many occasions, the worst loads for which the structures are designed are single time occurrences. It is only reasonable to use the concrete in tension for resisting moments under such unique and rare occasions. If tensile stresses permitted in concrete at transfer stage or at service load stage, it is possible to allow the centre of compression to be below the bottom kern point at transfer load stage and above the top kern point at service load stage. At transfer stage if the permissible tensile stress at top fibre is $-f_{cttr}$ the centre of compression can be located below the kern point by a distance s_2, so that moment created by the total compressive or tensile force through the lever arm of s_2 should generate a tensile stress $-f_{cttr}$ at top.

$$P_i s_2 = -f_{cttr} \cdot \frac{I_g}{y_{tg}}$$

$$s_2 = \frac{-f_{cttr} \cdot A_g k_b}{P_i} \qquad \text{.....(8.11)}$$

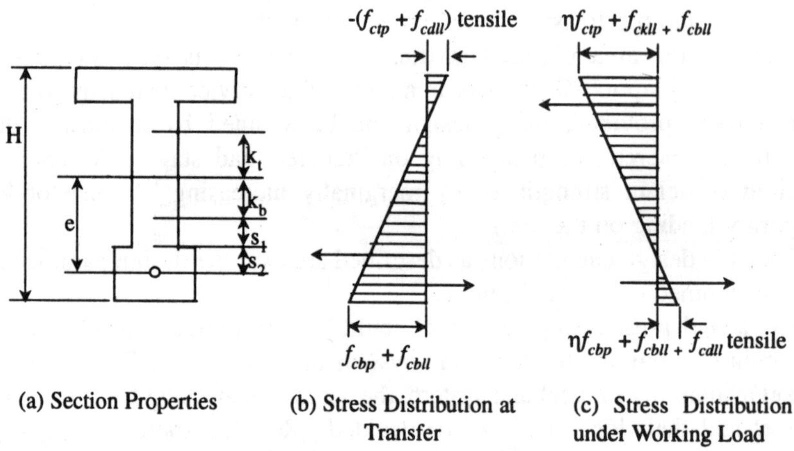

Fig 8.2 Stress Distribution Allowing and Considering Tension in Concrete

Hence for a given dead load moment M_g, the centre of steel can be located at a distance $(s_1 + s_2)$ below the kern point where $s_1 = M_g/P_i$,

$$s_2 = \frac{f_{cttr} \cdot A_g k_b}{P_i}$$

As soon as the prestress is transferred, the dead load starts acting and the centre of compression gets shifted by a distance s_1 from the steel level upwards. This position of centre of compression can be at distance s_2 below the bottom kern point. Hence the position of steel can be at a distance 'e' from the C.G. of section and e can be expressed as $k_b + s_2 + s_1$

$$e = k_b\left[\frac{f_{cttr}.A_g k_b}{P_i} + \frac{M_g}{P_i}\right] \qquad .....(8.12)$$

For an assumed initial prestressing force P_i the eccentricity can be calculated.

Similarly at service load stage, if tensile stresses are permitted at bottom fibre to a level of $-f_{cbs}$, the centre of compression can be above the top kern point by a distance s_3. This distance as lever arm, with the prestressing force P_∞ could develop a bending moment, which would lead to a tensile stress of $-f_{cbs}$ at bottom fibre.

$$\text{Hence } P_\infty s_3 = \frac{f_{cbs} I_g}{y_b} \qquad .....(8.13)$$

The total movement of centre of compression from the level of steel while resisting the total bending moment can be expressed as $[e + k_t + s_3]$

Hence $P_\alpha (e + k_t + s_3) = M_T$

$P_\infty (e + k_t) = M_T - P_\infty s_3$

$P_\infty (e + k_t) = M_T - f_{cbs} Ak_t$

$$P_\infty = \frac{M_T - f_{cbs} Ak_t}{e_t k_t} \qquad .....(8.14)$$

Using this expression, P_∞ can be evaluated. From the ratio of initial stress to final stress in steel or using a loss coefficient value, initial prestressing force P_i can be arrived at.

The concrete cross sectional area provided should be such that the stress level do not cross the limiting compressive stresses at bottom at transfer stage and at top during service stage

According the area of cross section required at transfer stage can be arrived from

$$f_{cbtr} = \frac{P_i}{A} + \frac{P_i e}{Z_b} - \frac{M_g}{Z_b} \qquad .....(8.15a)$$

$$f_{cbtr} = \frac{P_i}{A} + \frac{P_i e}{Ak_t} - \frac{M_g}{A_{kt}}$$

$$A_i = \frac{P_i}{f_{cbtr}} + \frac{P_{ie}}{f_{cbtr}} - \frac{M_g}{Ak_t . f_{cbs}} \qquad .....(8.15b)$$

Area of concrete required at service load stage can be arrived from

$$f_{cts} = \frac{P_\alpha}{A} + \frac{P_\alpha e}{Ak_b} + \frac{M_T}{Ak_t} \qquad .....(8.16a)$$

$$A = \frac{P_\alpha}{f_{cts}} + \frac{P_\alpha . e}{f_{cts}} + \frac{M_T}{k_b f_{cbs}} \qquad .....(8.16b)$$

After finding out the area required to keep the compression stress within limit both at transfer stage and service load stage, the actual steel area to be used, the type of prestressing element, anchorage details, duct details etc., have to be worked out. Then the exact section properties of the practical section chosen have to be evaluated and stresses checked at top and bottom fibres both at transfer stage and service load stage. Later a check on the moment carrying capacity has also to be made and if necessary untensioned reinforcement may have to be added to achieve the required safety factor against collapse.

Example 8.1

A small precast prestressed concrete beam is to be designed to cover a span of 12 meters and to carry a super imposed load of 15kN/mR. The quality of concrete proposed to be used is M45. The permissible stress is compression can be 14N.mm^2 and in tension 1.4N/mm^2. Force loss in prestressing cables can be assumed to be 15% of the initial force during service load conditions. Design the cross section dimensions; prestressing force and eccentricity. Adopt stress range approach for arriving at the cross sectional dimensions.

Solution

Span $= 12$ m

Superimposed load $= 15 kN/mR$

Bending Moment due to super imposed load $= \dfrac{15 \times 12 \times 12}{8} = 270 kNm$

Provide for 20% additional moment due to self weight $= 54\ kN$

Total bending moment $= 324\ kNm$

Preliminary design

Adopt a depth of beam of $\dfrac{L}{16}$ $= \dfrac{12}{16} = 0.75m$

Prestressing force $= \dfrac{324 \times 10^6}{0.61 \times 750} = 708196.9N$

Area of concrete required $= \left[\dfrac{708196.9}{0.85}\right] \times \dfrac{2}{14}$

$= 119024.6\ sq.mm$

Since it is a precast unit a symmetrical I section is assumed, with top and bottom width of 300mm and flange depth of 150mm and web width of 120mm.

Area $= 144000\text{mm}^2$

Moment of Inertia $= 300 \times \dfrac{750^3}{12} - 180 \times \dfrac{450^3}{12}$

$= 1.0547 \times 10^{10} - 0.1367 \times 10^{10}$

$= 0.918 \times 10^{10}\ mm^4$

$$Z_b = Z_t = \frac{0.918 \times 10^{10}}{375}$$

$$= 24.48 \times 10^6 mm^3$$

$f_{cbrange} = 0.85 \times 14 - (-1.4) = 13.3 N/mm^2$

$f_{ctrange} = 14 - 0.85(-1.4) = 15.19 N/mm^2$

$$M_g = \left(\frac{144000}{10^6}\right)24 \times \frac{12^2}{8}$$

$$= 62.208kNm$$

$$Z_b = \frac{[270 + 0.15 \times 62.208]}{13.3} = 21.00 \times 10^6 mm^3$$

$$Z_t = \frac{270 + 0.15 \times 62.208}{15.19} = 18.39 \times 10^6 mm^2$$

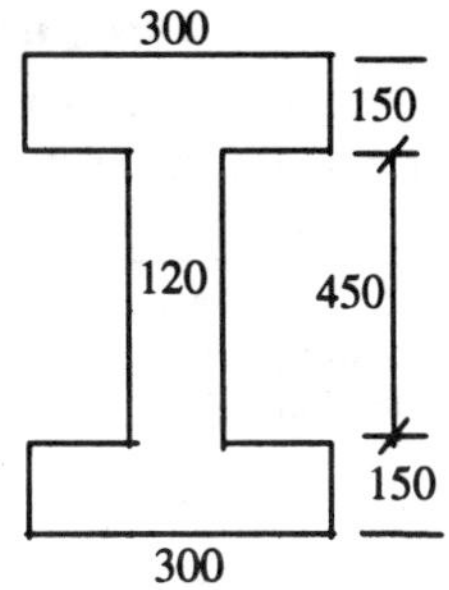

Z provided in more than what is required

Hence the section adopted is adequate

Prestressing force required at section load stage = P_∞

$$P_\infty = A\left[\frac{f_{cbp}Z_b + f_{ctp}Z_t}{Z_t + Z_b}\right]$$

$$f_{cbp} = \frac{-1.4}{0.85} + \frac{(270 + 62.208)10^6}{0.85 \times 24.48 \times 10^6}$$

$$- 1.647 + 15.966 = 14.32$$

$$f_{ctb} = -1.4 - \frac{62.208}{24.48} = -1.4 - 2.501 = -3.941$$

At transfer stage, with dead load effects the stress will be come less that 14N/mm^2 at bottom and less than 1.4N/mm^2 tension at top.

$$P_\infty = 144 \times 10^3 \frac{[14.32 - 3.941]24.48 \times 10^6}{2 \times 24.48 \times 10^6}$$

$$= 747.288kN$$

$$e = \frac{24.48 \times 24.48 \times 10^{12}}{144 \times 10^3}\left[\frac{14.32 - (-3.941)}{24.48 \times 10^6 \times (14.32 - 3.941)}\right]$$

$$= \frac{24.48 \times 10^6}{144 \times 10^3} \times \frac{18.261}{10.396} = 298.6mm$$

Provide 300mm eccentricity with a centre cover of 75 mm for steel 747.288kN of force can be got 24 nos of 7mm in 2 cable (12 × 7) stressed to 1000 N/mm^2 initially will give a force of 92.3161kN. With a percentage loss of 15% the final pressers will be 784.68 KN>747.288kN

Hence the design is adequate.

Example 8.2

A Prestressed concrete beam is to be designed to resist a bending moment of 500 *kNm* at service load conditions. The height or the depth of the beam is restricted to 900mm.High tensile steel of strength 2000 *N/mm*^{2} is available and it can be pre-tensioned upto 0.6f_p during initial prestressing, Concrete quality with f_{ck} of 45 *N/mm*^{2} is proposed to be used with a permissible compressive stress of 14 *N/mm*^{2} at service load conditions. The prestressing is proposed to be transferred only after the concrete attains the full strength of 45 *N/mm*^{2}. The dead load bending moment may be assumed to 100*kNm*. No tension is permitted in concrete at any stage (Type I - structure). Design the cross section and the details regarding prestressing, using Lyn's approach.

Solution

Preliminary design

Since no tension is permitted at any stage, the stress distribution could at most be triangular, Considering the service load stage.

$$P_\infty = \frac{M_T}{0.61h} = \frac{500 \times 10^6}{0.61 \times 900} = 910750N$$

Area of steel required, assuming the stress in steel as 1000 N/mm^2 after losses.

$$A_{sp} = \frac{910.75 \times 10^3}{1000} = 910.75mm^2$$

Assuming an average compressive stress of 7 N/mm^2 on concrete approximate area of concrete required

$$A_c = \frac{910.75}{7} \times 10^3$$

$= 130.107 \times 10^3\ mm^2$

Since Mg is small when compared to the total service load bending moment, bulbed I section is preferred.

The section as shown in Figure is assumed.

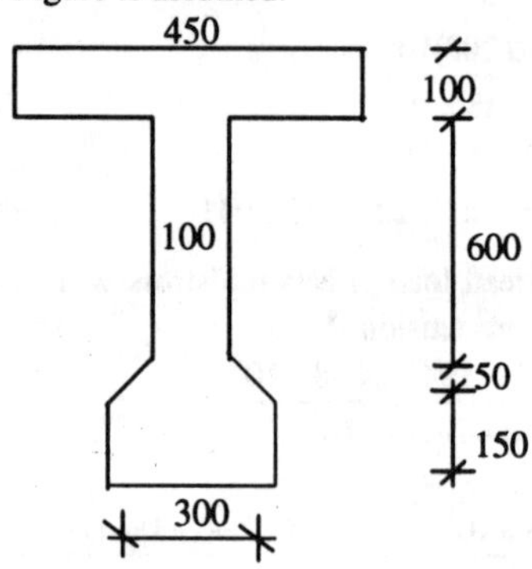

Area of Cross section = 152500mm^2

Taking moment of areas about the bottom soft, the distance of centroid from the soffit = y_b

$$y_b = \begin{pmatrix} 450 \times 100 \times 850 + 600 \times 100 \times 500 \\ + \frac{1}{2} \times 50 \times 100 \times 166.67 + 300 \times 150 \times 75 \end{pmatrix} \div 152500$$

$= 472.40mm$

$y_t = 427.60mm$

$I = 1.76 \times 10^{10}\ mm^4$

$r^2 = 1.76 \times 10^{10} / 152500$

$r^2 = 115409.8mm^2$

$$k_t = \frac{r^2}{y_b} = \frac{115409.8}{472.40} = 244.30mm$$

$$k_b = \frac{r^2}{y_t} = \frac{115409.8}{427.60} = 269.90mm$$

$P_\infty = 910.75kN$

$$P_i = 910.75 \times \frac{1200}{1000} = 1093kN$$

$$e - k_b = \frac{M_g}{P_i} = \frac{100 \times 10^6}{1093 \times 10^3} = 91.5mm$$

$e = 91.5 + 269.90 = 361.40mm$

This gives a facility of a minimum cover of 100mm and hence can be adopted

Final prestress $P_\infty = \dfrac{M_T}{e + k_t}$

$$= \frac{500 \times 10^6}{(361.40 + 260.90)}$$

$$= 792.6kN$$

Initial prestress $P_i = 1.2 \times 792.64 = 951.17kN$

Area of Concrete required is revised based on limiting compressive stresses at bottom and top

$$f_{cbi} = \frac{P_0}{A_g} + \frac{P_0.e}{Z_b} - \frac{M_g}{Z_b}$$

$$A_g = \frac{P_0}{14} + \frac{P_0.e}{14k_t} - \frac{M_g}{14k_t}$$

$$A_g = \frac{951.17 \times 10^3}{14} + \frac{951.17 \times 10^3 \times 361.40}{14 \times 244.30} - \frac{100 \times 10^6}{14 \times 244.30}$$

$Ag = (67.94 + 100.506 - 29.24)10^3$

$Ag = 139.206 \times 10^3$

Area provided is 152500mm^2

Hence the section provided is more than adequate.

Since the permissible stresses at service load stage is also 14N/mm^2 the section provided will be adequate for service load conditions also

The initial prestressing force to be provided is 951.17 kN

Area of streel required $= \dfrac{951.17 \times 10^3}{1000}$

$$= 951.17mm^2$$

2 Nos of 12 × 7 cable will give an area of 923.11mm^2 and it could be prestressed to 1030.5 N/mm^2. Even by elastic shortening losses, the stress level could come down to 1000 N/mm^2, or the eccentricity could be increased by 3% say from 361.40 to 375.0.Even then there will be a cover of 92.5mm which is adequate.

2 × 60mm ducts could be used for accommodating the 2nos of 12 × 7mm cables. This area could be subtracted and the net area and section properties can be calculated, Similarly the effective section properties could also be evaluated.

Net section properties with P_i can be used for evaluation of stresses at transfer stage. Effective section properties with P_∞ could be used for evaluating the stresses at service load stages. The stresses have be finally checked against permissible stresses. For illustration please refer example 8.3

After checking the stresses at service load conditions the ultimate carrying capacity of the section has to be evaluated as per procedures given in chapter 6. The same must be checked against the required capacity using partial safety factor for loads (γ_l).

Example 8.3

A prestressed concrete beam is to be designed to resist a total bending moment of 600 *kNm* at service load conditions and the depth of the beam is restricted to 1000*mm*. The concrete quality proposed to be used has a strength of 45 *N/mm*^{2} and permissible compressive stress and tensile stress at service load are 14 *N/mm*^{2} and 2 *N/mm*^{2} respectively. The prestressing is done with high tensile steel of capacity of 2000 *N/mm*^{2} and can be stressed upto 1200 *N/mm*^{2} at transfer stages. After all losses the stress can be assumed to be 1000*N/mm*^{2}.

Design the concrete cross section, prestressing force and eccentricity check the adequately of the section to resist the moment. Assume a dead load bending moment of 100 *kNm*. Adopt Lyn's method of approach for the design.

Solution

As a first step, it could be assumed that the stress distribution is triangular during service load. Hence the prestressing force required to resist the bending moment can be worked out as P_∞

$$P_\infty = \frac{M_t}{0.61h} = \frac{600 \times 10^6}{0.61 \times 1000} = 983.6kN$$

Assuming $P_\infty = 1000\ kN$

Area of concrete required to resist the force $= A_c = \dfrac{1000 \times 10^3}{0.5 \times 14}$

$= 142857.14\ mm^2$

Since dead load bending moment is relatively small, I section is preferred. Assuming a section as detailed to give the required are a

$A_g = 450 \times 100 + 750 \times 100 + 50 \times 100 + 300 \times 150$

$= 170000 mm^2$

Depth of centroid from the soffit $= y_b$

$$y_{bg} = \frac{450 \times 100 \times 950 + 750 \times 100 \times 525 + 100 \times 50 \times 166.67 + 300 \times 150 \times 75}{170000}$$

$y_{bg} = 507.84mm$

$y_{tg} = 1000 - 507.84 = 492.16\ mm$

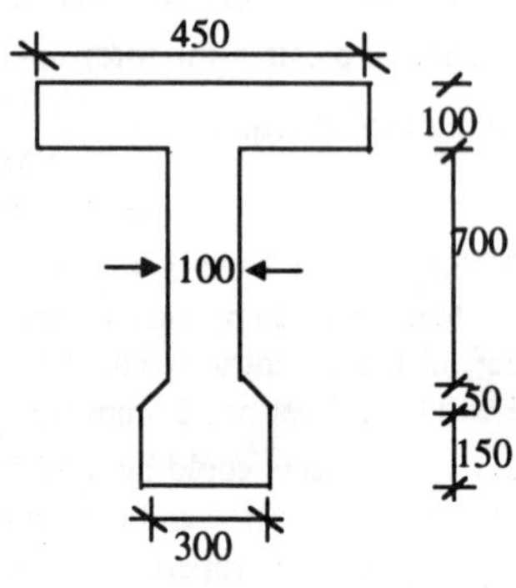

Moment of inertia of cross section I_g

$I_g = 2.33 \times 10^{10}\ mm^4$

$Ar^2 = 2.33 \times 10^{10}$

$r^2 = 137058.8mm^2$

Kern point distances k_b and k_t are calculated as follows

$$\frac{r^2}{y_{bg}} = k_{tg} = \frac{137058.8}{507.84} = 269.89mm$$

$$\frac{r^2}{y_{tg}} = k_{bg} = \frac{137058.8}{492.16} = 278.48mm$$

For the assumed final prestress force $P_\infty = 1000kN$ the initial prestressing force P_i is evaluated as

$P_i = 1.2 \times 1000 = 1200kN$

Considering the transfer load stage and allowing the permissible tensile stress at top fibre, $2N/mm^2$

The eccentricity is evaluated as follows

$e = s_1 + s_2 + k_b$

where $s_1 = \dfrac{M_g}{P_i} = \dfrac{100 \times 10^6}{1200 \times 10^3} = 83.33mm$

and

$$s_2 = \frac{\bar{f}_{ctI} I}{P_o y_t} = \frac{2 \times 2.33 \times 10^{10}}{1200 \times 10^3 \times 492.16} = 78.90m$$

Hence

$$e = 83.33 + 78.90 + 278.48$$
$$= 440.71mm$$

This eccentricity gives a cover of 67mm only which is considered in adequate, Providing a 'e' value of 425mm gives facility of clear cover of more than 50mm if a duct of 60mm dia meter is used. Considering the service load stage and restricting the tensile stress at the fibre not to exceed $2N/mm^2$, P_∞ is evaluated with an eccentricity of 425mm only.

$$p_\infty = \frac{600 \times 10^6 - 2 \times 170000 \times 269.89}{425 + 269.89}$$
$$= 850.25kN$$

From the final prestress force, the initial prestressing force can be evaluated as

$$Pi = 850.25 \times 1.2 = 1020.3kN$$

To see that the compressive stresses at bottom and top fibre do not cross the limiting value A_g required at transfer stage

$$= \frac{1020.3 \times 10^3}{14}\left[\frac{1 + 440.71 - \dfrac{100 \times 10^6}{858.264 \times 10^3}}{269.89}\right]$$

$$= 160421.35mm^2$$

A_g required at service load stage,

$$= \frac{P_\infty h}{f_{ctw} y_b - f_{ctl} \times y_b}$$

$$= \frac{850.25 \times 10^3 \times 1000}{4 \times 507.87 - 2 \times 507.84}$$

$$= 139520.65mm^2$$

At both the stage, area of concrete section required is less than area provided

The finalised design details are

A_g, $= 170000\ mm^2$

y_{bg}, $= 507.84\ mm$

y_{tg}, $= 492.16\ mm$

Z_{bg}, $= 4.589 \times 102mm^3$

Z_{tg}, $= 4.735 \times 107mm^3$

P_i, $= 1020.3\ kN$

P_∞, $= 850.25\ kN$

$e = 425mm$

It is necessary to check for stresses adopting proper section properties at transfer stage and service load stage.

$$P_\infty = 850.25kN \qquad P_i = 1020.3kN$$

Adopting 24nos of 7mm wires, in 2ducts of $60mm$ dia, area prestressing the wires to a stress level of $1105N/mm^2$, the total prestress forces of 1020.3kN can be achieved.

Assuming that the two ducts of 60mm diameter are placed at 80mm from the soffit, there will be a clear cover of 50mm for the ducts, before grouting, when the prestress is transferred, only nett section properties will be available. Hence stresses have to be calculated based on nett section properties at transfer stage.

Let the shift of the centroid from the gross section position be 'y_s'

Taking moment about the new position of the centroid

$170000\, y_s = 2 \times 60 \times 0.785(507.84 - + y_s)$

$164348\, y_s = 2418151.68$

$y_s = 14.7 mm$

$A_n = 164348 mm^2$

$= 232.65 \times 10^8$

$y_{bn} = 507.84 + 14.71$

$= 522.55 mm$

$y_{tn} = 492.16 - 14.71$

$= 477.45 mm$

$$Z_{bn} = \frac{232.65 \times 10^8}{522.55}$$

$= 4.452 \times 10^7 mm^3$

$$Z_{tn} = \frac{232.65 \times 10^8}{477.45}$$

$= 4.873 \times 10^7 mm$

When the ducts are granted, the effective section properties will be available for resisting the external moment.

Let the shift of centroid of effective section be y_s

$$170000 y_s = 24 \times \frac{\pi}{4} \times 49 \times 5(507.84 - 80 - y_s)4615.8(427.84 - y_s)$$

$1746158 y_s = 1974823.88$

$y_s = 11.3 mm$

$y_{be} = 507.84 - 11.3 = 496.54$

$y_{te} = 492.16 + 11.3 = 503.46$

$I_e = 241.233 \times 10^8$

$$Z_{be} = \frac{241.233 \times 10^8}{496.54} = 4.86 \times 10^7 mm^3$$

$$Z_{te} = \frac{241.233 \times 10^8}{503.96} = 4.792 \times 10^7 mm$$

Stresses at transfer

at top $\dfrac{1020.3 \times 10^3}{164348} + \dfrac{1020.3 \times 10^3 \times 442.55}{4452 \times 10^7} - \dfrac{100 \times 10^6}{4.452 \times 10^7}$

$= 6.280 + 10.14 - 2.246 = 14.102 N/mm^2$

at bottom $\dfrac{1020.3 \times 10^3}{164348} + \dfrac{1020.3 \times 10^3 \times 442.55}{4.873 \times 10^7} - \dfrac{100 \times 10^6}{4.873 \times 10^7}$

$= 6.208 - 9.26 + 1.032 = - 1.032\ N/mm^2$

Service load stage

$$\frac{850.25 \times 10^3}{174615.8} + \frac{850.25 \times 10^3 \times 416.54}{4.84 \times 10^7} - \frac{600 \times 10^6}{4.85 \times 10^7}$$

$= 4.87 + 7.302 - 12.37 = - 0.198 N/mm^2$

$$\frac{850.25 \times 10^3}{174615.8} + \frac{850.25 \times 10^3 \times 16.54}{4.792 \times 10^7} - \frac{600 \times 10^6}{4.792 \times 10^7}$$

$= 4.87 - 7.39 + 2.52 = 10\ N/mm^2$

Stresses at bottom during transfer is marginally larger than 14N/mm² (permissible). But this would be coming with in permissible limits immediates after friction loss which is not accounted for in calculation.

All other stress are with in permissible limit. After checking the stresses it is mandatory to check for the ultimate moment carrying capacity to see that the section is adequate to carry the total moment of 600 kNm with the partial safety factor for loads (γ_e).

For this purpose the reader is advised to refer to chapter 6. and evaluate the capacity. If the capacity is in adequate untentioned steel can be provided at the bottom flange to make good of any inadequacy.

8.5 Magnel's Approach

In the Lyn's approach, the design is monitored based on the behavioral aspects of the section at various load stages following the movement of centre of compression with application of load. Hence it is more suited for on the spot corrections and modifications. But with advent of computers and standardisation of sections, it is possible to have designs in terms of non-dimensional parameters connected with section properties, presstressing force, its eccentrictiy etc. Mathematical optimisation of design is also easily possible with the designs dealing with only non-dimensional parameters. Such an approach was first advocated by Prof. Magnel and hence called Magnels approach.

In any design approach, at the service load stages, the main considerations shall be that the basic stress constraints relating the actual stresses and limiting stresses at salient level (top and bottom fibre levels) at extreme load stages (governed primarily by the transfer load stage and service load stage) referred in equation 8.1a to 8.1d should be satisfied. These inequilities should be converted in terms of non-dimensional parameters.

The actual stresses that may occur on the sections, and limiting stress are first expressed in terms of the concrete strength f_{ck} with coefficients as detailed with Table 8.1.

Table 8.1

Details of stresses	Actual	Permissible
Compressive stress in concrete at transfer.	$c_t f_{ck}$ f_{cbtr}	$c_t' f_{ck}$ $\bar{f}_{cbtr}$
Tensile stress in concrete at transfer.	$a_t f_{ck}$ f_{cttr}	$a_t' f_{ck}$ $\bar{f}_{cttr}$
Compressive stress in concrete at service load.	$c f_{ck}$ f_{cts}	$c' f_{ck}$ $\bar{f}_{cts}$
Tensile stress in concrete at service load.	$a f_{ck}$ f_{cns}	$a' f_{ck}$ f_{cbs}

The basic stress inequalities to be satisfied in service load stress design as design as discussed in art 8.2.

$$f_{bpi} + f_{bdl} = c_t f_{ck} \le c_t' f_{ck} \ (comp. + ve) \quad .....(8.17a)$$

$$f_{tpi} + f_{tdL} = a_t f_{ck} \le a_t' f_{ck} \ (ten. + ve) \quad .....(8.17b)$$

$$f_{bpf} + f_{bdL} + f_{bLL} = af_{ck_1} \le a' f_{ck} \ (ten. + ve) \quad .....(8.17c)$$

$$f_{tpf} + f_{tdL} + f_{tLL} = cf_{ck} \le c' f_{ck} \ (comp. + ve) \quad .....(8.17d)$$

These co-efficients (c_t', c', a_t',a') can be had from the relevant code of practice. I.S. code 1343–1980 defines these coefficients in clause 22.8.1.1. The same has also been given in chapter 2 on materials.

The inequalities 8.17a to d can be expressed in terms of initial prestressing force P_i, its eccentricity e, the loss factor the section properties like moment of inertia (I), distance of extreme fibres at top & bottom (y_t, y_b) from centroid of section and the applied bending moment the least radius of gyration (r), dead load bending moment (M_g).

$$\frac{P}{A_n}\left[\frac{ey_t}{r^2}-1\right]-\frac{M_g y_t}{I_n}=a_t f_{ck}\leq a'_t f_{ck} \quad .....(8.18a)$$

$$\frac{P}{A_n}\left[\frac{ey_b}{r^2}+1\right]-\frac{M_g y_b}{I_n}=c_t f_{ck}\leq c_t f_{ck} \quad .....(8.18b)$$

$$-\eta\frac{P}{A_t}\left[\frac{ey_t}{r^2}-1\right]+\frac{M_t y_t}{I_t}=c f_{ck}\leq c' f_{ck} \quad .....(8.18c)$$

$$-\eta\frac{P}{A_t}\left[\frac{ey_b}{r^2}+1\right]+\frac{M_t y_b}{I_t}=a f_{ck}\leq a' f_{ck} \quad .....(8.18d)$$

Even in this approach, the main unknowns are kept as the prestressing force and eccentricity while the concrete section is chosen from standardised sections which will have the parameters satisfying the basic equations expressed in terms of such non dimensional parameters.

If by a preliminary design, the height required the concrete area and the moment of inertia are known roughly, one of the standard section which is commonly adopted can be chosen and its exact section properties can be arrived at. For such a chosen section graphs can be drawn between eccentricity and $\frac{I}{P}$ represent the above four equations and the same could be used for design of sections as suggested by Bennet (5).

For drawing such a graph, the basic stress inequalities (8.18a to 8.18d) can be expressed in the following form

$$\frac{1}{P_i}\geq\frac{\frac{ey_t}{r^2}-1}{\left[\frac{M_g y_t}{I}+a'_t f_{ck}\right]A_n} \quad .....(8.19a)$$

$$\frac{I}{P_i}\geq\frac{\left[\frac{ey_b}{r^2}+1\right]}{\left[\frac{M_g y_b}{I}+c'_t f_{ck}\right]A_n} \quad .....(8.19b)$$

$$\frac{I}{P_i} \leq \frac{\left[\dfrac{ey_t}{r^2} - 1\right]\eta}{\left[\dfrac{M_T y_T}{I} - c' f_{ck}\right] A_t} \qquad \text{.....(8.19c)}$$

$$\frac{I}{P_i} \leq \frac{\left[\dfrac{ey_b}{r_2} + 1\right]\eta}{\left[\dfrac{M_t y_b}{I} - a' f_{ck}\right] A_t} \qquad \text{.....(8.19d)}$$

The shaded zone in the graph (Fig 8.3) is the zone from which P and corresponding 'e' can be chosen which would satisfy the stress requirements for a given section. The maximum acceptable eccentricity goes with minimum prestressing force identified by point A on the graph (Fig 8.3) and similarly the maximum prestressing force and the corresponding eccentricity (minimum) is identified by point 'B'.

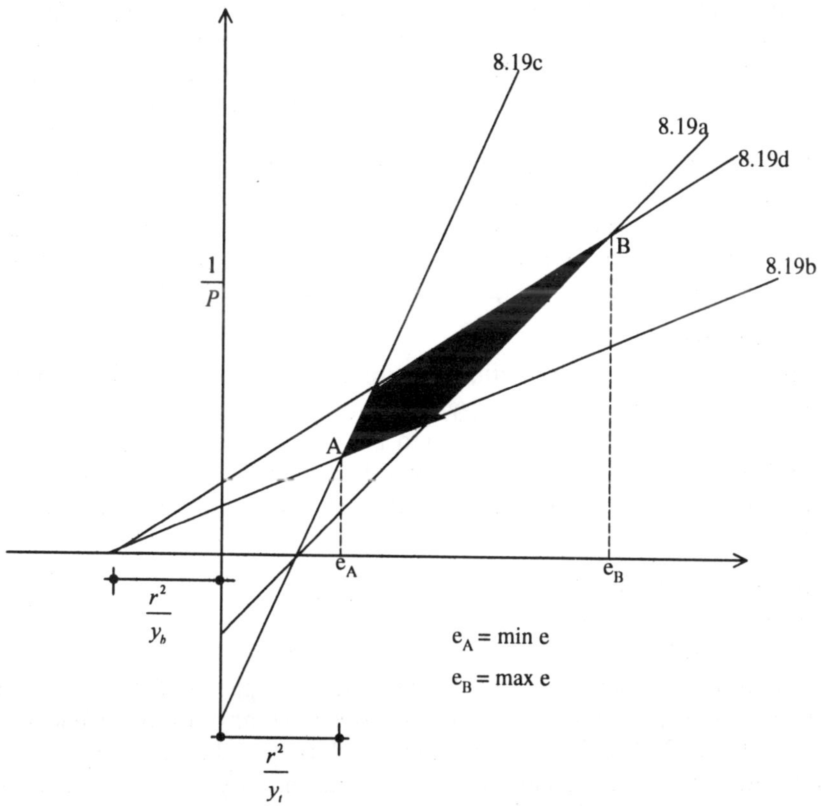

Fig 8.3 Least Weight Design Graphic Zone

8.5.1 Non dimensional variables

The non-dimensional variables with regards to a prestressed concrete section are defined as under. (i) depth factor (ω), (ii) efficiency factor (ρ), (iii) shape factor (Δ), (iv) moment ratio (R) (v) reinforcement ratio (m) and (vi) eccentricity ratio (∈). The factors have their own significances and the details are described below:

The depth factor (ω)

This is defined in terms of height, span, concrete strength and unit weight of concrete. $\omega = \dfrac{hf_{ck}}{\gamma L^2}$ This is controlled by architectural considerations, stiffness requirements, and weight per unit length, which is an essential requirement for handlability in case of precast construction. In practice, the factor varies between unity for very long spans to about 10 for short spans.

Efficiency factor (ρ)

This is a measure of distribution of areas with references the centroid of the area and is expressed as r^2/h^2. For flexure the areas should be away from the centroid so that the moment of inertia (second moment of area) is maximum. The theoretical maximum distance or the value of to r can be 0.5 h and hence the maximum efficiency factor can be 0.25, which is again only theoretical. For rectangular section, which is the least efficient section for flexure, it will be 0.0833. The practical range of ρ for prestressed concrete beams can be from 0.09 to 0.14. For 'T' section, it will be in the range 0.09 to 0.11 and for I section it will be 0.11 to 0.14. Hence in case of precast and cast in situ combination, for precast section, the value of ρ can be chosen around 0.09 to 0.10 but for built up, large flanged section it can go as high as 0.16.

Shape factor (Δ)

This factor refers to the position of extreme fibres from centroid for areas of the section and is defined as y_b/y_t. For symmetrical section it will be 1.0. For practical 'T' section it may go up to 1.6. For inverted 'T' section which will be used as a pre cast unit (in a precast-cast in situ combination) it will be around 0.8 to 0.9. This factor is helpful in rearranging the areas of section to satisfy the stress requirements.

Reinforcement ratio(m)

As in the case of reinforced concrete beams, where percentage steel is defined with reference to concrete area, in a prestressed concrete beam, the ratio of applied prestressing force to the concrete capacity of the section in compression is defined as $m = P/Af_{ck}$. The practical range for this value will be 0.15 to 0.40. The lower value can be chosen if the dead load when compared to total load is large where large eccentricity is possible. Higher value can be chosen if the dead load moment is low when compared to total moment due to smaller depth of cross section and the facility of eccentricity is also minimum.

Eccentricity ratio($\in$)

This ratio is a measure of effective utilitsation of the prestressing force and is defined as e/h. The moment to be resisted by the prestressing force depends on the facility of movement of centre of compression from the centroid of steel to the allowable limit above the top kern point. This facility is more if the eccentricity is more and the corresponding prestressing force required will be less. Practical limits for this ratio will be 0.25 to 0.35 for symmetrical section and it can go upto 0.55 for top heavy sections and limited to 0.25 for bottom heavy section. This can also be expressed as a function of shape factor Δ and cover 'g'. Eccentricity 'e' can be represented as (y_b-cover) and $\in$ may be expressed as $\frac{y_b - g}{h}$ where 'g' is cover $\in$ can be rewritten as

$$\in = \frac{y_b}{y_b + y_t} - \frac{g}{h} = \frac{\Delta}{\Delta + 1} - \frac{g}{h} \qquad \text{.....(8.20)}$$

g/h will normally be kept in the range of 0.05 to 0.1.

Moment ratio (R)

This is the most important design parameter. This is defined as the ratio of moments caused by applied loads to moment caused by self weight. The design should be such the self-weight (cross sectional area) required to resist certain applied moment. M_a should be minimum. Hence M_a/M_g should be as large as possible for a economical design. R is also defined as a function of span and unit weight of concrete.

$$R = \frac{8M_a}{A\gamma L^2}$$

The design parameters should be so chosen that they satisfy the stress requirements at service load and also lead to maximum possible value of R. This variable is inversely proportional to area of section. Now it is necessary to get relationship of R with other parameters so that R can be maximized within acceptable values of the other parameters, which are governed by the stress constraints.

The most efficient section is one where the stresses are reaching their limiting values and not exceeding it at any load stage. Hence the coefficient c_t, a_t, c and 'a' must be reaching their limiting values of c_t',a_t', c' and a' in any section.

The stresses restraints and the inequalities to be satisfied for the same for an acceptable design can now be expressed interms of the of the dimensionless parameters as follows:

$$m\left[\frac{\in}{\rho(1+\Delta)} - 1\right] - \frac{1}{8\omega\rho(1+\Delta)} = a_t \leq a'_t \qquad \text{.....(8.21a)}$$

$$m\left[\frac{\in\Delta}{\rho(1+\Delta)} + 1\right] - \frac{\Delta}{8\omega\rho(1+\Delta)} = c_t \leq c'_t \qquad \text{.....(8.21b)}$$

$$-\eta m\left[\frac{\in}{\rho(1+\Delta)}-1\right]+\frac{1+R}{8\omega\rho(1+\Delta)}=c\leq c' \qquad .....(8.21c)$$

$$-\eta m\left[\frac{\in\Delta}{\rho(1+\Delta)}+1\right]+\frac{\Delta(1+R)}{8\omega\rho(1+\Delta)}=a\leq a' \qquad .....(8.21d)$$

The dimensionless terms namely ω, m, R, ρ, ∈ can be expressed as a function of the stress coefficients and a few parameters.

From equations 8.21a and 8.21b, m can be expressed as

$$m=\frac{c_t-\Delta a_t}{1+\Delta} \qquad .....(8.22a)$$

Similarly from equation 8.21c and 8.21d m can be expressed as

$$m=\frac{\Delta c_r-a}{\eta(1+\Delta)} \qquad .....(8.22b)$$

Using the two values of m as expressed above, D can be evaluated as

$$\Delta=\frac{\eta c_t+a}{\eta a_t+c} \qquad .....(8.23)$$

Multiplying equation and by h and adding than to equation and R can be expressed as

$$R = 8\rho\omega[\eta(a_t+c_t)+(a+c)]-(1-\eta)] \qquad .....(8.24a)$$

Substituting value of Δ in term of stress co-efficients in equations 8.22a and 8.22b m can be rewritten as

$$\text{m} = \frac{c_t c - a_t a}{(a+c)+\eta(a_t+c_t)} \qquad .....(8.25)$$

Adding equation 8.21a and 8.21b and ∈ can be expressed on

$$\in = 1/m[\rho(a_t+c_t)+1/8\omega] \qquad .....(8.26)$$

Equation 8.23, 8.24, 8.25, 8.26, details out the concrete section properties required and and define the prestressing force required and the corresponding eccentricity.

Eliminating (ηa_t+c) and $(\eta\in_t+a)$ between equations 8.21a and 8.21b or 8.21c and 8.21d and R can also be expressed as

$$R = 8\rho\omega(\eta c_t+a)(1/\Delta+1)-(1-\eta) \qquad .....(8.24b)$$

$$R = 8\rho\omega(\eta a_t+c)(\Delta+1)-(1-\eta) \qquad .....(8.24c)$$

By eliminating ρ between 8.24b and 8.24c R can also be expressed as

$$\text{R} = \frac{8\omega \in(c_t c - a_t a)-\left[(a+c)+\eta(a_t+c_t)\right]}{a_t+c_t}-(1-\eta) \qquad .....(8.24d)$$

With these equations R, Δ, m, ∈, it is possible to arrive at an optimal section with optimum prestressing force.

For a minimum weight design, the parameter R has to be maximum and the value of co-efficient at, c_t, a, c should reach their limiting values and accordingly the shape factor D reaches its optimal value Δ_e

$$\Delta_e = \frac{\eta c'_t+a'}{\eta a'_t+c'} \qquad .....(8.23a)$$

This is purely dependent on permissible limiting co-efficients. In any design the first attempt should be to achieve a shape factor$\Delta = \Delta e$ In such a case, all the four basic stress inequialities are satisfied exactly. But this will be a rare design. This means that the four requirements can be exactly satisfied only if Δ reaches its optimal value of Δ_e. For a given set of allowable stresses Δ_e can be calculated and it is possible to choose a section to have $\Delta = \Delta_e$. For long spans, it is necessary to make $\Delta > \Delta_e$ to provide for sufficient cover for prestressing force.

It can be seen from the expression Δ (equation 8.23a).that for values of Δ other than Δ_e one or more of the stress coefficients will differ from the limiting coefficients marginally on the conservative side. If $\Delta > \Delta_e$ it can be seen from equation 8.24b that in order to obtain the maximum value R, or minimum area of concrete, the following conditions have to be fully satisfied.

$c_t = c'_t \quad a = a'$

Similarly, for a given value of $\Delta < \Delta_e$, it can be seen from equation 8.23a that in order to obtain the maximum value R, corresponding to minimum area of concrete the following conditions have to the fully satisfied.

$a_t = a'_t \quad c = c'$

When $\Delta > \Delta_e$ the requirements governed by equations 8.21b (stress livel at bottom fibre under initial load conditons) and 8.21d (stress level at bottom fibre under service load conditions) are to be satisfied exactly, while requirements goverened by 8.21a and 8.21c (stress level at top fibre at the initial and service load stages) are to be marginally satisfied.

When $\Delta < \Delta_e$, the requirements governed by equations 8.21a (stress level at top fibre under initial load conditions) and 8.21c (stress level at top fibre under service load conditions) are to be satisfied exactly, while requirements governed by 8.21b and 8.21d (stress level at bottom fibre at initial and service load stages) are to be marginally satisfied. When $\Delta = \Delta_e$ all requirements governed by 8.21a to 8.21d are to be exactly satisfied.

8.5.2 The amount of steel (steel requirement)

The amount of steel is defined by the reinforcement ratio $m = P/Af_{ck}$ and it is a measure of prestressing force. For given values of concrete area and permissible stresses, the steel area required is defined by the value of m. The minimum value of A (area of concrete) is discussed in the earlier sub-heading as a function of Δ. Similarly the minimum value of m will be discussed here for the same variations in Δ as given in the earlier article.

In case of a section, with given h, ρ, ω values and $\Delta > \Delta e$, the least concrete area is obtained, if $c_t = c_t'$ and $a = a'$. The magnitude of m varies from a maximum to a minimum depending on the values of a_t and c as discussed earlier. From equation 8.22a it can be seen that m decreases with increase of a_t (since $c_t = c_t'$ for given value of Δ). Higher values of a_t yield smaller values for m and hence lower area of prestressed steel at should be made equal to its maximum value of a'_t for minimum area of steel.

On the other hand the maximum value of m will be reached if a_t is smallest. Hence the maximum amount of prestressing force will result for a given when $c_t = c'_t$, $a = a'$ and $c = c'$.

Similarly when Δ is < Δe, for a given ρ, h, w values the least concrete area is obtained if $a_t = a'_t$ and $c = c'$. By similar arrangement given above, minimum prestressing steel area is possible if $a = a'$ and maximum prestressing steel area is required if $c_t = c'_t$. These details are summarised in Table 8.2 below.

Table 8.2 Least Weight Criteria

No	D	Area of concrete	Area Steel	Eccentricity	Conditions to be satisfied
1.	$\Delta > \Delta_e$	Min (w, given)	Min	Max	$a_t = a'_t$ $c_t = c'_t$ $a = a'$ $c < c'$
2.	$\Delta > \Delta_e$	Min	Max	Min	$c_t = c'_t$ $a_t < a'_t$ $a = a'$ $c = c'$
3.	$\Delta > \Delta_e$	Min	Min	Max	$a_t = a'_t$ $c = c'$ $a = a'$ $c_t < c'_t$
4.	$\Delta > \Delta_e$	Min	Max	Min	$a_t = a'_t$ $c = c'$ $c_t = c'_t$ $a < a'$

8.5.3 Discussions

Increasing Δ from Δ_e results in the increase of the required concrete area. It provides means to obtain sufficient cover for placing prestressing steel. If criterion of Table 8.2 is applied, minimum prestressing steel is got with possible maximum eccentricity. If criterion 2 is applied, prestressing force is increased to a maximum and eccentricity is kept to a minimum. This also facilitates greater cover, but with a higher percentage of steel. Decreasing Δ from Δ_e also increases the concrete area but does not, in any case, give better facility for cover whether criterion 3 or 4 is applied.

Hence. if $\Delta = \Delta_e$ could not be achieved, as a first choice $\Delta > \Delta_e$ with criterion 1 may be tried. If that does not give feasible result, criterion 2 may be adopted. After getting maximum ρ and minimum e, a via media solution be tried by increasing ρ.

As done in example 8.4 correct section properties may be calculated, stresses at transfer stage and at service load stage may be checked and the moment carrying capacity, may be checked before the design is declared as completed.

The idealised sections for design for flexure are

- 'I' section with bottom bulb
- 'T' section
- Box section with cantilever top flange
- Inverted I section with top flange as a bulb
- Double 'T' section

The relationship between the various section dimensions such as thickness of flanges, web, width of top flange, bottom flange, height of web and the corresponding values of ρ, ω, Δ etc., are available in standard literature which could be made use of for choice of sections. In the choice of sections, the dead load to live load ratio plays a predominant role. As the dead load increases the facility of movement of eccentricity downwards must be increased. This is possible only in thin but long webbed sections, like single 'T'. But if the live load is large the distance available above the centroid must be more. By studing the effect of variation of one parameter on the other, and by experiences of designers, the choice of parameters like ρ,ω and Δ can be made. A few guidelines are given here based on discussion made by Katchaturien and Gurfinkel (6). They are only guidelines and need not be adhered to strictly. They can be or may have to be violated under specific conditions.

- For span upto 30m, ω can be chosen as 1.0
- For spans > 30m, ω is limited to 1.5 and
  $\Delta = \Delta_e$ if sufficient cover is available
  For cases when $1.5 < \omega < 3.0$, Δ can be greater than Δ_e by 10%
- $\omega > 3.0$, Δ should be kept as close to Δ_e as possible
- It is desirable to go in for higher value of ρ.
- Choose ρ any where between 0.9 to 0.1 for large span when dead load to live load ratio is large and ρ any where between 0.11 to 0.13 for smaller span ρ
- For additional guide lines refer Table 8.3

Table 8.3 Limitation Values of Non-dimensional Parameters

No.	Section	h	ρ	Δ	R	m
1.	I with heavier top flange	$\frac{L}{15}$ to $\frac{L}{25}$	0.12 to 0.17	1.0 to 1.8	0.5 to 2.0	0.15 to 0.30
2.	I with heavier bottom	$\frac{L}{15}$ to $\frac{L}{25}$	0.10 to 0.14	0.6 to 1.5	1.0 to 5.0	1.5 to 0.30
3.	T section	$\frac{L}{15}$ to $\frac{L}{25}$	0.08 to 0.10	1.2 to 1.8	0.5 to 2.5	0.15 to 0.30
4.	Inverted T	$\frac{L}{15}$ to $\frac{L}{25}$	0.08 to 0.10	0.6 to 0.9	0.5 to 2.5	0.15 to 0.40

Example 8.4

A prestressed concrete beam covers a span of 17m and is to carry a load of 20 kN/m. Design the beam satisfying the I.S code regulations using Magnel's method. High tensile steel wire of 5mm, 7mm diameter are available for use ultimate strength of the wires can be assumed to 1500N/mm^2 and they are allowed to be stressed to a level of not greater than 1200N/mm^2 at transfer load stage and not greater 1020 N/mm^2, at service load stage. Concrete of strength 42N/mm^2 at 28 days may be assumed to be used. Assume a loss coefficient of 0.85, Prestressing to be transferred after 7 days after casting the concrete unit weight of concrete and as = 24kN/m^3.

Solution

Stress coefficient as per I.S. 1343 – 1980

$a'_t = 0$ $\quad\quad c'_t = 0.8 \times 0.472 = 0.378$

$c' = 0.312$ $\quad\quad a' = 0$

Assuming $\frac{1}{h}$ ratio of 17, h = 100cm

$$\omega = \frac{hf_{ck}}{\gamma L^2} = 5.05$$

Since the span is relatively small for a prestressed concrete beam ρ can be assumed to be 0.13

$$\Delta_e = \frac{c_t + a'}{a_t' + c'}$$

$$= \frac{0.85 \times 0.378 + 0}{0 + 0.312} = 1.03$$

Assuming $\Delta = \Delta_e = 1.03$ and $\rho = 0.13$, R can be evaluated as

$$R = 8\rho w[(a + c) + \eta(a_t + c_t)] - (1 - \eta)$$

$$= 1.04 \times 5.05\ [(0.00 + 0.312) + 0.85\ (0.378)] - 0.15$$

$$= 1.04 \times 1.03 \times 5.05$$

$$= 3.325 - 0.15$$

$$= 3.175$$

$$R = \frac{M_a}{M_g} = \frac{20000}{24000}\frac{1}{A} = 3.175$$

$$\Delta = 262468 \text{ mm}^2$$

From R value of 3.175, Area of concrete of the proposed section works out to be 262468mm^2

The percentage reinforcement (prestressing force) and eccentricity can be arrived by evaluating m, and $\in$.

$$m = \frac{c_1 c - a_1 a}{(a + c) + \eta(a_t + c_t)} = \frac{0.378 \times 0.312 - 0}{0.312 + 0.85 \times 0.378}$$

$$= \frac{0.1179}{0.6333} = 0.186$$

$$\in = \left[\rho(a_t + c_t) + \frac{1}{8w}\right]\frac{1}{m}$$

$$= \frac{1}{0.1861}(0.04914 + 0.0248)$$

$$= \frac{0.07394}{0.1861} = 0.3970$$

Assuming a possible cover of 0.1 h, ∈ (possible) is evaluated as

$$\in_{\text{possible}} = \frac{\Delta}{\Delta+1} - \frac{g}{h} = 0.5073 - 0.10$$

$$= 0.4073 > 0.3970 \text{ required}$$

Hence ∈ = 0.3970 be adopted

The position of C.G of section is evaluated from value of Δ

$$\Delta = \frac{y_b}{y_t} = \frac{y_b}{100 - y_b} = 1.03 \qquad y_b = 50.7\text{cm}$$

Cover available = 50.7 – 39.7 = 11cm

This cover is adequate for the prestressing force to be provided

Prestresses force required

$$P = m \, . \, Af_{ck} = 0.1861 \times 262468 \times 42$$

$$= 2051502.3\text{N}$$

$$= 2051.5\text{kN}$$

The design parameters arrived at are

$A = 20625 \times 10^5 mm^2$, $\omega = 5.05$, $\Delta = 1.03$, $\rho = 0.13$, h = 100cm,

P = 2051.5kN, e = 39.7cm.

From standard sections whose properties are described in standard text book, I section leading very nearly to these parameters can be chosen as shown in Figure.

$A = 265600 mm^2$

$y_b = 508$

$y_t = 492$

$I = 3465.805 \times 10^7 mm^4$

$Z_b = 682.245 \times 10^5 mm^3$

$Z_t = 704.43 \times 10^5$

$P = 2051.5\text{kN}$

$e = 375mm$

$l = 17m$

$Liveload = 20kN/mR$

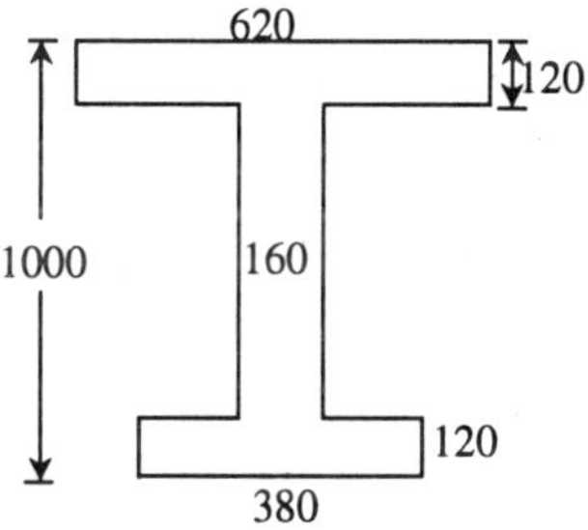

$$\text{Dead load} = \frac{265600}{1000^2} \times 24 = 6.37 kN/mR$$

$$M_{dl} = \frac{6.37 \times 17^2}{8} = 230.27 kNm$$

$$M_{ll} = \frac{20 \times 17^2}{8} = 722.5 kNm$$

This cross section will be checked for adequacy of stresses at transfer load stage and service load stage.

At transfer load stage

$$\text{Stress at bottom} = \frac{P}{A} + \frac{P_e}{Z_b} - \frac{M_{dl}}{Z_b}$$

$$= \frac{2051.5 \times 10^3}{265600} + \frac{2051.5 \times 10^3 \times 375}{682.245 \times 10^5} - \frac{230.27 \times 10^6}{682.245 \times 10^5}$$

$= 7.724 + 11.276 - 3.375$

$= 15.625N/mm^2 < 15.85N/mm^2 = c_t' f_{ck}$

Stress at top $= \frac{P}{A} - \frac{P_e}{Z_t} + \frac{M_{dL}}{Z_t}$

$$= \frac{2051.5 \times 10^3}{265600} + \frac{2051.5 \times 10^3 \times 375}{704.43 \times 10^5} + \frac{230.27 \times 10^6}{704.43 \times 10^5}$$

$= 7.724 - 10.92 + 3.27 = 0.074N/mm^2$

At service load stage condition

Stress at bottom $= \frac{\eta P}{A} + \frac{\eta P_e}{Z_b} + \frac{(M_{dL} + M_{ll})}{Z_b}$

$$= 0.85 \times \frac{2051.5 \times 10^3}{265600} + \frac{0.85 \times 2051.5 \times 10^3 \times 375}{682.245 \times 10^5} + \frac{(230.27 + 722.5) \times 10^6}{682.245 \times 10^5}$$

$= 6.5654 + 9.58 - 13.965 = 2.18$ (Residual çompression)

Stress at top $= \frac{\eta P}{A} + \frac{\eta P_e}{Z_t} + \frac{(M_{dL} + M_{ll})}{Z_t}$

$$= 0.85 \times \frac{2051.5 \times 10^3}{265600} + \frac{0.85 \times 2051.5 \times 10^3 \times 375}{704.43 \times 10^5} + \frac{(230.27 + 722.5) \times 10^6}{704.43 \times 10^5}$$

$= 6.5654 - 9.282 + 13.525 = 10.808N/mm^2 < 13.1m < c'f_{ck}$

The stresses are with in acceptable limit. The design has to be checked for safety against limit state of collapse. The moment carrying capable of the sections may be calculated as per procedures mentioned in chapter 6 and checked against the required factored moment carrying capables.

Example 8.5

Design a prestressed post tensioned concrete beam to span 40m and to carry an equivalent uniform live load of 20kN/m. The depth of beam is restricted to 2.0m due to service conditions at site 7mm, 5mm, stranes are available with a characteristic strength of $1500N/mm^2$. Concrete of strength $45N/mm^2$ is proposed to be used. Assume $E_s=2.1\times10^5N/mm^2$ $E_c=3.82\times104N/mm^2$; $\gamma=24kN/m^3$. Initial prestresses force may be transferred after the concrete attains a strength of 80% of tensile strength. Adopt Magnel's method of design.

Solution

Stress coefficients [From I.S 1343-1989] for M 45 concrete

$c'_t = 0,.364$ $\quad a'_t = 0.00$

$c' = 0.305$ $\quad a' = 0.00$

Stress level in prestressing steel at transfer f_s = limited to $0.7f_p$

Stress level in prestresses steel at service load f_s is limited to $0.6 f_p$

Since the span is large, dead load moment will be large. Accordingly a T section will be more suitable. Hence choose $\rho = 0.11$ h = 200mm. Assume a loss factor of 0.8

$$\omega = \frac{hf_{ck}}{\gamma L^2} = \frac{2000 \times 45 \times 10^9}{24000 \times 1600 \times 10^6} = 2.34$$

$$\Delta_e = \frac{\eta c'_t + a'}{\eta a'_t + c'} = 0.95$$

with $\rho = 0.11$, $\omega = 2.34$, $\Delta = 0.955$, the concrete area can be evaluated after getting R. The prestress force and eccentricity may be got after getting m and ϵ.

$R = 8\rho w\,[(\eta at' + c')] + (\Delta + 1) - (1- \eta)$

$= 8 \times 11 \times 2.34\,[(0 + 0.308)]\,(1.955) - 0.2$

$= 1.03$

$$m = \frac{c_t c - a_t a}{(a+c) + \eta(a_t + c_t)} = \frac{0.364 \times 0.305 - 0}{(0 + 0.305) + 0.8(0 + 0.364)}$$

$$= 0.1865 \approx 0.187$$

$$\in = \left[0.11(0.364)\frac{1}{8 \times 2.34}\right]\frac{1}{0.187}$$

$$= [0.04004 + 0.0534]\frac{1}{0.187} = 0.4996 \approx 0.5$$

$$\in_{possible} = \frac{\Delta}{1+\Delta} - \frac{g}{h} = 0.388 < \in_{required}$$

Hence this is not a feasible design. Design has to be done higher cover facilities. This can be achieved by increasing Δ greater than Δ_e and varying ρ values. Choosing $\Delta = 1.5$ and h = 200cm and (ω) > 2.34, any one of the two criteria for design can be adopted namely Minimum P, maximum e, or maximum P and minimum e. First adopt criterion 1.

Criterion - 1

Minimum P and maximum e.

$a_t = a'_t = 0$ $\quad\quad a = a' = 0$

$c_t = c'_t = 0.364$ $\quad\quad c < c'$

$$m = \frac{c_t - a_t}{1+\Delta} = \frac{0.364 - 1.5 \times 0}{2.5} = 0.1456$$

c, can be calculated from equation 8.5

$$1.5 = \frac{0.85 \times 0.364 + 0}{0 + c}$$

$c = 0.2062$

$\in$ is calculated as

$$\in = \frac{1}{0.1456}\left[0.11(0.364) + \frac{1}{8 \times 2.34}\right]$$

$$= (0.0934)\frac{1}{0.1456} = 0.6418$$

$$\in_{possible} = \frac{1.5}{2.5} - 0.1 = 0.5$$

This shows the prestressing force and the correspoding 'e' has been drastically reduced and this could not be achieved. Hence choosing the next criteria namely e minimum and P maximum.

$c_t = c'_t = 0.364$ $\quad\quad a = a' = 0$

$c = c' = 0.305$ $\quad\quad a_t < a'_t = -$

$$m = \frac{\Delta c - a}{\eta(1+\Delta)} = \frac{1.5 \times 0.305 - 0}{(0.8(1.5))} = 0.22875$$

Coefficient at from equation 8.

$$= \frac{0.80 \times 0.364 + 0}{0.8 \times a'_t + 0.305} = 1.5$$

$1.5[0.8a'_t + 0.305] = 0.2912$

$a'_t = 0.14$

$$\in = \frac{1}{m} = \left[\rho(a_t + c_t) + \frac{1}{8w}\right.$$

$$[0.11 \times (0.14 + 0.364)] + \frac{1}{8 \times 2.34} \times \frac{1}{0.22875}$$

$$[0.11 \times 0.504 + 0.0534] + \frac{1}{0.22875}$$

$$[0.05544 + 0.0534] \times \frac{1}{0.22875}$$

$= 0.476$

$\in_{possible} = 0.5$ Hence this is a possible design. Higher efficiency factor may be tried. Assume $\rho = 0.12$

$$\in = \frac{1}{m}\left[0.12(0.14 + 0.364) + \frac{1}{8 \times 2.34}\right]\frac{1}{0.22975}$$

$$= \left[0.12 \times 0.504 + \frac{1}{8 + 0.234}\right]\frac{1}{0.22875}$$

$\in = 0.487$

$$_{possbile} = \frac{\Delta}{\Delta + 1} - \frac{g}{h} = 0.5$$

Hence it is an acceptable design. Choose a section to suit the following reqirements

$\rho = 0.12, \quad m = 0.22875, \quad w = 2.34, \quad \Delta = 1.5, \quad h = 2.0m$

$R = 8\,\rho w\,[(0.8 \times 0.14 + 0.305)]\,(2.5) - 0.2$

$= 2.246\,[0.4417 \times 2.5] - 0.2$

$= 2.342 - 0.2 = 2.142$

$$R = \frac{M_a}{M_g} = \frac{20000}{24000A}$$

$A = 389044.5$

$3890 cm^2$

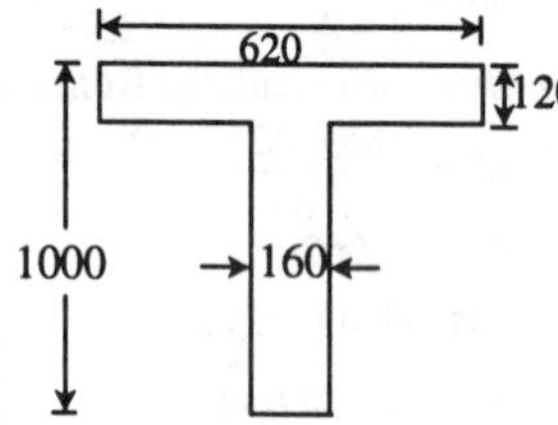

As done in example 8.4 correct section properties may be calculated and stresses at transfer stage and service load stage may be checked and the moment carrying capacity checked.

Reference

1. EVANS, R.H. and BENNETT E.W., (1958) **Prestressed Concrete,** Chapman and Hall, London, .
2. LIN, T.Y., (1960) **Prestressed Concrete Structures,** John Wiley and Sons, Ins, .
3. IS: 1343-1980, **Indian Standard Code of Practice for Prestressed Concrete,** (first revision), Indian Standards Institution, New Delhi.
4. MAGNEL, G., (1948) **Prestressed Concrete,** Concrete Publications, London, .
5. BENNETT, E.W., (1959) "A graphical method for the design of prestressed beams," **Concrete and Constructional Engineering,** Vol. 53, November, pp 399-403.
6. KATCHATURIEN and GURFINKEL (1960) **Prestressed Concrete** - Mc Graw - Hill, Company New York.

9. Shear and Torsional Behaviour

9.1 Concept of Shear

Any element in a structure subjected to a system of loading, experiences principle stresses at varying angles at different places. The principle stress-trajectories for beams subjected to uniformly distributed load is shown in Fig. 9.1.a. These principal stress trajectories could also be conveniently represented as normal and tangential stresses on any two orthogonal planes. The designers prefer that these stresses are represented as longitudinal and shear stresses. Hence the shear stress is only a nomenclature for defining the direction and magnitude of principle stresses. It is a convenient way of expressing the same in the element and has no specific extra significance. The equilibrium of forces at any particular cut section of a beam results in two orthogonal force systems, namely longitudinal forces and shear forces. They in turn lead to longitudinal (flexural) and shear stresses, which vary across the depth of the section.

The shear stresses are evaluated using the expression $\tau = \frac{V}{Ib}A\overline{y}$ and flexural stresses are evaluated as $\frac{M_y}{I} = f_x$ The principal stresses are evaluated using the expression,

$$f_I f_{II} = \frac{f_x + f_y}{2} + \sqrt{\left(\frac{f_x - f_y}{2}\right)^2 + \tau^2}$$

f_y is assumed to be zero in the theory of bending.

9.2 Mechanism of Shear Resistance in Concrete Beams

9.2.1 Uncracked beam

The expression for f_I & f_{II} given above are valid in a beam of homogeneous material, Prestressed concrete is considered as homogenous till it is cracked. A prestressed concrete member subjected to flexure and shear (Variable bending moment over its length) has the principal stress trajectories as shown in Fig 9.1 (b). It clearly shows the relative intensities of flexural and shear stresses. In zones where the slopes of the trajectories are flat, the influence of vertical forces (shear force) is minimum and that of flexural stresses are predominant. In the zones where the trajectories are steeper, the influence of flexural forces is minimal and in that of shear forces is predominant.

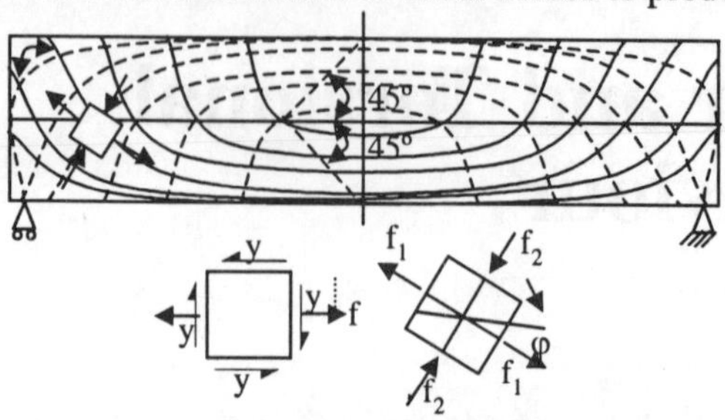

Fig. 9.1.a. Trajectories of Principal Stresses in a Homogenous Isotropic Uncracked R.C. Beam

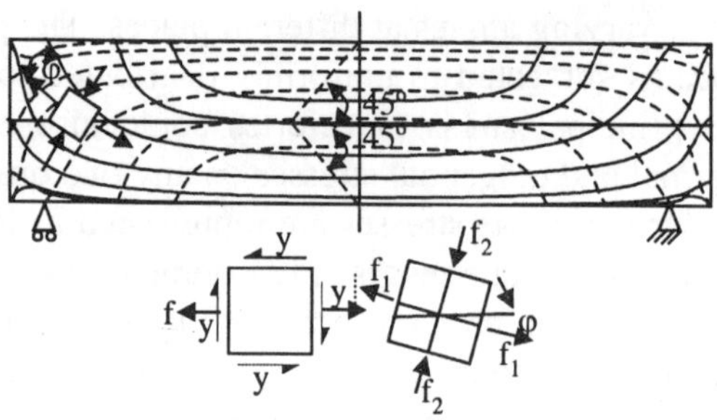

Fig. 9.1.b. Trajectories of Principal Stresses in a Homogenous Isotropic Uncracked Prestressed Concrete Beam

In a prestressed concrete beam, because of effect of prestress, the principal stress trajectories are more horizontal when compared to a Reinforced concrete beam and hence shear stresses are less predominant. These formulation are valid upto cracking load stage. The relationship between orthogonal stress ($f_x f_y$, τ) and principal stresses ($f_I f_{II}$) is also depicted by Mohr's circle, illustrated already in chapter 1(Fig.1.1.) It is shown that prestressing a concrete beam helps in increasing the shear resisting capacity of the beam. In a prestressed concrete beam subjected to flexure, f_x should include compressive stresses induced by pre- compression (prestressing) in addition to the compression/tensile stresses caused by bending due to loads. The component due to prestress makes longitudinal stresses f_x more compression predominant. f_y is considered as zero. Hence the principal tensile (f_{II}) stress are of very low value. At times both f_I and f_{II} may also be only compressive stresses under service load conditions. This eliminates the risk of diagonal tension cracks in prestressed concrete beams under service loads. ***The mechanism to resist these stresses (Principal stresses, or orthogonal stresses with shear stress) in an uncrcked beam is by the material capacities in tension and compression.*** But once the beam gets cracked in any place, the mechanism of resistance is different and it is also not uniform over the length of the beam.

9.2.2 Cracked beam

As the external loading is applied, the bending moment increases and the prestressed concrete beam reaches the stage of decompression in the tension zone. Later when the tensile stresses cross the flexural cracking capacity of the material concrete, cracks start developing in the tension zone. Similarly if the principal tensile stresses cross the tensile stress capacity of the material, anywhere in the beam, cracks start developing normal to the direction of such principal tensile stresses.

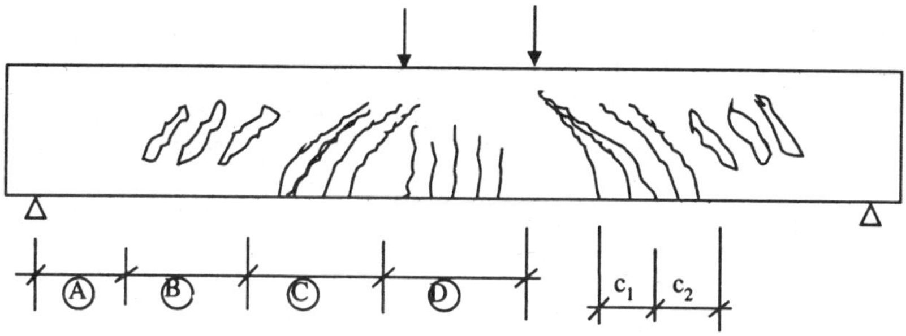

(A) No cracks zone
(B) Web shear creack zone
(C) Flexural shear creack zone
(D) Flexural crack zone
(C_1) Zone with parallel cracks
(C_2) Zone with coverging cracks

Fig. 9.2. Classification of Zones In a Cracked Beam (2)

Bachmann and Thurlimann [1] have clearly classified different zones (Fig 9.2) where cracks of different types, appear. The same has been adopted by CEB-FIP [2] also in their code. As per that classification, the zone where the direction of the principal trajectories and the flexural stresses are one and the same, [pure bending zone] cracks are developed perpendicular to the axis of the beam and progress ahead with increase in load. This zone is referred as zone of pure flexure (zone D).

In zones where flexure predominates, the cracks start as flexural cracks first. As they proceed, the influence of shear forces is felt and the cracks turn towards the zone of higher bending moment. This zone is referred as flexural shear crack zone (zone C). In the zone where the vertical force (equilibrating force) in the section predominates, the cracks develop in web somewhere in the middle where shear stress intensity is high. This is due to principal tensile stresses and they are called diagonal cracks. This zone is called web shear crack zone (Zone B) Here the flexural tension (bottom) chord is practically uncracked. Such a zone is practically small or can be considered non existent in a reinforced concrete beam.

Principle of limit state design against collapse is commonly adopted now days, because such a design is based on reality. Accordingly the demarcation of zones is also based on principal tensile stress at limit stage of collapse as suggested by Leonhardt [3] The ultimate load is governed by load factors and safety factors as discussed in chapter 7. At ultimate load stage, the tensile strength of concrete is compared with principal tensile stress that occurs in concrete. At that point of time principal compressive stresses are also higher and effective. CFB-FIP committee for shear has combined zone A & B together and fixed the demarcation between zone B and zone C with stress limit on the extreme tensile fibre $\sigma_{x.u}$ to be greater than equal to 0.55 $f^{2/3}_{ck}$ (kg/cm^2). But in the design practice, the design is done for safety against failure by shear both in zone B & zone C independently and the required material like steel or concrete (greater of the requirements for the two zones namely for zone B or zone C) is provided in both the zones. Hence the demarcation of zone is more of academic and research interest.

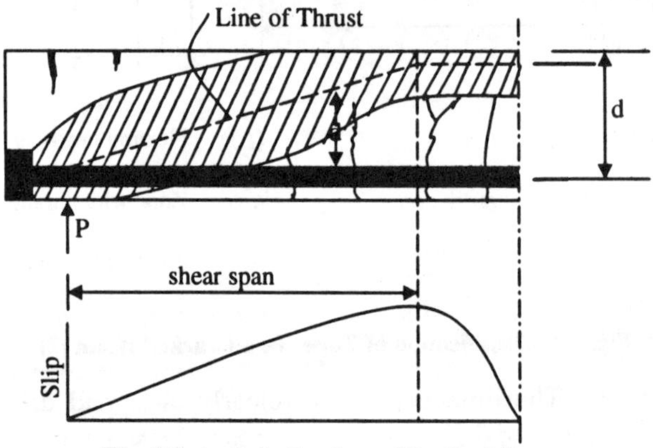

Fig. 9.3. Arch Action in an Idealized Beam

The Resistance to shear has been modelled and explained in many ways as reported by ASCE-ACI task committee 426[4]. The structural action of the beam is compared to that of a tied arch (Fig 9.3) in a beam where there are no shear reinforcements. Kani[5] explained the resistance to shear by means of tooth action of the concrete between the cracks. (Fig 9.4) Consecutive teeth are also considered as part of a frame with the concrete between the cracks acting at struts[4]. Moersch [6] and Baker and Ahamed [7] considered the behavior of a beam as that of a truss or a modified Truss where top and bottom chords resist flexural forces by forming a couple and the web portion represented with diagonal chords resist shear (Fig 9.5). With introduction of prestressing in beams, the above structural models are modified taking into consideration the effects of compressive stresses caused by prestress and vertical resisting component by draping of tendous.

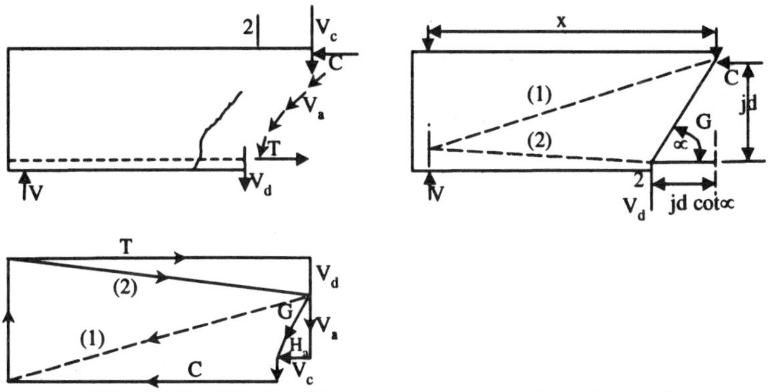

Fig. 9.4 Equilibrium Requirements in the Shear Span of Beam

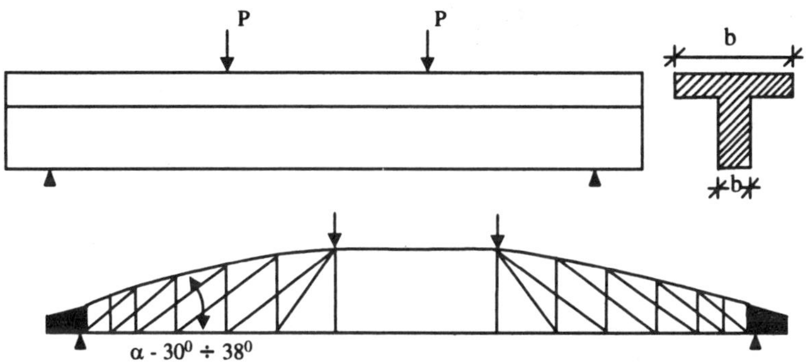

Fig. 9.5. Modified Truss-Analogous Beam

Detailed explanations on the various aspects mentioned above are available in the report of ASCE - ACI Task committee, 426 [4] McGregor and Hanson [8] have also discussed the effect of prestressing on the shear behavior of Prestressed concrete Beams. For the purposes of design for shear in pretressed concrete, knowledge on possible modes of shear failure is desirable and they are explained before the design procedure for shear reinforcement are detailed.

9.2.3. Modes of failure

The modes of failure due to shear, depend on the geometry of the beams and the possible loading pattern. Since loading pattern is not a constant, in any beam, an enveloping loading as equivalent uniformly distributed load or equivalent enveloping bending moment diagram is considered for shear. Under such conditions the geometry of beam especially the depth of the beam, thickness of the web, and stiffness of the tension chord (percentage longitudinal steel and the bond between concrete & steel) play major role.

For beams with M/Vd ratios ranging from 3 to 6, the cracks start as flexural cracks. As these cracks develop, the influence of shear is felt on these cracks and they turn towards the load point (toward the zone of greater bending moment) At one stage, with further increase in load, the compression zone becomes unstable. The cracks suddenly penetrate and cut into the compression zone causing failure. This type of a failure is called ***diagonal tension failure*** (Fig 9.6 a)

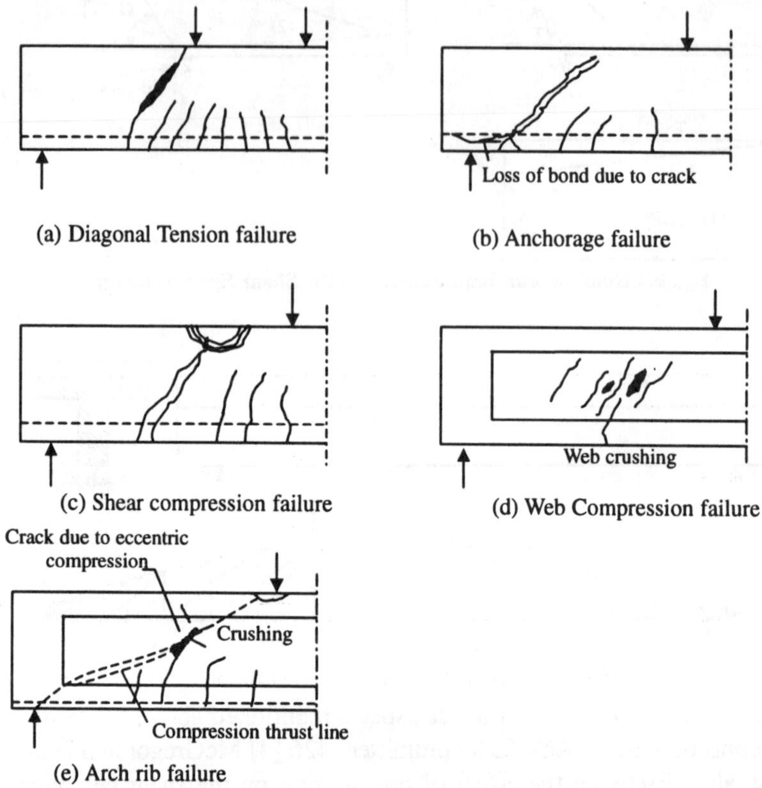

Fig. 9.6 Different Failure Patterns under Shear

In case of beams with *M/Vd* ratios 1 to 2.5, the shear takes a predominant role and triggers off two additional modes of failure purely based on shear. Shear crack which are developed from flexure, on increase of loads, widen causing higher stress level in the shear reinforcements. These stresses in the shear reinforcement reach their limiting values causing large deformation. At that point of time these shear cracks propagate on either direction both upwards & downwards. When it prorogate downwards, dowel forces are mobilised on longitudinal steel. If the tension chord is stiff, the cracks divert their direction and develop along the longitudinal reinforcement towards the support causing loss of bond. The main reinforcement slips causing an ***anchorage failure*** (Fig 9.6.b). Some times if the tension reinforcement is small and the tension chord is not stiff, the cracks cut across the tension chord in the same angle causing failure of the tension chord. This is called ***shear tension failure***. The shear crack when it propagates upwards, the compression zone gets reduced and gets crushed. It is known as ***shear***

compression failure. (Fig 9.6.c) In case of wide flanged beams the crack may not penetrate through the compression zone, but move longitudinally along the junction of flange and web moving towards higher bending moment zone leading to instability in the compression zone causing failure.

In the case of beams with thin webs such as I-beams or T-beams the shear stresses in the web zone play a predominant role leading to web failure either in tension or in compression. The ***web compression failure*** (Fig 9.6.d) is due to the principal compressive stress exceeding the limiting value and this occurs when the web is well protected against tension failure with reinforcement. This type of failure is rather sudden and violent. To avoid this type of failure, the percentage of web reinforcement is limited to a maximum value so that the compression capacity of the web is never reached. This is indirectly achieved in various national codes regulations by fixing up an upper limit of shear stress up to which alone shear could be allowed to be resisted by reinforcement. Any shear beyond that limit is not permitted to be covered by steel reinforcement. In such cases the section has to be redesigned. In case of beams experiencing larger tension in the web, the cracks in the web zone open out leading to buckling of the compression diagonals which further leads to eccentric compression and secondary crushing of the web. This failure is referred as ***arch rib failure***. This type of failure (Fig.9.6.e) could be avoided by providing a sufficiently thick web. This is the cause for introduction of minimum web thickness for concrete beams in different codes. ***These types of failures on the web are more commonly felt in prestressed concrete beams.***

It is necessary that a design should take care to see that these failures do not occur and enough factor of safety is also provided against such failures using proper material (concrete and steel) which would not be stressed beyond their limiting values taking into consideration material safety factors. After cracking of the beam, the shear is resisted by

- The vertical component of the inclined compressive force in concrete in the compression zone.
- Dowel action of the reinforcement bars and the coarse aggregates in tension chard.
- Aggregate interlock along the shear crack.
- Bending resistance of concrete teeth (between 2 cracks) or the compressive strut action of that portion of concrete (frame action)
- Shear resistance of uncracked concrete in tension chord in zone B
- Shear reinforcement, and
- Vertical component of draped tendons in case of prestressed beam with draped tendons a cracked section with all these reactive forces is shown in Fig.9.7.

When the cracks develop well, aggregate interlocks play no role because of separation of surfaces. Dowel action plays very little role in case of beams

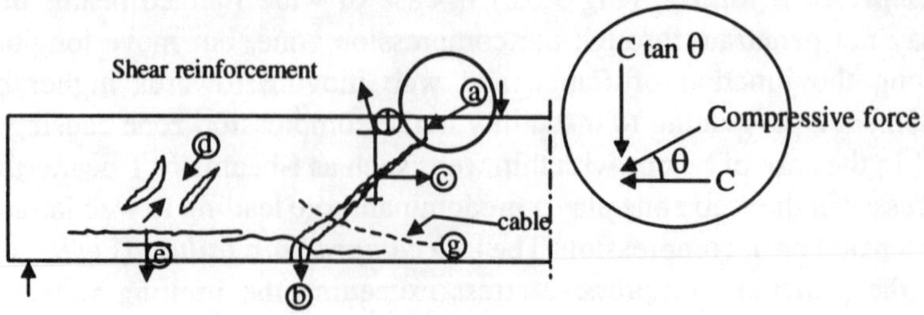

Fig. 9.7. Possible Shear Resistances in Cracked Beam

with web reinforcement since any dowel effort will be transferred to stirrup. The contribution of resistance to shear by frame action is negligible. Hence external forces represented as shear are resisted to a large extent only by a) Concrete contribution in compression and b) Shear reinforcement. But of course in zone B, the uncracked concrete is tension zone also contributes. The contribution to shear resistance expected from reinforcement can be expressed as

$$V_{st} = V - V_c - V_{dt}$$

where V is the total shear in the cross section after cracks have developed.

V_c is the shear resisted by concrete in compression zone(component of inclined compression force in compression chord)

V_{dt} is the shear resisted by draped tendons.

9.3. Design for Shear is Prestressed Concrete Beams.

The philosophy of cracking, classification of zones and types of failure are have been well recognised. Provision of material (Design) for safety against failure is also based on the contribution for shear resistance listed in art.9.2.3 above Quantifying the contributions is different in various national codes. Considering zone A, and referring to Fig 9.1 b, it is noted that the compression stress trajectories flow towards the support and they are directed towards the support. Hence the beam in this portion acts as a column with compression forces inclined. Hence there are no flexural or shear stresses. There are no cracks, hence no shear reinforcement is necessary. Zone A is extended upto a distance equal to half the depth of the beam ***along*** its axis in prestressed concrete beam because the compression force flow is more steep at the ends of the beams to reach to the support.

In zone B, the flexural stresses are small hence the tension chord is uncracked. The beam is considered homogeneous, though there could be cracks in the web. Analysing the beam element, principal stresses can be calculated by principles of mechanics. The principal stresses are due to flexural and shear stress. When principal tensile stress crosses the tensile strength of concrete, the corresponding flexural and shear stresses and the corresponding shear force can be evaluated. Any shear beyond the shear, occurring at the cross section in zone B should be resisted by shear reinforcement. The shear that is carried by concrete just before the principal tensile stress reaches the tensile capacity of concrete is designated as V_{co} in Zone B, Zone C is a zone where flexural cracks appear first and develop as shear cracks. The shear that is resisted by the beam up to formation of this crack is referred as V_{cr}. Once the beam is cracked, the shear forces are resisted by various structural actions mentioned in art. 9.2.3

Flexural cracks appear in a reinforced concrete beam in the tension zone immediately after loading because concrete is supposed to have no tensile capacity. But in prestressed concrete beams, cracking due to flexure happens only after the existing pre compression in concrete extreme fibre (due to prestress) is first nullified by tensile stresses developed due to bending caused by external loads and later extreme fibre stress in concrete reaches the flexural tensile capacity of material, The moment that is resisted upto a formation of a flexural crack is defined by M_{cr}. The corresponding shear at the cross section in that zone (which is considered for design) forms part of shear resisted by concrete [V_{c1}]. After cracking, on any further loading, shear is resisted only by the vertical component of the inclined compression trajectories in the compression chord. The total compression which form a couple with tensile force to resist the flexure at that cross section is referred as 'C' and the inclination of compressive force to the horizontal as θ as shown in Fig. 9.7. Then the shear resisted by that compression chord can be evaluated as $C \tan \theta$. This portion of shear resisted by concrete is designated as V_{c2}. This value of shear resisted by concrete compression chord '$C \tan \theta$ is normalised and expressed average stress $k\tau_c$ over the cross section, where τ_c is calculated as $C \tan \theta / bd$ and 'τ_c' is a function of concrete strength (f_{ck}) and the depth of compression block (x_u). The depth of compression block is affected by crack propagation and equilibrium of forces (namely the total compression equal to total tension.) Both crack propagation and the equilibrium of forces are functions of stiffness of tension chord expressed in terms of reinforcement (A_s) Hence τ_c is represented as a function f_{ck} and A_s / bd in literature. It may be noted that τ_c is only a representation of the shear resisted by concrete and it is not really the shear stress across the section.

The principal compression stresses trajectories are more horizontal in prestressed concrete beams when compared to those in reinforced concrete beams as explained earlier. Hence the value of V_{c2} [$C \tan \theta$] will be smaller in prestressed concrete beam and the corresponding normalised τ_e will be smaller. Instead of having different values of τ_c for reinforced and prestressed concrete beams with same f_{ck} and A_s / bd; a factor (k) less than one is used for prestressed concrete. This factor is function of prestressing level in steel. Hence V_{c2}, the contribution of compression chords for shear resistance is evaluated as $k\tau_e bd$. The total contribution by concrete is resulting shear will be $V_{c1} + V_{c2} = V_{cr}$

The factor k mentioned above is derived and formulated in different ways in various national codes. Out of the total shear to be resisted by a cross section in zone, the shear resisted by various components have to be subtracted and the left over shear alone should be resisted by reinforcement to be designed. Hence the forces to be resisted by shear reinforcement can be expressed as V_{st}

$$V_{st} = V - (V_{c1}+V_{c2}) - V_{d\ raped\ tendons}$$

The reinforcement is designed for zone 'C' and zone 'B' separately. In zone C, normally there could be more designs, there by providing different steel reinforcement in that zone. Various natural codes have formulated the design expressions in different ways. For protection against different types of failures, (ie) failures caused by shear compression, shear tension, diagonal tension etc shear reinforcement are designed on the basis detailed above. For protection against web compression failure, maximum shear permitted in the beam at any cross section inclusive of the facility of resisting by shear reinforcement is restricted so that the compression strut in the web does not fail by compression. For protection against arb rib failure which may be due to buckling, of the web, a minimum thickness of the web is called for. This is specially important in prestressed concrete beam because the prestressing tendons which normally pass though the web, occupies space reducing the effective concrete resisting arch rib failure. Various codes also stipulate the values for maximum permitted shear stress (nominal) and minimum thickness for the web.

9.3.1. I.S. 1343 - 1980 Code provisions (10)

The contribution from concrete for resistance to shear in Zone B, just before cracking in the web in defined as V_{co} It is function of principal tensile stress and its express as

$$V_{co} = 0.67bD\sqrt{ft^2 + 0.8 f_{cp} f_t}, \quad \text{where}$$

- b = breadth of the member which for T, I and L beams should be replaced by breadth of the rib bw,
- D = overall depth of the member,
- f_t = maximum principal tensile stress given by $0.24\sqrt{f_{ck}}$ taken as positive where fck is the characteristic compressive strength of concrete, and
- f_{cp} = compressive stress at centroidal axis due to prestress taken as positive.

In flanged members where the centroidal axis occurs in the flange, the principal tensile stress should be limited to $0.24\sqrt{f_{ck}}$ at the intersection of the flanged web; in this case, 0.8 of the stress due to prestress at this intersection may be used, in calculation V_{co}. For a section uncracked in flexure and with inclined tendons or vertical prestress, the component of prestressing force normal to the longitudinal axis of the member may be added to V_{co} as V_d for design shear reinforcement. The contribution from concrete for resistance to shear in Zone C is defined as V_{cr}.

$$V_{cr} = V_{c1} + V_{c2}$$

$$V_{cr} = k\tau_c bd + M_{cr}\left(\frac{V}{M}\right)$$

$$V_{cr} = \left(1 - 0.55\frac{f_{pe}}{f_p}\right)\tau_c bd + M_0\frac{V}{M}; \text{ where}$$

f_{pe} = effective prestress after all losses have occurred, which shall not be put greater than $0.6f_p$,

f_p = characteristic strength of prestressing steel,

τ_c = ultimate shear stress capacity of concrete obtained fromTable 9.1

b = breadth of member, which, for flanged section, shall be taken as the breadth of the web b_w,

d = distance from the extreme compression-fibre to the centroid of the tendons at the section considered,

M_0 = moment necessary to produce zero stress in the concrete at the depth, given by:

$M_o = 0.8 f_{pt}\frac{I}{y}$ Where f_{pt} is the stress due to prestress only at the soffit, y is the distance of soffit from the centroid of the concrete section, I is the moment of inertia of the Cross section, and V and M are shear force and bending moment respectively, at the section considered due to ultimate loads.

V_{cr} should be taken as not less than $0.1bd\sqrt{f_{ck}}$

τ_c is function of A_p / bd and f_{ck} and is given in Table 9.1

Table 9.1. Design Shear Strength of Concrete. τ_c

$100\frac{Ap}{bd}$	Concrete Grade		
	M 30	M 35	M 40 and Above
(1)	(2)	(3)	(4)
0.25	0.37	0.37	0.38
0.50	0.50	0.50	0.51
0.75	0.59	0.59	0.60
1.00	0.66	0.67	0.68
1.25	0.71	0.73	0.74
1.50	0.76	0.78	0.79
1.75	0.80	0.82	0.84
2.00	0.84	0.86	0.88
2.25	0.88	0.90	0.92
2.50	0.91	0.93	0.95
2.75	0.94	0.96	0.98
3.00	0.96	0.99	1.01

Note- A_p is the area-prestressing tendon.

The value of V_{cr} calculated at a particular section may be assumed to be constant for a distance equal to $d/2$, measured in the direction of increasing moment, from that particular section. For a section cracked in flexure and with inclined tendons, the component of prestressing forces normal to the longitudinal axis of the member should be ignored.The shear to be carried by shear reinforcement is decided as

$$V_{st} = V - [V_{cr} \text{ or } V_{co}] - V_d$$

The lower values of two namely V_{cr} & V_{co} should be used for design

9.3.1.1 Shear reinforcement

The shear resisted by concrete has to be taken as the smaller V_{cr} and V_{co} calculated above. If this shear (V_c) is greater than the total shear (V) to be resisted at any cross section, theoretically no shear reinforcement is necessary. But to avoid sudden collapse just after a shear crack, minimum reinforcement is suggested in IS1343-1980. The minimum reinforcement should be provided in the form of stirrup, such that the shear reinforcement (A_{sv}) satisfies the following conditions.

$$\frac{A_{sv}}{bs_v} = \frac{0.4}{0.87 f_y}$$

where

A_{sv} = total cross-sectional area of stirrup legs effective in shear,
b = breadth of the member which for T,I and L beams should be taken as the breadth of the rib, b_w.
s_v = stirrup spacing along the length of the member, and
f_y = characteristic strength of the stirrup reinforcement which shall not be taken greater than 415 N/mm².

However, shear reinforcement need not be provided in cases where V is less than 0.5 V_c, and in member of minor importance.

When V exceeds V_c, shear reinforcement shall be provided such that:

In rectangular beams, at both corners in the tensile zone, a stirrup should pass around a longitudinal bar, a tendon or a group of tendons having a diameter not less than the diameter of the stirrup. The depth d is taken as the depth from the extreme compression fibre either to the longitudinal bars or to the centroid of the tendons whichever is greater.

The spacing of stirrups along a member should not exceed 0.75 d_t nor 4 times the web thickness for flanged members. When V exceeds 1.8 V_c, the maximum spacing should be reduced to 0.5 d. The lateral spacing of the individual legs of the stirrups provided at a cross section should not exceed 0.75 d. To avoid web compression failure I.S 1343 suggested maximum permissible shear stress in a beam a shown in Table 9.2

Table 9.2. Maximum Shear Stress τ_c max

Concrete Grade	M 30	M 35	M 40	M 45	M 50	M 55 & over
Maximum Shear Stress N/mm²	3.5	3.7	4.0	4.3	4.6	4.8

To avoid arch rib failure Indian Roads Congress. specifies a minimum thickness of web to be provided as 250 mm or (250 + diameter of ducts in one horizontal section) in case I beam and box girder.

9.3.2 BS 8110 -1985 Code provisions (11)

BS 8110 ratio adopts similar design procedure as given in I.S code provisions in that, the shear resisting capacity of concrete in Zone B and Zone 'C' are first calculated. The evaluation of V_{co} & V_{cr} and the provision of reinforcement are explained below.

The design ultimate shear resistance of a section uncracked in flexure, V_{co} corresponds to the occurrence of a maximum design principal tensile stress f_t at the centroidal axis of the section equal to $0.24\sqrt{f_{cu}}$. In the calculation of V_{co}, the design value of the prestress at the centroidal axis should be taken as $0.8f_{cp}$. The value of V_{co} is given by the following equation

$$V_{co} = 0.67 b_v h \sqrt{\left(f_t^2 + 0.8 f_{cp} f_t\right)}$$

Values of $V_{co} / b_v h = \tau_{co}$ obtained from the above equation are given in Table 9.3 for applicable values of f_{cp}. In flanged members where the centroidal axis occurs in the flange the principal tensile stress should be limited to $0.24\sqrt{f_{cu}}$ at the intersection of the flange and web.

In this calculation, 0.8 of the stress due to prestress at this intersection should be used in calculation V_{co}.

For a section uncracked in flexure and with inclined tendons or compression flange, their effects should be combined algebraically with the external design load effects.

Table 9.3 Values of Design Shear Strength τ_{co}

f_{cp}	Concrete Grade			
	30	40	50	60
N/mm^2	N/mm^2	N/mm^2	N/mm^2	N/mm^2
2	1.30	1.45	1.60	1.70
4	1.65	1.80	1.95	2.05
6	1.90	2.10	2.20	2.35
8	2.15	2.30	2.50	2.65
10	2.35	2.55	2.70	2.85
12	2.55	2.75	2.95	3.10
14	2.70	2.95	3.15	3.30

In a pre-tensioned member the critical section should be taken at a distance from the edge of the bearing equal to the height of the centroid of the section above the soffit. In case this section occurs within the prestress development length, the compressive stress at the centroidal axis due to prestress to be used f_{cp} may be calculated from the following relationship:

$$f_{cpx} = \frac{x}{l_p}\left(2 - \frac{x}{l_p}\right) f_{cp}$$

where, f_{cp} is the design stress at the end of the prestress development length l_p.

The design ultimate shear resistance of a section cracked in flexure, V_{cr}, may be calculated using

$$V_{cr} = \left(1 - 0.55\frac{f_{pe}}{f_{pu}}\right)\tau_c b_v d + M_0 \frac{V}{M}$$

where τ_c is a function of total reinforcement (prestressing steel and untensioned steel together) and the effective depth in mm. The values can be taken from Table 9.4

Table 9.4 Values of τ_c, design concrete shear stress

$\frac{100A_s}{b_v d}$	Effective depth (in mm)							
	125	150	175	200	225	250	300	≥ 400
	N/mm²	N/mm²	N/mm²	N/mm²	N/mm²	N/mm²	N/mm²	N/mm²
≤ 0.15	0.45	0.43	0.41	0.40	0.39	0.38	0.36	0.34
0.25	0.53	0.51	0.49	0.47	0.46	0.45	0.43	0.40
0.50	0.67	0.64	0.62	0.60	0.58	0.56	0.54	0.50
0.75	0.77	0.73	0.71	0.68	0.66	0.65	0.62	0.57
1.00	0.84	0.81	0.78	0.75	0.73	0.71	0.68	0.63
1.50	0.97	0.92	0.89	0.86	0.83	0.81	0.78	0.72
2.00	1.06	1.02	0.98	0.95	0.92	0.89	0.86	0.80
≥ 3.00	1.22	1.16	1.12	1.08	1.05	1.02	0.98	0.91

Note 1. Allowance has been made in these figures for a γ_m of 1.25

For characteristic concrete strengths greater than 25 N/mm², the values may be multiplied by $\left(f_{cu}/25\right)^{\frac{1}{3}}$. The value of f_{cu} should not be taken as greater than 40.

The value of V_{cr} should be taken as not less than $0.1\, b_v d\sqrt{f_{cu}}$ The value of V_{cr} calculated using the above equation at a particular section may be assumed to constant for a distance equal to *d*/2, measured in the direction of increasing moment, from particular section. For a section cracked in flexure and with inclined tendons or compression chords, the design shear forces produced should be combined with external design load effects where these effects are increased.

Design of shear reinforcement is based on the following details. No shear reinforcement need to be provided if the shear calculated is less than $0.5V_c$ or if the member is declared as of minor importance.

In case the calculated shear exceeds $V_c + 0.4b_v d$, links should be provided to satisfy the equation $\frac{A_{sv}}{s_v} = \frac{V - V_c}{0.87 f_{yv} d_t}$, where s_v is the spacing of links.

The spacing of links along a member should not exceed $0.75d_t$ or four times the web thickness for flanged members. When V exceeds $1.8V_c$, the maximum spacing should be reduced to $0.5d$. The lateral spacing of the individual legs of the links provided at a cross section should not exceed d.

9.3.3. ACI 318-1998 Code provisions (12)

ACI also follows the concept of resistant to shear as explained in art.9.2 The values of shear resistance capacity of Zone B or Zone C are expressed in different format as detailed below. ***It may be noted that all units adopted in ACI specifications are FPS units.*** What is represented as V_{co} in, I.S specification is referred as V_{cw} in ACI specification.

$$V_{cw} = (3.5\sqrt{f_c} + 0.3f_{pc})b_w d + V_p$$

Alternatively, V_{cw} shall be computed as the shear force corresponding to dead load plus live load that results in a principal tensile stress of $4\sqrt{f_c^{1}}$ at the centroidal axis of members, or at the intersection of flange and web when the centroidal axis is in the flange. In composite members, the principal tensile stress shall be computed using the cross section that resists live load. Where

f_c' is the cylinder strength of concrete in *psi*
b_w is the width of web of beam
V_p is the prestress forces in tension (P_∞) in pound
f_{pc} is the stress in concrete at steel fibre level *psi*
d is the distance of extreme compression fibre from cantroid of prestressing steel or 0.8 which ever is greater

Shear resistance provided by concrete in Zone C is represented as V_{ci} and it is equivalent to V_{cr} of I.S code $V_{ci} = 0.6\sqrt{f_c b_w d} + V_d + \frac{V_1 M_{cr}}{M_{max}}$ where

V_d = Shear force at section due to unfactored dead load, but V_{ci} need not be taken less than $1.7\sqrt{f_c'}\, b_w d$, where $M_{cr} = (I / y_t)(6\sqrt{f_c'} + f_{pc} - f_d)$

When f_d is stress to unfactored dead load at extreme tensile fibre and values of M_{max} and V_I shall be computed from the load combination causing maximum moment to occur at the section.

As suggested in other codes the lower of V_{ci} or V_{cw} should be considered for evaluating the shear resistance to be provided by reinforcement. A minimum area of shear reinforcement shall be provided if the shear force to be resisted at limit state stage is more than 0.5 times the shear resisted by concrete (V_{ci} *or* V_{cw}) Minimum reforcement to be provided is expressed as

$A_{sv} = \frac{50b_w s_v}{f_y}$, where b_w, s_v are in inches and f_y in psi

If the shear force to be resisted at limit stage is more than that could be resisted by concrete the area of shear reinforcement is calculated as

$$A_{sv} = \frac{V_s s_v}{f_y(\sin\alpha + \cos\alpha)d}$$

Where V_s is the shear to be resisted by reinforcement α is the inclination of stirrups, to the horizontal and d is the effective depth.

The maximum shear that is permitted to be resisted by the beam is not greater than $3.5\sqrt{f_c'}\, b_w d$. If it is more, the width of the rib has to be increased.

9. 4. Shear in Flanged Beams

Shear as mentioned earlier is only an expression of indirect expression, along with flexural forces, to define the principal stress trajectories. In flanged beams the principal compressive forces flow through the flanges at pure bending moment region and the force have to flow towards the support where there is a variation in bending moment explained as shear in beam bending theory. The compression forces in the flange width should narrow down to web width and flow towards the support as principal compression trajectories. That is how the effective flange width at support section is taken as equal to web width only. More over for resisting bending moment, which is considered along the middle plane of the beam, all compressive forces should be brought together. Hence the cantilever portion of the flange should be connected to the web. This is done by using horizontally closed stirrup running for the full width of flange.

With the concept of shear resistance, in a cross section, the flexural shear flow in an uncracked cross section of a flanged beam can be described as shown in Fig.9.8. The shear flow also starts from free end of the flange and turns into the web to flow upto the bottom and ends. Hence the shear flow in the cantilever portion of the flange has to be resisted by means of reinforcement in the flange and the shear flow should be lead into the web. The quantity of shear flow in the cantilever portion of the flange can be arrived on the following lines. Since the flexural shear itself is caused by the inclined compression, the shear resisted by the concrete in the flange portion will be proportional to the compression force carried by that portion of the flange in relation to the total compression as detailed out by Leonhardt(3).

$C = 0.36 f_{ck}\, b_w\, x_u + 0.45 f_{ck}\,(b_f - b_w)\, D_f = (C_1 + C_2) + C_3$

V = Total shear resisted by concrete in compression in portion 1,2,3 of the Fig.9.8.

V_3 = Shear resisted by concrete in compression in portion 3

$V_{1,2}$ = Shear resisted by concrete in compression in portion 1 & 2

V_3 is proportional to C_3

V_1 & V_2 is proportional to C_1 & C_2 $\quad C_1 = C_2 \quad V_1 = V_2$

Hence shear carried by the cantilever portion of the flange on either side of the web $V_1 = V_2 = \left[\frac{C_1 or C_2}{C}\right]V$

Where $V = C \tan\theta = V_{cr}$ or V_{co} as the case may be

Once V_1 or V_2 is known, stirrups to carry the shear into the web portion can be designed as $Af_s = \frac{V_1}{s_v}$

where A_{svt} is the area of 2 legged stirrups in the flange portion.

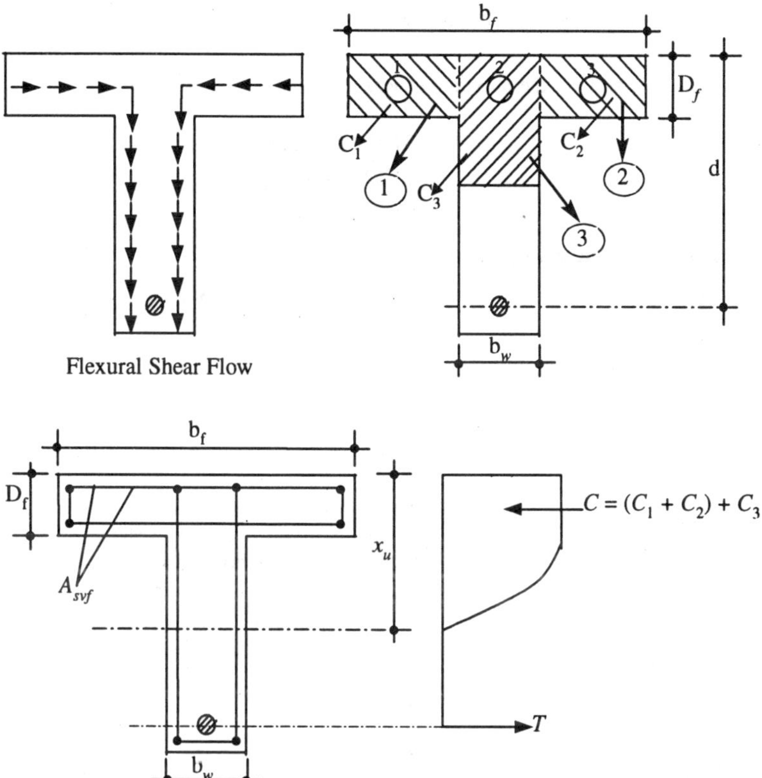

Fig. 9.8. Flexural Shear Flow and Shear Reinforcement in Flange

Example 9.1.

A prestressed concrete beam of rectangular section of size 250 × 600mm is prestressed with a straight cable with constant eccentricity of 100mm. The beam spans 10m and carries a load of 20kN/m and the prestressing force in the cable is 1000kN. Concrete has a density of 24kN/m^3 Calculate the maximum principal stresses at a cross section 300mm away from support.

Solution.

Section properties

Area 600 × 250 $= 150000\text{mm}^2$

Moment of Inertia $250 \times \frac{600^3}{12} = 4500 \times 10^6 mm^4$

Modulus of section $250 \times \frac{600^2}{6} = 15 \times 10^6 mm^3$

Self weight of the beam $\left[\frac{600 \times 250}{10^6} \times 24\right] kN = 3.6 kN/m$

Live load = $20 kN/m$

Total load = $23.6 kN/m$

Shear at 300mm from support $\frac{WL}{2} - W \times 0.3$

= 110.92kN

Maximum Shear stress at neutral axis level at that section $\tau_v = \frac{V}{Ib}(A\bar{y})$

For rectangular section, the maximum shear intensity $\tau_v = \frac{3}{2}\left(\frac{V}{bh}\right)$

$$= \frac{3}{2}\left(\frac{110.92 \times 10^3}{250 \times 600}\right)$$

= 1.1092N/mm²

Principal stresses at that element

Prestress effect $f_x = \frac{P}{A} = \frac{1000 \times 10^3}{250 \times 600} = 6.67 N/mm^2$

Maximum and Minimum principal stresses $= \frac{f_x}{2} \pm \frac{1}{2}\sqrt{(f_x)^2 + 4\tau_v^2}$

$$= \frac{6.67}{2} \pm \frac{1}{2}\sqrt{(6.67)^2 + 4(1.1092)^2}$$

= 3.33 ± 3.5146

= 6.844N/mm²(*compression*)

= –0.1846N/mm² (*tensile*) on a plane at a angle θ to the horizontal.

$$= 2\theta = \tan^{-1}\frac{2\tau_v}{f_x - f_y}$$

$$= \tan^{-1}\frac{2 \times 1.1092}{6.67 - 0}$$

$$\theta = 9.2°$$

Example 9.2.

A prestressed concrete beam of rectangular section of size 250×600mm is prestressed with parabolic cable having an maximum eccentricity of 100mm at the center of the span, reducing to zero at the supports. The beam spans 10m and carries a load of 20 kN/m and the prestress force on the cable is 1000kN. Concrete has a density of 24kN/m³. Calculate the maximum principal stress at the cross section 300mm away from support.

Solution

Section properties

Area = 600×250=150000mm²

Moment of Inertia $= 250 \times \frac{600^3}{12} = 4500 \times 10$

Modulus of section $= 250 \times \frac{600^2}{6} = 15 \times 10^6 mm^3$

Self weight of the beam = (0.6 × 0.25 × 24) = 3.6*kN/m*

Live load $= 20kN/m$

Total load W $= 23.6kN/m$

Shear at 300mm from support $= \frac{WL}{2} - W \times 0.3$

$= 110.92\text{kN}$

Slope of the cable at 300mm from support $= \frac{4e}{l^2}(1-2x)$

$$= \frac{4 \times 0.1}{(10)^2}(10 - 2 \times 0.3)$$

$= 0.0376$

Vertical component of prestressing force $= 1000 \times 0.0376$

$= 37.6kN$

Net shear force at 0.3m from the support $= 110.92\text{-}37.6$

$= 73.32\text{kN}$

Maximum Shear stress at neutral axis level $\tau_v = \frac{V}{Ib}(A\bar{y})$

$$\tau_v = \frac{3}{2}\left(\frac{V}{bh}\right)$$

$$\tau_v = \frac{3}{2}\left(\frac{V}{bh}\right) = \frac{3}{2}\left(\frac{73.32 \times 10^3}{250 \times 600}\right)$$

$= 0.7332N/mm^2$

Direct stress due to prestress

$$\frac{P}{A} = \frac{1000 \times 10^3}{250 \times 600} = 6.67N/mm^2$$

Maximum principal stresses

$$= \frac{6.67}{2} \pm \frac{1}{2}\sqrt{6.67^2 + 4 \times 0.7332}$$

$= 3.33 \pm 3.443$

$= 6.773N/m^2$ *(compression)*

$= -0.113N/m^2$ *(Tension)* acting

a plane at angle θ to the horizontal

$$2\theta = \tan^{-1}\frac{2\tau_v}{f_x - f_y}$$

$$= \tan^{-1}\frac{2 \times 0.7332}{6.67 - 0}$$

$\theta = 6.2°$

Example 9.3.

A concrete beam having a rectangular section 250mm wide and 600mm deep, is prestressed by a parabolic cable having an eccentricity of 100mm at the center of span, reducing to zero at the supports. The span of the beam is 8m. The beam supports a live load of 20kN/m. Determine the effective force in the cable to balance the dead and live loads on the beam. Estimate the principal stresses at a 300mm from a support.

Solution.

Self-weight of the beam $= (0.25 \times 0.6 \times 25) = 3.75\text{kN/m}$

Total load W= $(3.75+20) = 23.75\text{kN/m}$

Eccentricity of cable at centre of span $= 100\text{mm}$

Using the concept of load balancing, the prestressing force P can be evaluated by equating the external moment to that created by prestressing

$$P \times e = P \times 100 = \frac{23.75 \times 8000^2}{8}$$

$$P = 1900 \times 10^3\text{N} = 1900kN$$

Slope of cable at 300mm form support $\frac{4e}{l^2}(l-2x)$

$$= \frac{4 \times 100}{(8000)^2}(8000 - 2 \times 300)$$

$$= 0.04625$$

Vertical component of prestressing force $= 1900 \times 0.04625$

$$= 87.875kN$$

Shear force at 300mm from the support due to dead load and live load

$$= \frac{Wl}{2} - W \times 0.3$$

$$= \frac{23.75 \times 8}{2} - 23.75 \times 0.3$$

$$= 87.875kN$$

Therefore net shear force at a section 300mm from the support = 0

Horizontal prestress at that section $= \frac{1900 \times 10^3}{250 \times 600} = 1267N / mm^2$

∴ Principal stress = 12.67 N/mm² (Compression)

Example 9.4.

A T-beam section has the following section properties. Top flange-width and thickness are 500mm and 200mm, the depth and thickness of the web are 600mm and 200mm respectively. The beam is prestressed with H T steel of area of 2000mm². The cable profile used over an effective span of 16m, is parabolic with an eccentricity of 600mm at the centre of the span and 300mm at the support section. The effective prestress in the tendons is 1000 N/mm². Estimate the ultimate shear resistance at the support section, and the maximum permissible uniformly distributed working load on the beam using an overall load factor of 2.

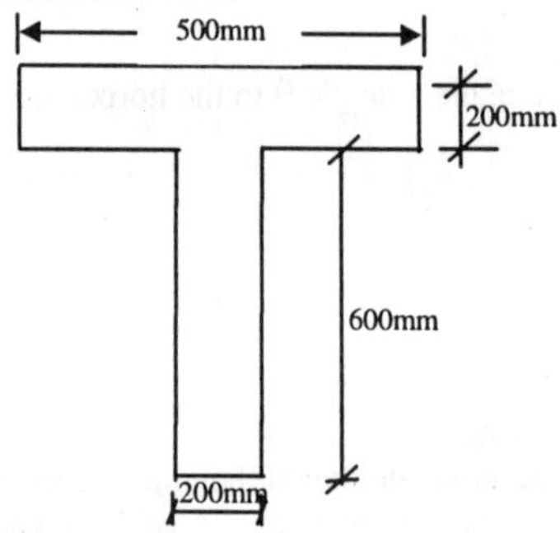

Solution.

Section properties

Moment of inertia $I_{xx} = 1.266 \times 10^{10} mm^4$

Gross cross-section area $A = 0.22 \times 10^6 mm^2$

$$f_t = 0.24\sqrt{f_{ck}}$$

$$f_t = 0.24\sqrt{40}$$

$$f_t = 1.518 N/mm^2$$

The ultimate shear resistance of the support section failing due to web-shear cracking mode is given by the expression.

$$V_{cw} = \left[b_w \left\{ \frac{I}{S} \right\} \sqrt{f_t^2 + f_{cp} f_t} + \eta P \sin\theta \right.$$

The first moment of area is obtained as:

$$S = A\bar{y} = [(600 \times 200 \times 218.18) + (200 \times 118.18 \times 59.09)]$$

$$= 0.275 \times 10^8 mm^3$$

$$\left(\frac{I}{S}\right) = \left(\frac{1.266 \times 10^{10}}{0.275 \times 10^8}\right) = 460.36mm$$

$\eta P = (2000 \times 1000) = 2000 \times 10^3 N$ Considering no loss in pressers

$f_{cp} = 9.09\ N/mm^2$

The slope of the parabolic tendon profile at support $\theta = \left(4\frac{e}{l}\right)$

Where e = dip of the cable between the support and the center of span,

$$\theta = \left(\frac{4(600-300)}{16\times 1000}\right) = 0.075$$

$\eta P.\sin\theta = (2000 \times 10^3 \times 0.075) = 150 \times 10^3\ I$

$$V_{cw} = \left[200 \times (460.36)\sqrt{(1.518)^2 + (9.09 \times 1.518)} + 150 \times 10^3\right]$$

$= 519.47 \times 10^3 N = 519.47kN$

The Maximum shear force under working loads

$$= \frac{519.47}{2} = 259.73kN$$

If q = Uniformly distributed working load over a span of 16m

$$q \times \frac{16}{2} = 259.7$$

$$q = 32.4669 kN/m$$

Example 9.5

The cross section of a prestressed concrete beam is an un-symmetrical T-section with an overall depth of 1200 mm. Thickness of web is 200mm. Width and depth of flange are 1000mm and 200mm respectively. At a particular section the beam is subjected an ultimate moment, are 2000kN m and a shear force is 250kN. Design the shear reinforcement necessary for zone near that section. The other details of the beam are as follows

Effective depth . d = 1100mm

Cube strength of concrete f_{ck} = 40N/mm²

Effective prestress at the extreme tensile face of the beam f_{ep}

= 19.3N/mm²

Moment of inertia = 7.533×10¹⁰mm⁴

Area of steel in the section A_p = 2310mm²

Tensile strength of tendons f_p = 1500N/mm²

Effective stress in tendons after all losses f_{pe} = 900 N/mm²

Solution.

Area of gross cross-section = $0.4 \times 10^6 mm^2$

Moment of inertia I_{xx} = $7.553 \times 10^{10} mm^4$

$\bar{y}_b$ = 800mm

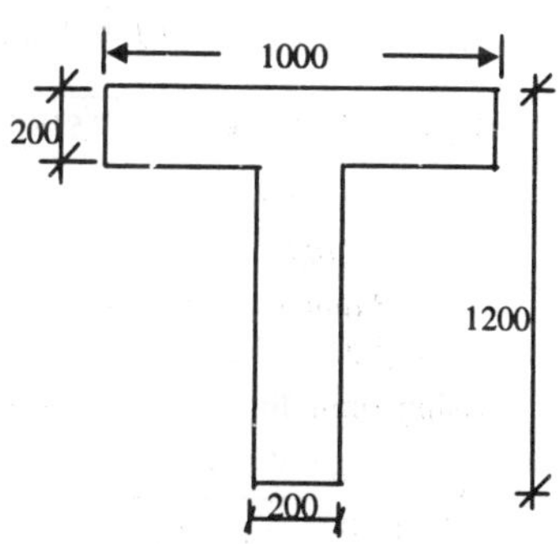

The flexural stress at the section where

Bending moment is 2000kNm is

$$f_x = \frac{2000 \times 10^6}{7.553 \times 10^{10}} \times 800 = 21.18 N/mm^2$$

This is greater than precompression available namely 19.3 N/mm². Hence it would have cracked. Hence estimitating the flexure-shear resistance of the section using Indian or British code regulations.

$$M_0 = \left[\frac{0.8 f_{ep} I}{y_b}\right] = \left[\frac{0.8 \times 19.3 \times 7.553 \times 10^{10}}{800}\right]$$

$= 145.77 \times 10^7 Nmm$

$$\left[\frac{100 A_p}{b_w d}\right] = \left[\frac{100 \times 2000}{200 \times 1100}\right] = 0.909 \qquad \tau_c = 0.65088$$

The flexure-shear resistance of the section is

$V_{cr} = (1 - 0.55 f_{pe} / f_p)\, \tau_c b_w d + (M_0 / M) V$

$$= \left(\frac{1 - 0.55 \times 900}{1500}\right) \times 0.65088 \times 200 \times 1100 + \left(\frac{145.77 \times 10^7}{2000 \times 10^6}\right) \times 250 \times 10^3$$

$= 278.15 \times 10^3 N$

$= 278.15 kN > 250 kN$ (actual shear)

Therefore minimum shear reinforcement alone is required name $A_{sv} = \dfrac{0.4 \times b s_v}{0.87 f_y}$

Aopting a spacing of 450mm for stirrup steel $A_{sv} = \dfrac{0.4 \times 200 \times 450}{0.87 \times 415} = 99.70 mm^2$ 8mm^2 legged stirrup will give an area = 100.48mm^2

Provide 8mm 2 legged stirrup at 450mm spacing.

Example 9.6.

The support section of a prestressed concrete beam, 150mm wide and 300mm deep, is required to support an ultimate shear force of 100kN. The compressive prestress at the centroidal axis is 5N/mm^2. The characteristic cube strength of the concrete is 40 N/mm^2. The cover to the tension reinforcement is 45mm. Design suitable reinforcements at the section using the Indian standarded code I.S:1343 recommendations.

Solution

$b_w = 150$mm $\qquad f_{cp} = 5$ N/mm^2

$D = 300$mm $\qquad f_{ck} = 40$ N/mm^2

$d = 255$mm $\qquad V = 100$ kN

Assume characteristic tensile strength of steel in stimups is 250 N/mm^2.

or the support section uncracked in flexure.

$$V_c = V_{cw} = 0.67 b_w D \sqrt{f_t^2 + 0.8 f_{cp} f_t}$$

$$f_t = 0.24\sqrt{f_{ck}} = 0.24\sqrt{40} = 1.518 N / mm^2$$

$$V_c = 0.67 \times 150 \times 300 \cdot \sqrt{(1.518)^2 + (0.8 \times 5 \times 1.518)}$$

$= 87.26 \times 10^3 N$

$= 87.26 kN$

$\therefore$ *Balance shear* $= (V - V_c)$

$= (100 - 87.26)$

Using 8mm diameter two-legged stirrups, the spacing of the stirrups

$$S_v = \left[\frac{A_{sv} 0.87 f_y d}{(V - V_c)}\right] = \frac{2 \times 50.26 \times 0.87 \times 250 \times 255}{12.74 \times 10^3}$$

$= 437.65 mm$

Maximum permissible spacing = 0.75d = (0.75 × 255)=191.25mm

Therefore 8mm diameter two-legged stirrups at 190mm centre to centre may be provided.

Example 9.7.

A prestressed concrete T beam as given in example 9.5 is to be designed for shear at a cross section where the factored moment is 2500kNm and the factored shear is 300kN. The fibre stress at extreme tensile face due to prestress is 10N/mm². All other details are as given in example 9.5. Design the flexural shear reinforcements using British code regulations.

Solution.

Data from the example 7.5 are

Over all depth of T-beam section (h)	1200mm
Thickness of web (b_v)	200mm
Width of flange (b_f)	1000mm
Depth of flange (d_f)	200mm
Distance of top fibres from centroid (y_t)	400mm
Distance of bottom fibres from centroid (y_b)	800mm
Factored moment at section xx (M)	2500kNm
Factored shear force at section xx (V)	300kN
Effective depth of section (d)	1100mm
Cube strength of concrete (f_{cu})	40N/mm²
Effective prestress at the extreme tensile face of beam (f_{cp})	10N/mm²
Second moment of area (I)	7.553x1010mm⁴
Area of steel in the section (A_p)	2310mm²
Tensile strength of prestressing tensions (f_{pu})	1500N/mm²
Effective prestress in tendons after all losses (f_{pe})	900N/mm²

According to British code, ultimate flexure shear resistance of section is given by

$$V_{cr} = \left(1 - 0.55\frac{f_{pc}}{f_{pu}}\right)\tau_c b_{vd} + M_0\left(\frac{V}{M}\right)$$

Where,

$$M_0 = M_0 = 0.8 \times f_{cp}\left(\frac{I}{y_b}\right)$$

$$= \frac{0.8 \times 10 \times 7.553 \times 10^{10}}{800}$$

$$= 755kNm$$

From Table 9.3 design shear strength (τ_c) for $f_{cu} = 40N/mm^2$

$$\tau_c = 2.55N/mm^2$$

$$V_{cr} = V_{cr}\left(1 - 0.55 \times \frac{900}{1500}\right) \times 2.55 \times 200 \times 1100 + \frac{755 \times 10^6 \times 300 \times 10^3}{2500 \times 10^6}$$

$$= 466.5\text{kN} > 300\text{kN}\ (actualshear)$$

provide minimum shear reinforcements

$$\frac{A_{sv}}{S_v} = \frac{0.4b_v}{0.87f_{yv}}$$

Adopting a spacing of 450mm for stirrup steel

$$A_{sv} = \frac{0.4 \times 200 \times 450}{0.87 \times 415}$$

$$A_{sv} = 99.70mm^2$$

8mm two legged stirrups will give an area = 100.48mm². Provide two legged 8mm ϕ stirrups at 450mm c/c.

Example 9.8.

Design the flexural shear reinforcements for the beam given in Example 9.6. Span of the beam is 66.66 feet. Use American code specification ACI – 318 – 1998 Hints: Convert all units are in FPS system data

Solution.

From the example 9.6 the data are converted into FPS units and the details are given below

$M_{max} = 22 \times 10^6 lb$

$V_l = 66 \times 10^3 lb$

$f_c^l = 4400 lb/in^2$

$f_{pe} = 1375 lb/in^2$

$y_b = 32 in$

$y_t = 16 in$

$b_w = 8 in$

$d = 44 in$

$A = 640 in^2$

$I = 11 \times 933 \times 10^3 in^4$

$f_c^l = 0.8 f_{ck} = 0.8 \times 5500 = 4400 lb/in^2$

Shear resistance of the section is given by

$$V_{ci} = 0.6\sqrt{f_c b_w d} + V_{ci} + \frac{V_1 M_{cr}}{M_{max}}$$

$$V_d = \frac{W_d l}{2}$$

unfactored dead load of the section (W_d)

$$W_d = [40 \times 8 \times 40 \times 8] \times 0.0825 = 52.8 lb/in$$

span of beam = 800in

$$V_d = \frac{52.8 \times 800}{2} = 21120 lb$$

unfactored total design load

$= 52.8 \times 800 = 42240 lb$

$$f_d = \frac{42240}{640} = 66 lb/in^2$$

$M_{cr} = 11.84 \times 10^6 lbin$

$$V_{cv} = 0.6\sqrt{4400} \times 8 \times 44 + 21120 + 11.84 \times 10^6 \left(\frac{66 \times 10^3}{22 \times 10^6}\right)$$

$= 70.65 \times 10^3 lb > 66 \times 10^3 lb$

calculated shear force is more than the actual shear. So, provide minimum shear reinforcement in form of the stirrups,

$$A_{sv} = \frac{50 b_w S_v}{f_y}$$

$f_y = 57062.5 lb/m^2 \quad b_w = 8 in$

Assuming $= S_v = 18 in$

$$A_{sv} = \frac{50 \times 8 \times 18}{57062.5} = 0.13 in^2$$

provide 2 legged 8mm F stirrups at 18in c/c $A_{sv} = 0.160 in^2$

9.5. Concept of Torsion

Torsion in concrete structures can be categorised under two groups, namely equilibrium torsion and compatibility torsion (refer fig. 9.9.) Equilibrium torsion is one for which provisions have to be made in the design of the structural element compulsorily for resistance and stability of the structure. Compatibility torsion is one, which occurs in a structural element to make that element coexist with other elements compatible with each other. If provisions are not made for such a torsion in an element, the structure may not become unstable, but yield to such a torsion by means of twist and it will be reflected as deformations and cracks. But when these deformations and cracks become excessive the structural may become unserviceable. A main beam supporting a canti lever beam is an example of an element subjected to equilibrium torsion. Various elements in a grid structure subjected to loading are examples of structural units subjected to compatibility torsion. Design for the torsion is better under stood if the behavior of the structural elements under torsional loading and the possible modes of failure is are clear

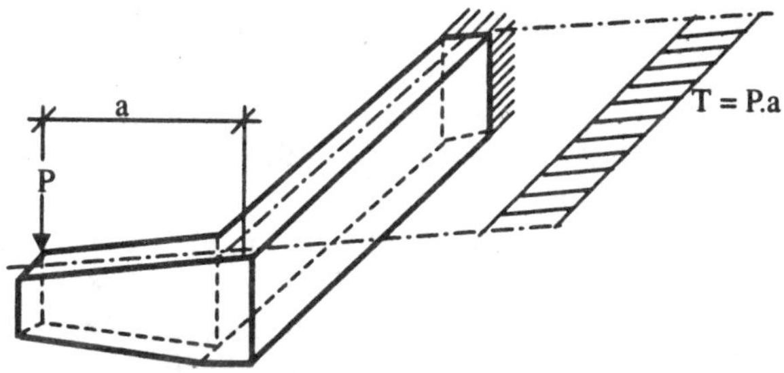

Fg. 9.9.a. Equilibrium Torsion (Primary)

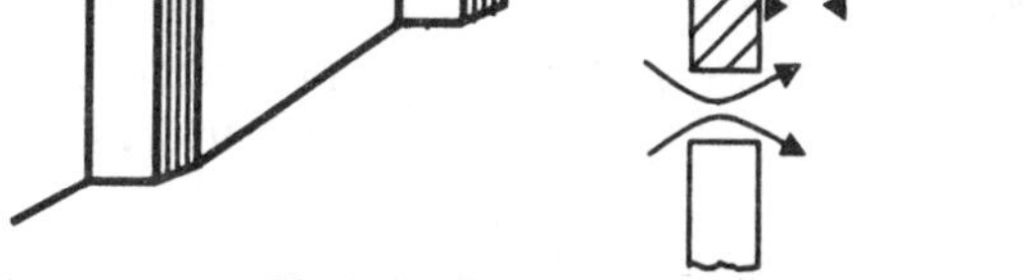

Fig.9. 9.b. Equilibrium Torsion (Secondary)

9. 6. Behavior of Structural Elements under Torsion

Torsion introduces shear stresses and it is normally proportional to the distance of the point under consideration from the axis about which torsion is applied. It is also dependent on the shape of the cross section. An axi symmetrical cross section such as circular section, resists torsion with shear stresses in the plane of the section and does not lead to a change in the shape of the cross section and the shear stresses are evaluated as $\left(\frac{T}{J}r\right)$ where T is the applied torsion, J is the polar moment of inertia of the section and r is the distance of the point from the centre of the section. A non-circular section like rectangular section also resists torsion with shear stresses occurring at the face of the section. But the section face does not remain in the same plane. It goes out of shape called warping. The shear stresses generated both in circular and non-circular section is caused by St. Venant's torsion. In noncircular section, in addition to these shear stresses, the change of shape of cross section causes longitudinal stresses called warping stresses.

They will not be of uniform through out the length of the beam. Even in the face of the section they will not be of same nature. They will change from compressive stresses to tensile stresses. Diagonally opposite points go through tension or compression depending on boundary conditions of the elements. There may be varying along the length of the speciments which may introduce warping shear stresses. Table 9.4. shows various cross-sections, their polar moments of inertia and maximum torsion shear stresses when subjected to a Torque 'T'. The values suggested are not exact but for the case of circular sections. The values are got, based on elastic analyses due to St Venant (13) and Bach (14) and presented by Seely and Smith (15). These shear stresses, combined with other stresses, if any, occuring in the cross section form principal tensile and compressive stresses. Principal tensile stresses cause cracking of the section. After cracking the torsional stiffness gets reduced and the formulae for shear stresses are not any further applicable. Prestressing introduces a pre-compression thus influences the formation of principal tensile stresses and delays cracking and failure. Lambert & Thurlimam (16) after through experimental investigation have found that after cracking, only comparatively thin concrete shell at the periphery resists the torsional load and rectangular sections behave as thin walled box sections. The thickness of the shell layer depends on the longitudinal reinforcement and its portion. This is also seen in the formula suggested by Bredt (17) for evaluation of shear stresses. The behaviour of the prestressed concrete elements differs from reinforced concrete elements only in the way that the cracking is delayed because of exisisting pre-compression in concrete and so also change in principal tensile stresses. Incase of non-circular section, prestressing also plays a role in modifying the warping stresses which are longitudinal in nature.

9.7. Failure of Concrete Elements under Torsion

Failure of concrete element under torsion depends on whether it is a plain concrete element or reinforced or prestressed and also on the percentage of reinforcement. Basically there are two schools of thought of behavior of concrete elements under torsion developed by Rausch (18) in Europe and Hsu (19) in the American continent. Rausch developed a space Truss model for failure. Torsion cracks are helically developed on the surface of the speciman and penerfrate inside the specimen forming struts and ties diagonally. (Fig.9.10.) The speciman is supposed to fail under compression while the tensile forces are resisted by reinforcement. Lessig and Yudin (20) further modified this model for reinforced concrete beams and suggested that the helical cracks proceed on 3 sides of a beam. But before it goes to the 4th side helically, it reverses its direction and closes at the starting point on the fourth side. This gives a skew/warped surface of failure. The equilibrium of forces and moments on this warped surface is studied. Hsu (19) converted this warped surface into a plane surface and formed the equilibrium of forces and moments. This is called skew bending approach. There are three modes of failures as shown in Fig.9.11.a,b,c. They are briefly discussed below. Mode one type of failure has the compression face on the top, which is very similar to the flexural failure. This type of failure happens when flexure is predominant, and torsion adds up to the flexural shear and warping stresses. Mode two type of failure creates compression face on the sides of the beam showing that it is a shear predominant failure.

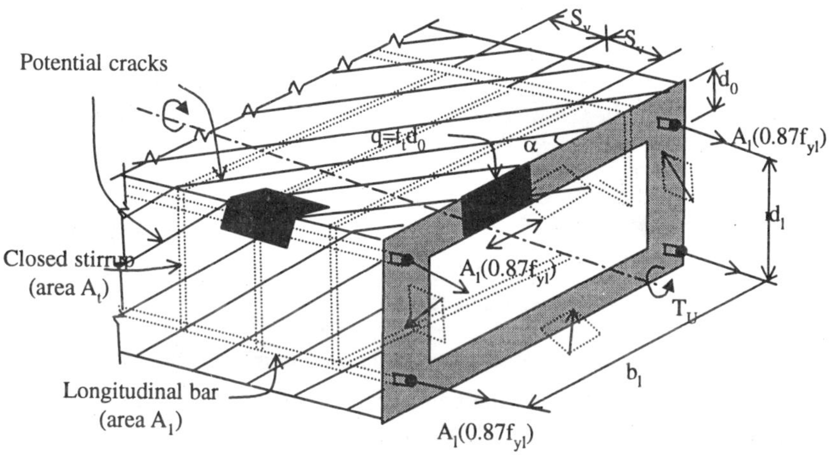

(a) space-truss model

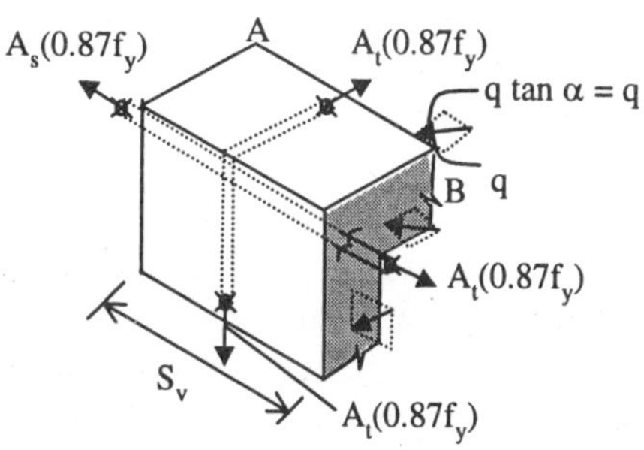

(a) detail 'X'

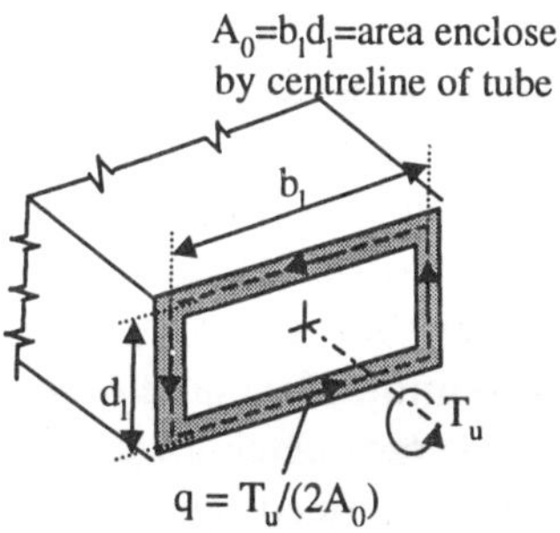

(a) thin-walled tube (box section)

Fig. 9.10. The Idealised Space-Truss Model

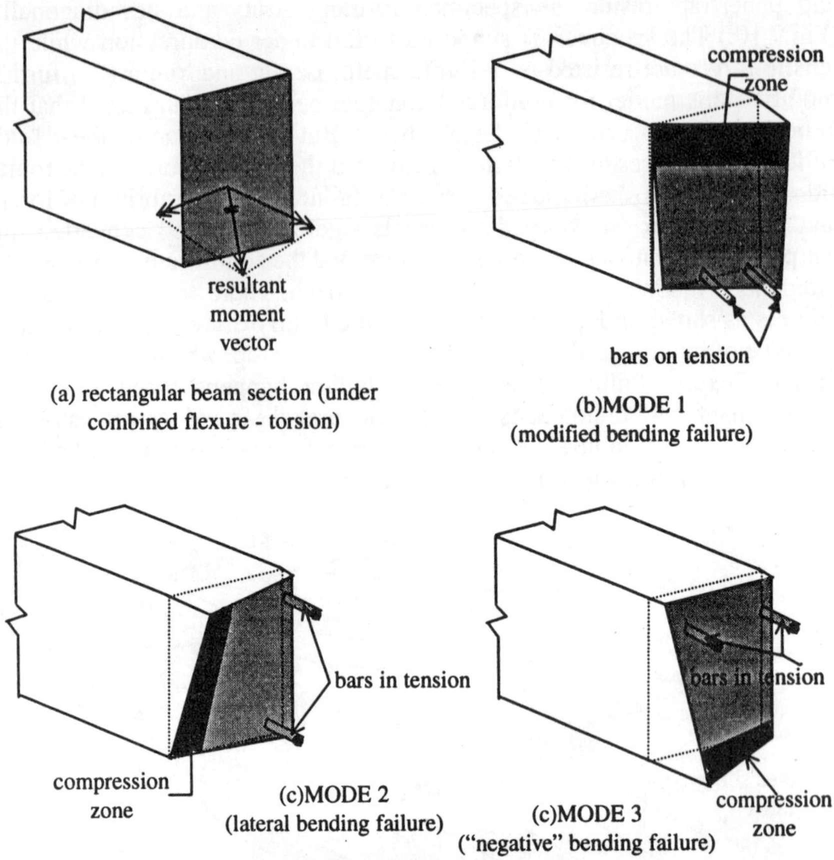

Fig. 9.11. Failure Modes for Combined Flexure and Torsion

This shear predominance happens because of higher amount of torsion compared to bending. The mode three type of failure is rather uncommon and the compression face is on the bottom of the beam where tension occurs due to flexure. This type of failure happens when torsion is predominant and even in torsion the warping stresses are more because of the shape of the cross section. The designers are to take care to dimension the cross section avoid this kind of failure The space Truss model was later developed by Lampert and Collins (21). Elfgreen (22) Collins and Mitchell (23) developed another model based on compression field theory which is a special case of space true model.

Experiences show that pure skew bending theory is applicable in case of marginally reinforced concrete beam under torsion. Reinforced concrete beams with average percentage of reinforcement subjected to torsion take a failure mode as defined by space tress analogy for low values of the ratio M/T. If the moment is highly predominant in structure again skew-bending theory is more applicable for predicting the failures.

The reinforcements to be provided for resisting torsion can be only stirrups in case of circular section but in case of non-circular section, the torsion has to be resisted both by shear and longitudinal stresses.

Necessary reinforcements have to be provided in the form of stirrups as well as longitudinals. As mentioned earlier in the Bredt's formula, the outer peripheral concrete only resists the torsion, the reinforcements have to be provided only in that zone. The longitudinal can be equally spaced around the periphery. But experimental evidences show that the corner reinforcements are sufficient to take care of warping stresses.

In case of space truss analogy model which is developed as a cracked concrete model the pure torsional the capacity of section is evaluated as

$$T_u = 2b_1d_1\left(\sqrt{\left(\frac{A_tf_y}{s_y}\right)\frac{A_1f_y}{2(b_1+d_1)}}\right)$$

where

A_t = is one leg of vertical and Al is longitudinal reinforcement

f_y = yield / limits stress in steel

s_v = spacing of stirrups

b_1, d_1 = are the dimension between the longitudinal steel in the corners.

In this formula there is no facility to have an increased capacity due to prestressing since prestressing does not contribute to Torsional Capacity after cracking. In case of skew bending approach the torsional capacity of the beam is defined as the torsion capacity of the beam of the same dimension without reinforcement (T_c) plus the torsional capacity of the reinforcements provided (T_{st}). The torsional capacity of plain concrete beam is defined as the torsion resisted by that beam up to cracking point. This would automatically include the effect of prestressing. With the introduction of reinforcement, this capacity is reduced to 40%. The torsion capacity of the reinforcements is the concrete is considered additionally.

It has been shown by Hsu (24) that the torsional capacity can be expressed as:

$$T = T_c + T_s = T_c + \frac{\Omega A_0 A_t f_y}{s}$$

where

T = permissible torque

T_c = allowable torque carried by concrete inclusive of effect of prestressing based on the elastic theory reduced to 40% i.e. $0.4\ T_{cr}$

T_s = allowable torque carried by web steel

A_t = cross-sectional area of one leg of web steel

f_y = allowable stress in web steel

A_0 = area of core enclosed by stirrup

s = spacing of stirrups

Ω = a constant varying from 1.6 to 2.0

The value of T_c can be taken as $\Sigma\left(\frac{1}{3}\right)b_1^2 d.f_{t,}$ when f_t is tensile strength of concrete. The component T_c takes care of prestressing in prestressed concrete elements. f_t is taken as then the tensile strength of prestressed concrete which is a function of applied prestress. The tensile strength of prestressed concrete is represented by $\lambda_p f_t$ and $\lambda_p = \left[1 + \frac{12 f_{cp}}{f_{ck}}\right]^{\frac{1}{2}}$

The combined action of Torsion, bending shear etc., have to be considered using interaction expressions. These aspects are dealt in specialist literatures on Torsion. The various national and international codes follow these philosophies for providing materials for torsional design. The code provisions as given IS:1343, BS:8110, ACI:318 are presented below. The code provisions take care of the possible failures and provides material for safety against such failure.

9.8. I.S. 1343 - 1980 Code provisions (10)

9.8.1. General philosophy

Design recommendations in I.S. 1343 - 1980 for prestressed concrete members subjected to torsion are based on the torsional behavior explained earlier. The code provisions are in the form of minimum dimensions, prestressing and reinforcement to take care of the different modes of failure described in art 9.7. These provisions follow similar provisions made in Australian code A.S. 1481(25) developed by Rangan and Hall (26).

If a prestressed concrete element is subjected to torsion, torsional shear stresses and longitudinal warping stresses are developed. Torsion is resisted as helical forces in the element. But once cracking starts, plain prestressed concrete element without passive reinforcement fails violently. To avoid such a failure, minimum reinforcement to take care of tensile forces in the helical form has to provided I.S. code specifies minimum reinforcement for this purpose as equivalent vertical and horizontal reinforcement.

If the prestressed concrete element is already reinforced for bending and shear in addition to prestressing and it is subjected to torsion in addition to bending, the combined effects of bending and torsion is considered.

Torsion produces longitudinal tensile or compression forces due to warping and also shear stresses. Hence torsion is converted as equivalent bending moment and equivalent shear in addition to the bending moment and shear already existent on the elements. The equivalent bending caused by torsion may be additive or subtractive to already applied bending moment depending on the mode of failure. The mode of failure is dependent on M/T ratio. The equivalent shear due to torsion is always be additive to the flexural shear. But for calculating equivalent shear, full applied torsion is not considered. The torsion resisted by concrete before cracking reduced by a factor is subtracted from the applied torsion.

The reinforcements longitudinal, as well as vertical are calculated on this basis as could be seen in the clauses described below.

9.8.2. Application of design rules for torsion

The design rules mentioned are applicable to: a) beams of solid rectangular cross section ($D > b$), b) hollow rectangular beams with $D > b$ and with a wall thickness t $\geq b/4$, and c) T - beam and I - beams.

In all these cases the average intensity of prestress in the concrete shall be less than $0.3 f_{ck}$.

9.8.2.1. Longitudinal reinforcement

The longitudinal reinforcement shall be designed to resist an equivalent ultimate bending moment M_{el} given by

$M_{el} = M + M_t$, where

M = applied ultimate bending moment at the cross-section acting in combination with T,

$M_t = T\sqrt{\left(1+\frac{2D}{b}\right)}$, the sign of M_t being the same as that of M,

D = overall depth of the beam, and

b = breadth of the member which for T and I beams shall be taken as the breadth of the web, b_w

Where the numerical value of M is less than that of M_t, the member shall also be designed to withstand a moment M_{e2} given by $M_{e2} = M_t - M$, the moment M_{e2} being taken as acting in the opposite sense to the moment M.

Where the numerical value of M is less than or equal to that of M_t, the beam shall be designed to withstand an equivalent transverse bending moment M_{e3} (not acting simultaneously with M_{el}), given by

$$\text{by } M_{e3} = M_1\left(1+\frac{x_1}{2e}\right)^2\left(\frac{1+\frac{2b}{D}}{1+\frac{2D}{b}}\right)$$

and acting about an axis at right angles to the axis of M, where x_1 is the smaller dimension of a closed hoop used as torsional shear reinforcement and e is as defined below

9.8.2.2. Transverse reinforcement

Torsional moment and shear carried by concrete is evaluated as follows. The reduced torsional moment carried by the concrete T_{c1} is given by:

$$T_{c1} = T_c\left(\frac{e}{e+e_c}\right), \text{ where}$$

$$T_c = \Sigma 1.5 b^2 D\left(1+\frac{b}{30}\right)\lambda_p\sqrt{f_{ck}}$$

$$e = \frac{T}{V}$$

$$e_c = \frac{T_c}{V_c}$$

$$\lambda_P = \sqrt{\left(1 + \frac{12 f_{cp}}{f_{ck}}\right)}$$

In the above expression,

T_c = torsional moment carried by concrete,

b = breadth of the member, which for T and I beams shall be taken as the breadth of the web, b_w,

D = overall depth of beam,

f_{ck} = characteristic compressive strength of concrete,

T = torsional moment applied to a cross-section under ultimate load conditions,

V = shearing force at a cross section calculated for the specified ultimate loads,

V_c = theoretical shear strength at a cross section, assuming the most unfavorable conditions for inclined cracking, that is, smaller of V_{co} and V_{cr}, and

f_{cp} = average intensity of effective prestress in concrete.

The shear force carried by the concrete V_{c1} is given by:

$$V_{c1} = V_c\left(\frac{e}{e + e_c}\right), \text{ where}$$

V_c = smaller of V_{co} and V_{cr} described. in art 9.3.1.

The transverse reinforcement should take care of shear caused by variation in bending moment and also torsion. The area of cross-section, to total transverse reinforcement A_{sv} necessary is provided in the form of closed stirrup enclosing the corner longitudinal bars A_{sv} shall be taken as the larger of the following two values:

$$A_{sv} = \frac{M_t s_v}{1.5 b_1 d_1 f_y}$$

and $A_{sv} = A_v + 2A_t$

where $$A_v = \frac{(V - V_{c1})\, S_v}{0.87 f_y d_1}$$

$$A_t = \frac{(T - T_{c1}) S_v}{0.87 b_1 d_1 f_y}$$

In the above expression,

s_v = spacing of the stirrup reinforcement,

b_1 = centre to centre distance between corner bars in the direction of the width,

d_1 = centre to centre distance between corner bars in the direction of the depth,

f_y = characteristic strength of shear reinforcement,

M_t, V, V_{c1}, T, T_{c1} are as defined in art 9.8.2.1 and 9.8.2.2.

Minimum reinforcement is provided to avoid sudden and brittle collapse. This amount of reinforcement A_{sv} shall not be taken less than that given by:

$$\frac{A_{sv}}{bs_v} = \frac{0.4}{0.87f_y}$$

There shall be at least one longitudinal bar not less than 12mm in diameter in each corner of the stirrups.

9.8.2.3. Distribution of torsion reinforcement

When a member is designed for torsion, torsion reinforcement shall be provided following the stipulations given below

a) All transverse reinforcement provided for torsion shall be in the form of closed stirrups perpendicular to the axis of the members;

b) The spacing, s_v, of the stirrups shall not exceed $(x_1 + y_1)/4$ or 200 mm whichever is smaller, where x_1 and y_1 are respectively short and long dimensions of the stirrup;

c) Each end of the bar forming the stirrup shall be anchored in accordance with IS:456-2000 and

d) Torsional reinforcement shall be continued to a distance not less than $(D + bw)$ beyond the point at which it is no longer than theoretically required, where D is the overall depth and b_w is the effective width of the web of a flanged member.

9.9. British Code Provision B.S. 8110 - 1985 (11)

Design provisions for torsional resistance of a prestressed Concrete elements are based on the works of Lin and Chow (29). Though the capacity of a prestressed concrete element is more than that of reinforced concrete, ***British code has not taken advantage of the same***. The code provisions take care of protection against the possible failure due to torsion, or bending and torsion by any of the modes of failure already explained.

When a prestressed concrete element is subjected to torsion, principal tensile and compressive stress trajectories are generated orthogonally. Sudden failure in tension is avoided by providing minimum reinforcement as stirrups and longitudinals. Sudden failure under compression is avoided by defining the maximum shear stress that the element may be subjected to. This shear is inclusive of shear caused by variation of bending moment.

When the prestressed concrete element is also reinforced longitudinally and vertically, the behavior of the element is considered analogous to a space truss and the reinforcements are designed accordingly. The various clauses given below are developed based on the above philosophy.

In case of rectangular section the torsional shear stress Vt at any section should be calculated assuming a plastic stress distribution and may be calculated from equation:

$$v_t = \frac{2T}{h_{\min}^2\left(h_{\max} - \dfrac{h_{\min}}{3}\right)}$$

where

h_{max} is larger dimension of the section.

h_{min} is smaller dimension of the section.

T is the torsional moment at failure load in the segment.

T, L or I sections are divided into component rectangles; these are chosen in such a way as to maximize $h^3_{min}h_{max}$ in the following expression. The torsional shear stress v_t carried by each of these component rectangles may be calculated by treating them as rectangular sections subjected to a torsional moment of $T\left(\frac{h^3_{min}h_{max}}{\Sigma\left(h^3_{min}h_{max}\right)}\right)$ Box and other hallow sections in which wall thickness exceed one-quarter of the overall thickness of the member in the direction of measurement may be treated as solid rectangular sections. For other sections, specialist literature should be consulted.

9.9.1. Limit to shear stress

In no case should the sum of the shear stress resulting from shear force and torsion $(v+v_t)$ exceed v_{tu} given in Table 9.5. nor, in the case of small sections where $y_l < 550$mm, should the torsional shear stress v_t exceed $v_{tu}\, y_l/550$.

9.9.2. Reinforcement for torsion

Where the torsion shear stress v_t exceeds v_t,*min* in Table 9.5., reinforcement should be provided. Recommendations for reinforcement for combinations of shear and torsion are given in Table 9.6.

TABLE 9. 5. VALUES OF $v_{t,\,min}$ AND v_{tu}

Concrete grade	$v_{t,min}$ N/mm²	v_{tu} N/mm²
25	0.33	4.00
30	0.37	4.38
40 or above	0.40	5.00

NOTE 1. Allowance is made for γ_m

NOTE 2. Values of $v_{t,m\text{-}in}$ and v_{tu} (in N/mm²) are derived from the equtions:

$v_{t,\,min} = 0.067\sqrt{f_{cu}}$ vfcu but not more than 0.4 N/mm²

$v_{tu} = 0.8\sqrt{f_{cu}}$ but not more than 5 N/mm²

TABLE 9. 6. REINFORCEMENT FOR SHEAR AND TORSION

	$v_t \leq v_{t,min}$	$v_t > v_{t,min}$
$v \leq v_c$	Nominal shear reinforcement; no torsion reinforcement	Designed torsion reinforcement only.
$v > v_c$	Designed shear reinforcement; no torsion reinforcement	Designed shear and torsion reinforcement.

9.9.3. Area of torsional reinforcement

Torsion reinforcement should consist of rectangular closed links together with longitudinal reinforcement. This reinforcement is additional to any requirements for shear or bending and should be such that:

$$\frac{A_{sv}}{s_v} > \frac{T}{0.8x_1y_1(0.87f_{yv})}$$

$$A_s > \frac{A_{sv}f_{yv}(x_1+y_1)}{s_vf_y}$$

f_y and f_{yv} should not be taken as greater than 460 N/mm^2

9.9.4. Spacing and type of links

The value s_v should not exceed the least x_1, $\frac{y_1}{2}$ or 200mm. The links should be of the closed type.

9.9.5. Arrangement of longitudinal torsion reinforcement

Longitudinal torsion reinforcement should be distributed evenly round the inside perimeter of the links. The clear distance between these bars should not exceed 300mm and at least four bars, one in each corner of the links, should be used. Additional longitudinal reinforcement required at the level of the tension or compression reinforcement may be provided by using larger bars than those required for bending alone. The torsion reinforcement should extend a distance at least equal to the largest dimension of the section beyond where it theoretically ceases to be required.

Arrangement of links in T, L or I-sections

In the component rectangles, the reinforcement cages should be detailed so that they interlock and tie the component rectangles of the section together to felicitate shear flow. Where the torsional shear stress in a minor component rectangle does not exceed $v_{t,min,}$ no torsion reinforcement need to provided in that rectangle.

9.10. American Code Specification ACI - 318 - 1998 (12)

American code, as usual assumes a factor φ greater than 1 to the applied torsion for design purposes. The sudden failure of a prestressed concrete element subjected torsion is prevented

a) by a fixing a limit on shear stresses indirectly by defining the cross section dimensions as a function of concrete strength and
b) by fixing minimum torsion reinforcement as a combination of longitudinal and vertical reinforcement.

For members subjected to bending and torsion design of reinforcement is arrived on based on space truss analogy with the angle of diagonal struts as 37.5^0. The code also defines minimum prestress in the elements for this assumption. The relevant clauses for design are given below. These recommendation accepted by prestressed concrete Institute Chicago (28).

9.10.1. Notation

All units adopted are F.P.S. units and the notation are as per ACI speafications.

A_c	=	area of concrete section resisting shear transfer, in^2.
A_{cp}	=	area enclosed by outside perimeter of concrete cross section,in^2
A_l	=	total area of longitudinal reinforcement to resist torsion, in^2.
A_{oh}	=	area enclosed by centerline of the outermost closed transverse torsional reinforcement, in^2.
A_t	=	area of one leg of a closed stirrup resisting torsion within a distance s, in^2.
b_w	=	web width, or diameter of circular section, in^2.
$f_c^{`}$	=	specified compressive strength of concrete, psi
f_{pc}	=	compressive stress in concrete (after allowance for all prestress losses) at centroid of cross section resisting externally applied loads or at junction of web and flange when the centroid lies within the flange, psi. (In a composite member, f_{pc} is resultant compressive stress at centroid of composite section, or at junction of web and flange, due to both prestress and moments resisted by precast member acting alone)
f_{yv}	=	yield strength of closed transverse torsional reinforcement, psi
f_{yl}	=	yield strength of longitudinal torsional reinforcement, psi
T_n	=	nominal torsional moment strength
T_u	=	factored torsional moment strength
V_c	=	nominal shear strength provided by concrete
V_u	=	factored shear force at section
θ	=	angle of compression diagonals in truss analogy for torsion.
ρ_h	=	ratio of horizontal shear reinforcement area to gross concrete area of vertical section
Φ	=	strength reduction factor.

9.10.2. Design torsion.

In prestressed members, sections located less than a distance *h/2* from the face of a support shall be designed for not less than the torsion T_u computed at a distance *h/2*. If a concentrated torque occurs within this distance, the critical section for design shall be at the face of the support.

9.10.3. Allowable torsion without reinforcement.

In case of prestressed members it shall be permitted to neglect torsion effects when the factored torsional moment T is less than $T_{u,min}$

$$T_{u,min} = \Phi\sqrt{f_c^l}\left(\frac{A_{cp}^2}{\rho_{cp}}\right)\sqrt{1+\frac{f_{pc}}{4\sqrt{f_c^l}}}$$

9.10.4. Torsional moment strength.

The cross-sectional dimensions shall be such that:

(a) for solid sections:

$$\sqrt{\left(\frac{V_u}{b_w d}\right)^2 + \left(\frac{T_u \rho_h}{1.7A_{oh}^2}\right)^2} \leq \phi\left(\frac{V_c}{b_w d} + 8\sqrt{f_c^1}\right)$$

b) for hollow sections:

$$\left(\frac{V_u}{b_w d}\right) + \left(\frac{T_u \rho_h}{1.7A_{oh}^2}\right) \leq \phi\left(\frac{V_c}{b_w d} + 8\sqrt{f_c^1}\right)$$

If the wall thickness varies around the perimeter of a hollow section, shall be evaluated at the location where the left-hand side of is a maximum.

If the wall thickness is less than A_{oh}/p_h, the second term or the left hand side expression in shall be taken as, $\left(\frac{T_u}{1.7A_{oh}^t}\right)$, where t is the thickness of the wall of the hollow section at the location where the stresses are being checked.

Design yield strength of nonprestressed torsion reinforcement shall not exceed 60,000 psi. The reinforcement is designed for torsion T_n and T_u shall be determined from $\phi T_n \geq T_u$

The transverse reinforcement for torsion shall be designed using the formula $T_n = \frac{2A_o A_t f_{yv}}{s} cot\,\theta$ where A_o shall be determined by analysis except that it shall be permitted to take A_o equal to $0.85A_{oh}$; θ shall not be taken smaller than 30 deg nor larger than 60 deg. It shall be permitted to take θ equal to 37.5 deg for prestressed members with an effective pressure force not less than 40 percent of the tensile strength of the longitudinal reinforcement.

The additional longitudinal reinforcement required for torsion shall not be less than

$$A_l = \frac{A_t}{s} \rho_h \left(\frac{f_{yv}}{f_{yt}}\right) cot^2\,\theta$$

where θ shall be the same value used for design of transverse reinforcement and A_t and s are transverse reinforcement and its spacing. The reinforcement required for torsion shall be added to that required for shear, bending moment and axial force which act along with torsion.

In prestressed beams:

(a) the total longitudinal reinforcement including tendons at each section shall resist the factored bending moment at that section plus an additional concentric longitudinal tensile force equal to $A_l f_{yl}$, based on the factored torsion at that section, and

(b) the spacing of the longitudinal reinforcement including tendons shall satisfy the requirements mentioned in art 9.10.7.

9.10.5. Details of torsional reinforcement

Torsion reinforcement shall consists of longitudinal bars or tendons and one or more of the following:

(a) closed stirrups or closed ties, perpendicular to the axis of the member, or
(b) a closed cage of welded wire fabric with transverse wires perpendicular to the axis of the member, or
(c) in non prestressed beams, spiral reinforcement.

Transverse torsional reinforcement shall be well anchored. Longitudinal torsion reinforcement shall have enough development lenght at both ends.

For hollow sections in torsion, the distance from the centerline of the transverse torsional reinforcement to the inside face of the wall of the hollow section shall not be less than $0.5A_{oh}/\rho_h$.

9.10.6. Minimum torsion reinforcement

A minimum area of torsion reinforcement shall be provided in all regions where the factored torsional moment T_u exceeds the minimum values specified in 9.10.3. Where torsional reinforcement is required, the minimum area of transverse closed stirrups shall be computed by:

$$(A_v + 2A_t) = \frac{50 b_w s}{f_{yv}}$$

The minimum total area of longitudinal torsional reinforcement shall be computed by: $A_{l,\min} = \dfrac{5\sqrt{f_c^1}\,A_{c0}}{f_{yl}} - \left(\dfrac{A_t}{s}\right)\rho_h \dfrac{f_{yv}}{f_{yl}}$

where A_t/s shall not be taken less than $25b_w/f_{yv}$

9.10.7. Spacing of torsion reinforcement.

The spacing of transverse torsion reinforcement shall not exceed the smaller of $\rho_h/8$ or 12 in.

The longitudinal reinforcement required for torsion shall be distributed around the perimeter of the closed stirrups with a maximum spacing of 12 in. The longitudinal bars or tendons shall be inside the stirrups. There shall be at least one longitudinal bar or tendon in each corner of the stirrups. Bars shall have a diameter at least 1/24 of the stirrup spacing, but not less than a No.3 bar.

Torsion reinforcement shall be provided for a distance of at least (b_t+d) beyond the point theoretically required.

Example 9. 9.

A pretensioned girder of a box section with top & bottom flanges 300mm wide and 50mm thick has an overall depth of 600mm. The thickness of the two webs are 50mm each. The prestress is equally distributed on the flange to create an average compressive stress of 12N/mm^2 through out the section. The maximum shear stress due to transverse load is 3N/mm^2. Determine the increase in the principal tensile stress when the box section is subjected to a maximum torque force of 3kNm.

Solution.

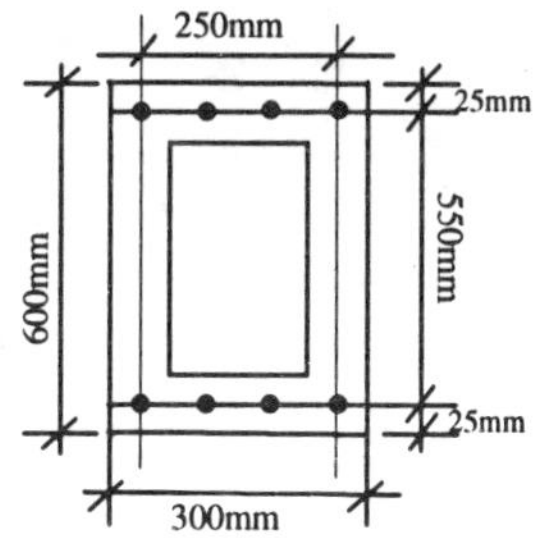

Minimum principal tensile stress is given by the expression

$$f_{min} = \frac{f_x + f_y}{2} - \frac{1}{2}\sqrt{(f_x - f_y)^2 + 4\tau^2}$$

where: $f_x = 12$ N/mm² $\quad f_y = 0 \quad \tau = 3$ N/mm²

$$f_{min} = \frac{12}{2} - \frac{1}{2}\sqrt{12^2 + 4 \times 3^2}$$

$$= 6 - 6.708$$

$$= -0.708 \text{ N/mm}^2 \text{ (tension)}$$

Shear stress due to torque at the centre of the web in box section can be calculated as follows.

For box section,

$$\tau_{max} = \frac{T}{2A_0 t_i}$$

where

$A_0 = b_0 h_0$ [area covered by the centre line of the box shell]

$b_0 = 250$mm; $h_0 = 550$mm; $t_i = 50$mm

$$\tau_{max} = \frac{3 \times 10^6}{2 \times 250 \times 550 \times 50} = 0.22 N/mm^2$$

Total shear stresses inclusive of the shear due to transverse loads

$$= 3 + 0.22$$

$$= 3.22 \text{N/mm}^2$$

Principal tensil stress with torque is given by

$$f_{min} = \frac{12}{2} - \frac{1}{2}\sqrt{12^2 + 4 \times 3.22^2}$$

$$= -0.809 N/mm^2 \ (tension)$$

Increase in principle tensile stress due to torque is

$$= 0.809 - 0.708$$

$$= 0.101 N/mm^2$$

Example 9.10.

Evaluate the maximum torque that could be resisted by the prestressed box section given in Example 9.9. If the concrete strength is 40 N/mm² and there are no other reinforcements in the section.

Solution.

Max principal tensile stress permitted for concrete of strength of 40 N/mm².

$$= 0.24\sqrt{f_{ck}}$$

$$= 0.24\sqrt{40}$$

$$= -1.52 N/mm^2$$

The section has a compressive stress f_x = 12N/mm². If it can take only 1.52 N/mm² as principal tension, the shear stress permitted on the section τ_{max} can be calculated as follows

$$f_{min} = \frac{f_x + f_y}{2} - \frac{1}{2}\sqrt{(f_x - f_y)^2 + 4\tau^2}$$

$$-1.52 = \frac{12}{2} - \frac{1}{2}\sqrt{12^2 + 4 \times \tau^2}$$

$$\sqrt{12^2+4\tau^2} = 15.04$$

$$4\tau^2 = 226.20 - 144 \qquad \tau^2 = 20.55$$

$$\tau = 4.53 N/mm^2$$

If the maximum permissible shear stress is 4.53 N/mm², the torsion that could be applied on the box section is given by.

$$T = \tau_{max} \times 2A_{ti}$$
$$= 4.53 \times 2 \;\; 250 \times 550 \times 50$$
$$= 62.29 kNm$$

Different empirical formulations in codes give also procedures to evaluate the torsional capacity of a prestressed concrete section as torsional capacity of the plain concrete sections multiplied by a factor greater than one. This factor is a function of applied prestress and strength of concrete I.S. code gives this value as

$$\lambda_p = \sqrt{1+\frac{12f_{cp}}{f_{ck}}} \qquad \lambda_p = \sqrt{1+\frac{12\times 12}{40}}$$

$$= 2.144$$

Maximum torque that could be resisted by the section if it were a plain concrete section is evaluated as follows.

Maximum principal tension permitted = -1.52 N/mm²

Since $f_x = f_y = 0$ is on the beam $f_{min} = \tau$

Hence $f_{min\,\alpha} = 1.52 N/mm^2$ *(tension)*

The torque that could generate is shear stress of 1.52 N/mm².

$$T = 1.52 \times 2 \times 250 \times 550 \times 50$$
$$= 20.9 kNm$$

Hence the increase torque that could be permitted on the section if it is prestressed to 12 N/mm²

$= 20.9 \times \lambda_p$

$= 20.9 \times 2.144 = 44.82$ as against the correct value of 62.29. The codes always suggest conservative values for resisting capacities.

Example 9.11

A prestressed concrete beam of size 400*mm* × 800*mm* is subjected to following service load conditions *M*=300kNm *V*=120kN and *T*=150kNm. It is prestressed with an effective force of 1000kN acting at an eccentricity of 220mm. Area of prestressing steel is 940mm². Design the longitudinal and transverse reinforcement necessary as per I.S. specification to resist the service load forces in the section using IS:1343-1980 code recommendations. Assume cover of 50mm f_{ck}= 40N/mm²; f_y= 415N/mm² f_p= 1600N/mm² take effective stress in steel = $0.6f_p$.

Solution.

At ultimate load conditions,

$P_u = \gamma_m \times P$ and $e = 220$mm

$= 0.9 \times 1000$

$= 900$kN

$M_u = 1.5 \times 300 = 450$kNm

$T_u = 1.5 \times 150 = 225$kNm

$V_u = 1.5 \times 120 = 180$kN

Area of prestressing steel = $940 mm^2$
Effective depth = $800 - 50 = 750 mm$
Area of concrete = 400×800
= $32 \times 10^4 mm^2$

Second moment of Area (I) = $\frac{400 \times 800^3}{12} = 17.07 \times 10^9 mm^4$

(i) Design of longitudinal reinforcements

Equivalent moment as per I.S. specifications

$$M_{el} = M + M_t$$

$$M_t = T\sqrt{1 + \frac{2D}{b}}$$

$$= 225\sqrt{1 + \frac{2 \times 800}{400}}$$

$$= 503.11 kNm$$

$$M_{el} = 450 + 503.11 = 953.11 kNm$$

Longitudinal steel section is designed for ultimate moment carrying capacity of 953.11kNm. Ultimate moment capacity is given by

$$M_u = f_{pu} A_p (d - 0.42 x_u)$$

$$\frac{f_p A_p}{b d f_{ck}} = \frac{1600 \times 940}{400 \times 750 \times 40} = 0.13$$

from Table 6.2 and using interpolation

$$\frac{f_{pu}}{0.87 f_p} = 1.0$$

$$\frac{x_u}{d} = 0.28$$

$$\therefore d = 750 mm$$

$$x_u = 0.28 \times 750 = 210 mm$$

$$f_{pu} = 0.87 \times 1600 = 1392 N/mm^2$$

$$M_u = 1392 \times 940 (750 \times 0.42 \times 210)$$

$$M_u = 865 kNm < 953.11$$

The additional moment beyond 865kNm should be resisted by untentioned steel

Additional moment to be resisted = 953.11 – 866
= 87.11kNm

Assuming Fe 415 steel, Ast required is evaluated using the formula

$$M_u = 0.87 f_y A_{st} d \left[1 - \frac{A_{st} f_y}{b d f_{ck}}\right]$$

$$87.11 \times 10^6 = 0.87 \times 15 \times A_{st} \times 750 \left[1 - \frac{A_{st} \times 415}{400 \times 750 \times 40}\right]$$

Solving

$$A_{st} = 326 mm^2$$

Provide 2 bars of 16mm ϕ each at an effective depth of 750mm. This will give an area of $402 mm^2$

(ii) Design of transverse reinforcement :-

Average intensity of prestress at the centroid portion of section f_{cp}

$$f_{cp} = \frac{P}{A}$$

$$= \frac{900 \times 10^3}{32 \times 10^4}$$

$$= 2.81 N/mm^2$$

Tensile strength of concrete

$$f_t = 0.24\sqrt{f_{ck}}$$

$$= 0.24\sqrt{40}$$

$$= 1.51 N/mm^2$$

The reduced torsional moment carried by the concrete T_{c1} is given by

$$T_{c1} = T_c\left[\frac{e}{e + e_c}\right]$$

where $T_c = 0.15b^2D\left[1 - \left(\frac{b}{3D}\right)\lambda_p\sqrt{f_{ck}}\right]$

$$\lambda_p = \sqrt{1 + \frac{12 f_{cp}}{f_{ck}}}$$

$$= \sqrt{1 + \frac{12 \times 2.81}{40}} = 1.36$$

$$T_c = 0.15 \times 400^2 \times 800\left[1 - \frac{400}{3 \times 800}\right] \times 1.36 \times \sqrt{40}$$

$$T_c = 137.62 kNm$$

$$e = \frac{T}{V} = \frac{225}{180} = 1250 mm$$

$$e_c = \frac{T_c}{V_c}$$

where V_c is smaller of V_{co} and V_{cr}

$$_{co} = 0.67bD\sqrt{f_t^2 + 0.8 f_{cp} f_t}$$

$$= 0.67 \times 400 \times 800\sqrt{1.51^2 + 0.8 \times 2.81 \times 1.51}$$

$$= 510 kN$$

$$V_{cr} = \left(1 - 0.55\frac{f_{pe}}{f_p}\right)\tau_c bd + M_o\frac{V}{M}$$

$\frac{f_{pe}}{f_p} = 0.6$ Effective steel area $= 940 + \frac{A_t f_y}{f_p}$

$$A_p = 940 + 402 \times \frac{415}{1600}$$

$$A_p = 1044 mm^2$$

$$\frac{100A_p}{bd} = \frac{100 \times 1044}{400 \times 750} = 0.35$$

From Table.9. Using interpolation

$$\tau_c = 0.432 N/mm^2$$

$$M_o = 0.8 f_{pt}\left(\frac{I}{y}\right)$$

$$\frac{900 \times 10^3}{400 \times 800} + \frac{900 \times 10^3 \times 220 \times 400}{\dfrac{400 \times 800^3}{12}}$$

$$f_{pt} = 7.45 N/mm^2$$

$$M_0 = 0.8 \times 7.45 \times \frac{17.07 \times 10^6}{400}$$

$$M_0 = 254.3 kNm$$

$$V_{cr} = (1 - 0.55 \times 0.6) \times 0.432 \times 400 \times 750 + 254.3 \times \frac{180 \times 10^3}{450}$$

$$V_{cr} = 188.5 kN$$

from that smaller values of V_{co} & V_{cr} $V_c = 188.5 kN$

$$e_c = \frac{T_c}{V_c} = \frac{137.62}{188.5} = 0.73m = 730m$$

$$T_{cl} = 31.87 \text{ kNm}$$

$$V_{cl} = V_c\left(\frac{e}{e + e_c}\right) = 188.5 \times \left(\frac{220}{220 + 730}\right)$$

$$= 43.65 kN$$

$$V\text{-}V_{cl} = 180 - 43.65 = 136.35 kN$$

$$T - T_{cl} = 225 - 31.87 = 193.13 kNm$$

The average steel as closed stirrups enclosing the longitudinal bas is taken as the larger of following two values.

$$A_{sv} = \frac{M_t s_v}{1.5 b_1 d_1 f_y} \quad \text{or} \quad A_{st} + 2A_T$$

Where

$$A_{st} = \frac{(V - V_{cl})s_v}{0.87 f_y d_1} \quad A_T = \frac{(T - T_{cl})s_v}{0.87 b_1 d_1 f_y}$$

Assume

$$s_v = 75mm \quad = \frac{136.35 \times 10^3 \times 75}{0.87 \times 415 \times 735}$$

$$A_{st} = 38.53mm^2$$

$$A_T = \frac{(T - T_{cl})S_v}{0.87 b_1 d_1 fy}$$

$$A_T = \frac{193.13 \times 10^6 \times 75}{0.87 \times 335 \times 735 \times 415}$$

$$= 162.93mm^2$$

$$A_{sv} = A_{st} + 2A_T$$

$$= 38.53 + 2 \times 162.93$$

$$= 246.18mm^2$$

larger value of $A_{sv} = 364.4\text{mm}^2$ is to be adopted in the design 12mm ϕ two legged stirrups at a spacing of 75mm ($A_{sv} = 452\text{mm}^2$) would be sufficient.

$$x_1 = 320mm;\ y_1 = 720mm$$

$$\frac{x_1 + y_1}{4} = \frac{320 + 720}{4} = 260 > 75mm$$

provide, 12mm Φ 75mm c/c -2 two legged stirrups.

Example 9.12

A prestressed concrete beam of size 400mm×800mm is subjected to following forces at ultimate load conditions M_u = 450kNm, V_u = 180kN and T_u = 225kNm. It is prestressed by an effective force of 1000kN acting at an eccentricity of 220mm. Area of prestressed steel is 940mm². Design the longitudinal and transverse reinforcements of given section, using **Space-Truss Analogy** method.

Solution.

Size = 400mm × 800mm

M_u = 450kNm

T = 225kNm

V = 180kN.

The yield force in the prestressing steel

$$P = f_p \times A_p$$

$$= 1600 \times 940$$

$$P = 1540\text{kN}$$

providing top and bottom covers of 50mm and side covers of 25mm.

$$b_1 = 400 - 2 \times 25 = 350\text{mm}$$

$$d_1 = 800 - 2 \times 50 = 700\text{mm}$$

distance from the top longitudinal bars to the prestressing steel, allowing 75mm from bas

$$= 700 - 75 = 624\text{mm}$$

Therefore, moment carrying capacity of beam,

$$= 1504 \times 0.625$$

$$= 940\text{kNm}$$

This exceeds the required capacity of M = 450kNm so longitudinal bar are to be designed to resist the torque and shear only.

Shear flow due to torsion is given by

$$q_T = \frac{T}{2A_0}$$

$$\frac{225 \times 10^6}{2 \times 350 \times 700}$$

$$= 459.18 N/mm$$

Shear flow due to flexural shear force is obtained from the relation

$$q_v = \frac{V}{2d_1} = \frac{180 \times 10^3}{2 \times 100} 128.57 N/mm$$

Maximum shearflow could be

$$= 459.18 + 128.57$$

$$= 587.75 \text{N/mm}$$

The force in stirrups is obtained from expression

$$s = q\, s_v \tan\alpha$$

where,

q – maximum shear flow

s_v – spacing of vertical stirrups

α – Angle of imaginary concrete strut (Assume as 45°)

$$s = 587.75 \times s_v \times \tan 45^\circ$$

$$= 587.75\,(s_v)\ \text{N}$$

The required stirrups reinforcement is

$$\frac{A_{sv}}{S_v} = \frac{587.75}{415} = 1.416 mm^2/mm$$

Adopt spacing of 75mm

$$A_{sv} = 1.416 \times 75 = 106.2 \text{mm}^2$$

Provide 12mm Φ two logged stirrups at 75mm centre to centre.

$$A_{sv} = 226 \text{mm}^2$$

Additional longitudinal reinforcement required at bottom left and right hand corner to resist the torsion and shear

$$A_{l1} = \frac{1}{2f_y}\left[\frac{V}{2} + \frac{T}{2A_0}(b_1 + d_1)\right]$$

$$= \frac{1}{2f_y}\left[\frac{180 \times 10^3}{2} + \frac{225 \times 10^6}{2 \times 350 \times 100}(350 + 100)\right]$$

$$= 689.33 mm^2$$

$$A_{lr} = \frac{1}{2f_y}\left[\frac{T}{2A_0}(b_1 + d_1) - \frac{V}{2}\right]$$

$$= \frac{1}{2 \times 415}\left[\frac{225 \times 10^6}{2 \times 350 \times 700}(350 + 700) - \frac{180 \times 10^3}{2}\right]$$

$$= 472.4 \text{mm}^2$$

box sake of symmetry, provide $A_{sl} = A_{sr}$ hence provide 2 bars of 25mm dia leading to $A_{sl} = 981 \text{mm}^2$ at the left and right hand bottom corners.

Example 9.13

A prestressed concrete beam of size 200mm × 400mm is subjected to the following forces M_u = 40kNm; V_u = 60kN and T_u = 15kNm. It is prestressed by an effective force of 300kN after all losses with 2 stands with C.G located at 75mm above the soffit. Area of prestressing steel is 300mm². Design the shear reinforcement to take care of tension and flexured shear. Assume the cover of 40mm at center of reinforcements at corners f_{ck} = 50N/mm²; f_p = 1600 N/mm²

Solution.

Equivalent ultimate bending moment M_u is given by

$$M_u = M + M_t$$

Where

$$M_t = T\sqrt{1+\frac{2D}{b}}$$

$$15\sqrt{1+\frac{2\times 400}{200}}$$

$$= 33.54\text{kNm}$$

$$M = 40\text{kNm}$$

$$M_{el} = 40 + 33.54 = 73.54\text{kNm}$$

$$M_{el} = 73.54\text{kNm}$$

In this case

$$d = 400 - 75 = 325\text{mm}$$

$$\frac{A_p}{bdf_{ck}} = \frac{300\times 1600}{200\times 325\times 50} = 0.15$$

from Table 6.2 for post tensioned case,

$$\frac{f_{pu}}{0.87f_p} = 1.0$$

$$f_{pu} = 0.87 \times 1600 = 1392\text{N/mm}^2$$

corresponding

$$\frac{x_u}{d} = 0.316$$

$$x_u = 0.316 \times 325$$

$$= 102.7mm$$

ultimate moment capacity

$$M = f_{pu} \times A_p\,(d - 0.42 \times 4)$$

$$= 1392 \times 300 \times [325 - 6.42 \times 102.70]$$

$$= 117.70\text{knM} > \text{M}_{\text{c1}}$$

Hence section is adequate

[Note - If M_{c1}>ultimate moment capacity, the section has to be redesigned]

$$f_{ck} = 50N/mm^2\ f_t = 0.24\sqrt{50} = 1.697$$

Average intensity of prestress (2 stands)

$$= \frac{300\times 1000}{200\times 400}$$

$$= 3.75N/mm^2$$

$$V_{co} = 0.67bd\sqrt{f_t^2 + 0.8f_{cp}f_t}$$

$$V_{co} = 0.67\times 200\times 400\sqrt{1.697^2 + 0.8\times 3.75\times 1.697}$$

$$= 151.326kN$$

$$V_{cr} = \left(1 - 0.55\frac{f_{pe}}{f_p}\right)\tau_c bd + M_o\left(\frac{V}{M}\right)$$

where $M_0 = 0.8 f_{pt}\left(\frac{I}{y}\right)$

$$f_{pt} = \frac{P}{A} + \frac{P_e(y)}{I}$$

$$\frac{300 \times 10^3}{200 \times 400} + \frac{300 \times 10^3 \times 125 \times 200}{\left(\frac{200 \times 400^3}{12}\right)}$$

$$= 3.75 + 7.031$$
$$= 10.78 \text{ N/mm}^2$$

$$M_0 = 0.8 \times 10.78\left(\frac{\frac{200 \times 400^3}{12}}{200}\right)$$

$$M_0 = 46 kNm$$

$$\frac{100A_p}{bd} = \frac{100 \times 300}{200 \times 325} = 0.46$$

From Table 9.

τ_c = $0.489 N/mm^2$ (*for M*50)
f_{pe} = $0.6 f_p$

$$V_{cr} = [1 - 0.55 \times 0.6]0.489 \times 200 \times 325 + \frac{46 \times 10^6 \times 60 \times 10^3}{40 \times 10^6}$$

$$V_{cr} = 90.3kN$$

V_{cr} should be taken as not less than

$$0.1bd\sqrt{f_{ck}}$$

$$= 0.1 \times 200 \times 325\sqrt{50}$$
$$= 45.96kN < V_{cr}$$

Torsional moment

$$T_{cl} = T_c\left(\frac{e}{e + e_c}\right)$$

$$T_c = 1.5b^2 d\left[1 - \left(\frac{b}{3D}\right)\right]\lambda_p f_{ck}$$

$$\lambda_p = \sqrt{1 + \frac{12 f_{cp}}{f_{ck}}}$$

$$= \sqrt{1 + \frac{12 \times 3.75}{50}}$$

$$\lambda_p = 1.378$$

$$T_c = 1.5 \times 200^2 \times 400\left[1 - \frac{200}{3 \times 400}\right] \times 1.378\sqrt{50}$$

194.88kNm

$$e = \frac{T}{V} = \frac{15}{60} = 0.25m = 250m$$

$$e_c = \frac{T_c}{V_c} = \frac{194.88}{90.30} = 2.158m$$

$$T_{cl} = 194.88\left(\frac{250}{250+2158}\right)$$

$$= 19.423\ kNm$$

Taking s_v = 75mm and b_l-135mm and d_l = 350

$$A_{sv} = \frac{M_t S_v}{1.5 b_1 d_1 f_y}$$

$$= \frac{33.54 \times 10^6 \times 75}{1.5 \times 135 \times 350 \times 250}$$

$$A_{sv} = 141.96mm^2\ (f_e\ 250)$$

$$V_{cl} = V_c\left(\frac{e}{e+e_c}\right)$$

$$= 90.3\left(\frac{250}{250+2158}\right)$$

$$= 9.375kN$$

$$V - V_c = 60 \times 10^3 - 9.375 \times 10^3 = 50.625N$$

$$A_{sv} = \frac{(V - V_{cl})S_v}{0.87 \times f_y d_1} = \frac{50.625 \times 10^3 \times 75}{0.87 \times 250 \times 350}$$

$$= 49.87mm^2$$

$T - T_{cl}$ *is – ve* i.e 15 – 22.99

$A_{sv} = 141.96mm^2$

minimum

$$A_{sv} = \frac{b_{sv}(0.4)}{0.87 f_y}$$

$$\frac{200 \times 75 \times 0.4}{0.87 \times 250}$$

$$= 27.58mm^2$$

hence assumed $A_{sv} = 141.96mm^2$ is o.k.

for Fe = 415 and A_{sv} = 141.96mm² assume s_v = 125mm

For Fe 415 steel

$$A_{sv} = \frac{200 \times 125 \times 0.4}{0.87 \times 415}$$

$$= 27.70mm^2$$

hence $A_{sv} = 141.96mm^2$ is o.k.

$$\text{maximum spacing} = \frac{x_1 + y_1}{4}$$

$$= \frac{120 + 320}{4}$$

$$= 110mm < 125mm$$

so provide spacing 75mm for Fe 250 and 110mm for Fe 415

Now recalculate A_{sv} with s_v = 110mm

$$A_{sv} = \frac{33.54 \times 10^6 \times 110}{1.5 \times 135 \times 350 \times 415}$$

$$= 125.43mm^2$$

$$A_{sv} = 125.43mm^2$$

provide 12mmΦ 2 legged stirrups at a spacing of 110mm (A_{sv} = 226mm²) (Fe 415 steel)

Reference

1. H. BACHMANN U. B. THURLIMANN: (1966) **"Schubbemessung von Balken and Platten aus Stahlbeton, Stahlbeton mit Spannzulagen and Spannbeton".** Schweizerische Bauzeitung 1966, p. 583u. 599.
2. COMITE EUROPEEN DU BETON U.(1970) **"Federation Internationale de la Precontrainte: International Richtlinien zur Berechnung and Ausfuhrung von Betonbauwerken. Veroffentlicht zum 6."** FIP-Kongress, Prag
3. LEONHARDT, F. : (1970)**"Shear and torsion in prestressed concrete".** VI. FIP Congress, Prague.
4. Joint ASCE-ACI Task Committee 426. (1973)**"The Shear Strength of Reinforced Concrete Members".** Journal of the Structural Division, ASCE, Vol.99. ST6. June 1973.pp.1091-1187.
5. KANI, G.N.J.: (1969)**"A rational theory for the function of web reinforcement".** Journal of the American Concrete Institute, March.
6. E.MORSCH, (1908) **"Der Eisenbeton. Seine Theorie andAnwendung,"** Wittwer, Stuttgart.
7. A.L.L. BAKER U. S. AHMAD: (1962) **"A statically indeterminate truss model for reinforced concrete beams failing by shear".** CEB, Bulletin No.37, Paris, 1962, p.204.
8. McGREGOR, J.G. and J.M. HANSON, (1969) **"Proposed changes in shear provisions for reinforced and prestressed concrete beams,"** Proceedings of the American Concrete Institute, vol.66, No.4, pp 276-288.
9. REGAN. P.E., (1971) **"Behaviour of reinforced concrete and prestresed concrete subjected to shear force"**, Proceedings, Inst. of Civil Engineers, Paper 7441 S, 1971 Supplement (XVII), pp. 337-374.
10. IS 1343-1980. Indian Standard **"Code of Practice for Prestressed Concrete"**, Bureau of Indian Standards, New Delhi.
11. BS 8110-1985. **"British Standard for Structural use of Concrete"** British Standards Institution, London WIA 2BS.
12. ACI 318-1998. **"Building Code requirements for Structural Concrete"**,American Concrete Institute, Forminton Halls, HI 48333, USA.
13. TOD HUNTER, I, and K. PEARSON, (1960) **"A History of the Theory of Elasticity and of the strength of Materials"**, Part 1. Dover Publications, New York, pp 1-69.
14. BACH, C., ELASTIZITAT and FESTIGKEIT, (1911) Springer verlag, berlin, 1911, pp. 312-320.
15. SELLY FRED, B AND O. SMITH JAMES, (1952) **"Advanced Mechanics of Materials"**, 2nd edition, Wiley International, John Wiley, 1952, New York, pp 266-294.

16. P. LAMPERT u. B. THURLIMANN: (1969) **"Torsions - and Torsions Biegeversuche an Stahlbetonbalken. ETH Zurich, Institut fur Baustatik, Berichte Nr".** 6506-2 and 6506-3, 6506-4.

17. LEONHARDT. P, u.a, **Torsionsversuche an Stahlbetontragern. DafSt.,**

18. RAUSCH, E., (1929) **"Berechnung des Eisenbetons gegen verdrehung and Abscheren (Design of reinforced Concrete for Torsion and shear),"** Springer Verlag, Berlin.

19. HSU, T.T.C. (1968) **"Torsion in structural concrete - uniformly prestressed members without web reinforcement,"** Journal of the Prestressed Concrete Institute, Vol.13, pp 34-44.

20. YUDIN, V.K., (1962) **"Determination of the load bearing capacity of R.C. elements of rectangular cross-section under combined torsion and bending",** Beton No.6, June 1962, pp. 265-268.

21. LAMPERT, P. and M.P., COLLINS, (1972) **"Torsion, bending and confusion - An attempt to establish the facts,"** Journal of ACI. Vol. 69, No.8, August 1972 pp 500-504.

22. ELFGREN, L., (1971) **"Reinforced concrete beams loaded in combined torsion bending and shear - A study of the ultimate load-carrying capacity,"** Publications No.71:3, Chalmers University of Technology, Goteborg, second edition, p.230

23. COLLINS, M.P. and D. MITCHELL, (1980) **"Shear and torsion design of prestressed and non-prestressed concrete beams, "** Journal of the Prestressed Concrete Institute, Vol.25, pp 32-100.

24. HSU. T.T.C. (1968) **Ultimate torque of reinforced rectangular beams.** Journal of structural division ASCE. Vol.94. No ST2. Proceedings paper 5814 Feb 1968.

25. AS:1481-1974, Australian Prestressed Concrete Code.

26. RANGAN, B.V and A.S. HALL, (1975) ***"Design of prestressed concrete to beams subjected to combined bending, shear and torsion,"*** A.C.I. Journal Proceedings, Vol.72, No.3, March 1975.

27. LIN, T.Y. and P. CHOW, (1983) **Looking into the Future** - A Handbook of Structural Concrete, edited by KONG F.K., EVANS, R.H., COHEN, E and ROLL, F. PITMAN, London and McGraw-Hill, New York, Chapter 1. Warner R.F. Prestressed Concrete and Partially Prestressed Concrete, Chapter 19 pp 19-1 to 19-52.

28. **Prestressed Concrete Institute** (1985) PCI design Handbook, Chicago, Prestressed Concrete Institute.

10. Transmission and Anchorage Zone (Analysis and Design)

10.1 General

Earlier it has been discussed that the prestressing force acting on a structural member may be transmitted by an internal force system developed through bond between steel and concrete in case of a pretensioned prestressed concrete element, and an through external anchorage system in a post tensioned prestressed concrete element. The zone within which the applied prestressing force on steel is fully dissipated in to concrete and made to act as a distributed force is known as transmission zone in pretensioned elements and anchor zone in post tensioned elements. Transfer of force from steel to concrete by bond in pretensioned members is improved by wobbling, bends, loops, indentation, etc on steel. Anchor plates, lead cones and other steel elements are used to reduce the stress concentration effects and allow a smooth transfer of force in post-tensioned elements. During the transmission of force, different types of stresses are generated in concrete and hence that portion of the structural element has to be designed for all those stresses.

10.2 Pretensioned Prestressed Concrete Elements

In case of pretensioned elements the force is transversed by means of bond and the bond can be of different types such as friction bond, adhesion bond, shear or mechanical bond. The distance through which the force in the steel is transmitted to concrete is defined as transmission length and it is marginally higher than the bond length as could be seen in Fig 10.1. An empirical relationship has been established between the bond length and transmission length and codified in DIN specification [1] The factors which affect the bond length or the transmission length are the shape, surface characteristics, and ductility of steel in addition to the strength of concrete and the magnitude of prestress. A combination of characteristics like surface friction in steel, strength of concrete ductility of steel and applied prestressing force leads to Hoyer effect [2] (Fig. 10. 2). As the wire returns to original diameter on release of force at the ends in a pretensioning bed, radial compressive forces are exerted on the surrounding concrete. This increases the frictional bond force and reduces the transmission length. This is an important factor in the design of pretensioned prestressed concrete element especially in short pretensioned elements.

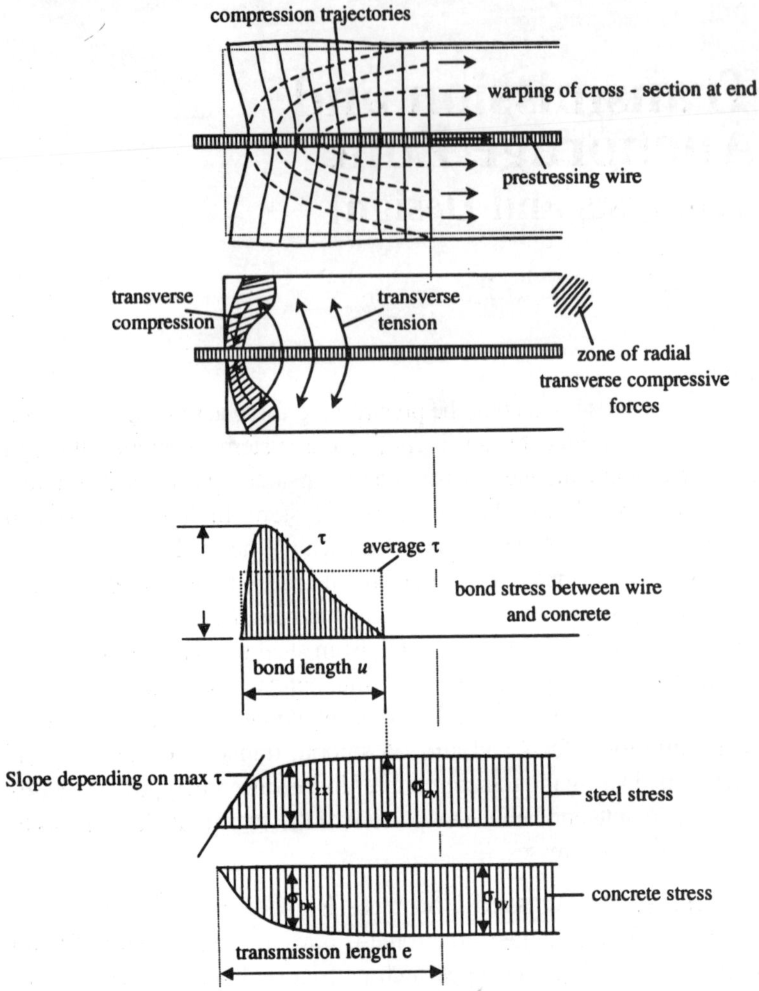

Fig 10.1 Bond Stress Distribution at the End of a Bond Anchorage of a Pre-Tension Wire

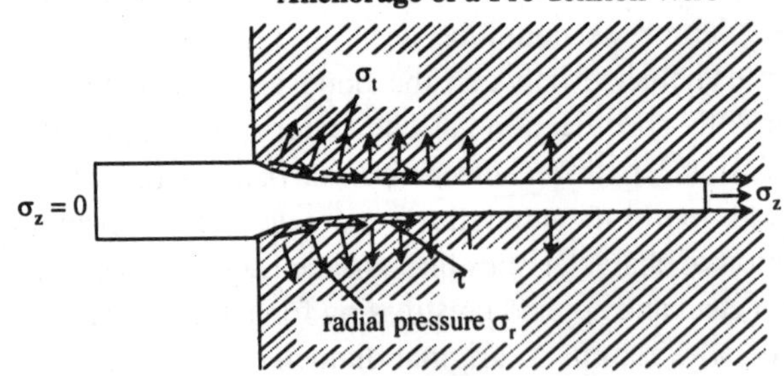

Fig 10.2 The Hoyer Effect

Only beyond the transmission length the prestress can be assumed to be uniformly applied on to concrete. Anywhere in the transmission zone the effect of prestress on concrete is less and it affects the allowable moment and shear in that zone. This bond or transmission length is also linked to the slip, which is considered as failure of bond. There are various expressions for defining the bond length. Based on elastic and theoretical analysis of Jenny [3] and Guyon [4], Marshall [5] has suggested the following formula for bond length.

$$_b = \left[\frac{3\varphi\left(1+\frac{E_s}{E_c}(1+\gamma_c)\right)}{\gamma_s\cdot\mu}\right] \qquad \text{....(10.1)}$$

Where φ = diameter of bar/wire

E_s = modulus of elasticity of steel

E_c = modulus of elasticity of concrete.

γ_c = Poisson's ratio for concrete.

γ_s = Poisson's ratio for steel.

μ = Friction coefficient e assumed as 0.3.

This is an improved version of expression developed by Hoyer[2]

Later, Marshall and Krishnamoorthy[6] proposed a simplified version for evaluating the bond length as

$$L_b = \sqrt{\frac{\sqrt{f_{ck}} \times 10^3}{k_1}} \qquad \text{....(10.2)}$$

assuming slip as proportional to square root of concrete strength. k_1 is a constant depending on the diameter of the wire/strand. (vide Table 10.1)

It may be noted that as per equation 10.2, the bond length works out to be more for an increased concrete strength, which does not appear reasonable. A series of tests on pretensioned concrete specimens were conducted by Rôs[7] at EMPA (Material Testing Laboratory in Zurich) and an empirical relationship has been developed to define the transmission length in form of diameter of the wire. This is universally accepted nowadays. This relationship holds good for concrete of quality M 45. For concrete of different strengths, the transmission length is modified by a factor $\sqrt{\frac{45}{f_{ck}}}$. The results from EMPA are given in the Table 10.2.

Table 10.1 Value of Constant k_1 in Equation 10.2

Serial number	Details of wire of stand	k_1
1	2mm dia.wire	0.144
2	5 mm dia. Wire	0.0235
3	7 mm dia.wire	0.0174
4	10 mm dia., 7 wire strand	0.144
5	12.5 mm dia., 7 wire strand	0.058
6	18 mm dia., 19 wire strand	0.0235
7	19 mm dia., 7 wire strand	0.0235
8	Twin twisted wires or 6.25 mm dia., 7 wire strand	0.077

Table 10.2 Transmission Length in Pretensioned Elements

Type of wire	Dia. mm	Average bond stress N/mm²	L_t mm in terms of dia.
Round drawn	1.5	32.5	140 - 90 to 95
" "	2.0	17.5	340 - 170
" "	3.0	12.5	720 - 240
" "	5.0	10.0	1500 - 300
Deformed wire	3.0	50.0	180 - 60
Strand	4 Nos. of 2.6 mm	30.0	520 - 200 of 2.6mm
Twisted strand (square) 45a	a = 45	20.0	680 -150

DIN specification [1] after considering various research works gives an expression linking Transmission length and bond length as

$$L_t = \sqrt{(L_b)^2 + D^2} \qquad \text{....(10.3)}$$

Where

L_t is the transmission length

L_b is the bond length

D is the diameter of bar

Time dependent properties like creep in concrete and relaxation in steel lead to an increase in the transmission length. This normally affects the smaller pretensioned elements in their functioning especially when they are subjected to dynamic or fatigue loading. Repeated loads lead to slip and horizontal cracking and finally bond failure. For such cases, a few prototype elements have to be tested for ascertaining the actual partial safety factor available under such loading conditions.

Having known the transmission length, it is necessary to know about the bond stress distribution in concrete, the tensile stress in steel, and transverse tensile stress in concrete in the transmission zone. The bond stress distribution can be assumed to be parabolic, with the maximum value near the free end of the rod reducing to zero stress at the end of the transmission length, the parabola having a decreasing slope with distance. Marshall[8] proposed an expression for bond stress based on the experiments conducted at universally of leads as follows:

$$\tau_x = \tau_{max}\, e^{-Bx} \qquad ....(10.4)$$

τ_{max} can be assumed to 7.42N/mm^2. B is a factor dependent on diameter of steel and change in stress in steel along the transmission length. It can be assumed as 0.029 divided by diameter in mm units; x is the distance (in mm) measured from free end. Similarly the tensile stress in steel at a distance 'x' measured from the free end is expressed as a function of permitted slip and the applied maximum stress.

$$f_{px} = f_{s\,max}\,(1 - e^{x/\lambda}). \qquad ....(10.5)$$

$$x = \frac{E_s}{f_{s\max}}.\Delta$$

Δ = Maximum slip permitted.

This is further simplified as

$$f_{px} = f_{s\max}\left[1 - e^{-Bx}\right] \qquad ....(10.6)$$

Where B = .029/diameter of the wire in mm

The transverse tensile stress distribution at a distance 's' on a horizontal section is evaluated as

$$\sigma_v = \frac{5M}{bc^2}\left[-1 + 12\frac{S^3}{c^3} + 16\frac{S^3}{c^3}\right] \qquad ....(10.7)$$

Where 'c' is the length of transmission zone (L_t)

'b' width of section in transmission zone

M is the resultant of the bending moments of the prestressing force and internal stresses in concrete at the end of transmission zone about the horizontal section considered and 'S' is the horizontal distance of the point considered measured from centre of transmission length, considered positive towards the end of the beam.

The above expression has been developed by Marshall [8] and Krishnamoorthy [9] subsequently and simplified as

$$\sigma_v = \frac{10M}{b.h.L_t}\left(1 - \frac{x}{Lt}\right)e^{-3.5xLt} \qquad ....(10.8)$$

Where 'h' is the height of the section and other terms as defined earlier.

The maximum tensile stress is expected to occur very close to the end ($0.05L_t$) and the value could be got by substituting x = 0 in the equation 10.8 and $\sigma_v = \dfrac{10M}{bhL_t}$

The transverse tensile distribution can be assumed to be a cubic parabolic with a decreasing slope as shown in Fig. 10.3.

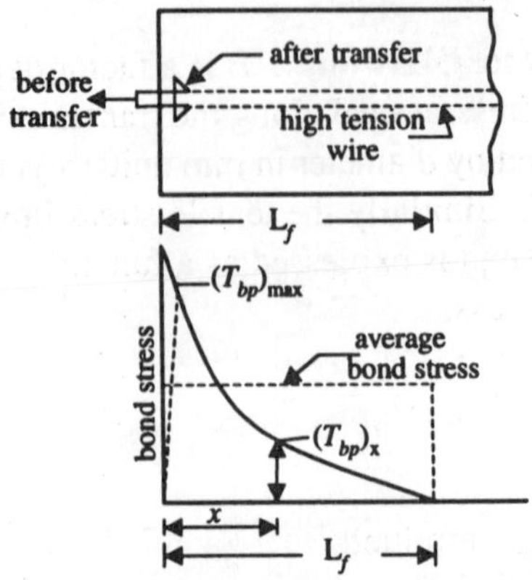

Fig 10.3 a & 10.3 b Bond Stress in Pretensioned Beams

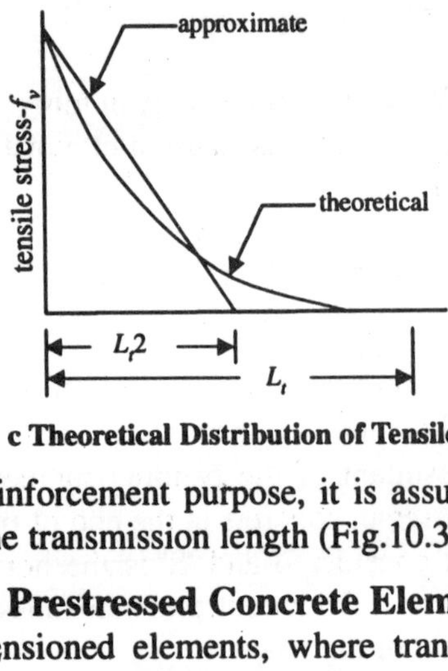

Fig 10.3 c Theoretical Distribution of Tensile Stress

For detailing of reinforcement purpose, it is assumed to vary linearity from the end to half the transmission length (Fig.10.3 c.)

10.3 Post tensioned Prestressed Concrete Elements

In case of post tensioned elements, where transmission is primarily through external anchorages, high compressive stresses are generated just behind the anchorages, and after a small distance, tension is developed in a direction normal to the direction in which the compression force is transmitted. The high compressive force behind the anchor plate plays an important role in taking decision on the size of the anchor plate and the strength of concrete to be adopted for the anchor zone. The tensile forces developed are to be resisted by provision of adequate and proper reinforcement in that zone. These tensile forces within the anchor zone are called splitting forces. Similar tensile forces are also developed on the zone away from the anchor zone and they are termed as spalling forces. The force flow from the point or the area of application of the prestressing force to the zone where it gets fully distributed can be seen as the stress trajectories shown in Fig. 10.4.

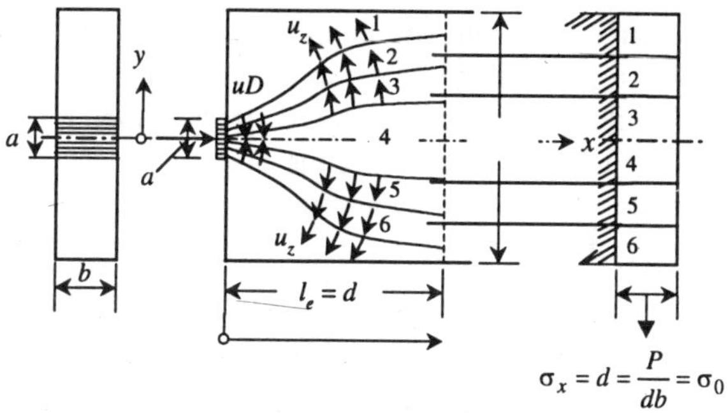

Fig. 10.4 Approximate Pattern of the Lines of Compression, and the "Radial" Forces Associated therewith, Produced by the Transmission of a Concentrated Force into a Concrete Member

If a concentrated force is applied normal to the surface on a small area on a semi infinite body, the lateral expansion of the punched area is being resisted by the concrete surrounding it, causing internal hoop forces (hoop tension) and triaxial state of compression in the punched area. Hence just behind the anchorage's, there is a heavy compressive force in the all directions. This would be taken care by proper choice of concrete of good strength or by confining the concrete in that zone so that its capacity is increased. It is known that the capacity of concrete under triaxial compression is more than the capacity of the same concrete under uniaxial compression. But a little away from bearing zone there will be a compressive stress flow only along the direction of application of force and tensile force normal to that direction. Accordingly provisions have to be made in the design of anchor zone. National and International codes define the pressure limit behind the anchorage under various conditions as bearing stresses. Provisions made in this regard in IS 1343 is discussed later in this chapter.

But the splitting tensile forces inside the structural element, a little away form the points of application along the direction of application are to be evaluated more critically and reinforcement should be designed to take care of these tensile stresses. There are different methods of approach of analysis for solving this problem.

Morsch (10) has arrived at the total splitting forces based on the stress calculations as in the case of a rocker bearing. Magnel (11) has considered the anchorage zone as a deep beam subjected to concentrated forces (prestressing forces) at one end with distributed reaction at the other end of the anchor zone and evaluated the principal tensile stress along longitudinal sections on different elements and accordingly suggested reinforcement in that zone. Guyon (12) considered the anchor zone as a semi-infinite plate and anlaysed the same for tensile force on particular plane. The boundary conditions of the long edge are approximately satisfied and those at the smaller edges are not considered. Guyon has also given influence lines for evaluating splitting forces at salient sections. Iyengar (13) has considered the anchor zone as the semi infinite Plate with exact boundary condition and has evolved a solution for tensile forces on a particular plane. Sargious (14) made photo elastic studies of the anchor zone and confirmed the stress values got by Iyengar. All these solutions are arrived considering the problem as a planar problem. The dimensions of the width of the beam has not been considered because of the complex nature of the problem

10.3.1 Behavioral pattern

A concentrated force applied at a point plasticises a small area till the stress level is reduced to failure stress. Then the force gets distributed over a small area called anchor area. Just behind this area the concrete is in a state of compression (triaxial). But within small distance compression stresses are felt only along the direction of application of force but, normal to that direction, tensile stresses are generated.

The behaviour is very similar to that of a split tension test on a concrete cylinder. The force flow is also similar to the flow of water through a nozzle or a jet till the energy is dissipated and does not cause damage to the medium in which it flows. In case of the end blocks, while on one side, a concentrated force is applied, on the other side it is fully supported for the full height by more or less a uniformly distributed force. Hence, the tension developed along the line joining the point of application of the concentrated force and the centre of the opposite face will not be uniform as in the case of split tension test.

The stress along the middle line of the end block will be as shown in Fig 10.5.

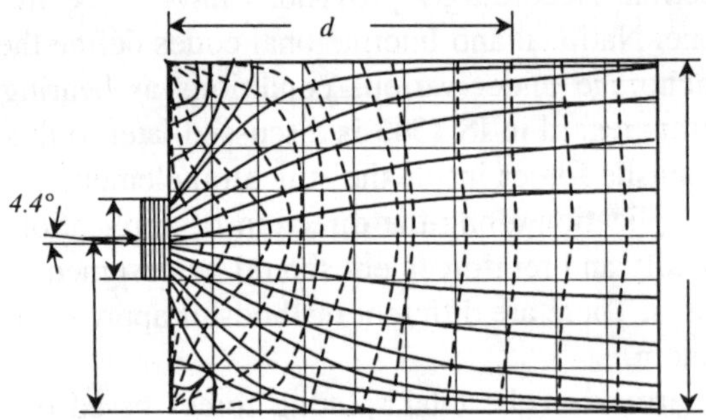

Fig 10.5a Principal Stress Trajectories for a Concentrated Force Applied at 1/2d, as determined from Photo-Elastic Investigations by *Sargious*. In this Case the Tendon Axis is Sloped at 4.4^0

If the prestressing force at the point of application fully distributed over the full depth of the beam (i.e. a = d) (namely the size (dimensions) of the anchorage is such that it sits on the full cross section of the beam) then the end block is supported on either side with equally distributed forces. This would not lead to any tension along the line joining the centres of these two faces.

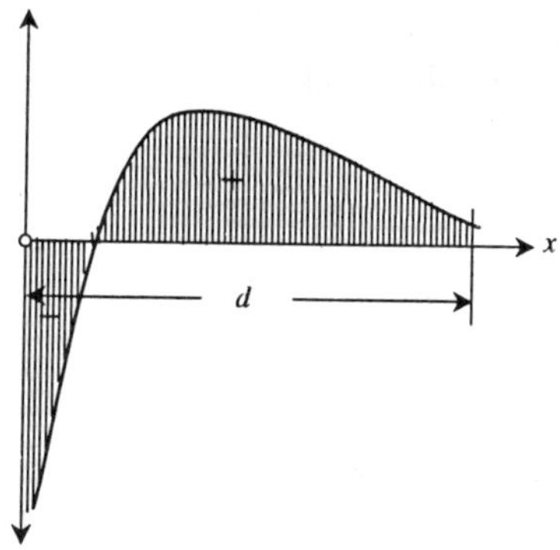

Fig 10.5b Distribution of the Transverse Stresses σ_Y on the Section Y = 0 for a Certain Ratio a/d

From this it is clear that the ratio of the height of distribution of concentrated force to the depth of the section (a/d) plays a role in the tensile stresses developed on a plane connecting the middle level of both faces.

If a/d is zero (hypothetical), this tensile stress will be maximum. If a/d is 1, there will not be any tensile stress at all. Fig 10.6 gives the pattern of tensile stress for different a/d ratios along the middle plane.

In addition to the tensile stress mentioned above which are defined as splitting tension, there are possibilities of tensile stress being developed at the corners (away from the anchorages) of the end zone in the horizontal direction. Referring to the fig 10.5 on stress flow and trajectories, it may be noted that the principal tensile trajectories which are vertical at the middle level of the end block turns towards horizontal on the top and the bottom portion being orthogonal to compressive stress trajectories. These trajectories are causing tensile forces called spalling forces. The spalling forces also slowly vanish as a/d ratio increases from zero to one.

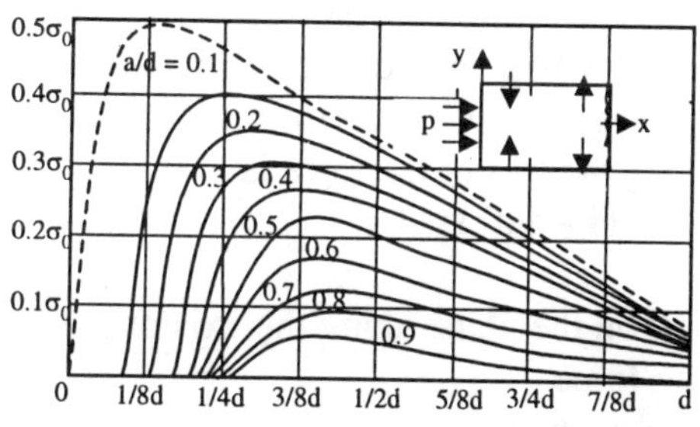

Fig 10.6 Distribution of the Transverse Tensile Stresses σ_Y for Various Ratios a/d as a Function of σ_0 = P/bd for the Axis y = 0 (from Ref 13)

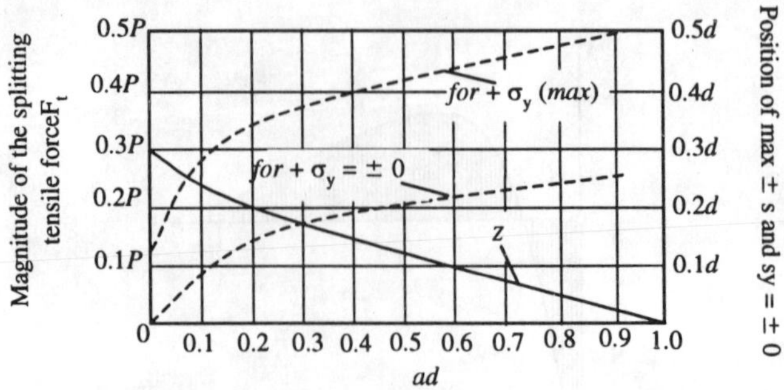

Fig 10.7 Magnitude of the Splitting Force as a Proportion of P, and istance for Max, + σ_y and for σ_y = 0 (from Ref 13)

If these splitting and spalling force could be arrived at accurately, reinforcements could be provided for resisting the same. The different methods mentioned earlier are various attempts to arrive at the design splitting tension and spalling tension forces.

10.3.2 Iyengar's Approach

Iyengar analysed the end zone of a prestressed concrete beam of width 'b', with a concentrate prestressing force at centre, in one plane as a semi infinite disc with forces on one face 'p' distributed over a length 'a' on one face, and over the full height of 'd' at the other face at infinity. The top and bottom surfaces of the beam being free, have zero normal forces in the plane of the disc and also zero shear force along the full length equal to 'd'. The same problem was solved for different a/d ratios. The results are plotted in fig 10.7. This figure gives the value of the total tensile force on the horizontal plane along the axis of the prestressing force, the position of maximum intensity of the tensile stress, and also the position of zero tensile stress. Knowing the total quantum of tension, the amount of reinforcement to be provided is decided. This reinforcement has to be distributed over the length; starting at the place around 0.05D where the tensile stress is possibly zero and to the end distance equal to the depth of the beam. The reinforcement should be more closely spaced around the zone where the tensile stresses are maximum. In short the spacing shall be based on the pattern of tensile stress distribution given earlier. The total tension on the plane along the axis of the force is given by

$$F_{bst} = 0.3\ [1 - a/d]\ P$$

Where

F_{bst} = total tension on the plane along the axis of prestressing force.
P = Prestressing force
a = the height of anchorage.
d = depth of the beam.

This formed the basis of arriving at the splitting tensile force (total tension). For different combinations, the same figure is used and combined effects are arrived at. If the width of anchor plate is very much less than the width of the beam 'b', horizontal spread of the tensile forces with horizontal plane also has to be recognised and these splitting forces can be calculated using b'/b in place of a/d ratio, where 'b' is the width of anchorage. These results have been confirmed by the photo elastic study carried out by Sargious[14].

10.3.3 Magnel's approach

Magnel considered the end zone of the prestressed concrete beam as similar to a deep beam subjected to a concentrated prestressing force on one face and distributed pressure forces on the inner face of the end zone in addition to the reaction force on the bottom face. The inclined cable force, (if any) may lead to tangential force at these areas on the end face of the beam. The anchorage zone is considered effective to a distance equal to the depth of the prestressed beam. The equilibrium of forces on this portion, considered as deep beam, leads flexural shear and normal stresses at various points. They could be converted as principal stresses. Reinforcements are designed to take care of principal tensile stresses. Since the behaviour of this zone is idealised to that of a deep beam, which is a shear predominant element, the critical section (horizontal plane) is considered as that plane which has maximum shear force (horizontal).

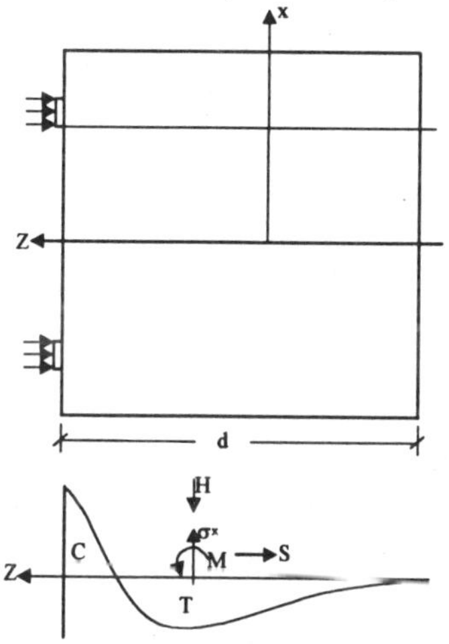

Fig 10.8.

The forces acting on this section are identified as a bending moment M, shearing force S and a normal force H, normal to that plane as shown in Fig.10.8. Defining the direction normal to that plane as 'x' direction, the axis direction as 'z' and the width direction as 'y', the stresses acting on that plane (zy plane) are due to moment M and normal force H. The stress due to M are flexural stresses and it is assumed to vary as a cubic parabola in 'z' direction i.e., along the length of the main beam. The $\sigma_{x(1)}$ caused by bending moment M is expressed as . $Az + Bz^2 + Cz^3 + Dz^4$. The boundary or edge conditions of the structural element (deep beam) may define the value of the coefficients A, B, C and D. From the boundary conditions it is noted that,

$$\frac{d\sigma_{x(1)}}{d_z} = 0 \text{ at } z = 0$$

$$\sigma_x = 0 \text{ at } \quad z = -d/2$$

From the equilibrium of forces, it is noted that

$$\int_{-d/2}^{+d/2} \sigma_{x1}.b.d.z = 0$$

From the equilibrium of Moments, it is noted that

$$\int_{-d/2}^{+d/2} \sigma_x z.b.d_x = M$$

Using the boundary conditions and equilibrium conditions, the constants are evaluated as. $A = \frac{-5M}{bd^2}, B = 0, C = \frac{60M}{bd^4}, D = \frac{80M}{bd^5}$.

Hence one portion of transverse tensile stress is expressed as,

$$\sigma_{x1} = \frac{5M}{bd^2}[-1 + \frac{12z^2}{d^2} + \frac{16z^3}{d^3}] = K_1 \frac{M}{bd^2}$$

The stress due to normal force 'H' in 'x' direction is also assumed to be varying parabolically in 'z' direction. That component is similarly evaluated as

$$\sigma_{x2} = \frac{2H}{bd}[1 + \frac{3z}{d} - \frac{6z^2}{d^2} - \frac{6z^3}{d^3}] = K_2 \frac{H}{bd}$$

Then the total transverse tensile stress is

$$\sigma_x = \sigma_{x1} + \sigma_{x2}$$

$$= K_1 \frac{M}{bd^2} + K_2 \frac{H}{bd}$$

Considering two such horizontal planes, the variation of σ_x is expressed as shear stress. Taking the equilibrium of forces on the element strip contained by two adjacent horizontal plane shear stresses in the sides being s, it is seen (Fig 10.9)

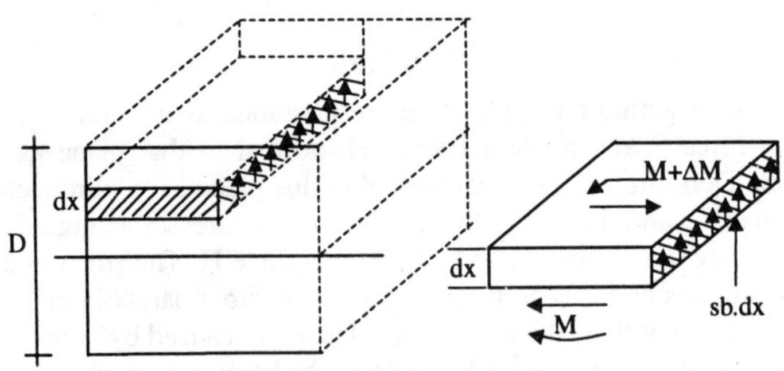

Fig 10.9

$$s.b.dx = \int_{z}^{d/2} \frac{d\sigma_{x1}}{dx}.dx.b.d_z$$

(σ_{x2} on both faces are the same)

$$\frac{d\sigma_1}{d_x} = \frac{K_1}{bd^2}\frac{dM}{dx} = \frac{K1}{bd^2}(S)$$

$$s.b.dx = \frac{S}{bd^2}\int_{z}^{d/2} k_1 dz.b.dx$$

$$s = \frac{S}{bd^2}\int_{z}^{d1/2} K_1 d_z = \frac{5S}{bd}[\frac{1}{4} + \frac{z}{d} - \frac{4z^3}{d^3} - \frac{4z^4}{d^4}]$$

$$s = K_3\frac{S}{bd}$$

The values of K_1, K_2 and K_3 are defined for various values of z as a function of d namely z = 0.1 d and so on as shown in Table 10.3.

Table 10.3 Magnel's Stress Coefficients for Normal Stress in a Plane.

Z	K_1	K_2	K_3
0.00	20.0	-2.0	0.00
0.10	9.72	0.0	1.458
0.20	2.56	1.28	2.048
0.30	-1.96	1.96	2.058
0.40	-4.32	2.16	1.728
0.50	-5.00	2.00	1.250
0.60	-4.48	4.60	1.768
0.70	-3.24	1.08	0.378
0.80	-1.76	0.56	0.128
0.90	-0.52	0.16	0.018
1.00	0.00	0.00	0.00

Considering an element on that plane of maximum shear it is subjected to normal stress of ($\sigma_{x1} + \sigma_{x2}$) in 'x' direction shear stress of 's' and another normal stress σ_z which is evaluated as follows:

The prestressing force P is assumed to be dispersed with an angle of 45°. The force flow axis is defined as the line joining the centre of the depth of dispersion at any section. Till such time the dispersion centre does not cut an outer face or inter face with dispersion cone of another force, the axis of flow will be along the direction of the prestressing force at starting point. Once the dispersion cone cuts an outer face, the line of joining the centre of depth of dispersion cone will not be the same along the direction of prestressing force at starting point. Considering the position of element, the eccentricity of the force flow line, with respect to the original direction of prestressing force, the stress in z direction is written as

$$\sigma_z = \frac{P}{bd'}[1 + \frac{12(e')^2}{(d')^2}]$$

Where e' is the eccentricity of the element form the force flow line and d' the depth dispersion cone at that section.

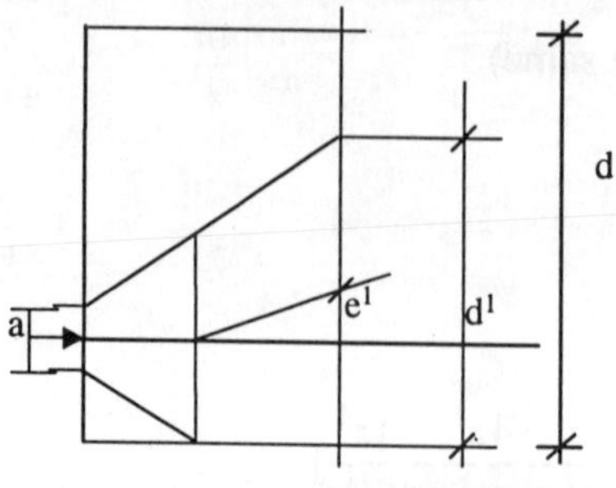

Fig 10.10

With a system of forces of σ_x, s,σ_z on the element the principal stresses at any point on the horizontal section are calculated as

$$\sigma_{I,II} = \frac{\sigma_x + \sigma_z}{2} \sqrt{\frac{(\sigma_x - \sigma_z)^2}{4} + S^2}$$

The principal tensile stresses are calculated at various points along the plane of maximum shear force 'S' and the reinforcement is provided to resist the principal tensile forces. The same is explained with an illustrative example.

Example 10.1

A post tensioned prestressed concrete beam has a rectangular end zone with dimensions 250 × 500mm. Two prestressing cables having forces 500KN in each are anchored at the end face, one cable is anchored at 100 mm from bottom and another 100 mm from top. The top cable has a slope of 1 in 10 at the anchoring zone up to a distance is 500 mm. This beam, due

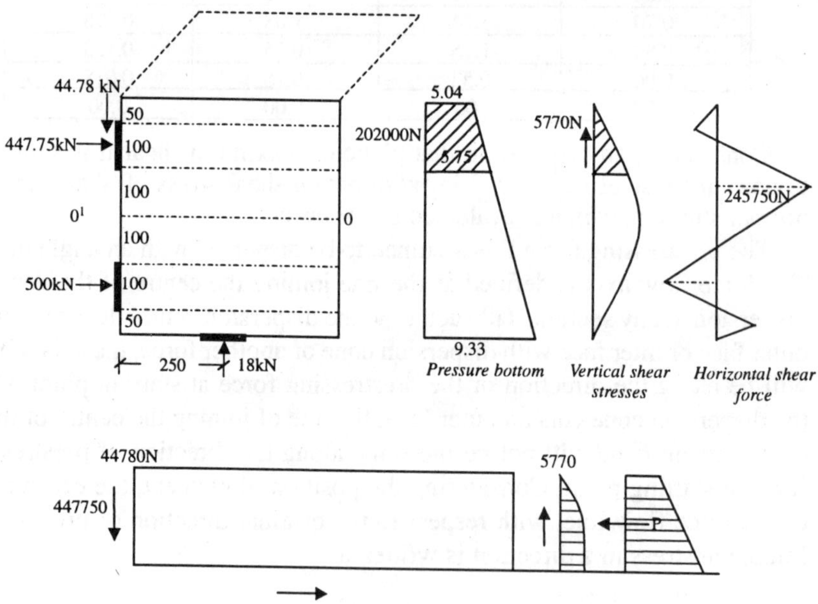

Fig 10.11

to its self-weight, has a reaction of 18 kN on a support located at 200 mm from the end face of the beam. The anchor plates are of dimension 100mm × 100 mm. Evaluate, the critical stress in the anchor zone to design for anchor zone reinforcement. Use Magnel method of approach and plot the stress variation along the plane of maximum shear.

Solution

The direct stress (pressure distribution) at the vertical section at the distance of 500 mm from the end face [end of anchor zone] varies from 5.04 M mm^2 at the top to 9.33 N/mm^2 at the bottom as could be calculated by as axial force componen plus the component due to moment at point o at the end sections. The shear along the various horizontal planes have also been evaluated as shown in Fig 10.11 The shear on the vertical face at the end of anchor zone is parabolically distributed and shear on that face up to any level can be evaluated.

Shear force diagram shows that the maximum shear (horizontal) occurs along the plane AB at 150 mm from top plane. The stress due to axial force and moment at that level is 5.75N/mm^2

Total direct force at the inner face $= \dfrac{5.4 + 5.75}{2} \times 150 \times 250 = 202000N$

Shear force on the vertical plane at the end of anchor zone up to the level AB

$$= 250 \int_{100}^{250} \frac{44780 - 18000}{260000 \times 10^4 \times 250} \int_{y}^{250} (250.y.dy)dy$$

$$= \frac{26780}{26 \times 10^8} \int_{100}^{250} 250[\frac{250}{2} - \frac{Y^2}{2}]dy$$

$$= \frac{26780}{26 \times 10^8}[62000 \times \frac{250}{2}.y - \frac{250}{2}.\frac{y^3}{3}]_{100}^{250}$$

$$= \frac{26780}{26 \times 10^8}[62000 \times \frac{250}{2} \times 150 - \frac{250}{6}(250^3 - 100^3)] = 5770\text{N}.$$

Calculation of principal stresses on plane AB using coefficients from Table 10.4.

Table 10.4

Distance from end mm	Direct stress N/mm^2 $\sigma_x = \frac{K_1 M_1}{BD^2} + K_2 \frac{H}{BD}$	f_h	Shear stress (s) N/mm^2 K_3S/BD	Min. Principal stress (or) N/mm^2
0.0	$\frac{20 \times 5740000}{250 \times 500 \times 500}$ $\frac{-2 \times 39010}{250 \times 500}$ $1.835 - 0.625 = 1.21$	17.5	0	1.15
0.1D = 50	0.891– 0 = 0.891	0	2.85	0.035
0.2D = 100	0.235 + 0.4 = 0.635	6.84	4.00	–1.125
0.3D = 150	-0.18 + 0.61= 0.43	6.72	4.02	–1.320
0.4D = 200	-0.397 + 0.674 = 0.277	6.65	3.38	–1.015
0.5D = 250	-0.460 + 0.674 = 0.163	6.57	2.45	–0.566
0.6D = 300	-0.411 + 0.500 = 0.89	6.50	3.46	–0.203
0.7D = 350	-0.298+ 0.337 = 0.39	6.40	0.74	–0.028
0.8D = 400	-0.161 + 0.180 = 0.19	6.32	0.25	0
0.9D = 450	0.048 + 0.05 = 0.098	6.25	0.035	0
1.0D = 500	0 + 0 = 0	6.08	0	0

Moment about center of the horizontal plane of all direct forces

$$= 250[5.73 \times 150 \times \frac{150}{2} - (5.73 - 5.04)\frac{150}{2} \times \frac{150}{3} \times 2]$$

$$= 1550000 \text{ Nmm}$$

Total bending moment in the plane on consideration

$$B.M = 44780 \times 250 + 15500000 + 5770 \times 250 - 447750 \times 50$$

$$= 5740000 \text{ Nmm}$$

Direct force $= 44780 - 5770 = 39010$N

Shear force $= 447750 - 202000 = 245750$ N

The principal tensile stress distribution above the length of horizontal sections is calculated as shown in Table 10.4 and plotted in magnitude only as shown in fig.10.12.

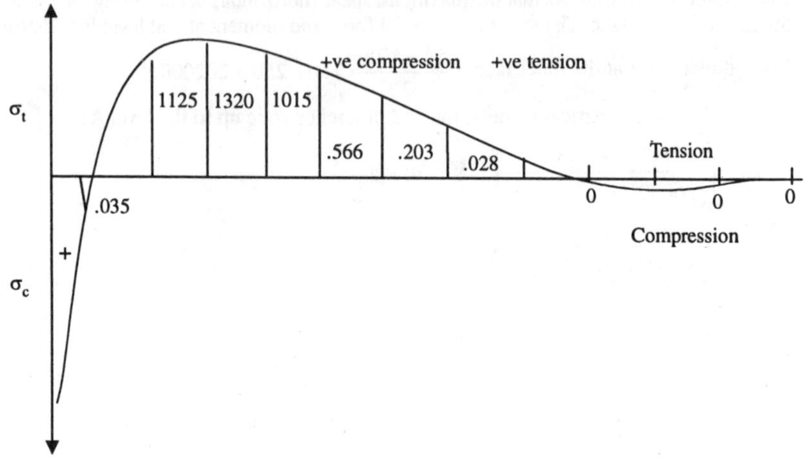

Fig. 10.12

10.3.4 Guyon's Approach for analysis of end zone of a prestressed concrete beam

Guyon considered the anchor zone as a plate with in plane forces acting as concentrated forces at the boundaries. This plate was considered as a semi-infinite plate and the boundary condition along the longitudinal edges were fully satisfied. The problem was considered as a two-dimensional planar problem. The solution was aimed at to get the splitting forces. The values are given in the form of a tabular statement or influence line. The problem was considered and results were presented under two different headings. a) End anchor zone with evenly distributed forces. b) Anchor zone with eccentric forces.

10.3.5 Anchor zone with evenly distributed forces

In case of beams where the pressure distribution at the end of anchor zone is such that the line of action of concentrated forces passes through the centroid of the corresponding pressure diagram, the anchor zone is assumed to have evenly distributed forces. In such a case the average concrete pressure f_{rp} was considered to be equal to the total force/BD. Each concentrated prestressing force is assumed to have an influence over a effective prism depth defined as '2g'. The value of 'g' is defined as the smallest of a) the distance of the free face from the centre of prestressing force. b) half the distances of the force from the next adjoining force. c) half the depth of the beam.

The splitting force is calculated considering a/2g ratio and f_{rp}. Table 10.5 gives the maximum tensile stress as a ratio of f_{rp}, the position of maximum tensile stress and the position of zero tensile stress. Reinforcement will be provided for the maximum tensile stress for the length starting from the point of zero tensile stress to the length equal to depth of beam.

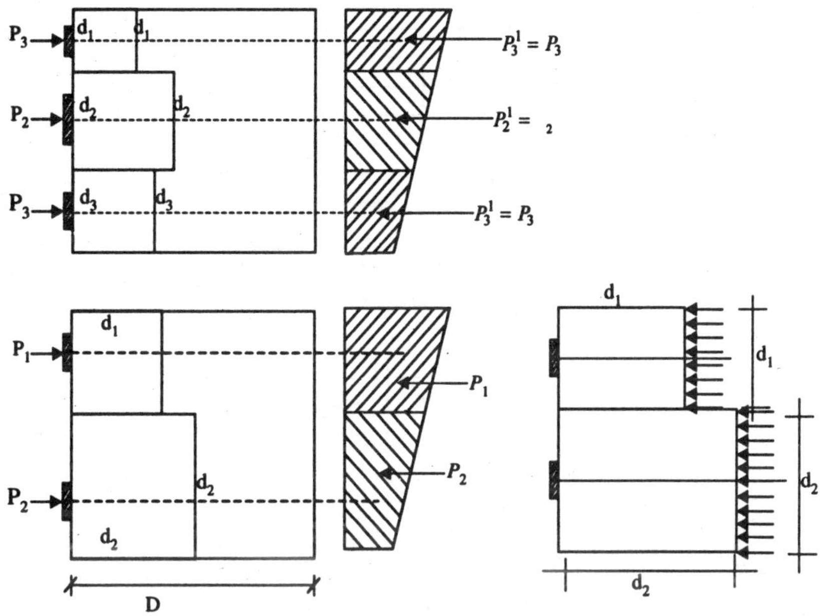

Fig 10.13

Fig 10.13 shows the case of Anchor zone with uniformly distributed forces.

10.3.6 Anchor zone with unevenly distributed forces:

In this case of beams, the pressure distribution at the end of anchor zone will be such that the line of action of the concentrated prestressing force is not collinear with the centre of pressure of the force acting on corresponding prism of influence. Here Guyon has suggested analysis of every individual prism with the corresponding forces, and then consider two forces together with the corresponding prism of influence and so on. This is known as successive resultant approach. In each analysis of an individual prism for one force or a prism of influence for a combination of forces, the tensile stressed developed at any layer is worked out using the coefficient given. The coefficients are given in Tables 10.6 & 10.7 for unit force intensity acting at eccentricities of 1/8th, 1/4th, 3/8th, 1/2th the depth of the prism on either side of the plane under consideration. Hence the prism under consideration will be divided into 8 equal parts and the concentrated forces on one side of prism will be applied on equivalent concentrated forces at 1/8th points on the equivalent prism and the effect of each such force at 1/8th points on the plane under consideration will be evaluated.

Table 10.5 Distributed Axial Force

Distribution Ratio	Position of Zero stress	Position of maximum stress	Maximum tensile stress
0.0	0.00	0.17	0.50
0.1	0.09	0.24	0.43
0.2	0.14	0.30	0.36
0.3	0.16	0.36	0.33
0.4	0.18	0.39	0.27
0.5	0.20	0.43	0.23
0.6	0.22	0.44	0.18
0.7	0.23	0.45	0.13
0.8	0.24	0.46	0.09

Table 10.6 Concentrated Eccentric Force

Eccentricity of force, e/D	Distance of stress from end of beam, x/D						
	0	$\frac{1}{12}$	$\frac{1}{6}$	$\frac{1}{4}$	$\frac{1}{3}$	$\frac{1}{2}$	$\frac{3}{4}$
+1/2	–2.187	–0.913	–0.428	–0.014	+0.307	+0.399	+0.192
+3/8	–1.222	–0.601	+0.125	+0.192	+0.250	+0.242	+0.122
+1/4	–0.758	–0.025	+0.238	+0.154	+0.062	–0.024	–0.016
–1/8	–0.566	+1.004	+0.074	–0.144	–0.266	–0.262	–0.128
0	0.000	–0.448	–0.500	–0.462	–0.423	–0.314	–0.161
–1/8	–0.566	+1.004	+0.074	–0.144	–0.266	–0.262	–0.128
–1/4	–0.758	–0.025	+0.238	+0.154	+0.062	–0.024	–0.016
–3/8	–1.222	–0.601	+0.125	+0.192	+0.250	+0.242	+0.122
–1/2	–2.187	–0.913	–0.438	–0.014	+0.307	+0.399	+0.192

Table 10.7 Concentrated Stress Force

Eccentricity of force, e/D	Distance of Stress from end of Beam, x/D						
	0	$\frac{1}{12}$	$\frac{1}{6}$	$\frac{1}{4}$	$\frac{1}{3}$	$\frac{1}{2}$	$\frac{3}{4}$
+1/2	-	0	0	0	0	0	0
+3/8	+5.66	+2.96	+0.87	+0.18	–0.05	–0.14	–0.07
+1/4	+4.00	+3.10	+1.52	0.44	–0.22	–0.20	–0.10
+1/8	+5.66	+2.96	+0.87	+0.19	–0.05	–0.14	–0.07
0	-	0	0	0	0	0	0
–1/8	–5.66	–2.96	–0.87	–0.19	+0.05	+0.14	+0.07
–1/4	–4.00	–3.10	–1.52	–0.44	+0.22	+0.20	+0.10
–3/8	–5.66	–2.96	–0.87	–0.19	+0.05	0.14	+0.07
–1/2	-	0	0	0	0	0	0

The tensile stresses normal to the horizontal plane are available at specific points at distances 0, 1/12th, 1/6, ¼, 1/3, ½ and ¾ the depth of the beam from the anchor end of the beam. The effect of all the concentrated forces at various

$1/8^{th}$ points are added and the final stress at the specificpoints are arrived at. Similarly even shear forces (or the vertical component of an inclined prestressed force) are divided between the 1/8th points on the equivalent prism and their effects on the horizontal plane under consideration at specific points as described earlier can be evaluated.

These calculations are completed for each prism along its axis and the forces are combined in twos and threes and so on and corresponding prisms of influence are analysed. The reinforcement was provided for the worst effect namely the highest tensile stress at any point. The details of calculations can be seen from the illustrative example.

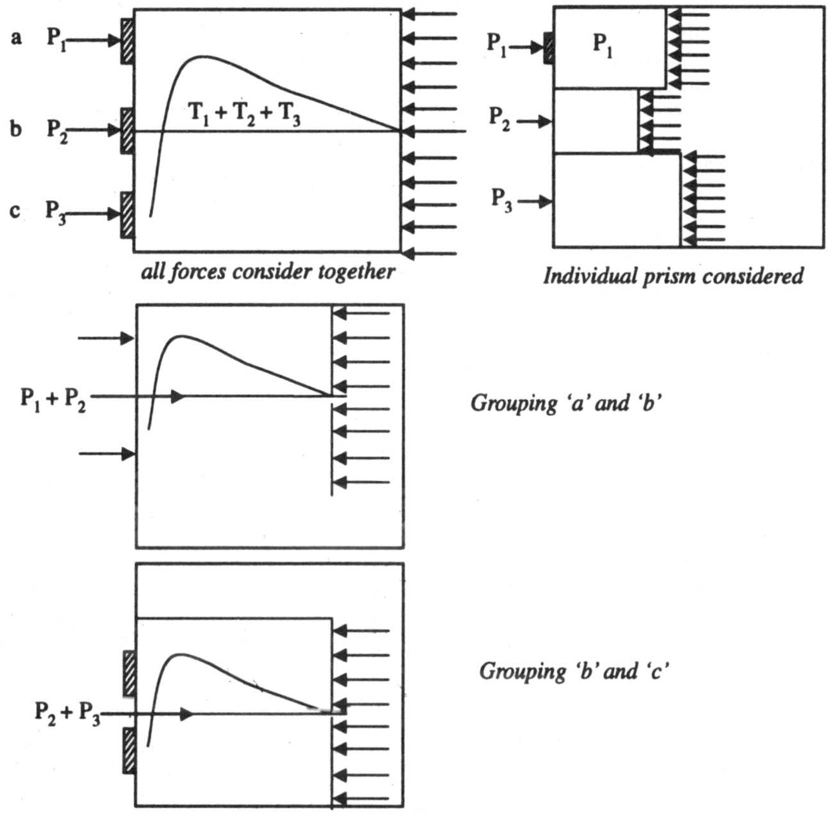

Fig 10.14.

Example10.2

A post tensioned prestressed concrete Beam has an end block of dimension 500mm × 250mm. Two cables with 450kN of force in each are anchored at 100 mm from top and bottom. The anchor plates have a dimension of 100 × 100 mm. The top cable has a slope of 1/10. Evaluate the anchor zone stresses along the centre line of the Beam using Guyons' approach. Plot the stress pattern variation on along the axis of the beam in the end zone.

Solution

Since the system of forces acting on the beam is very nearly symmetrical with respective the middle plane of the beam, both the forces are considered together and the equivalent prism will have the full depth of the beam as its depth and the prestressing forces put together will be considered as nine concentrated loads at $1/8^{th}$ points at the distributed end as shown in the fig. 10.15.

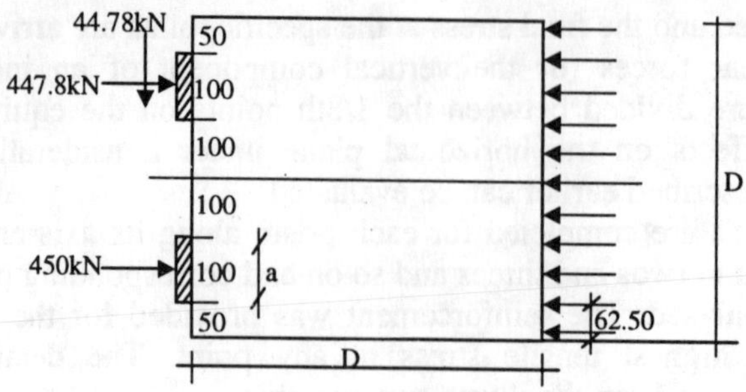

Fig 10.15

The equivalent concentrated force at these 9 nodal points is calculated first as R1 to R9. Their effect as tensile stresses along the middle surface plane are evaluated using the coefficients given in table 10.6.

Referring to fig R1 to R9 are calculated as detailed

$R_1 = 4.478 \times 1.25 \times \frac{1}{10} = 0.56$ 5.6kN R_5

$R_2 = 14.0$

$R_{2L} = 4.478 \times 1.25 \times \frac{9}{10} = 5.03$ $\Bigg\}$ $R_2 = 190.3kN$

$R_{eR} = 4.478 \times 25 \times \frac{5}{6.25} = 8.956$

$R_{3L} = 4.478 \times \frac{6.25}{2} = 14.0$ $\Bigg\}$ $R_3 = 229.5kN$

$R_{4R} = 0$

$R_{4L} = 4.478 \times 2.5 \times \frac{1.25}{6.25} = 2.239$ $\Bigg\}$ $R_4 = 22.39kN$

$R_5 = 0$ $R_6 = 22.5kN$

$\overline{R_{6L}} = 0$

$\overline{R_{6R}} = 4.5 \times 2.5 \times \frac{1.25}{6.35} = 2.25$ $\Bigg\}$

$\overline{R_{7R}} = 4.5 \times 2.5 \times \frac{5}{6.25} = 9$

$\overline{R_{7L}} = 4.5 \times \frac{6.25}{2} = 14.08$ $\Bigg\}$

$\overline{R_{8R}} = 14.08$

$\overline{R_{8L}} = 4.5 \times 1.25 \times \frac{9}{10} = 5.05$ $\Bigg\}$ $R_8 = 19.13kN$

$\overline{R_9} = 4.5 \times 1.25 \times \frac{1}{10} = 0.56$ $R_9 = 5.6$ kN

These forces R_1 to R_9 are applied at 1/8th points at the innerface of the end block and their effects as tensile stresses on the middle horizontal plane are evaluated as shown in Table 10.8

Table 10.8. Tensile stresses on middle surface plane

Position e/D	Force kN	Average Stress N/mm²	Stress at (x mm) Distance from end						
			0	41.6	83.3	125.0	167.7	250.0	375.0
½	5.6	0.045	–0.098	–0.041	–0.0192	–0.0063	+0.0138	0.0779	+0.0168
3/8	190.3	1.520	–1.860	0.915	+0.190	+0.292	+0.380	+.369	+0.186
¼	229.5	1.835	–1.390	–0.046	+0.427	+0.283	+0.114	–0.044	–0.029
1/8	224.0	0.179	–0.10	+0.180	+0.0132	–0.026	–0.048	–0.047	–0.023
0	0	0	0	0	0	0	0	0	0
-1/8	225.0	0.180	–0.102	+0.180	+0.013	–0.027	–0.048	–0.047	–0.023
-1/4	230.8	1.840	–1.385	–0.046	+0.438	0.284	+0.115	0.045	–0.030
-3/8	191.3	1.530	–1.870	–0.920	+0.0191	+0.292	+0.387	+0.37	+0.187
-1/2	5.5	0.0448	–0.099	–0.041	0.020	–0.006	0.014	+0.078	0.017
			–6.914	–1.649	+1.2716	+1.0857	+0.9232	+0.7109	+0.307

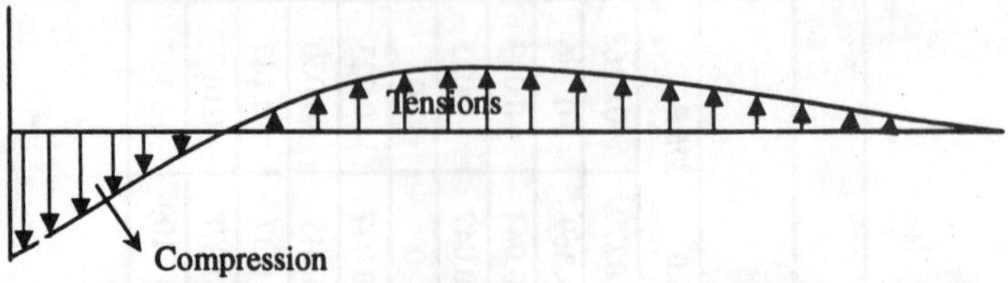

Fig 10.16

10.4 Indian Standards Code Provisions (!5)

Indian Standards Code IS 1343 - 1980 requires the following design criteria to be followed.

i) Bearing Stress

The bearing stress on concrete after accounting for all losses due to relaxation of steel, elastic short running, creep of concrete, slip and seating of anchorage shall not exceed.

$$0.48\ fci\ \sqrt{\frac{Abearing}{Apunching}}$$ or 0.8 f_{ci} whichever is smaller where f_{ci} is

the concrete strength at transfer and $A_{bearing}$ is the bearing area and $A_{punching}$ is the punching area. The punching area is defined as a contact area of the anchorage devices. In case of circular anchor plate, the area shall be replaced by an equivalent square area.

The bearing area shall be the maximum area of that part of the member, which is geometrically similar and concentric to effective punching area. The side of the bearing area ($\sqrt{A_{bearing}}$) is limited to the least of the following

- 2 times the side of the punching area ($\sqrt{A_{punching}}$)
- 2 times the distance of the centre of the anchorage to the nearest free surface.
- Drawing the 45° dispersion lines for adjacent anchorage's, two times the distance of the intersection of the dispersion line from the centre of the punching area.

When there is an already existing compressive stress on the bearing area as in the case of anchorage in places inside the body of the structure, the total stress shall not exceed the limiting values given above. If the anchorages are embodied in concrete, the bearing stress shall be investigated after accounting to surface friction between the anchorages and the concrete.

ii) Bursting Tensile Stresses:

The bursting tensile stress/force, which is also called splitting force in the anchorage zones of bonded, post tensioned members shall be on the basis of tendon jacking load. For un-bonded members, the bursting tensile forces shall be assessed on the basis of the tendon jacking load or the load on tendon at limit state of collapse, whichever is greater. This bursting tensile forces F_{bst} in an individual square end block symmetrically placed on a square anchor plate may be expressed as below

$$\frac{F_{bst}}{P_k} = 0.32 - 0.3\frac{y_{po}}{y_o}$$

Where,

F_{bst} = bursting tensile force,
P_k = load in the tendon assessed as above,
y_{po} = side of loaded area, and
y_o = side of end block.

The force F_{bst} will be distributed in a region extending from 0.1 y_o to y_o from the loaded face of the end block. Reinforcement provided to sustain the bursting tensile force may be assumed to be acting at its design strength (0.87 times characteristic strength of reinforcement) except that the stress should be limited to a value corresponding to a strain of 0.001 and the cover to the reinforcement is not less than 50 mm.

In rectangular end blocks, the bursting tensile forces in the two principal directions should be assessed on the basis of this clause. When circular anchorage or bearing plates are used the sides of equivalent square area should be used. Where groups of anchorages or bearings plates occur, the end blocks should be divided into a series of symmetrically loaded prism and each prism treated as independently end block itself. For designing the end blocks having a cross-section different in shape from that of the general cross-section of the beam, reference should be made to specialist literature.

Should the anchorages be eccentrically located, the side of the end block will be restrained by the depth of the prism of influence. The prism of influence is defined by its depth which is to be determined as the least of the following:

- 2 times the distance of the free surface from the centre of the anchorage.
- 2 times the distance of the centre of the anchorage to intersection of dispersion line of two adjacent anchorages.

Based on the above splitting tension/bursting tensile force will be calculated and the forces are supposed to be distributed over the length from 0.1 times the depth of the prism to full depth of the prism. The philosophy of detailing of reinforcement and actual procedure to be adopted is explained later.

10.5 FIP Recommendations - 1984 Specifications (16)

The FIP recommendations are more general in nature and it should be complimented with the individual national recommendations for concrete structures. But the philosophy of design as presented in FIP for anchorage zone are based on the fact that when any concrete element is subjected to a concentrated force at one end to be dispersed at a distance does undergoes splitting and spalling tension and also high compressive stresses just behind the anchorages. Provisions have to be made to take care of these stresses in terms of the concrete strength and reinforcement. The related clauses of design for anchorage zone of tendons as given in FIP are reproduced below

10.5.1 Anchorage zones of tendons

10.5.1.1 Local force

Checking of local effects in the concrete and of the transverse reinforcement should be carried out for an applied force

$P_k = 1.35\, A_{sp} f_{po}$

The local resisting force from concrete should be determined as follows

$$F_{Rdu} = f_{ck} / \gamma_m \sqrt{(A_{cl} A_{co})} \neq 3.3 \frac{f_{ck}}{\gamma_m A_{co}}$$

Where A_{co} is the loaded area and A_{cl} is a geometric enlargement of A_{co} with the same centre of gravity, lying totally within A_c (Fig. 10.17).

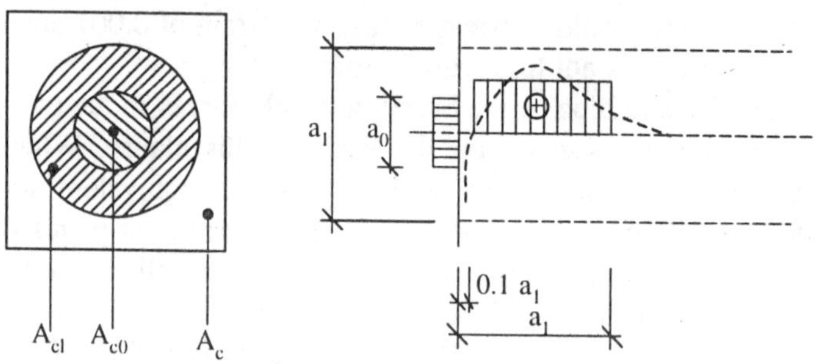

Fig 10.17 Definitions

The corresponding transverse tensile force may be assumed to be constant over a distance measured between 0.1 a_1 and a_1 and should be calculated in each direction using

$$F_{bst} = 0.3 P_k \left(1 - \frac{a_0}{a_1}\right)$$

Reinforcement has to be provided for resists then splitting tensile force.

The values of a_o and a_1 generally differ from one direction to another; normally, examination of two directions at right angles suffices.

10.5.1.2 Dispersion of the prestress and detailing

In addtion to the local reinforcement according to section 10.4.1.1, transverse reinforcement associated with the general dispersion of the prestress should be provided within the anchorage zones of tendons (see Fig. 10.18). This reinforcement may be calculated either by elastic theory or by a sample truss analogy (see Fig 10.19). The range of preferred nominal diameters ϕ (mm) for bars is as follows: 5, 6, 10, 22, 16, 20, 23, 32, 40, 50, 60. For $\phi > 32$ the permissible stress value has to be reduced. Hence large diameter bars are normally avoided.

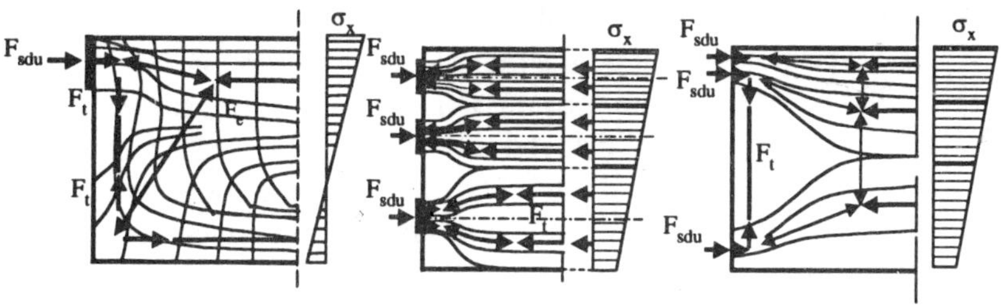

Fig 10.18 Dispersion of Pressers in a Plane (or web)

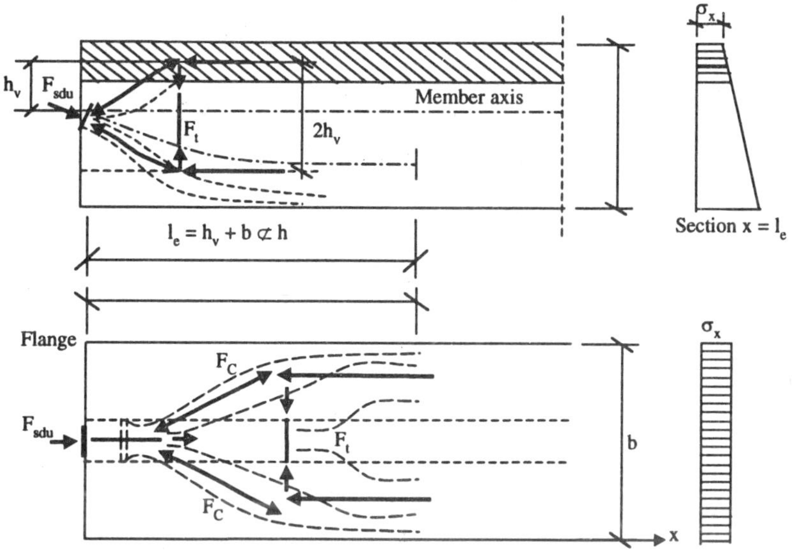

Fig 10.19 Dispersion of Pressers in T-beam

10.5.2 Basic bond values in case of pretensioned products

The bond stress τ_{bd} may be assumed to act with a constant value the bars, and the determination of anchorage lengths is based on limiting values of bond stress under failure condition.The bond conditions are considered to be good for

- all bars placed in elements whose depth in the direction of concreting does not exceed 250 mm
- all bars embedded in elements with a depth greater than 250 mm which are, when concretting is completed, in the lower half of the element (or of any part cast during the same stage of concreting) or at least 300 mm from the top.
- all bars with an inclination of 45° - 90° to the horizontal.

The basic length necessary for the straight anchorage of a bar is

$$L_b = \frac{Q}{4} \frac{(f_y / \gamma_m)}{\tau_{bd}}$$

For other details the readers are referred to F.I.P. recommendations [16]

10.6 BS 8110 - 1985 Specifications (17)

The BS Specifications are on similar lines as the provisions given in IS 1343 – 1980 and defines the transmission length in pretensioned members and design provisions for the end blocks for post-tensioned members separately. While for the pre-tensioned members the transmission length alone is defined, for post-tensioned members its design reinforcement in the transmission zone for splitting and spalling tension are described. The transmission length for post-tensioned members is taken is equal to depth of the stress block as defined by Guyon and the reinforcement for resisting tension is expressed as a function of applied prestressing force and the ratio of the loaded area to the equivalent stress block area.

Transmission length in pretensioned members is defined as a function of strength concrete, size of the tendon, and the surface characteristics of the tendons. It could vary from 50 -160 times the diameter of the tendons. Experimental evidence has been given greater importance. In the absence of experimental evidences the transmission length L_t is evaluated as

$$L_t = \frac{K_t \phi}{\sqrt{f_{ci}}}$$

Where f_{ci} is the concrete strength at transfer; ϕ is the nominal diameter of the tendon; K_t is a coefficient for the type of tendon and is selected from the following:

(a) plain or indented wire (including crimped wire with a small wave height): $K_t = 600$;
(b) Crimped wire with a total wave height not less than $0.15\ \phi : K_t = 400$;
(c) 7-wire standard or super strand: $K_t = 240$;
(d) 7-wire drawn strand: $K_t = 360$.

10.6.1 End blocks in post-tensioned members

10.6.1.1. General

In the design of end blocks, consideration should be given to:

(a) bursting forces around individual anchorages;
(b) Overall equilibrium of the end block;
(c) Spalling of the concrete from the loaded face around anchorages

Information on item (b) and (c) is to be taken from specialist literature.

10.6.1.2 Serviceability limit state

At the serviceability limit state, the design bursting tensile force, F_{bst} in an individual square end block loaded by a symmetrically placed square bearing plate, may be derived from Table 10.8 on the basis of the tendon jacking load. With rectangular anchorages and /or rectangular end blocks, the bursting tensile forces in the two pricipal directions should be assessed in relation to the value of Y_{PO}/Y_o for each direction where

Y_o is half the side of the end block;
Y_{PO} is half the side of the loaded area;
P_o is the tendon jacking force.
Circular bearing should be treated as square plates of equivalent area.

	Table 10.9 Design Bursting Tensile Forces in End Blocks					
Y_{po}/y_o	0.2	0.3	0.4	0.5	0.6	0.7
F_{bst}/P_o	0.23	0.23	0.20	0.17	0.14	0.11
	Note. Intermediate values maybe interpolated.					

This force, F_{bst}, will be distributed in a region extending from $0.2Y_o$ to $2Y_o$ from the loaded face, and should be resisted by reinforcement in the form of spirals or closed links, uniformly distributed throughout this region, and acting at a stress of 200 N/mm^2.

When a large block contains several anchorages it should be divided into a series of symmetrically-loaded prisms and each prism treated in the above manner. However, additional reinforcement will be required around the groups of anchorages to ensure overall equilibrium of the end block. Special attention should also be paid to end blocks having a cross section different in shape from that of the general cross section of the beam.

10.6.1.3 Ultimate limit state

For members with unbonded tendons the design bursting tensile force, F_{bst}, should be assessed from Table 10.9 on the basis of the characteristic tendon force; the reinforcement provided to sustain this force may be assumed to be acting at its design strength ($0.87f_y$). No such check is necessary in the case of members with bonded tendons.

10.6.2 Limitations on area of prestressing tendons

The size and number of prestressing tendons should be such that cracking of the concrete in the end block would precede failure of the beam in flexure. This requirement may be considered to be satisfied if the ultimate moment of resistance in flexure exceeds the moment necessary to produce a flexural tensile stress in the concrete at the extreme tension fibers equal to $0.6\sqrt{f_{ck}}$. In this calculation the prestress in the concrete may be taken as the value after all losses have occurred.

10.7 Detailing of Reinforcement in General

The reinforcement of transmission zone should be disposed transverse to the flow of the prestressing force and the distribution pattern should be in accordance with the direction of principle tensile stress trajectories. The flow of prestressing force is in the form of trancated come opening out fully to size of the circle, which could be accommodated in the body of the beam. Hence the reinforcement should be in the radial direction to take care of the principle tensile trajectory. Since such a layout of reinforcement pattern is difficult to be provided, it is managed by providing reinforcement in orthogonal directions in any cross section. To keep the tensile trajectories contained, it is necessary that there should be reinforcement along the flow direction along the periphery of the cone of flow of forces. This is achieved in practice by providing its longitudinal reinforcement. While the tension domain is continuous and the reinforcements provided are discrete it is preferred with they are placed as closely as possible with smaller diameter bars commensurate with the facility of flow of concrete. The present day prestressing system to a large extent by a leading cone of metal. The present international specifications of providing only helical reinforcement of increasing diameter from the point of applicantion of the forces considers that the tensile capacity of concrete could be raised by confinement. But a judicious compromise of providing reinforcement in 3 directions and also providing helical reinforcement around the ducts would assure a smooth transmission of flow from steel to concrete. It is hastened to add that the concrete mass must be continuous without break or honeycomb and that could be assured only by providing enough place for the concrete to flow and choosing good workable high performance concrete mix.

The total Quantum of principle tensile force generated over the length of the tensile zone from point of application to the cross section where uniform distribution takes place is taken care of by properly spaced reinforcement. By experimental studies and experience details of quantum of reinforcement to be distributed is given in Fig. 10.20.

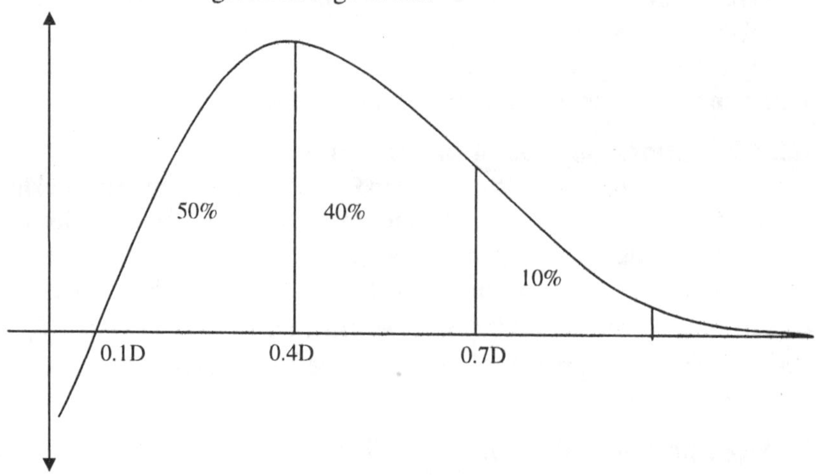

Fig 10.20

This will define variations in spacing of reinforcement at various zones through the length equivalent to the depth of the section from the end face.

In addition, reinforcement have to be provided for what are called as spalling forces. Spalling tensile stresses are based on the profile of the principle tensile trajectories as shown in Fig.10.5a.

If the principal tensile trajectories have a tendency to turn horizontally, which will be specifically the case in anchorages of eccentric forces, provisions have to be made for spalling tension. The spalling tensile reinforcement has to be empirically decided. By experience, it is found to be not greater than 10 percent of the prestressing forces or 33 percent of the total splitting tensile forces. The reinforcement will run for the full length of the end bock. The cross section areas of steel required will be distributed into top quarter and the bottom quarter of the equivalent stress block.

The reinforcement in general meant for splitting as well as for spalling tension cannot be placed in a single vertical plane. They should be distributed in more number of vertical planes not less than 3 depending on the thickness of the end block (width of the beam).

References

1. DEUTSCHER AUFSCHUSS FUR STAHLBETON: (1973) **Bemessung und Ausfuhrung von Spannbetonbauteilon unter Beruchsichtigung** von DIN Ausgabe 1, 72, Fassung June 1973.
2. HOYER E and E. FRIEDRICH.,- (1939)" **Beitrag zur frage der Hatspannung in Eisen betonbauteilen",** Beton and eisen (Berlin), vol.38, No.6, pp 107-110.
3. JANNEY R J - (1954) **"Nature of Bond in Pretensioned Prestressed Concrete",** Journal of the American Concrete Institute, May 1954, Vol.50, pp.717-736.
4. GUYON Y , (1953)- **"Prestressed Concrete",** Contractors Record Limited, London, pp.175-212.
5. MARSHALL W T, (1966)- **"A Theory for End-zone Stresses in Pretensioned Concrete Beams",** Journal of the Prestressed Concrete Institute, April 1966, Vol.11, pp.45-51.
6. MARSHALL W T.,AND D KRISHNAMURTHY., (1970) **"Design of end zone reinforcement for pre-tensioned concrete beams",** Proceedings of the Sixth Congress of the Federation Internationale de la Precontrainte, Prague, June 1970.
7. ROS, M R EMPA BERICHT Nr.155. (1946) Vorgespannter Beton, Zurich, Marz.
8. MARSHALL G (1949) **"End anchorage and bond stresses in prestressed concrete,"** Magazine of Concrete Research, vol.1, No.3, pp 123-127.
9. KRISHNA MURTHY, D., (1971) **"A method of determining the tensile stresses in the end zones of pre-tensioned beams".** The Indian Concrete Journal, Vol.45, No.7, July 1971, pp 286-297.
10. MORSCH E (1924) **Uber die Berchnung der Gelenkquader, Beton u, Eisen** 23 H.12, S 156-161.
11. MAGNEL G (1944) **Le fluage des aciers et son importance en beton preconstraint.** Science et Technique, H 10.
12. GUYON Y (1951) **Constraints dans les pieces prsmatiques soumises a des forces appliquees sur leurs bases, au voisinage de ces bases.** IVBH Abh.11 S.165-226.
13. SUNDARA RAJA IYENGAR K. T. (1960) **Der Spannungazustand in einem elastiachen Halbstreifen und seine technischen Anwendungen.** Diss. Techn. Hochsch. Hannover.
14. SARGIOUS M. (1960) **Beitrag zur Ermittlung der Hauptzugspannungen am Endauglager vorgespannter Betonbalken.** Diss. Techn Hochsch, Stuttgart. Or Hauptzugkrafte am Endauglager vorgespannter Betonbalken. Bautechnik 38 (1961) H.3. S. 91-97.

11. Composite Construction

11.1 General

Composite Construction in general refers to use of two materials of construction to form one structural unit, But the term composite construction in prestressed concrete works mostly refers to a construction in which a prestressed precast member acts in combination with a cast in situ concrete element. Prestressing the precast member has the advantage of having the unit as small and slender as possible to facilitate easy transport. In small elements of composite constructions full prestressing (which is eventually required for service load conditions), is transmitted in one stage, on the precast element itself. But in bigger units, prestressing is done in two or more stages. As a first stage, prestressing is done only to the level, that it takes care of self-weight of precast unit and to some extent the additional loads coming on the precast element. This is normally referred as first stage of prestressing. After the cast in situ concrete is poured and set, both the elements act together monolithically under service loads. Generally, the precast prestressed element is made of high strength concrete. The superlaid in-situ concrete is usually of lower compressive strength. They act together at service loads. To keep the stresses within permissible limits at various leading stages after the cast in situ concrete is poured, the section is prestressed additionally, which is defined as second stage prestressing. The second stage prestressing may be again made in two steps. The main aim is that the stresses should be within permissible limits at salient points in the cross section at any stage. The stresses have a relation to section properties. As the section becomes a composite section with a precast beam and a cast in situ slab, additional prestress can be introduced as first step of the second stage prestressing to take care of the self weight moment introduced by the increase in weight due to weight of the slab. Later, further superimposed dead loads may came on the structure in addition to the live load for which the elements are designed. For example, in bridge deck construction, the wearing coat, kerb, handrails etc are the additional superimposed loads which may come after the composite action is achieved, in addition to the traffic load for which the deck is designed. For taking care of those loads, one more step of prestressing is done on the composite section. Hence composite construction with prestressed precast beams and cast in situ slab always goes with stagewise prestressing. The analysis and the design hence have to be done with stagewise prestressing in mind.

11.2 Need for Composite Constructions

Composite construction is normally warranted when the cast-in-situ construction is very difficult, if not impossible. Moreover, the composite construction is an automatic choice when quality construction is not assured at site and the speed of construction is a decisive factor. The advantages in using prestressed precast concrete with cast in situ slab in composite construction are

- Appreciable saving in cost of steel in the composite member when compared to a fully reinforced or fully prestressed cast-in-situ member.
- The prestressing can be in stages as explained earlier. This would facilitate providing only the required prestress to take care of service loads.
- In many cases, precast prestressed units serve as supports and dispense with the formwork required for placing the cast in situ concrete.
- Easy and fast execution of work, is possible particularly in bridge construction, where normal traffic cannot be disrupted for long.
- Proper utilisation of the properties of concrete in the composite member, such as use of high strength concrete in the precast prestressed unit, a part of which become the tensile zone after the composite action is established and use of medium strength concrete in the insitu portion which forms the compression zone of the composite member.
- Possibility of use of different materials like light weight concrete for the cast-in-situ portion resulting in reduction of dead load of the structure which would lead to overall economy.

11.3 Types of Composite Construction

The most commonly used cross sections of composite construction are shown in Fig. 11.1. The cross sections shown in Fig11.1 a, b are meant for short to medium span bridge girder decks. Where there is a restriction on depth, inverted T- sections placed side by side are used. They are connected by means of a continuous cast-in-situ slab, which fills the gaps between them as shown in Fig.11.1b. Transverse prestressing is provided for road widths providing for more than two lanes. Voided slabs could be achieved as shown in Fig.11.1c, which are adopted in multistoried buildings or culvert slabs. The cast-in-situ portion can be in lightweight concrete if proper connections are used. When the interface contact area is large, the composite action is achieved by means of bond that exists between the precast concrete portion and the in-situ concrete. When the interface contact area is small, provision of additional vertical steel at the interface is necessary for shear transfer and they are called shear connectors.

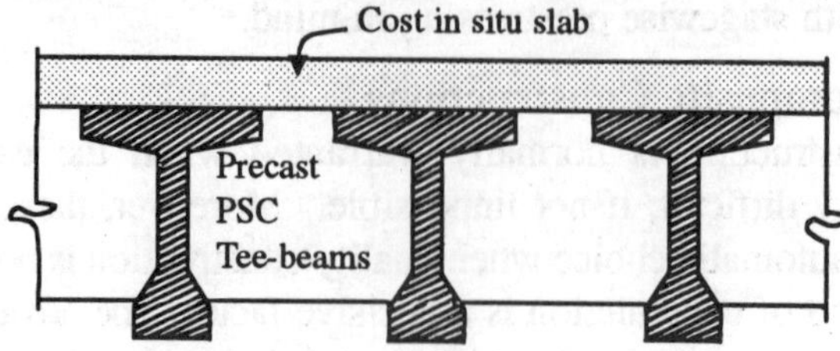

Fig 11.1.a

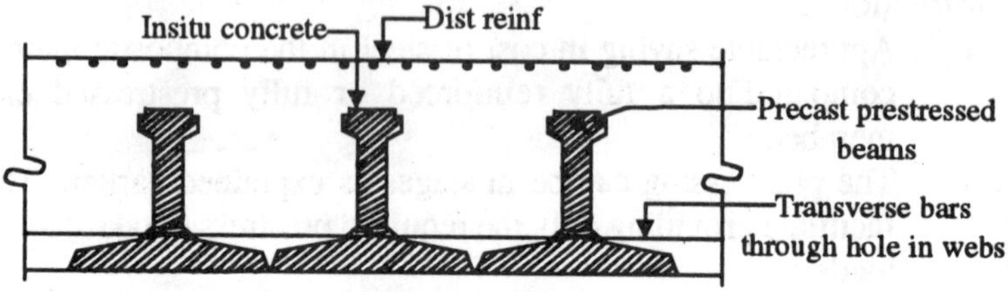

Fig 11.1.b

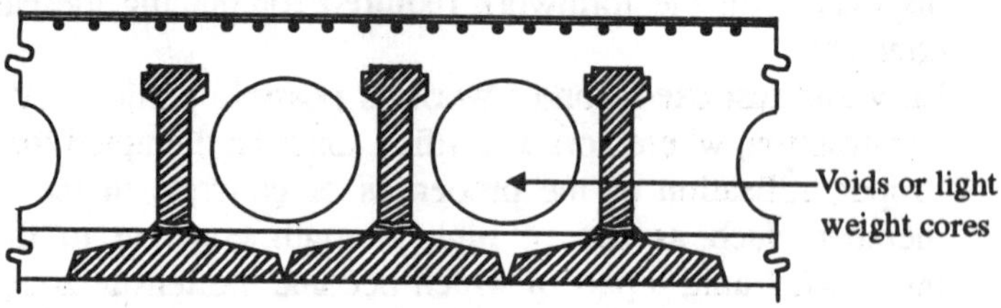

Fig 11.1.c

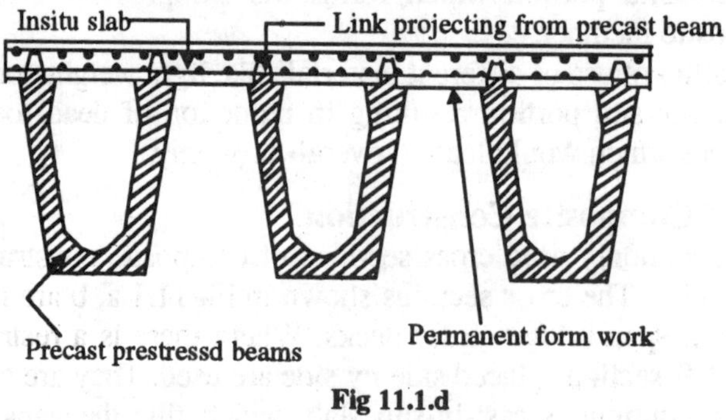

Fig 11.1.d

Fig11.1 Typical Precast Prestressed Section in Composite Section

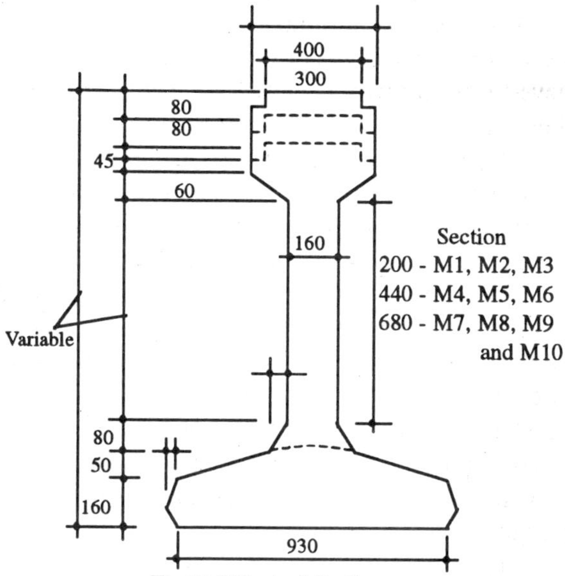

Fig 11.2 Typical Section

Table 11.1 - Typical Section						
Section properties of Typical Section						
Number	Overall depth mm	Area mm^2 above bottom fibre mm	Height of centroid	Section modulli mm^3 × 10^6 Top fibre	Bottom fibre	Dead load kN/m
M1	640	284,650	220	24.72	47.17	6.71
M2	720	316,650	265	35.64	61.04	7.46
M3	800	348,650	310	46.96	74.31	8.21
M4	880	323,050	302	43.41	85.95	7.61
M5	960	355,050	357	59.39	100.33	8.37
M6	1040	387,050	409	75.39	116.23	9.12
M7	1200	393,450	454	66.46	123.16	8.52
M8	1200	393,450	454	87.39	143.57	9.27
M9	1280	425,450	512	108.09	161.96	10.02
M10	1360	457,450	568	128.65	179.36	10.78

A typical inverted T sections as developed by the Ministry of Transport and the Cement and Concrete Association, U.K. for various spans which could be purchased from manufacturers of precast products are shown in Fig. 11.2. Channel sections as precast prestressed units are also used with cast in situ deck slab to form box girder bridge decks, which are commonly adopted in long span bridges.

11.4 Behaviour of Composite Construction

Precast prestressed concrete, with cast-in-situ concrete composite constructions are used mainly as flexural members. For analysis of such a composite unit as a beam, the well known simple beam bending theory with Naviers hypothesis is adopted. Accordingly the assumptions used in Naviers hypothesis have to be fulfilled, namely the plane section remains plane at various loading stages. In composite constructions, because of the likely hood of physical separation between the two materials, sliding of cast in situ concrete slab over precast prestressed beam may occur leading to tendencies for in-plane deformation causing longitudinal shear stress. Further, in a composite beam, if a portion of the span alone consists two layers of materials and the remaining portion has only the first layer of the material as shown in Fig.11.3, the loading on the top portion of the composite portion of the structural element may be transferred as reaction loads on the bottom layer. This leads to tendencies for vertical separation between the layers and hence normal stressses between the layers at the interface are introduced. Johnson (1) has explained the behaviour of composite beams with steel concrete composites which are equally applicable here also for stydying the structural behaviour.

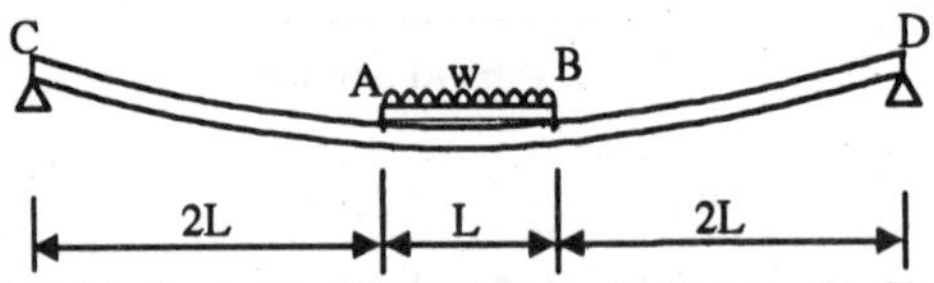

Fig 11.3 Cause for Uplift Forces in Composite Units

For achieving monolythic action of the composite construction the two composite materials used in a composite constructions have to be tied together to prevent the horizontal and vertical separation. The horizontal shear stress and vertical normal stresses at the interface layers have to be properly taken care of. This can be achieved by good bond properties of the materials at the interface or using reinforcement known as shear connectors. In the composite construction, normally the precast prestressed units are prepared in a factory with good quality control and the cast-in-situ concrete units are laid on top of prestressed, precast units with surface preparation at the interface or with shear connectors if necessary. If the precast prestressed unit is unpropped during the time of placing the cast insitu concrete, the insitu concrete is considered as dead load on the precast unit and the flexural stresses are calculated using the section properties of precast unit only. In case the precast prestressed units are propped while the insitu concrete is placed, the dead load of the cast-in-situ slab starts acting only after the props are removed. At that time the cast-in-situ concrete is well set and it acts monolithic with precast beam. Hence the flexural stresses are calculated using the section properties of the combined section. The distribution of flexural stresses at various stages of loading in propped and unpropped constructions are shown in Fig11.4. The composite constructions are normally used in bridge decks where the cast in situ units cover up the compression zone for number of precasting beams. For the purpose of calculations and design of composite section for the service load conditions the effective width of compression flange of T beams (interior beams) and edge beams (exterior beams) may be calculated as in the case of normal concrete sections as defined in IS456-2000(2). But precautions are taken in detailing the reinforcement to provide monolithicity of the slab with the beam.

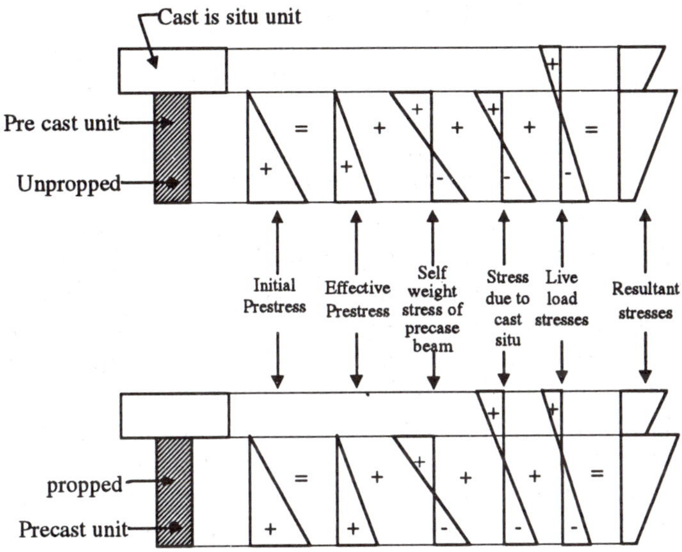

Fig 11.4 Stress Distribution in Unpropper and Propped Composite Construction

If T or L sections are considered specially in continuous beams over supports, the flange thickness shall not be less than one-half the width of the web and a total flange width not more than four times the width of the web. For effective stress transfer, it is desirable to splay the junction of the web and the flange so as to form an angle of not less than 110° as shown in Fig.11.5. Where the principal reinforcement is a slab which is considered as the flange of a T or L beam is parallel to the beam, transverse reinforcement shall be provided at the top of the flange. This reinforcement shall be equal to sixty percent of the main reinforcement of the slab at its mid span unless it is specially calculated. The length of such reinforcing bars shall be as indicated in Fig.11.6.

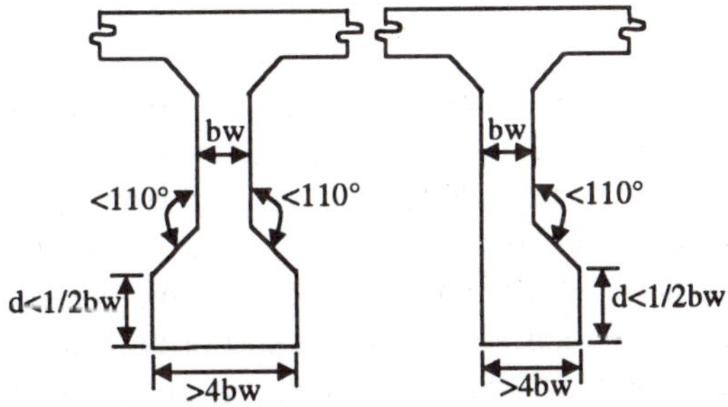

Fig 11.5 Splay in Precast Units

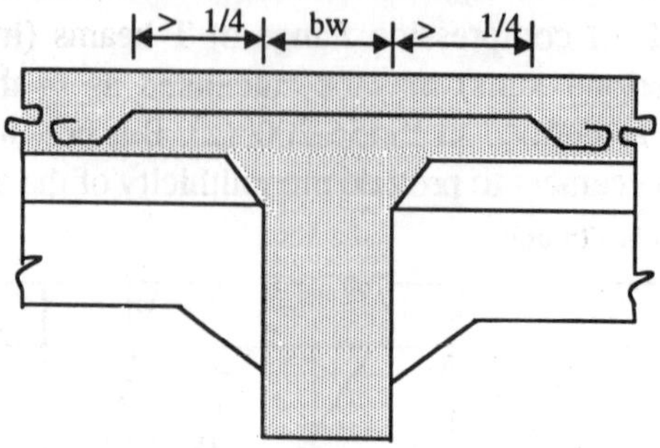

Fig 11.6 Reinforcement in the deck slab

11.5 Flexural Stresses

In most composite construction, where precast prestressed units and cast insitu concrete are used, the grade and properties of concrete used for the prestressed precast units and those used for the cast-in-situ portion are different. Since the behaviour of the composite sections depends on the compatibility at the interface, the stresses developed in both the cast in situ concrete portion and the precast concrete portion depend on the moduli of elasticity of concretes used for the precast and insitu portions. This aspect is taken into consideration in the design by multiplying the area of the precast prestressed unit by the ratio of the moduli of elasticity of the two concretes. The other properties of the composite section like the distance of the centroid from the face, modulus of section, moment of inertia etc are then calculated based on this transformed area and the stresses are subsequently evaluated using these sectional properties. This is illustrated in the examples worked out later.

11.6 Longitudinal Shear Transfer

As flexural loading is applied to a composite beam there is a tendency for the cast in situ slab to slip horizontally as shown in Fig.11.7b. The bottom face of the slab moves outward with respect to the top face of the precast web, which tends to displace inwards. If this slip is not prevented, the flange and the web start acting two separate beams each resisting its portion of the load independently by bending about its own centroidal axis as shown in Fig 11.7b.

If the slip is prevented the longitudinal shear across the interface between the two components of the composite members is generated. Resistance against the shear could be provided by natural adhesion and friction between the two concrete faces. It could be enhanced if the top of the precast unit is roughened deliberately rather than trawled smooth. For composite beams having a large contact (interface) area no extra provision for shear transfer needs to be made. For beams having a smaller contact area, vertical stirrups called shear connectors, designed to take the interface shear at that section (vertical), are to be provided.

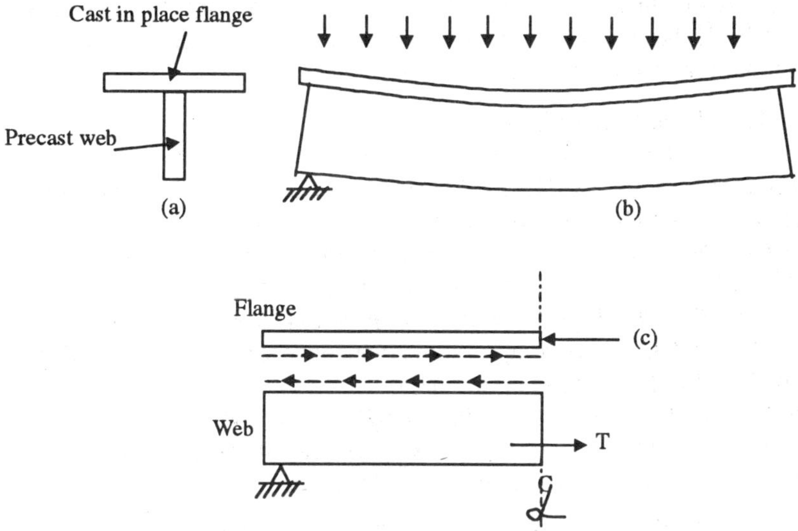

Fig 11.7 Composite Action in Tee-Beam (a) section. (b) Non Composite Behaviour (c) Shear Stresses on Interface

These stirrups are taken to the full depth by extending them in to the cast in situ portion. This provides a greater frictional resistance and also a dowel force. Tests have shown that after a slip 0.12 mm, the composite action will be lost. The intensity of shear at the interface is calculated using the principles of structural mechanics. This should not exceed the limiting shear stresses which is a function of concrete quality and surface conditions. Various codes give these values differently as discussed later.

The horizontal shear stress at any level across the cross section is calculated as the complimentary effect of vertical shear or equal to the compressive force carried by the concrete above the plane of interface at the time of failure.

As the effect of vertical shear, shear stresses are produced at various levels. At the interface it is expressed as $\frac{V}{I}(A\bar{y})$ where V is the total vertical shear, I is the moment of intertia of the cross section, A is the area of concrete above the interface plane and $\bar{y}$ is the distance centre of gravity of the area above the interface plane from the centroid of the composite section. As the effect compressive forces on the flange, there could be horizontal in-plane shear stress at the interface level. This would be equal to the compressive force above the interface level divided by the width (smaller) of section at the interface (namely top flange width of the precast unit). Enough resistance has to be provided by means of interface bond or shear connectors. For cracked concrete section, the shear intensity cannot be calculated based on the expression given above and hence only the effective compressive force above the interface is considered. Indian standard code IS 1343-1980[3] does not make specific recommendations for preventing such failure, along the interface due to longitudinal shear. But British standard BS 8110-1985 [4] have made certain recommendations which are used in professional practice now a days

11.6.1 British code provisions

British code has made provisions to be taken care in the design based on ***Ultimate Load*** conditions. The total horizontal shear force at the interface at ultimate load is either of the following

- ***where the interface is in the tension zone:*** the total compression (or tension) calculated from the ultimate bending moment; or
- ***where the interface is in the compression zone:*** the compression from that part of the compression zone above the interface, calculated from the ultimate bending moment.

The horizontal shear stress is calculated by dividing the compressive force above the level by the width of contact between precast and cast in either unit. The intensity of shear stress shall not exceed the value given in table 11.2.

Table 11.2 Design Ultimate Horizontal Shear Stresses at Interface(V_u)

Precast unit	Surface type	Grade of in situ concrete		
		25	30	40 and over
		N/mm^2	N/mm^2	N/mm^2
Withoutlinks	As-cast or as-extruded	0.4	0.55	0.65
	Brushed, screeded or rough-tamped	0.6	0.65	0.75
	Washed to remove laitance or treated with retarder and cleaned	0.7	0.75	0.80
With nominal links projecting into in situ concrete	As-cast or as-extruded	1.2	1.8	2.0
	Brushed, screeded or rough-tamped	1.8	2.0	2.2
	Washed to remove laitance or treated with retarder and cleaned	2.1	2.2	2.5

Note 1. The description 'as-cast' covers those cases where the concrete is placed and vibrated leaving a rough finish. The surface is rougher than would be required for finishes to be applied directly without a further finishing screed but not as rough as would be obtained if tamping, brushing or other artificial roughening had taken place.
Note 2. The description 'as-extruded' covers those cases in which an open-textured surface is produced direct from an extruding machine.
Note 3. The description 'brushed, screeded or rough-tamped' covers those cases where some from of deliberate surface roughening has taken place but not to the extent of exposing the aggregate.
Note 4. For structural assessment purposes, it may be noted that the appropriate value of material safety factor may be included.

Nominal links: When provided, nominal links should be of cross section at least 0.15% of the contact area. Spacing should not be excessive. The spacing of links in T-beam ribs with composite flanges should not exceed four times the minimum thickness of the in situ concrete nor 600mm. Links should be adequately anchored on both sides of the interface.

Links in excess of minimum: Where the horizontal shear stress exceeds the value given in the Table 11.2, all the horizontal shear force should be carried on reinforcement anchored either side of the interface. The amount of steel required, A_h (in mm²/m) should be calculated form the following equation $A_h = \dfrac{1000bv_h}{0.87f_y}$

11.6.2 American code provisions

American code provisions ACI 318-1998[5] demand that the full transfer of horizontal shear forces to be assured by contact surface of the inter connected elements at the interface. Limit state concepts are to be used and a capacity reduction factor times of 0.85 is recommended. The horizontal shear force is to be calculated as a factor times the vertical shear forces given as $\dfrac{V_u}{\mu\phi}$ and μ is assumed to be 1.0 if concrete is placed on already hardened concrete, 1.4 when concerning is done monolithically and 0.7 if concrete is placed over steel. Then the limiting value of horizontal shear stress is evaluated as horizontal shear forces divided by the width at interface. These values shall not exceed the limiting values given in Table 11.3. The limiting values are different for 3 different categories of surfaces. If the shear stress exceed the values reinforcement to be provided for resisting the full horizontal shear.

The area of steel is distributed over a unit length of the beam as shear connectors will be in accordance with the following equation $A_{vf} = \dfrac{V_u}{\phi f_y \mu}$

	Table 11.3	
Type	**Description of contact surface**	V_h
1	Clean and intentionally roughened with a full amplitude of approx. 6mm without ties	0.56Mpa
2	Clean but not intentionally roughened with ties satisfying minimum requirements given below	0.56Mpa
3	Clean and intentionally roughened with ties satisfying minimum requirements given below	2.45MPa

11.7 Transverse Shear

Transverse shear refers to the shear in the vertical plane running orthogonal to the cross section of the main girder. (refer section 1–1in Fig.11.9, running normal to the plane of the paper). The longitudinal shear along the length of the girder produces also complementary shear in a direction transverse to the length of the girder. While the shear connectors take care of the shear in the longitudinal plane, reinforcements are also required in the transverse direction to the length of the beam. Most of international codes do not give clear guidelines for assuming safety against such structural action since the flexural reinforcement provided in the cast in slab is more than adequate to take care of this structural action. But IRC 22 1986[6] gives certain guide lines

The shear force (V_L) per unit length, transformed from the prefabricated girder in to the slab through the shear plane, shall not exceed either of the following and the reinforcements will be accordingly provided.

i) $0.4L_s\sqrt{f_{ck}}$

ii) $0.7A_s\sigma_y + 0.08L_s\sqrt{f_{ck}}$ Where L_s is the length of shear plane under consideration in mm. Typical shear planes are shown in Fig. 11.9, f_{ck} is the Cube (characteristics) strength of concrete (Mpa) but not more than 45 Mpa, As is the sum of the cross section areas per unit length of beam of all reinforcing bars intersected by the shear plane (mm^2/mm) and y is the yield stress (Mpa) of the reinforcing bars intersected by the shear plane but not more than 450 Mpa.

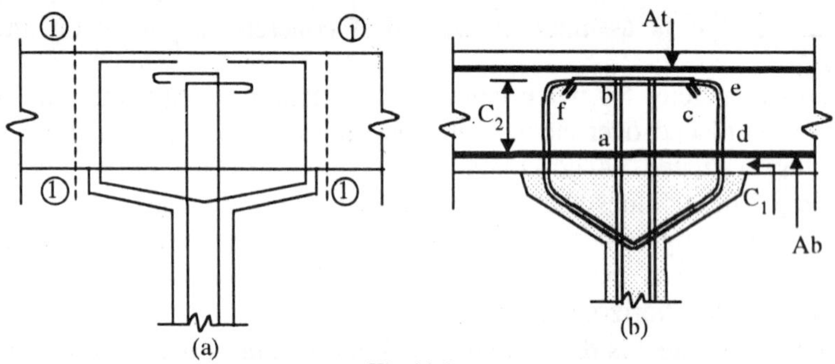

Fig 11.9

The reinforcement to resist longitudinal shear are provided as stirrups passing through the precast girder and cast-in-situ slab. In addition to these reinforcement, sufficient reinforcement is tobe provided to resist the transverse shear. These reinforcement shall be for unit length of the beam. But these are in horizontal plane to cut the shear plane 1-1 shown in Fig.11.9. There may be already similar reinforcement existent in the slab for flexural purposes. The area of such reinforcement which are intersected by the shear plane and if they are fully anchored on eitherside could be considered to resist with transverse shear. The quantity of the reinforcement is decided by the proportionate compressive force taken by the compressive zone above the interface. If this proportionate forces is C′ and the total force is C the transverse shear is evaluated as $\frac{C'}{C}\times$ (total shear in the cross section). This shear force in the shear plane has to be resisted by horizontal plane reinforcement. Further there is also a requirement of provision of minimum reinforcement given by $A_s = \frac{0.8L_s}{f_y}$ where A_s, L_s and f_y are as defined earlier.

I.R.C. 22-1966[6] discusses about the shear connectors and has proposed certain design requirements such as

- The shear connectors shall be designed to ensure integral action between the prefabricated unit and the in-situ concrete unit so that the two units act as composite structure. The shear connectors shall satisfy the two basic requirements namely (i) Shear along the contact surface is transferred without slip, and (ii) Separation of the prefabricated unit and the in-situ slab in the perpendicular direction is prevented.
- The shear connectors shall be of the type which permit a thorough compaction of concrete in order to ensure perfect contact of their entire surfaces with concrete.
- The load factor for design of shear connectors under ultimate load shall be 1.5 for dead load and 2.5 for live load. The dead load to be taken for calculating the ultimate horizontal shear shall be the dead load operating after composite action is effective.
- The ultimate longitudinal shear V_L per unit length at the interface shall be evaluated from the expression $\frac{V}{I}(A\bar{y})$ (b) by calculating ultimate vertical shear with the load factors.
- The ultimate shear resistance of an anchorage connector is given by $Q_u = A_s f_y \times 10^{-3}$

Where

Q_u = The ultimate shear resisting capacity of the anchorage connector in kN

A_s = The cross-sectional area of the anchorage connector mm^2

f_y = The ultimate tensile strength of steel of the anchor connector in Mpa. The anchor bars shall be placed either vertical or at an angle of 45°.

- The spacing of the shear connector shall be calculated from the formula

$$s_u = \frac{\Sigma Q_u}{V_L}$$

s_v = Spacing of shear connector

Q_u = Ultimate shear resistance of each connector

V_L = Ultimate longitudinal shear per unit length

- Notwithstanding the fact that full shear resistance is provided by anchor bars, the ultimate bond stress at the interface shall not exceed 2.1 Mpa. The interface shall always be made rough for effective bonding.

11.8 Creep and Shrinkage Effects in Composite Construction

The cast insitu portion is laid over the precast prestressed elements, which has already undergone creep and shrinkage to a great extent. But the in-situ slab starts creeping only after it is laid and that is larger because of wet concrete. Hence as both units start creeping together thereafter, the relative creep will be different. This differential creep and shrinkage causes stresses both in the cast insitu slab and in the precast beam. The differential strain is applied as shown in Fig.11.10. This is later converted to an equivalent and opposite forces. The couple is then transferred to an axial compression and bending moment on the composite section. The stresses caused due to these on the composite beam can be seen in Fig 11.10a and 11.10b.

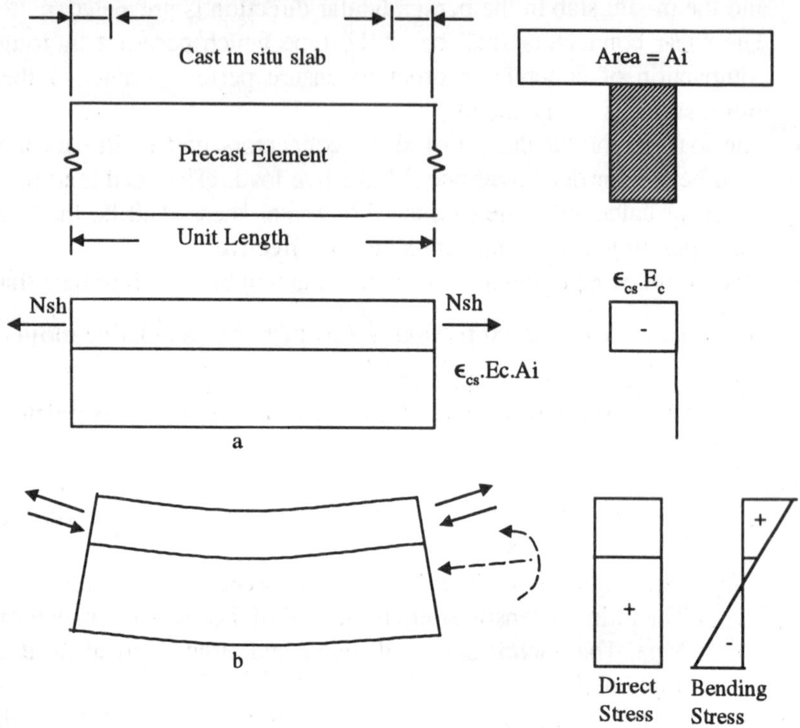

Fig 11.10

This approach is only on approximate one, as the creep coefficient is dependent on applied load which itself is not a constant with respect to time. Recent research has taken all this in to consideration and Trost (7) has provided a solution for such effects in composite structures.

11.8.1 Effects of Creep and Shrinkage - Trost's Approach

The structural effect due to differential creep and shrinkage is dependent on time and applied stress on the elements concerned. For getting the differential creep and shrinkage strain, the creep and shrinkage strain of individual units are to be evaluated. As mentioned in chapter 2 Trost has arrived at a general formula for strain in concrete at time 't' as:

$$\epsilon_c\ t) = \frac{\sigma_{co}}{E_c}(1+\varphi) + \frac{\Delta\sigma_c}{E_c}(1+\rho\varphi) + \epsilon_{s\infty}\frac{\varphi(t)}{\varphi_{(\infty)}}$$

Where σ_{co} = stress in concrete at the time of transfer

φ = Creep coefficient which is dependent on applied stress

$\Delta\sigma_c$ = The variation in concrete stress from the time at which the load is transferred to the time at which the creep strain is calculated.

ρ = A relaxation factor for creep coefficient.

It is well known that creep strain is a function of applied stress. The first term gives the effect of creep due to the initially applied stress. It should be noted that as the creep strain increases, the applied stress decreases and accordingly, there is a decrease in the rate of change of creep strain. This is represented by the second term where $\Delta\sigma_c$ is the reduction in applied stress (hence negative). But since the applied stress becomes smaller creep coefficient also should be reduced. That is reflected in the expression through in the creep relaxation factors 'ρ'. The relaxation factor takes care of the reduction in the rate of creep strain at any time due to the variation in the applied load. The last term of the equation refers to shrinkage strain at a time 't', which is idealised to a value propotional to creep strain at any point of time. It is expressed as shrinkage strain at time '∝' is multiplied by the creep factor at time 't' is divided by the creep factor at infinite time. Such calculated strain at the interface due to creep and shrinkage must be compatible.

Referring to Fig.11.11 $\Delta\sigma_{1t}$, $\Delta\sigma_{1b}$, $\Delta\sigma_{2t}$, $\Delta\sigma_{2b}$ are the stresses developed due to the differential creep and shrinkage at the interface in the precast and cast insitu portions. Since these stresses are developed under 'no load' condition, the following equations have to be satisfied

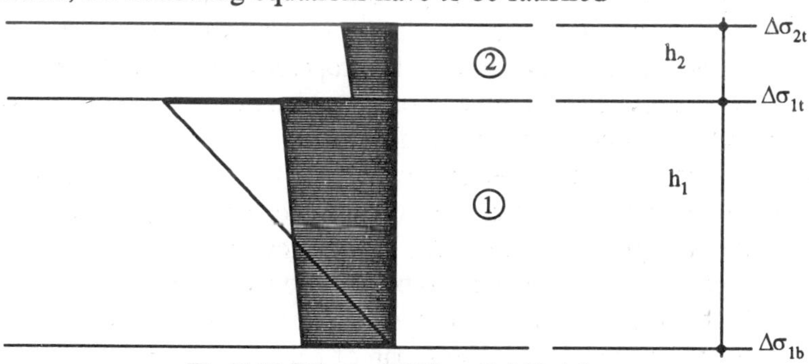

Fig 11.11 Effects of Differential Shrinkage

$\int_{A_1+A_2} \Delta\sigma.dA = 0$ by equilibrium of forces

$\int_{A_1+A_2} \Delta\sigma.y.dA = 0$ by equilibrium of moments

$\Delta\varepsilon_{1t} = \Delta\varepsilon_{2b}$ by compatibility of strains at surface remain plane through out the various loading shapes inclusive of service loads.

$\dfrac{\Delta\varepsilon_{1t} - \Delta\varepsilon_{1b}}{h_1} = \dfrac{\Delta\varepsilon_{2t} - \Delta\varepsilon_{2b}}{h_2}$ due to assumption that plane sections. remain plane all through loading stages.

The values of strains are evaluated using Trost's formula and stresses are evaluated using the constitutive relationship of material for precast and cast in situ concrete.
Solving the four equations given above, the values $\Delta\sigma_{it}$, $\Delta\sigma_{1b}$, $\Delta\sigma_{2t}$, $\Delta\sigma_{2b}$, namely the stresses due to differential creep and shrinkage are evaluated. An illustrative example (Example 11.9) gives details of the above procedure.

11.9 Stage Prestressing

As indicated earlier, the prestressing is mostly done or preferably done in two or three stages. The first stage is to take care of the self weight of the precast units with some additional load and the second and third stages to take care of the superimposed dead load and liveload respectively. While for stage one, the precast section properties are considered, for stage two and three, the composite section properties are to be used. Stage two and stage three are some times combined together as one stage as itself namely the second stage. While analysis the structure for stress resultants at various levels at various stages a few precautions as mentioned are worth while to be taken note of. They are

- The creep coefficient and shrinkage coefficient for different stages of prestressing have to be separately considered depending on age of the concrete when forces are applied.
- The elastic shortening of the beam and the corresponding losses in prestressing in stage I cables are to be considered when stage two prestresing is done.
- Even in first stage prestressing if the number of cables are more, there may be loss in the forces in the cables stressed earlier when a new cable is stressed which have to be properly accomodated for.
- These losses due to creep and shrinkage causes deformation problems specifially in indeterminate structure.
- The layout of cable, the grouping of cables the spacing between the cables and the anchorage positions of cables have to be carefully identified specifically in stagewise prestressing.
- The cables which are going to be stressed later have to have a profiles at higher elevation th an those which are to be prestressed earlier.
- It is preferable to grout the group of cables after prestressing all the individual cables in that group.
- The grouping of cables have to be done only in one vertical plane.
- Due to necessities of circumstances if in second level of stage two prestressing cables have to be anchored at the slab, back up anchorage reinforcements are to be used. The anchorage forces have to be tied back in the opposite direction by specifically additional reinforcement.
- The cables will be covered with water proofing material over which the layer of concrete slab or wearing coat will be laid.

Example 11.1

A precast pretensioned unit of rectangular section of size of 120mm × 240mm is used a part of a composite beam to span 6m. This unit is prestressed by tendons with their centroid coinciding with the bottom kern point. The initial force in the tendon is 240 kN. The loss of prestress may be assumed to be 15%. The unit is incorporated as the web of a composite beam by casting a slab of flange width of 480 and athickness of 40mm. On the top of theprecast unit the composite beam supports a live load 4.0 kN/m. Calculate the resultant stresses developed in the precast and cast insitu concrete assuming the pretensioned unit as unpropped while casting the insitu slab the ratio of modullii of elasticity between the precast unit and cast in situ slab is 1.25.

Solution

Section properties of precast unit

Area $= 120 \times 240 = 28800\ mm^2$

Z $= \dfrac{120 \times 240^2}{6} = 1152 \times 10^3 mm^3$

Initial prestressing force P_0 = 240 kN

Stress due to P_0 at top = $0 N/mm^2$

at bottom $= \dfrac{240 \times 10^3 \times 2}{28800} = 16.66 N/mm^2$

Final prestressing force $P\infty$ $= 240 \times 0.85 = 204 kN$

Stress at top = 0

at bottom = $14.161\ N/mm^2$

Self weight of precast beam $= 0.12 \times 0.24 \times 24 \times 10^3$

$= 691.25$ N/m

Moment due to self weight $= 0.125 \times 691.25 \times 6.25^2$

$= 3375.2$ Nm

Stress at top and bottom due to self Weight bending moment $= \dfrac{3375.2}{1152 \times 10^3}$

$= 2.298$ (Comp)

Self weight of cast-in-situ slab

$= 0.04 \times 0.48 \times 6.25^2 = 460.8 N/m$

Moment due to slab weight $= 0.125 \times 460.8 \times 6.25^2 = 2250 Nm$

Stress at top of the precast unit $= \dfrac{2250 \times 10^2}{1152 \times 10^3} = 1.953 N/mm^2$ Comp)

Stress at bottom of the

Precast unit $= 1.953\ N/mm^2$ (tension)

Section properties of composite section:

The centroid of the equivalent composite section is determined by taking moments by taking moments about the axis passing through the soffit of the beam.

If y is the distance of centroid of the beam from the soffit

$(19.28 + 1.25 + 28.8)10^3 . y = 19.28 \times 10^3 \times 260 + 1.25 \times 28.8 \times 10^3 \times 120$

$y = \dfrac{9332.8}{55.28} = 168.82 mm$

Second moment of inertia of equivalent composite section:

$$I_e = \left(\frac{480 \times 40^3}{12} + 480 \times 40 \times 91.18^2\right) + 1.25\left(\frac{120 \times 240^3}{12} + 240 \times 120 \times 48.82^2\right)$$

$= 420.78 \times 10^6 mm^4$

Live load moment: $4 \times 10^3 \times 6.0^2 \times 0.125 = 18000000 Nmm$

Stresses in cast insitu slab

at top $= \dfrac{18 \times 10^6}{420.78 \times 10^6} \times 111.18 = 4.756 N/mm^2 (comp)$

at bottom $= \dfrac{18 \times 10^6}{420.78 \times 10^6} \times 71.18 = 3.05\ N/mm^2 (comp)$

stress in precast beam

at top $= \dfrac{18 \times 10^6}{420.78 \times 10^6} \times 71.18)\ 1.25 = 3.80\ N/mm^2 (compression)$

at bottom $= \dfrac{18 \times 10^6}{420.78 \times 10^6} \times 168.8\ )\ 1.25 = 9.02\ N/mm^2 (tension)$

Final stresses

Cast in situ slab top = 4.756 N/m^2 (*Compression*)

Cast in situ slab bottom = 3.05 N/mm^2 (*Compression*)

Precast beam top = 2.928 + 1953 +3.80

= 8.6828 N/m^2 (*Compression*)

Precast beam bottom = –2.928 – 1.953 – 9.027 + 14.161

= +0.253 Nm^2 (*Compression*)

Example 11.2

A composite 'T' beam is made up of a pretensioned rib 300 mm wide and 1000 mm deep and a cast in situ slab of 200 mm thickness and 1500 mm width. The modulus of elasticity of cast in situ slab will be 28 KN/mm². The diffrential shrinkage and creep is 0.0001. Determine the stresses caused by this on the precast and cast in situ concrete.

Solution

Differential deformation in terms of strain = 10×10^{-5}

Area of in situ concrete =$1500 \times 200 = 30 \times 10^4 mm^2$

Uniform tensile stress inducted in cast-in-situ slab = $10 \times 10^{-5} \times 28 \times 10^3$

= 2.8 N/mm²

Force on cast-in-situ slab = $2.8 \times 30 \times 10^4$

Centroid of the composite section is located at 400mm from top and 300 mm from centroid of slab

Moment = $840 \times 10^3 \times 300 = 84 \times 3 \times 10^6 Nmm$

Second moment of inertia of composite section = $0.8 \times 1000^4 = 800 \times 100^4$

Section moduli at various fibre levels

$Z_t = 0.2 \times 1000^3$

$Z_b = 0.1 \times 1000^3$

$Z_j = \dfrac{800 \times 100}{200} = 0.04 \times 100^5$

Direct compression on composite section = $\dfrac{840 \times 10^3}{60 \times 10^4} = 1.4 N/mm^2$

Bending stresses on composite section

Top fibre level $= \dfrac{84 \times 3 \times 10^6}{0.2 \times 1000^3} = 1.12 N/mm^2$

Bottom fibre level $= \dfrac{84 \times 3 \times 10^6}{0.1 \times 1000^3} = 2.24 N/mm^2$

Junction level $= \dfrac{84 \times 3 \times 10^6}{0.04 \times 10^{10}} = 0.63 N/mm^2$

Differential shrinkage streses

a) Precast pretensioned beam
   At top : $1.4 + 0.63 = 2.03\ N/mm^2$
   At bottom : $1.4 - 2.24 = -0.84\ N/mm^2$

b) Situ slab
   At top : $1.4+1.12=2.52-2.8=0.28 N/mm^2$
   At bottom : $1.4+0.63=2.03-2.8=-0.77 N/mm^2$

Example 11.3

A precast pretensioned beam has a rectangular cross section over which a reinforced concrete slab is cast. The dimensions of the beam and slab elements are shown in figure (11.12). Both the elements are made of similar concrete and hence have same E value. The precast beam is subjected to precasting force od 300 tons and an external moment of 170 t.m. The prestressing force is applied at distance of 100 mm from soffit The section properties, creep coefficeinet, creep relaxation factor, shrinkage strain etc., are given below. Evaluate the stresses at top and bottom and interface at the two units due to creep and shrinkage under the stress conditions.

$A_1 = 0.3 \times 10^6 mm^2$	$A_2 = 0.3 \times 10^6$
$I_1 = 0.025 \times 10^{12} mm^4$	$I_2 = 0.001 \times 10^{12} mm^4$
$Z_1 = 0.05 \times 10^9 mm^3$	$Z_2 = 0.01 \times 10^9 mm^3$
Creep coefficient = 2	Creep Coefficient = 4
$\epsilon_{s1} = 0; \rho_1 = 1$	$\epsilon_{dz} = -0.0004; \rho_2 = 1$

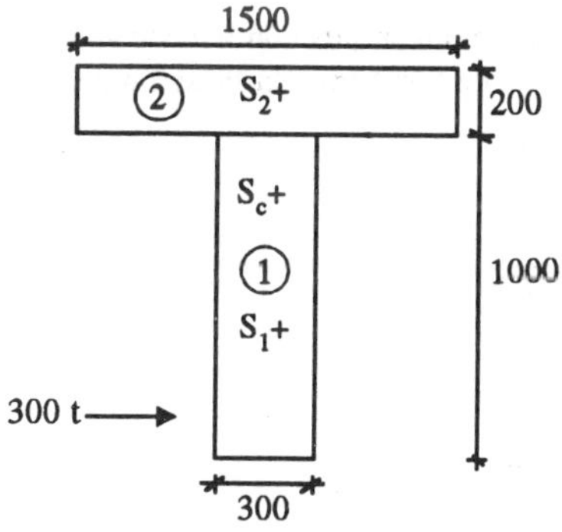

Fig 11.12

Composite section properties

$A = 0.6 \times 10^6 mm^2$	$I = 0.08 \times 10^{12} mm^4$
$Z_b = 0.1 \times 10^9 mm^3$	$Z_t = 0.2 \times 10^9 mm^3$

Solution

Stressses at time = 0 in section 1

$$\text{Bottom} = \frac{-300 \times 10^4}{0.3 \times 10^6} - \frac{300 \times 10^4 \times 400}{0.05 \times 10^9} + \frac{170 \times 10^3}{0.05 \times 10^9}$$

$= -10 - 24 + 34 = 0$

Top $= -10 + 24 - 34 = -20\text{N/mm}^2$ (compression)

1) Equilibrium equation for forces

$$\frac{[\Delta\sigma_{1t} + \Delta\sigma_{1b}].A_1}{2} + \frac{[\Delta\sigma_{2t} + \Delta\sigma_{2b}].A_2}{2} = 0$$

Since $A_1 = A_2$

$$\Delta\sigma_{1t} + \Delta\sigma_{1b} + \Delta\sigma_{2t} + \Delta\sigma_{2b} = 0 \qquad \text{.......(1)}$$

Moment equation about the interface shall be written as follows: (Ref. Fig. 11.11)

$$\frac{[\Delta\sigma_{2t} + \Delta\sigma_{2b}].A_2}{2} \times \frac{2(\Delta\sigma_{2t}) + \Delta\sigma_{2b}}{\Delta\sigma_{2t} + \Delta\sigma_{2b}}.\frac{h_2}{3}$$

$$= \frac{[\Delta\sigma_{1t} + \Delta\sigma_{1b}].A_1}{2} \times \frac{2(\Delta\sigma_{2t}) + \Delta\sigma_{2b}}{\Delta\sigma_{2t} + \Delta\sigma_{2b}}.\frac{h_1}{3}$$

$$= [2\Delta\sigma_{1t} + \Delta\sigma_{2b}]\, A_2 h_2 = [2\Delta\sigma_{1b} + \Delta\sigma_{1t}]\, h_1 A_1$$

$$= 2\Delta\sigma_{2t} + \Delta\sigma_{2b} - 10\Delta\sigma_{1b} - 5\Delta\sigma_{1t} \quad = 0$$

Moment equilibrium equation can be written as

$$-5\Delta\sigma_{1t} - 10\Delta\sigma_{1b} + 2\Delta\sigma_{2t} + \Delta\sigma_{2b} = 0 \qquad \text{.......(2)}$$

Compatibility of strain as the junction

$$\Delta\varepsilon_{1t} = \Delta\varepsilon_{2b}$$

$$\frac{\Delta\sigma_{1t}}{E} = \frac{\Delta\sigma_{2b}}{E}$$

$\frac{\Delta\sigma_{1t}}{E}$ due to creep and shrinkage is calculated using Trost's formula:

$$\frac{\Delta\sigma_{1t}}{E} = \frac{\sigma_{1t}}{E}(1+\varphi) + \left[\sigma_{1t} - \sigma'_{1t}\right]\left[1 + \rho\varphi\right] + \varepsilon_{sa}.\frac{\varphi}{\varphi_a}$$

Since the compatibility of strains at junction is concerning the strains after the elastic strains due to σ_{1t} has occurred the first term (elastic strain due to σ_{1t}) is omitted.

$$\frac{-20}{E} + 2 + \frac{\Delta\sigma_{1t}}{E}(1+2) + 0$$

$$\frac{1}{E}.(-20 \times 2 + 3\Delta\sigma_{1t}) = \frac{\Delta\sigma_1}{E}$$

$$\Delta\sigma_1 = (-20 \times 2 + 3\Delta\sigma_{1t})$$

Similarly

$$\frac{\Delta\sigma_2}{E} = \left[\frac{\Delta\sigma_{2b(1+4)}}{E} - 0.0004\right]$$

$$\Delta\sigma_2 = [5\Delta\sigma_{2b} - 10]$$

$$-40 + 3\Delta\sigma_{1t} = 5\Delta\sigma_{2b} - 10$$

$$3\Delta\sigma_{1t} - 5\Delta\sigma_{2b} = 30 \qquad \text{.......(3)}$$

$$\frac{\Delta\varepsilon_{1t} - \Delta\varepsilon_{1b}}{h_1} = \frac{\Delta\varepsilon_{2t} - \Delta\varepsilon_{2b}}{h_2}$$

$$\Delta\varepsilon_{1t} - \Delta\varepsilon_{1b} - 5\Delta\varepsilon_{2t} + 5\Delta\varepsilon_{2b} = 0$$

$$-20.2 + 3\Delta\sigma_{1t} - 3\Delta\sigma_{1b} - 25\Delta\sigma_{2t} + 25\Delta\sigma_{2b} = 0$$

$$3\Delta\sigma_{1t} - 3\Delta\sigma_{1b} - 25\Delta\sigma_{2t} + 25\Delta\sigma_{2b} = 40 \qquad \text{.......(4)}$$

solving four equations 1,2,3, and 4,

$\Delta\sigma_{1t} = 7.47 N/mm^2$

$\Delta\sigma_{1b} = -4.23 N/mm^2$

$\Delta\sigma_{2t} = -1.72 N/mm^2$

$\Delta\sigma_{2b} = -1.52 N/mm^2$

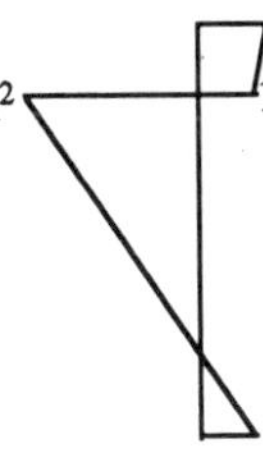

Fig 11.13

11.10 Additional Design Consideration

Normally composite construction forms part of a structural system like a bridge deck or floor system in large covered area with heavy load on top. There may be main beam precast and prestressed, and subsidary beams or diaphragm as cast-in-situ units alongwith slab. The shear transfer at the junctions of main beams and diaphragms has to be properly distinguished. If there are end diaphragms in a bridge deck with a gap slab, the analysis for design moments on the gap slab portion and the provision reinforcements on the slab have to be carefully done. These aspects are discussed by Leonhardt [8] as shear transfer in indirect supports (mainbeam and diaphragm meeting) and as the force flow at the continuous system for live loads on Bridge decks.

11.10.1 Shear transfer in indirect supports

When two members meet at the same level, shear has to be transferred from one member (called secondary member) to another member(called primary member) through interface shear or through reinforcements. In case of monolithically cast system (both primary and secondary member) interface shear does not play role and only shear reinforcement have to be provided in both the members to see that the shear is transferred from the secondary member to the primary member. Normally in composite construction the diaphragms/cross girders are secondary members and they meet the main girder at the junction. The main girders are acting as primary members. In case of end cross girders which are seated bearings the main girders are transferring the load to the end cross girders, then the end cross girder, becomes the primary members and the main girder becomes the secondary member. The force flow at the junctions of a primary member meet the secondary members are given in Fig 11.14.

Primary member, taking over the force from the secondary member, must have shear stirrups to take care of the full load of the coming from secondary member. These stirrups must be located within the zone equal to half the depth of the primary member on either side of the joint. Of course, the secondary member will have its own shear reinforcements at the junctions, which will cater for shear forces at that point.

It may be seen and noted from the Fig 11.14 that the forces are transferred from the secondary member to primary member by the inclined

Compression strut. Force is transferred at the bottom of the primary member and the bottom portion of the primary member will normally be in tension and may be cracked. Hence it is necessary that the primary member takes over the load from the secondary member only by stirrups to its compression zone.

The amount of reinforcement may also vary marginally if the depth of members meeting are different as could be seen from Fig11.14.

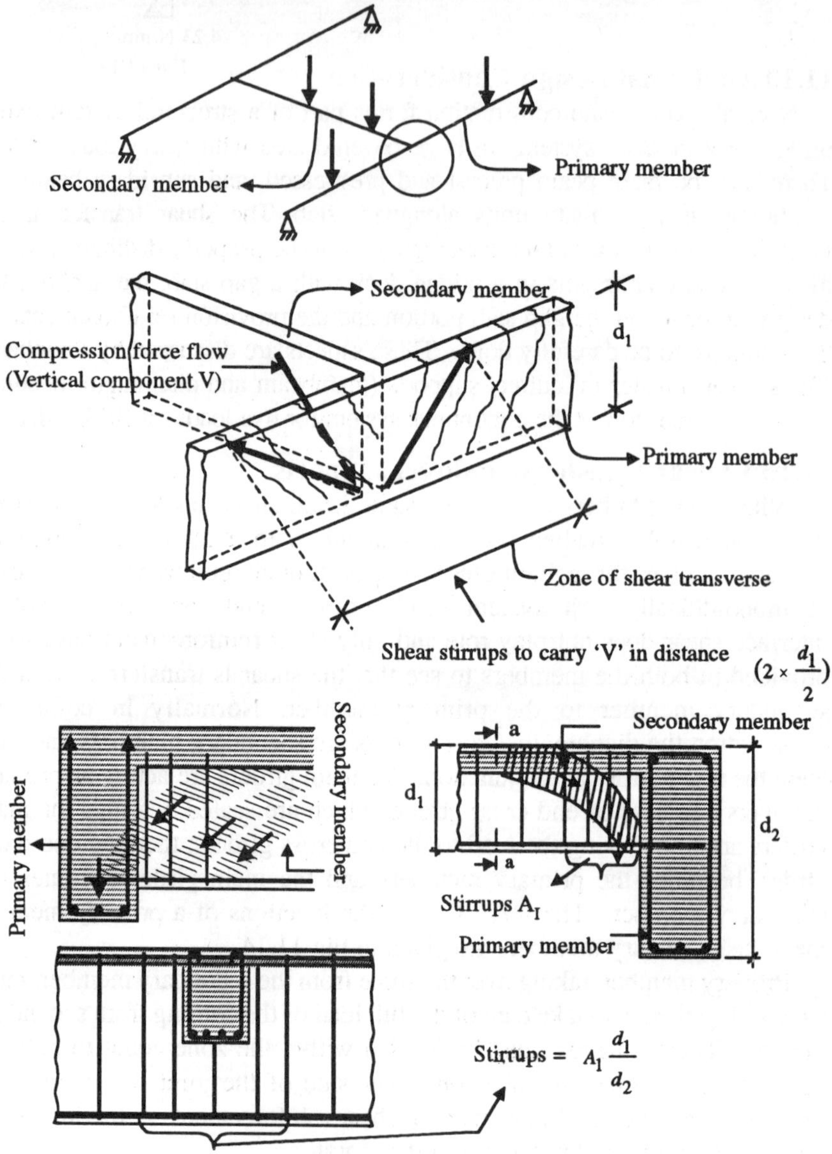

Fig 11.14 Shear Transer in Indirect Supports

11.10.2 Detailing at the continuous support zone for live loads

In the fast track approach for construction, prestressed concrete beams are cast at the casting yard, and brought to the site and placed over individual spans. Then the cast-in-situ slabs along with the diaphragms are cast to achieve in the monolithicity for the deck. This concrete may also be poured over the support so that the structure becomes continuous for loads which come on the structure after the monolithicity is reached. Hence the structure is considered as simply supported elements for self weight of the elements concerned and becomes a continuous system for the superimposed dead load and live load.

The spacing between the individual beams resting on the common support pier may be filled up with the diaphragms over which the cast-in-situ slab is also cast. This would mean that the end anchorages of the precast prestressed individual beams will be completely embedded and facility of inspection of the anchorages are not available. Alternately the individual end diaphragms for individual spans may be cast a little inwards from the anchored zone of individual beams. In this case there will be a gap slab which will be filling up the distance between the end diaphragms of two beams sitting over the pier. This gapslab will have tension at the top and may crack because of its smaller thickness. Hence the gap slab will be prestressed by a cap cable or a straight cable as shown in Fig.11.15 The analysis(local) to evaluate the moments occuring on the gapslab has to be done seperately by one of the sophisticated methods. This portion of the deck will act as a frame with diaphragms on eitherside and gapslab as a top boom. Hence the moment on the gapslab will be much reduced than what it would be when otherwise calculated as a portion of the continuous deck with a smaller span. Based on the actual stress resultants (moments acting on the slab), the cap cables or small straight cables have to be provided.

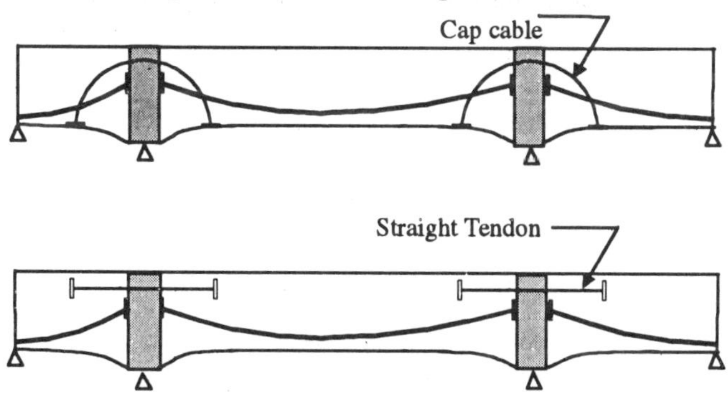

Fig 11.15 Cable Lay-outs of Continuous Beams

Detailing of this portion of the structure inclusive of the provision of cables has to be carefully done to avoid cracking on the slabs at the joints at the junctions of the two spans due to live load. Such detailing of cables, play a major role in profiling of cables in prestressed concrete continuos structures. Lin[9], Visvesvaraya[10] have studied these problems and reported in literature. Detailing of reinforcement at the junction of diaphragm and the main beam have to be done for shear transfer at the junction. Further if the end diaphragms are not supported on bearings at the points where the mainbeams join the diaphragm, and there is no further beam joining the diaphragm on the otherside, the diaphragms has to be checked for lateral bending caused by the horizontal component of inclined compression chord of the longitudinal girder.

References

1. R.P. JOHNSON, ***Composite Structure of Steel and Concrete (Volume 1),*** Blackwell Scientific Publication (Second Edition), U.K., 1994.
2. IS 456-2000 "Code of Practice for Plain and Reinforce Concrete", Indian Standard Institution, New Delhi.
3. IS 1343 - (1980). Indian Standard ***"Code of Practice for Prestressed Concrete",*** Bureau of Indian Standards, New Delhi.
4. BS 8110 - (1980). ***"British Standard for Structural use of Concrete"*** British Standards Institution, London WIA 2BS.
5. ACI 318 - (1998). ***"Building Code requirements for Structural Concrete",*** American Concrete Institute, Forminton Halls, HI 48333, USA.
6. IRC 22-1986, ***"Standard Specifications and Code of Practice for Roads Bridges"*** SectionVI-Composite Construction, New Delhi.
7. TROST. H (1967) ***"Implications of the superposition principle in creep and relaxation problems for concrete and prestressed concrete"*** Betons Stahlbetonbau 62(10) PP 230-238 & 62 (11) PP 261-169, (in German).
8. F. LEONHARDT ünd E.MÖNNING ***Vorlesungen über Massivbau Dritter Teil Grundlagen zum Bewehren im Stahlbetonbau*** Springer-Verlag Berlin. Heidelberg. New York 1974.
9. LIN T.Y., ***Prestressed Concrete Structures,*** John Wiley and Sons, Inc. 1960. Pp 288-292
10. VISVESVARAYA, H.C and N. RAGHAVENDRA, ***"Continuity in prestressed concrete construction,"*** Seminar on problems of prestressing, Madras, 1970, pp I-91-94.

12 Statically Indeterminate Structures

12.1 General

Any structure that is statically determinate could be made more efficient with respect to its structural functioning, concerning strength and serviceability by introducing certain restraints on deformations at the boundaries or anywhere in between. These introduced restraints make the structure statically indeterminate. A simplest form of a statically indeterminate structure is a continuous beam. By making the structure indeterminate or continuous, specific advantages are gained as detailed.

- The bending moments are more evenly distributed through the entire structure than to be concentrated at specific points.
- Accordingly the size of the members could be reduced which make the structure lighter.
- The ultimate load carrying capacity of the structure is based on the limit state of collapse which is more dependent on the number of hinges to be formed before failure. This is a function of the statical indeterminacy and also the phenomenon of redistribution of moments. Hence, the capacity of the structure at failure is much larger as compared to that of a determinate structure with the same cross sectional dimensions.
- The stability of the structure is improved.
- Continuity of members at joints or over supports reduces the rotation at these places and hence reduces the deflection.
- Continuity over joints provides good facility for erection of structures by segmental methods of construction using precast or cast in situ units and prestressing.
- Number of external anchorages and prestressing operations are less in continuous structures leading to reduction in cost and time of construction.

But there are also countable disadvantages in using an indeterminate prestressed concrete structure. These disadvantages are

- Concurrence of maximum moment and maximum shear at joints which reduces the moment carrying capacity at those sections calling for additional untentioned reinforcement.
- Non structural deformation (such as sinking of supports, in adequacy in structural dimension of connected units etc.,) may cause additional stresses in the system.
- Losses in prestressing force, caused by friction, wobbling etc., due to cable profiling (warranted because of continuity) are higher.
- There may be constructional difficulties while trying to achieve continuity.

Hence it is prudent for an engineer to look at the overall advantages and disadvantages before choosing determinate or indeterminate structure for a particular purpose.

12.2 Analysis of Prestressed Indeterminate Structures

Any structure, where the design forces like moments, shear etc., at any cross section, can not be determined explicitely using static equations of equilibrium, is defined as an indeterminate structure. If the bondary (support) reactions alone could not be determined using equations of equilibrium, they are said to be externally indeterminate; A propped cantilever, a continuous beam, a single bay portal etc., fall under this category. If the design forces could not be evaluated even after the boundary reactions are known, those structures are called internally indeterminate structures. There can be a combination of both, as in the case of continuous doubly curved shells etc.,. The scope of this chapter is limited to prestressed concrete beams and frames. Of course special structures like pipes, tanks are dealtwith separately in a later chapter. First, the aspects dealing with effect of prestressing on the analysis of continuous beams are discussed. This is followed by analysis of frames with prestressing effects.

12.3 Continuous Beams

A prestressed concrete continuous beam can be considered as a **plain concrete** continuous beam with prestressing as external force. In the elastic analysis, the moments due to external loads in an ordinary indeterminate beam are evaluated as the sum of two quantities namely the moment generated by external loads on corresponding determinate beam and the moment generated due to indeterminacies. In similar way the moment due to prestressing in an indeterminate beam also can be calculated considering the prestressing force on a corresponding determinate beam as a load and then superposing the effect of indeterminacy on the earlier calculated moment. Accordingly any one of the methods for solving indeterminate structures with external loadings such as consistent deformation method, slope deflection method (flexibility and stiffness approaches) or iterative procedures like moment distribution method can be used for analysing an indeterminate prestressed concrete beam.

In a simply supported (determinate) beam, the prestressing force 'P' acting at an eccentricity 'e' causes a moment P.e (M_p^o) and this moment causes internal stresses and the centre of compression and centre of tension (prestressing force) coincides. The effect of self weight is not considered. This system is a self equilibrating system and does not cause any external reactions.

Considering a two span continuous beam as shown in Fig.12.1 the prestressing force introduces a determinate moment at every cross section (P.e = (M_p^o)) which introduces a deformation. The point B at centre lifts up due to this structural action. But compatibility condition demands that the soffit of beam should be in contact with the support at point B. That means that a negative reaction (pulling down) force is warranted at that point B (R_B). This reaction force R_B cannot be evaluated using static equilibrium conditions and hence called the indeterminacy. This indeterminate reaction introduces moments which is called indeterminate moment at any section (M_p^1)

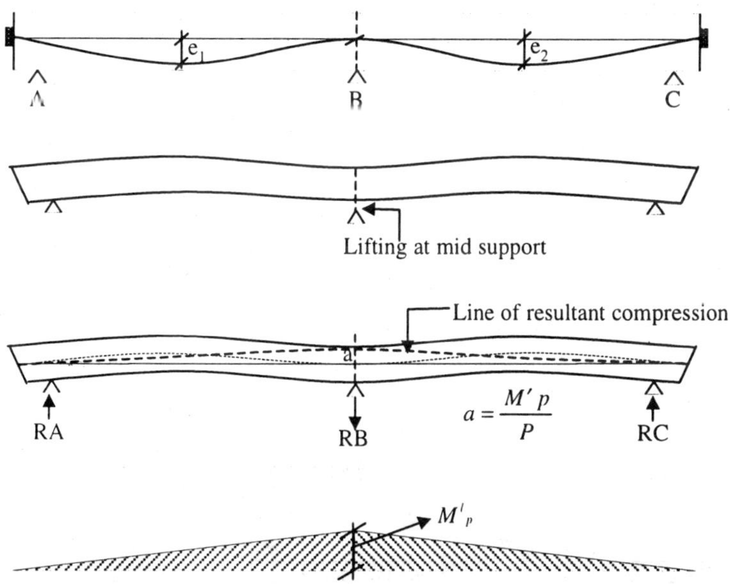

Fig 12.1 Effect of Prestress in Continuous Beams

This final moment due to prestressing M_p is evaluated as, $M_p^0 + M_p^l$ where M_p^0 = Moment on the determinate structure due to prestressing, and

M_p^l = Moment due to indeterminacy, This moment M_p^l at any section can also be equated to an equivalent eccentricity 'a' for the prestressing force 'P' and $a = \frac{M_p^l}{P}$. Now the compression line is shifted by a distance 'a' from line of action of tensile force. This is called centre of pressure line. The resultant final moment at any section (M_p) can be evaluated as $[P_o e + P_o a] = M_p^0 + M_p^l$. Anyhow, in the above analysis, it is assumed that the eccentricities are small when compared to the length, of the beam, friction loss along the beam is negligible, and the same tendon is continued for full length.

While there are different methods of analysis of indeterminate structure, as mentioned earlier, there is an additional but easier method of approach for arriving at the final moment due to prestressing alone. It is called equivalent load method.

12.2.1 Equivalent Load Method

First the beam is considered with cable force alone without the support reactions. It is a determinate system since the prestressing in the beam is a self equilibrating force. The moment at any cross section induced by the cable force is balanced by equivalent stresses developed in the section. The moment due to prestressing force (P.e) is plotted along the span. From this moment diagram, the shear force diagram is plotted along the span as it is only the differential of the bending moment at any point. From the shear force diagram the corresponding loading diagram is also plotted on the span,

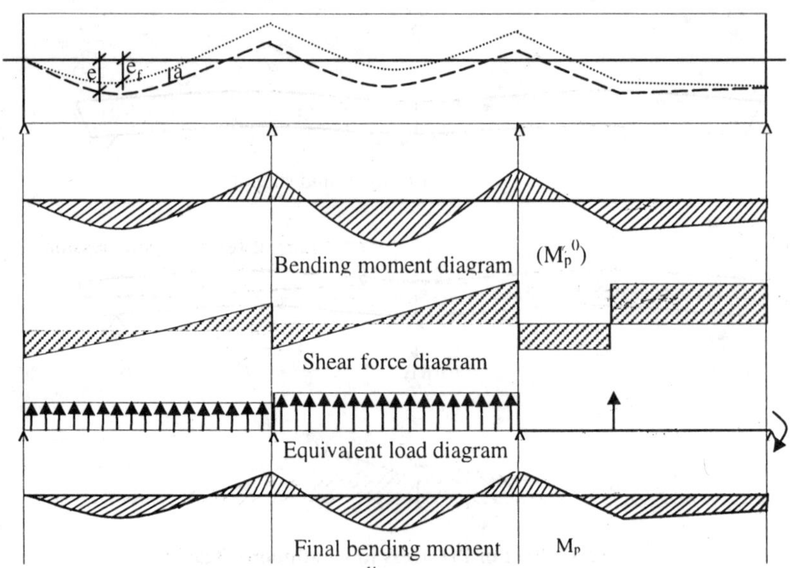

Fig 12.2 Derivation of Equivalent Loads

which again is formed by the differential of shear force. This loading diagram can now statically replace the prestressing cable force on the beam and **it is called the equivalent load** [1].

Now the continuous beam can be analysed with this equivalent load using any of the established methods of analysis and the final moment can be evaluated. This final moment due to prestressing (M_p) can be converted to an equivalent eccentricity (e_f) for the given original prestressing force (P_0) on the cable. Then the moment caused by the indeterminacy can be expressed as $M_p - M_p^{o} = P\,[e_f - e_o] = P.a$. The distance 'a' is called the shift of the centre of tension line to get the centre of pressure line. This shift is upward if the eccentricity is downward. It is equal to $a = \dfrac{M_p - M_p^{o}}{P}$

The equivalent load which represents the prestressing force can be arrived at by different methods, one of them being the above mentioned method. This procedure is schematically explained in Fig.12.2 .

Use of Equivalent method for solving the effect of prestressing in a continuous beam is illustrated by an example.

Example 12.1

A continuous prestressed concrete beam with two spans with bonded tendons has got a running cable as shown in figure 12.3. The cable consists of two linear profiles in the first span AB with a kink at a point 9m from support A and it has parabolic profile in second span BC. Analyse the beam for the effect of prestressing, using equivalent load method. Effect of dead load need not be considered. The prestressing force is 1200kN at the anchor points and the loss due to friction can be ignored.

Solution

Moment caused by prestressing force due to its eccentricity in the cross section is called primary moment and idenfied as M_p^{0} The primary moments at various salient sections are calculated first. Primary moment at sections

A = – 1200 × 60 = – 72000 kN mm

D = – 1200 × 240 = – 288000 kN mm

B = – 1200 × 120 = – 144000 kN mm

C = – 1200 × 270 = – 324000 kN mm

Bending moment diagram is drawn with the above primary moments at the relevant sections and the corresponding shear forces and loading diagram are drawn using principles of simple bending.

Slope of Bending Moment Diagram at = $A = \dfrac{-72000 + 288000}{9000} = 24$

Shear Force at A = 24 kN

Slope of Bending Moment Diagram at = $D = \dfrac{288000 + 14400}{6000} = 72$

Shear Force at D = 72 kN

Slope of Bending Moment Diagram at B towards D = 72

Slope of Bending Moment Diagram at

B towards E = $4 \times \dfrac{270 + 60}{15000} + \dfrac{120}{15000} \times 1200 = 116$

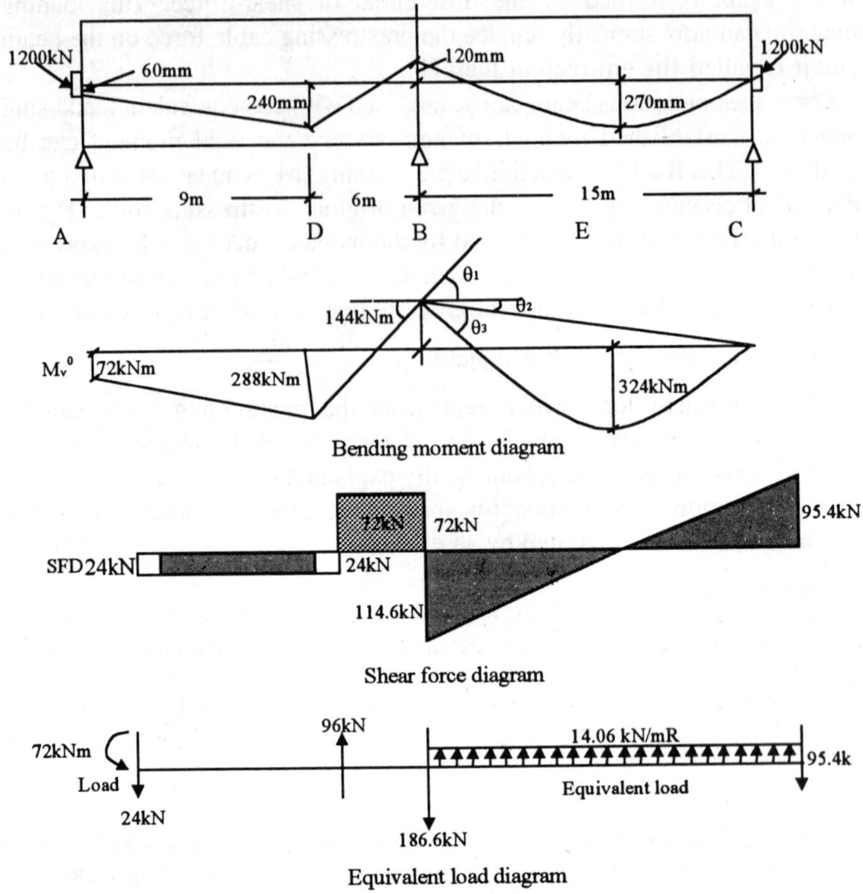

Fig 12.3. Derivation of Equivalent Loads for Example 12.1

Shear Force to the right B towards E = 116 kN

Let intensity of load on the span BC be ω/mR

The maximum bending moment at the centre of span BC

$$\left(324 + \frac{144}{2}\right) = 396kNm$$

$$= \frac{15 \times 15}{8} . w$$

Then $w = \dfrac{396 \times 8}{15 \times 15} = 14\ kN/m$

Reaction at support points B and C

$R_{B_1} = 72kN$

$R_{B_2} = 105 + 9.6 = 114.6kN$

$R_{C_2} = 105 - 9.6 = 95.4kN$

Loading diagram is shown in Fig. 12.3 itself

$R_A = 24\ kN$

$M_A = 72kNm$

R_B = 186.6kN = (114.6 + 72)

R_c = 95.4 kN

The final bending moment on the continuous beam due to prestressing is evaluated as the bending moment caused by equivalent load as calculated above, using moment distribution method for Analysis.

Fixed End Moments on spans AB, and BC

$$\text{Fixed End Moment at } A = \frac{96 \times 6^2 \times 9}{15^2} = 138.20\,kNm$$

$$\text{Fixed End Moment at B towards } A = \frac{96 \times 9^2 \times 6}{15^2} = 202.2\,kNm$$

$$\text{Fixed End Moment at B towards } C = \frac{14 \times 15^2}{12} = 262.5\,kNm$$

Fixed End Moment at C = 262.0*kNm*

Distribution Table

Joints	A	B		C
Distribution factor		0.5	0.5	
Fixed End Moment	+138.2	–202.2	+262.0	–262.0
	–210.2 →	–105.6	+131.0 ←	+262.0
Balancing and Redistribution of moments		–307.8 –42.6	+393.0 –42.6	0
Final Moments	–72.0	–350.4	+350.4	0

The Final Bending Moment Diagram is shown in Fig12.4.

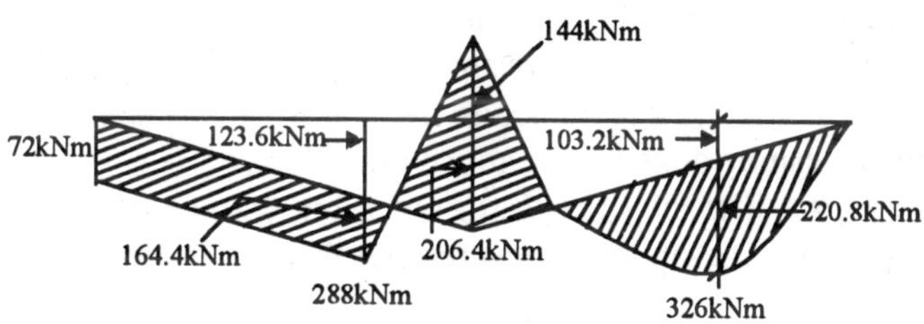

Fig 12.4. Final Bending Moment

The eccentricity of the line of resultant compression (e_f) can be computed by dividing the final bending moment values by prestressing force P^0. The final bending moment is algebric sum of primary bending moment and the indeterminate bending moment. The primary bending moment is represented by the C.G,.S. profile. Indeterminate bending moment could be represented by a distance 'a' called shift which is got by dividing the indeterminate bending moment by the prestressing force 'P'.

The indeterminate bending moment at B is 350.4 – 144.0 = 206.4 kNm.

The corresponding shift is $206.4 \times 10^3/120 = 172mm$

The indeterminate moment at supports A and e are zero and hence shift is zero. In between at all sections the shift can be calculated by linear interpolation. Once the shift is known the resultant eccentricity of compression line can be calculated at $[e_0\text{-}a]$ as illustrated below.

Position of resultant of compression at Section D

Shift of the centre of steel line a $a = (172)\frac{9}{15} = 103$

Resultant eccentricity e_f at D = 240 – 103 =137 mm.

Position of centre of compression at Section E

Shift of the centre of steel line a = (172) ½ = 86

Resultant eccentricity e_f at E = 270 – 86 = 184 mm.

These eccentricities (e_f) give the line of thrust (pressure) position for the given problem. The line of resultant thrust is other wise called C.G.C line and it is shown along with cable profile line in Fig.12.5.

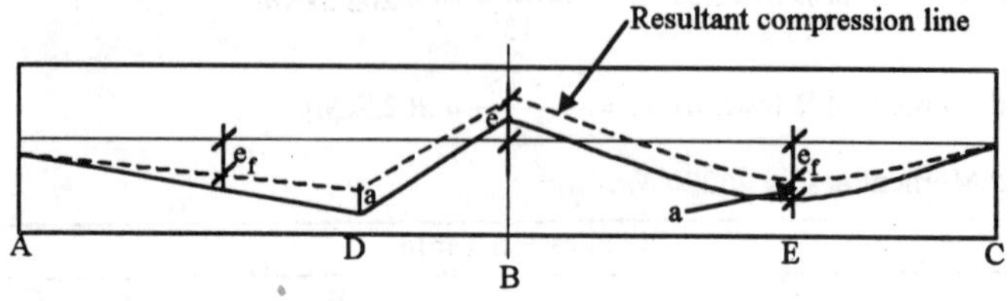

Location	A	D	B	E	C
e_f mm	0	137	292	184	0
a mm	0	103	172	86	0
e mm	0	240	120	270	0

Fig 12.5. Positioning of C.G.C (Line of Resultant Compression)

12.2.2 Load Balancing Method

It is possible to arrive at the equivalent load directly without drawing the bending moment diagram due to prestress alone. Such an approach is called load balancing method [2], If the loads are arrived at by this approach, and further analysis can be done by the above procedure.

Considering an elemental length (dx) of beam with an inclined cable as shown in Fig.12.6. The applied prestressing force is P and the change in

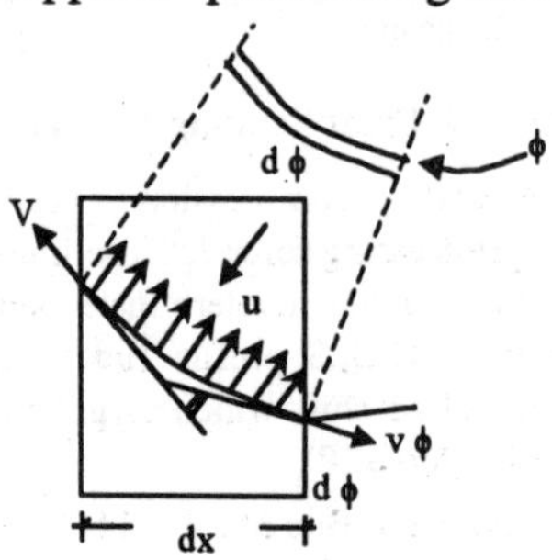

Fig 12.6

angle of the cable profile in a distance Δx be = dϕ. The slope of the cable at one end = ϕ The cable suspends an angle dϕ at centre of curvature. Let the equivalent load be 'u'.

By equilibrium of forces

$$u.ds = 2P\sin\frac{d\varphi}{2}$$

For small angles $\sin\frac{d\varphi}{2} = \frac{d\varphi}{2}$

Hence $u.ds = P.d\varphi$

$$u = P.\frac{d\varphi}{ds}$$

$$\frac{d\phi}{ds} = \frac{1}{R} = \frac{y''}{\left(1+y'^2\right)^{3/2}}$$

For small values of y

$$\frac{1}{R} = y''$$

$$u = P.y''$$

Considering ds = dx and neglecting higher order differential

$$\frac{d\varphi}{ds} = \frac{d\phi}{dx} = \frac{1}{R} = y''$$

For parabolic profile $y = \frac{4l}{l^2}\left(1x - x^2\right)$

$y'' = \frac{8f}{l^2}$. accordingly $u = P\left[\frac{8f}{l^2}\right]$

If y is considered positive downwards the equivalent load will be upwards. Hence it is represented as $u = P^{(-)}\left(\frac{8f}{l^2}\right)$ where P is the prestressing force creating compression in concrete.

If there is a kink in the straight cable profile there will be concentrated force acting at the place where there is a change in the profile. The value of the concentrated force W = $P^{(-)}(\varphi)$ where ϕ is the change of angle of the profile. The end reaction forces should not be neglected in the equivalent loading.

Example 12.2

Evaluate the equivalent loading diagram to replace the prestressing profile for the continuous beam given in example 12.1 using load balancing principle.

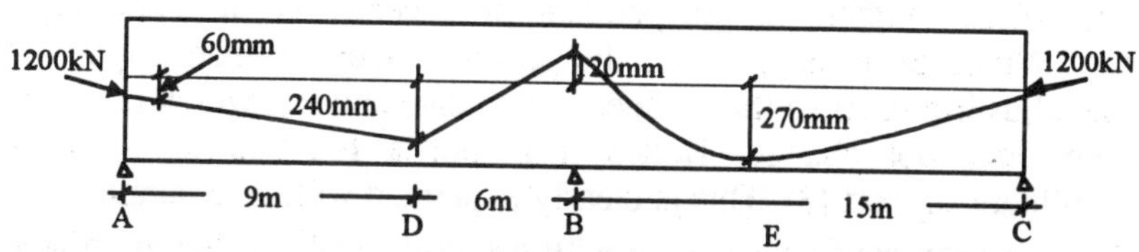

Fig 12.7.

Solution

Load on Span AB is evaluated based on Load balancing Principle. Profile is linear and it has a kink at D. This loads to a concentrated load at D = W, W = $P^{(-)}(\phi)$ where ϕ is change in angle of the profile at D.

$$Tan\phi = \left[\frac{24+12}{600} + \frac{24-6}{900}\right] = \frac{1440}{1800}$$

For small angles, $\tan\phi = \phi$ in radians

Hence $W = \frac{1440}{1800} \times 120 = 96kN$

Similarly the load on BC is also evaluated

Cable Profile is parabolic for zone BC. Hence it would be equivalent to u.d.l of intensity 'u'/m.

Loading $u = \frac{8f}{l^2}(P)^{(-)} = \frac{8 \times 33}{150 \times 150} \times 1200 = 14.06kN/m$

Force at support A $= P.\phi = P(\tan\varphi) = 1200\left[\frac{24-6}{900}\right] = 24kN$

Force at support B $= 1200\left[\frac{36}{600} + \frac{4 \times 33}{1500} + \frac{12}{1500}\right] = 187kN$

Similarly force at support C $= 120\left[\frac{4 \times 33}{1500} - \frac{12}{1500}\right]$

$$= \frac{120}{1500} \times 1200 = 96kN$$

The loading diagram is shown in Fig12.8

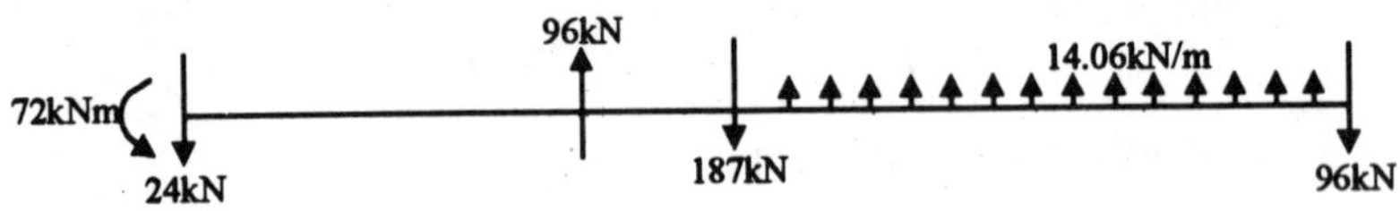

Fig 12.8.

12.2.3 Independence of Final moment M_P on mid support eccentricies.

It is well known in determinate structures the bending moment caused by prestressing can be evaluated as P.e where P is the prestressing force and 'e' is the eccentricity at the sections. But in indeterminate structure, the final moment at any cross section due to prestressing is not P.e but Pe_f where e_f is (e-a) and 'a' is the shift of the centre of tension line due to the effect of prestressing. Though e_f is defined as (e-a), it is not separately dependent on 'e' or 'a'. For a given intrinsic shape of the profile and an eccentricity at one end, the final moment is independent of eccentricity at indeterminate support. Whatever 'e' is given for the profile at indeterminate support, 'a' will be adjusting itself so that e_f does not change and $P.e_f$ which is the final moment does not change. Hence it is not a function of 'e' at the indeterminate support [3]. This is clearly explained with an example.

The two span continuous beam with parabolic cable profile as shown in Fig.12.9 could be considered as a continuous beam with equivalent load calculated based on load balancing principle. This cable profile can be replaced by equivalent load of u.d.l. of intensity u and concentrated forces at places where it changes its curvature abruptly (ie. at mid support and at ends).

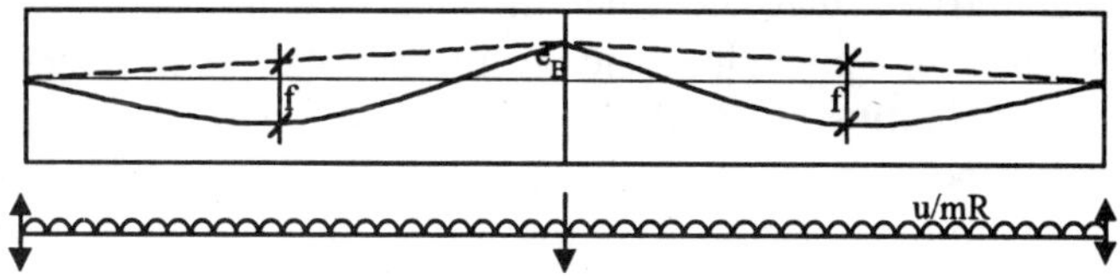

Fig 12.9.

The final bending moment at support section B can be evaluated using support moment distribution method.

Distribution Table				
Sections	A	B		C
Distribution Factor		0.5	0.5	
Fixed End Moment	$+\frac{ul^2}{12}$	$-\frac{ul^2}{12}$	$+\frac{ul^2}{12}$	$-\frac{ul^2}{12}$
Balancing Moment	$-\frac{ul^2}{12}$ C.O →	$-\frac{ul^2}{24}$	$+\frac{ul^2}{24}$	← C.O $+\frac{ul^2}{12}$
	0	$-\frac{3ul^2}{24}$	$+\frac{3ul^2}{24}$	0
Final Moment		$-\frac{ul^2}{8}$	$+\frac{ul^2}{8}$	

Final moment at B towards A is = $-\frac{ul^2}{8}$ $(bottomtension)$

Primary moment $M_P{}^0 = P^{(-)}.e_B{}^{(-)}$ bottomtension

Substituting for u as P.y" 'u' becomes equal to $-P(\frac{8f}{l^2})(upward)$

The final moment at support

$$M_p atB = P^{(-)}\frac{8f}{l^2}.\frac{l^2}{8} = Pf\text{(bottomtension)}$$

It may also be noted here ***that the final moment at the intermediate support is independent of eccentricities of the cable at this support and it is dependent on the dip of the parabola 'f'.***

$$M_p{}^0 = P.e_B(toptension)$$

$$M_p' = Pf - P.e_B$$

If M_p' is to be equal to zero 'f' should be equal to e_B

If the eccentricity at the mid support is the same as the dip of the cable (f) there will not be any effect due to indeterminacy of the structure on the effect due to prestressing.

It has been proved that the final moment due to prestressing in a continuous beam with a given intrinsic profile for the cable is independently mid span eccentricities, considering a specific case of spans with symmetrical parabolic profiled cables using load balancing method and moment distribution method for arriving at the final moments. The same point namely ***the final moment at a indetermediate support is independent of the eccentricity of the cable at that support is proved*** for a more general case using the principle of consistent deformation. By this approach the use of consistent deformation approach for solving indeterminate structure for the effect of prestressing is also explained.

12.2.4 Method of Consistent Deformation

Consider a two span continuous beam as shown in Fig.12.10 with cable profiles parabolic in nature with eccentricity a,b and c at supports A,B, and C. The dip of the cable in span AB is r_1 and that in span BC is r_2. The span length of AB=l_1 and that of BC = l_2. As per consistant deformation method, first the indeterminacy at B (moment resisting capacity at support B) is removed. This makes the structure determinate consisting of two spans AB and BC independent with a hinge at B. This is considered as a primary structure and the bending moment (M_P^0) on the primary structure is drawn. The indeterminacy at B namely the moment resisting capacity is introduced as unit moment first and the bending moment due to this unit moment on the span AB and BC are drawn. The exact value of indeterminate moment B is assumed to be (X_1)

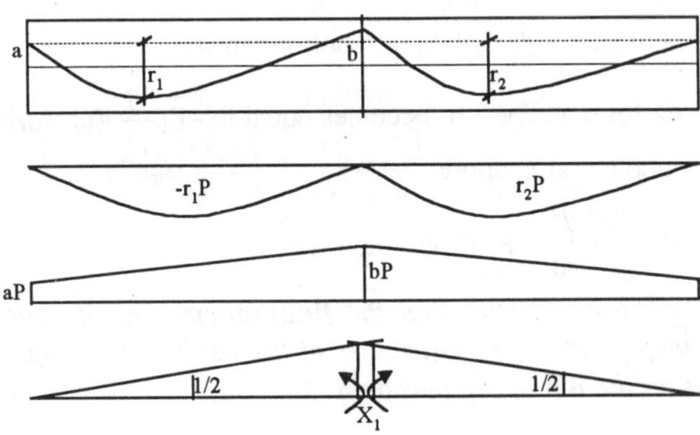

Fig 12.10.

The rotation at B due to prestressing alone is = $\int M_P^{\,0}.m_1.\dfrac{dx}{EI}$

The rotation at junction B due to unit moment acting in place of redundancy. $X_1 = \int m_1 m_2 \dfrac{dx}{EI}$ Compatibility conditions (namely the relative rotation at junction B should be zero) demands that

$$\int M_p^{\,0}.m_1.\frac{dx}{EI} + X_1\int m_1.m_1.\frac{dx}{EI} = 0$$

Integration of these bending moments on a single span (l) could be easily arrived at using Simpson's rule. If two figures as shown in Fig. 12.11 are to be integrated the integral value is given by

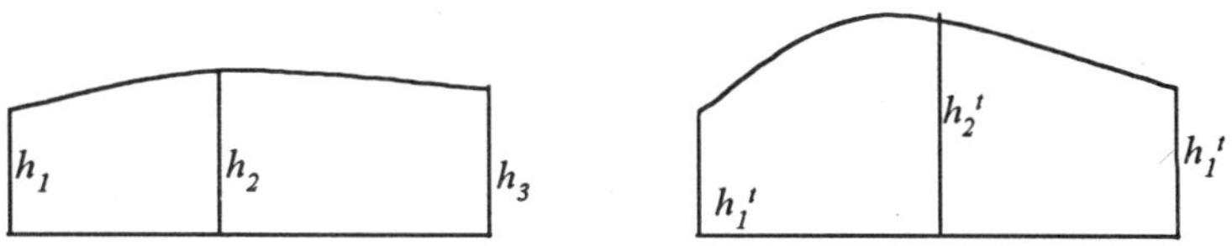

Fig. 12.11

$$\frac{l}{6EI}[h_1.h_1'+4h_2.h_2'+h_3.h_3']$$

Using the moment diagrams drawn on the primary structure and the Simpon's rule for integration.

$$\int M_p^{\,0}.m_1.\frac{dx}{EI} = \frac{l_1}{6EI}(4.P.r_1.\frac{1}{2}) + \frac{l_2}{6EI}(4.P.r_2.\frac{1}{2})$$

$$+\frac{l_1}{6EI}[aP.\frac{1}{2}.4 + aP.1]+$$

$$+\frac{l_2}{6EI}[cP.\frac{1}{2}.4 + cP.1]+$$

$$+\frac{1}{6EI}[bP - aP]2 + \frac{l_2}{6EI}[bP - cP]2$$

$$= -\frac{1}{6EI}\left[2r_1Pl_1 + 2r_2Pl_2\right]$$

$$+\frac{1}{6EI}\left[3aPl_1 + 3cPl_2\right]$$

$$+\frac{1}{6EI}\left[2bPl_1 - 2aPl_1\right]$$

$$+\frac{1}{6EI}\left[2bPl_2 - 2cPl_2\right]$$

$$= \frac{P}{6EI}\left[l_1(2r_1 - a) + l_2(2r_2 - c) - 2b(l_1 + l_2)\right]$$

$$\int \frac{m_1.m_1dx}{EI} = \frac{1}{3}(l_1 + l_2)\frac{1}{EI}$$

$$X_1 = \frac{{}^1\!/_6(2r_1 - a)l_1 + (2r_2 - c)l_2 - 2b(l_1 + l_2)}{\frac{1}{3}(l_1 + l_2)}.P$$

$$= \frac{P[2r_1 - a]l_1 + [2r_2 - c]l_2}{2(l_1 + l_2)} - bP$$

$$M = M_0 + M_1$$

$$= bP + P\frac{(2f_1 - a)l_1 + l_2(2f_2 - c)}{2(l_1 + l_2)} - bP$$

$$= \left[\frac{1_1(2r_1 - a) + 1_2(2r_2 - c)}{2(1_1 + 1_2)}\right]P$$

Now it seen that the final moment at B is independent eccentricity at B namely 'b'.

This is valid for any profile of tendon. The eccentricities at mid supportcan be chosen to get a good lever arm for resisting ultimate moment without affecting service load stage. Large rise '*f*' also give greater moment due to prestressing in the span as well as supports. It is economical to choose the rise 'r' of the parabola and eccentricity 'b' at the interior support as large as possible.

The product integrals of any two definite shapes of bending moment on a given span, can be evaluated using Simpson's rule. For the benefit of readers a few of them are summarised and presented in Table 12.1

It has been shown earlier that the final moments due to prestressing at the indeterminate supports are independent of the eccentricities at those supports. This means that the cable can be moved up or down at those supports with out change in the final bending moment. Any movement of eccentricity at mid support will move the eccentricities of the other points.

But the revised eccentricities can be arrived at by knowing the shift of eccentricities at various place using linear interpolation between the shift of eccentricities at the supports (proportionate shift with distances and it is only linear). By such movement of eccentricities, the intrinsic shape of the profile will not change and hence the equivalent loads will not change and also the effect of prestressing. Such a changed profile is called transformed profile. Since the eccentricities at all places are changed linearly with distance along span, it is called linearly transformed profile. This principle linear transformation of profile can also be explained from the bending moment diagram as could be seen from the solution of the example 12.1.

Table 12.1 Product Integral for Flexibility Influence Coefficients (Simpson's Rule)

$m_j \downarrow$ \ $m_i \rightarrow$	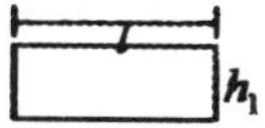		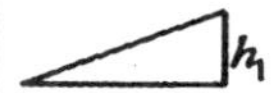			
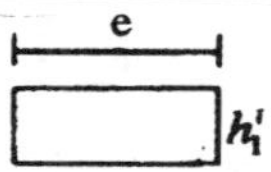	lh_1h_1'	$\frac{1}{2}lh_1h_1'$	$\frac{1}{2}lh_1h_1'$	$\frac{2}{3}lh_1h_1'$	$\frac{1}{2}lh_1h_1'$	$\frac{1}{2}l(h_1+h_2)h_1'$
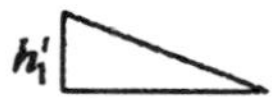	$\frac{1}{2}lh_1h_1'$	$\frac{1}{3}lh_1h_1'$	$\frac{1}{6}lh_1h_1'$	$\frac{1}{3}lh_1h_1'$	$\frac{1}{4}lh_1h_1'$	$\frac{1}{6}l(2h_1+h_2)h_1'$
	$\frac{1}{2}lh_1h_1'$	$\frac{1}{6}lh_1h_1'$	$\frac{1}{3}lh_1h_1'$	$\frac{1}{3}lh_1h_1'$	$\frac{1}{4}lh_1h_1'$	$\frac{1}{6}l(h_1+2h_2)h_1'$
	$\frac{2}{3}lh_1h_1'$	$\frac{1}{3}lh_1h_1'$	$\frac{1}{3}lh_1h_1'$	$\frac{8}{15}lh_1h_1'$	$\frac{5}{12}lh_1h_1'$	$\frac{1}{3}l(h_1+h_2)h_1'$
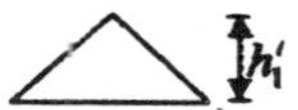	$\frac{1}{2}lh_1h_1'$	$\frac{1}{4}lh_1h_1'$	$\frac{1}{4}lh_1h_1'$	$\frac{5}{12}lh_1h_1'$	$\frac{1}{3}lh_1h_1'$	$\frac{1}{4}l(h_1+h_2)h_1'$
	$\frac{1}{2}lh_1(h_1'+h_2')$	$\frac{1}{6}lh_1(2h_1'+h_2')$	$\frac{1}{6}lh_1(h_1'+2h_2')$	$\frac{1}{3}lh_1(h_1'+h_2')$	$\frac{1}{4}lh_1(h_1'+h_2')$	$\frac{1}{6}l\{h_1(2h'_1+h_2)$ $+h_2(2h_2+h'_1)\}$

$\frac{l}{6EI}\left[h_1h_1' + h_2h_2'4 + h_3h_3'\right]$ *for continuous curves*

12.4 Linear Transformation and Concordancy of Cable Profiles

In the illustrative example 12.1 the resultant compression line (C.G.C.) has an eccentricity (e_f) = 292mm at support B, 137 mm at point C and 184 mm at point D. At support A and C the original position of the cable is maintained since these are no indeterminate moment. The upward movement (a^t) of the resultant compression line(C.G.C. line) at support B = 292 – 120 = 172 mm and at A it is zero. Correspondingly the upward movement of eccentricity at D works out to be $\frac{72 \times 9}{15} = 103$ and at E $\frac{172}{15} \times 7.5 = 86$ The final eccentricities at D and E are 240 – 103 = 137 and 270 – 86 = 184mm respectively.The eccentricity of centre of pressure(resultant compression line) e_f at point D and E have also been worked out from final M_P diagram as 137 mm and 184mm respectively. This shows that shifts (a') of 103 mm and 86 mm of the C.G.S. line upwards at D and E respectively are only linearly proportionate shift at D and E for a shift of 172mm at B.

Hence the position of the resultant compression line at various sections can also be located by evaluating the shift of cable profile line (C.G.S. line) due to a shift of C.G.S. line at B, without altering the position of C.G.S line at supports. Further, if the C.G.S. line (centre of tension) is shifted by 172 mm at B (which corresponds to resultant compression) without any change in position at A and C, the intrinsic profiles of the cable in both spans AB and BC have not changed and accordingly the equivalent loads, which could replace the prestressing, have not changed. This means the kink of point D will even now produce an upward force of 96kN as in earlier case, before shifting of C.G.S. line. Similarly the uniformly distributed load on span BC will remain as 14.05 kN per m/run. If the loading on the span AB and BC which replaces the prestressing are not changed, the final bending moment diagram on spans AB and BC, due to prestressing is also not affected. But by moving the support eccentricity at B from 120 to 292 mm, the cable profile gets lifted proportionately along the length of spans BA and BC. Such a transformation of the profile is called linear transformation of the cable profile. The final moment diagram due to prestressing (M_p) which comprises of M_P^0 and M_v^1 is unaltered by the linear transformation. If M_P^0 is altered because of change of eccentricity (C.G.S. line) of the cable, the indeterminate moment M_P^1 gets accordingly changed so as to keep M_P unaltered.

It is seen now that by linear transformation of cable profile (C.G. S. line) the final bending moment does not get changed. By the linear transformation of the profile it is possible to get the C.G.S. line (cable profile line) to coincide with C.G.C line. If the C.G.C. line and C.G.S. line are one end the same, it is known that the indeterminacy of the structure does not bring in indeterminate moment ($M^l_p = 0$). Such a profile where the indeterminate moment (M_P') is zero is called as concordant profile [2]. This may be noted that during linear transformation of cable profile the eccentricity at the end support and eccentricity at the cantilever end beam should not be changed. It may also be noted that the linear transformation may cause different forces at supports, different from what had been the case when the first profile of the able was adopted. Since these forces are directly transformed to supports, they do not cause any variation in the bending moment diagram M_p due to prestressing and hence in the C.G.C. line (e_f)

In conclusion it may stated

- When the position of C.G.S. line (cable profile) is moved over interior supports of a continuous beams without changing intrinsic shapes (curvature and bends equally) of the cable line are within the individual spans, the cable line said to be linearly transformed.
- The end eccentricity should not be altered.
- Each cable could individually be considered for linear transformation.
- By linear transformation of profile it is possible to achieve a particular profile for which the bending moment diagram due to prestressing is straight away given by determinate component $(M^0_p = P_e.e)$ **Such a profile is known as concordant profile.**

Hence concordant profile does not produce any secondary moment (indeterminate moment). Each cable in a **statically determinate** structure is a concordant profile because no indeterminate moment is involved and also no external reaction is produced, whereas, in a continuous beam a prestressing cable running over two or more spans introduces support reactions which causes indeterminate moment, otherwise known as secondary moment. Hence if by chance or by any method the prestressing force is made **not to cause any reaction** on supports such as profile is said to be concordant. A concordant profile does not produce any incompatibility at middle support by way of displacement.

Besides the fact that the concrodant cable line is easier for analysis, there is practically no gain using a concordant profile in the design. A non concordant profile may normally be better suited so that the indeterminate moment caused by prestressing may also act against the applied loads.

Any moment diagram for a continuous beam produced by any combination of external loads whether due to transverse loads or moments, plotted to any scale is one location of concordant profile. If this location falls out the beam, it can be linearly transformed to give more practical location without changing C.G.C. line.

Example 12.3

For a continuous prestressed concrete beam loaded as shown in Fig.12.12 obtain a concordant cable profile.

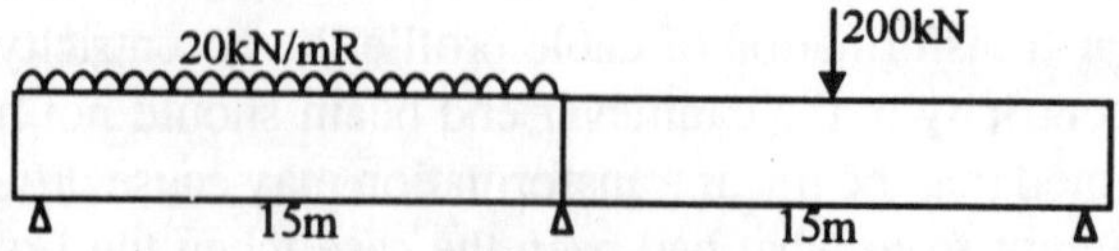

Fig 12.12.

Bending moment Diagram is first drawn for given loading. A cable profile is chosen to counter act the load by equivalent load. That will be a parabola in span AB and two straight with kink at D an BC.

If there is no eccentricity at any of the support, the cable force will cause a similar bending moment as shown in Fig 12.13b. An eccentricity is introduced so that, it will cause 562.5kNm bending moment of opposite nature at support B. Assuming 500kN of prestressing force the eccentricities at

A = 0

B = 1125mm

C = 937.5mm

D = 0

The profile and the shape of the Bending Moment Diagram with eccentricities as given below, will be a concordant profile for certain scale. If the facilities of eccentricities are not available, increasing the prestressing force to 1000kN. Smaller profile with eccentricities is obtained.

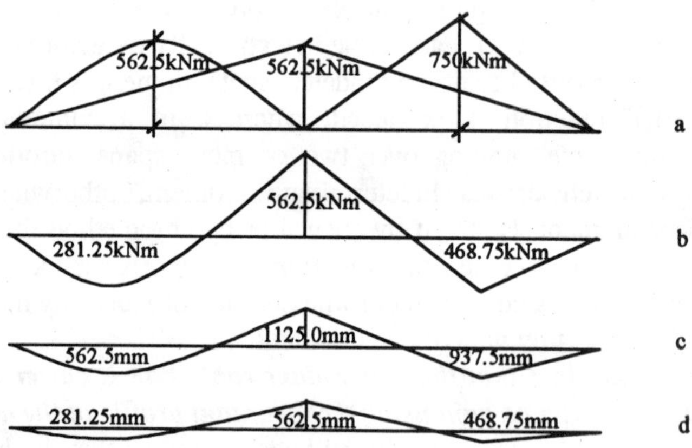

Fig 12.13.

12.3.1 Detailing at the continuity joint

It may be noted that the prestressing cables are given a profile in individual spans of a continuous beam and they are primarily anchored only at the ends of the continuous beam. To provide for continuity of the cable over the continuous supports, the cable profiles have to be smoothened over the supports and that leads to a uniformly distributed downward load at the support for over small region or additional cables are to be provided in the support region depending on the construction procedure.

For example, when precast construction units are used, sometimes continuity of the beam is brought out only for live load conditions and in such cases an additional cable is introduced at the continuity joint as a straight cable or as a cap cable [4]. Continuity is achieved by cable profiles shown in Fig 12.15

Such detailing of cables, play a major role in profiling of cables in prestressed concrete continuos structures. Lyn [5], Visvesvaraya [6] have studied these problems and reported in literature about the proper detailing of the structure at the joint.

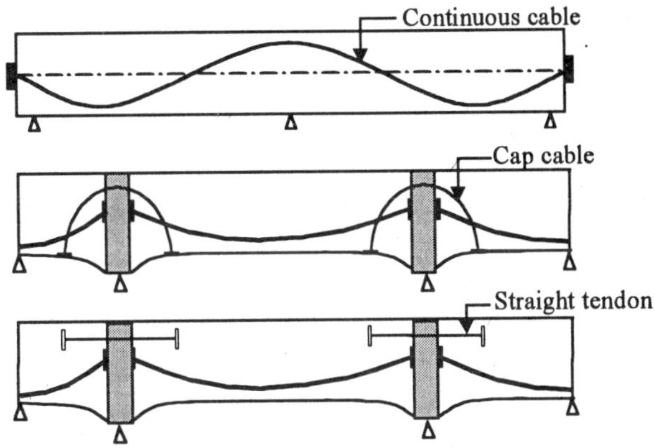

Fig 12.15 Cable Lay-outs of Continuous Beams

12.3.2 Limit State Design of Continuous Beams

In an indeterminate structure the limit state of collapse under flexure is not arrived at, based on the capacity of an individual section. Even after an individual section reaches its capacity, there can be a possible redistribution to other sections till some more hinges are formed and the structure becomes a determinate structure. After-wards when sufficient hinges are formed to create a determinate structure, any additional hinge formation leads to collapse of the structure. Generally the analysis is done on the basis of elastic behaviour. But it is not true after cracking of the section. Hence the moment to be resisted at limiting loads in continuous beams cannot be proportionately worked out from the moment calculated at the service load using elastic approach. But in practical cases these aspects are not given enough weightage and the sections are designed for the moments calculated using service load conditions multiplied by a load factor. This is an approximation.

Further to this approximation, in indeterminate structures the actual hinge formation and redistribution of moments in an indeterminate structure (till it becomes determinate structure) is not being carefully looked into. But a 15% redistribution is assumed in various national and international codes as per the experiences of engineers. Hence the indeterminate structure is said to have its limiting capacity if its critical section reaches its capacity, factored by 1.15 due to redistribution, It should be recognised that there are two stages of conservatism in this design, of which one state of conservatism is reduced by a assumption of 15% redistribution.

Baker (U.K.) [7], Prager (USA) [8] have researched on this area and developed a limit state analysis for steel structures where plasticity of material is well defined. But not much of accepted informations is available on the aspect of plastic hinges in reinforced concrete structures. The reduction of the moment of inertia after cracking, is not being considered in detail and an approximate emprical value for redistribution is assumed.

Prestressed Concrete Structures also behaves as reinforced concrete structures at the time of collapse, hence above assumptions mentioned are being adopted in prestressed concrete structures also and the capacity of a prestressing concrete continuous beam is decided accordingly.

12.5 Frames

Longspans of industrial structures are normally covered by portal frames. To make the structure slender, it is to desirable to prestress the different elements such as beams and columns. Since it is a frame, prestressing of anyone element affects the behaviour of all the elements and hence the structure has to be analysed fully for the effects of prestressing.

It has been clearly shown earlier that the effects of the prestressing can be considered similar to the effects of loads and any method of analysis such as consistant deformation method, slope deflection method, moment distribution method (flexibility and stiffness approaches) can be adopted for analysing such structures. Here for analysis of prestressed concrete portals and to study the effects of prestressing, moment distribution method is chosen.

For analysing the indeterminate structure such as a frame for the effect of prestressing using moment distribution method, the fixed end moment due to the prestressing in the individual elements such as columns, beams etc., are to be evaluated. Then they have to be redistributed to achieve moment equilibrium at the joints as per the philosophy of moment distribution developed by Hardy cross [9], [10], [13]. Evaluation of the fixed end moments at the ends of the individual elements, such as the beam or the column, caused by prestressed cables, itself is problem of an indeterminate structure. A typical case of a fixed ended prismatic beam prestressed with a parabolic cable with eccentricity 'o' at supports and 'e' at mid span is considered for arriving end moments caused by prestressing. The same is illustrated first. But, for other cases, the fixed end moments for prismatic elements with the different cable profiles and boundary conditions are given in Table 12.2. [3]. Fixed end moment on a beam of prismatic section with a parabolic cable profile as shown in Fig12.16 is evaluated as follows.

Final moment (Fixed end) $= M_P = M_P^{o} + M_P^{1}$

Indeterminate moment ($X_1 = X_2$) is solved by consistent deformation method. The structure is first assumed to be hinged at both ends and the indeterninate moment is introduced at hinged ends as unknown. Because of symmetry of load and structure these indeterminate moments at ends are one and the same. The moment diagram due to prestressing on the determinate structure is drawn as M_P^O diagram and the moment diagram

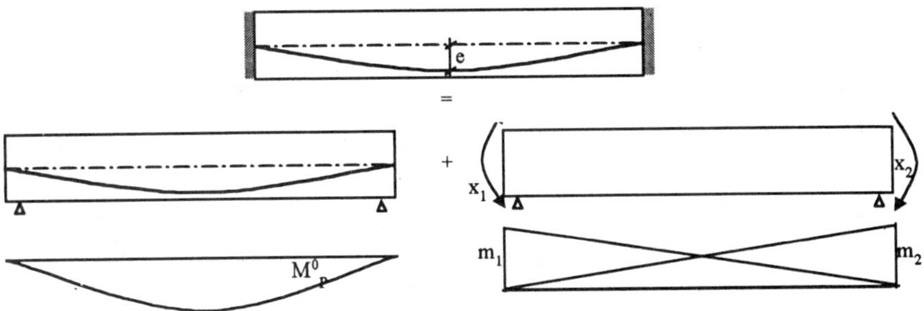

$$[M_p^0 + x_1 m_1 + x_1 m_2 = M_p]$$

Fig. 12.16.

due to indeterminate moment at ends are drawn as m_1 and m_2 diagrams as shown in Fig.12.16. Then the final rotation at the ends A or B is evaluated as rotation caused by the prestressing on determinate structure and the same caused by the moment's m_1 and m_2. This is expressed as

$$\int M_p^o m_1 \frac{dx}{EI} + x_1 \int m_1 m_1 \frac{dx}{EI} + x_2 \int m_1 m_2 \frac{dx}{EI} = 0$$

by symmetry $x_1 = x_2$

by Simpson's rule of integration

$$\int M_p^o m_1 \frac{dx}{EI} = \frac{L}{6EI}\left(p.e.4.\frac{1}{2}\right) = \frac{2PeL}{6EI}$$

$$\int m_1 m_1 \frac{dx}{EI} = \frac{L}{6EI}\left(1.1 + \frac{1}{2}.\frac{1}{2}.4\right) = \frac{2L}{6EI}$$

$$\int m_1 m_2 \frac{dx}{EI} = \frac{L}{6EI}\left(\frac{1}{2}.\frac{1}{2}.4\right) = \frac{L}{6EI}$$

The compatibility equation becomes

$$\frac{2PeL}{6} + \left(\frac{2L}{6EI} + \frac{L}{6EI}\right)x_1 = 0;\ 2Pe + 3x_1 = 0;$$

$$x_1 = -\frac{2}{3}Pe;\ x_2 = x_1 = \frac{-2}{3}Pe$$

This is the indeterminate moment. The final moment diagram can be got by super position.b

Table 12.2 Indeterminate Moments Due to Prestressing

Cable profile	Interminate moment
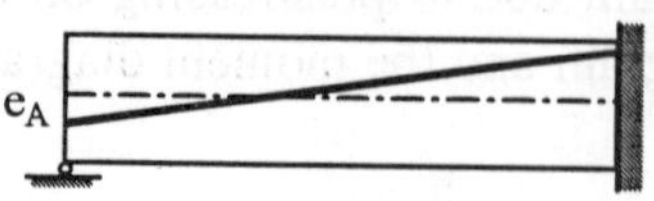	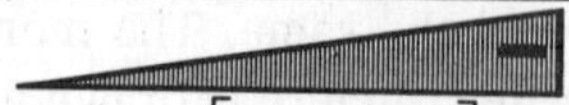 $M'_{PB} = -\left[e_B + \frac{1}{2}e_A\right]P^{(-)}$
	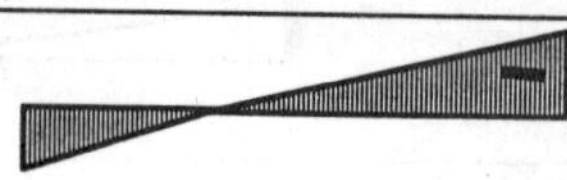 $M'_{PA} = -e_A P^{(-)}$ $M'_{PB} = -e_B P^{(-)}$
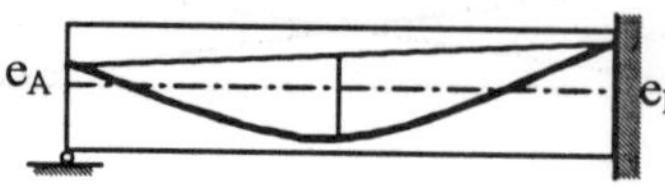	 $M'_{PB} = -\left[f + e_B + \frac{e_A}{2}\right]P^{(-)}$
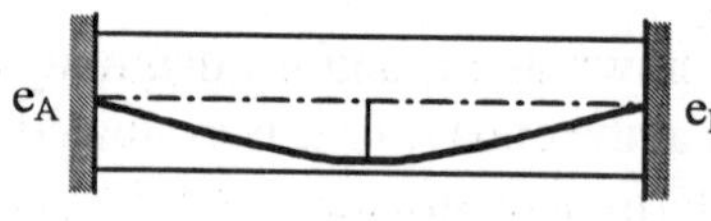	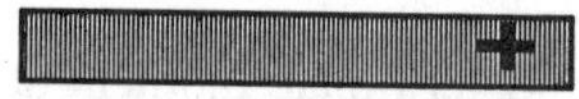 $M'_{PA} = M'_{PB} = \frac{2}{3} fP^{(-)}$
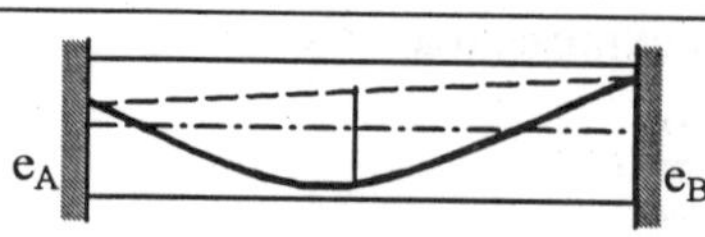 *e : downward, +ve upward, −ve.*	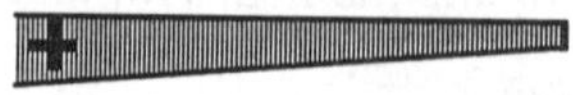 $M'_{PA} = -(\frac{2}{3}f + e_A)P^{(-)}; M'_{PB} = -(\frac{2}{3}f + e_B)P^{(-)}$
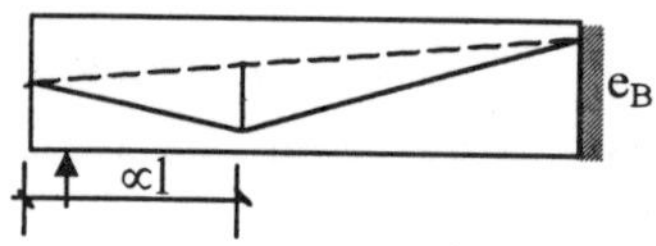	 $M'_P = -\left[e_B + \frac{1}{2}f(1+\alpha)\right]P^{(-)}$
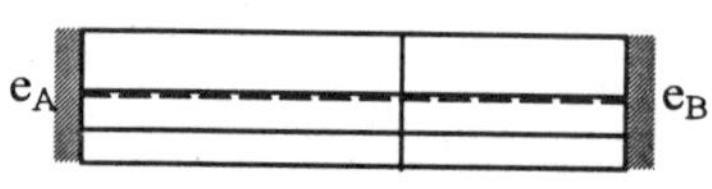	 $M'_{PB} = -e_B P^{(-)}$ $M'_{PA} = -e_A P^{(-)}$
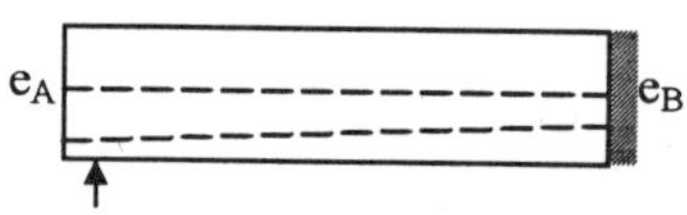	 $M'_{PB} = -\frac{1}{2}(e_A + e_B)P^{(-)}$
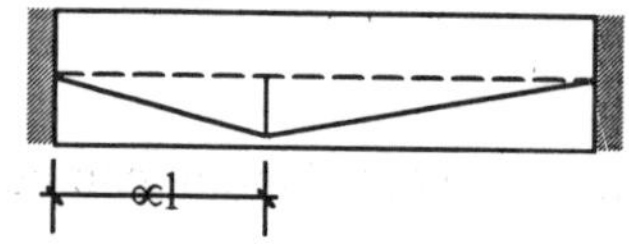	 $M'_{PA} = -f\alpha P^{(-)}$ $M'_{PB} = -f(1-\alpha)P^{(-)}$

Sign Convention : Bottom Tension Positive

As discussed earlier, a given frame with prestressing profiles could be analysed using moment distribution method. First, all the joints are assumed to be fixed without allowance for any rotation and the fixed end moments are evaluated as discussed for one typical case (values are available in Table 12.2 for various cable profiles). Once the joints are released there will be unbalanced moments and they would be distributed as per stiffnesses of members meeting at that joint and the distributed moments are carried out to the farther joints. This process is repeated till all the joints are having moments even after release and carried to the farther joint till the joint reach a state of balanced moment even after release.

It may be noted that the fixed end moment at every joint consists of a determinate moment (M_p^O) and in determinate moment M_p^1. The determinate moment at any joint is already balanced at the joint. Hence the unbalanced moment is caused only by the indeterminate component of the moment at each joint namely (M_P^1). This is also explained by Leonhardt [3] and Walther [11] how the self-equilibrating force (P.e) will not contribute to unbalanced moment at any joint even if the 'e' at the joint section for two adjacent spans are different. Hence in moment distribution process, only the unbalanced moment is considered for distribution. After getting the balanced moment at each joint due to the indeterminate moment component, the final moment at each joint is arrived as the determinate moment plus the balanced version of indeterminate moment. This is illustrated in a case of portal frame in example 12.4

Example 12.4

Analyse a portal frame of span 10m and height 5m with both support ends fixed. The moments of inertia of the beam element and the column element are in the same. The beam element alone is prestressed with a cable having a parabolic file with eccentricity 'e' (upward) at the ends and eccentricity 'e' (downwards) at mid span (Ref. Fig.12.17).

Draw the bending moment diagram on the portal due to applied prestressing force of *P*.

Solution

The fixed moment at A towards B = $M_{PAB}^F = 0$

B towards A = $M_{PBA}^F = 0$

B towards C = $M_{PBC}^F = 0$

C towards B = $M_{PCB}^F = 0$

C towards D = $M_{PCD}^F = 0$

Fixed moment $M_{PBC}^F = M_{PBC}^o + M_{PBC}^1$

(Ref. Table 12.2)

M_{PBC}^I is the indeterminate part

That alone causes unbalance at the joint.

$M_{PBC}^I = -[\frac{2}{3}.2e - e]P^{(-)}$

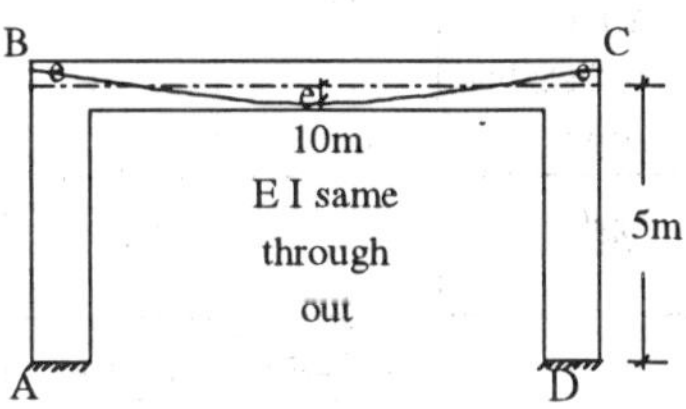

Fig. 12.17

$= -[\frac{1}{3} P.e]^{(-)}$

$= \frac{1}{3} Pe$ (bottom tension)

(Clockwise at B towards C)

Similarly

$M'_{PCB} = \frac{1}{3} P.e$ (bottom tension)

(anticlockwise at C towards B)

Because of symmetry the frame is considered fixed at mid point (E) of the top beam portion and the stiffness element. B towards E is considered half of the value of I/span for that element.

$$\frac{I}{L} forBE = \frac{I}{2 \times 10} = \frac{I}{20}$$

$$\frac{I}{L} forBA = \frac{I}{5} = \frac{I}{5}$$

Distribution ratio at B is 4:1 between column and beam

Moment Distribution Table

Joints	A	B		E
Member	AB	BA	BE	EB
Distribution Factor	-	0.8	0.2	-
M'_P	-	-	$\frac{Pe}{3}$	-
Distributed moment	-	$\frac{4Pe}{15}$	$-\frac{Pe}{15}$	-
Carried over moment	$-\frac{2Pe}{15}$			-
Final Moment	$-\frac{2}{15}Pe$	$-\frac{4Pe}{15}$	$+\frac{4Pe}{15}$	-

This balanced moment of the indeterminate component should be added to the determinate component of moment (M_P^O). This is presented in the tabular form with due sign convension [bottom tension positive and top tension negative].

Joint	A	B		C		D	E
Member		BA	BC	CB	CD		
M'_P	0	0	$\frac{Pe}{3}$	$\frac{Pe}{3}$	0	0	
M'_P after redistribu -tion	$-\frac{2}{15}Pe$	$\frac{4}{15}Pe$	$\frac{4}{15}Pe$	$\frac{4}{15}Pe$	$\frac{4}{15}Pe$	$-\frac{2}{15}Pe$	$\frac{4}{15}Pe$
M_P^0	0	0	$\frac{15}{15}Pe$	$\frac{15}{15}Pe$	0	0	$-\frac{15}{15}Pe$
M_p	$-\frac{2}{15}Pe$	$\frac{4}{15}Pe$	$\frac{19}{15}Pe$	$\frac{19}{15}Pe$	0	0	$-\frac{11}{15}Pe$

The final bending moment diagram for the frame due to the effects of prestressing is shown in Fig.12.18.

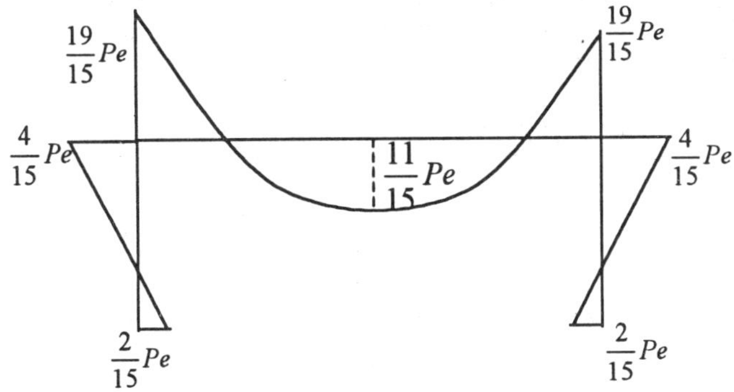

Fig. 12.18 Bending Moment Diagram drawn on compression side

12.6 Choice of Prestressing and Cable Profiles.

The choice of prestressing and cable profiling is a design problem. It is based on satisfactory structural performance to resist the effect of external loading, economy of materials required (H.T. steel) the prestressing operations involved and their costing. For explanation a portal frame is chosen and it is a flexure based system. The design forces are evaluated in terms of bending moment. [A shell structure is a membrane force predominant structure and hence the design forces are membrane forces]. Considering a single bayed portal, first the bending moments on various elements such as beam and columns, caused by external loading are evaluated and the resultant bending moment diagram is drawn. As a next step a few possible cable profiles are chosen. These are based on the shape of the Bending moment diagram due to external loading or based on the possible equivalent loads that could be generated by the cable profile which would counteract the given external loads. After choosing a set of cable profiles, the portal frame is analysed with those cable profiles for a given prestressing forcc(P) and the Bending moment diagram due to prestressing in drawn. Different prestressing forces can also be chosen for various elements such as P,P/2. 3P/2 etc. The bending moment diagrams for such cable profiles may also be drawn. These bending moment diagrams may be compared with the bending diagrams on the portal caused by external bending. The cable profiles which lead to maching (but opposite in nature) bending moment diagrams to that caused by external loading may then be identified. From amongst them, the profiles which gives the minimum prestressing force to resist the bending moment caused by external load may be chosen. Then, as a next stage, the profile which leads to minimum material cost and operational cost may be chosen. This procedure of choice of profile and prestressing is illustrated with an example.

Example 12.5

The portal frame given in example 12.4 is subjected to an uniformly distributed load of 120kN/m on the beam portion alone. This load includes self weight and the superimposed loads. Analyse the portal for the given loading and suggest suitable prestressing with cable profile so that the members could be made slender.

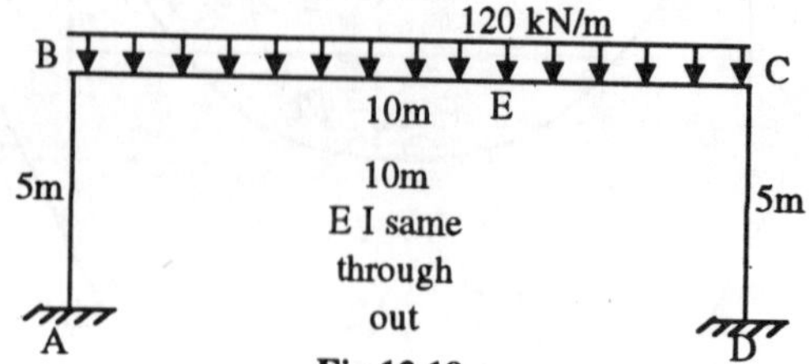

Fig 12.19.

Solution

The portal is first analysed for the given loading of 120kN/m on the beam using moment distribution method. The symmetry of the structure and the loading is made use of and hence only half the portal is considered as in the case of example 12.4. The portal is assumed to be fi xed at the mid point (E) of beam and the stiffness of the beam is reduced to half its value. Fixed end moments due to loading.

at A towards B = 0

B towards A = 0

$$\text{B towards E} = \frac{wl^2}{12} = \frac{120 \times 10 \times 10}{12} = 1000 kNm$$

Distribution factor at B towards A = 0.8

B towards E = 0.2

Distribution Table

Joint	A	B		E
Member	AB	BA	BE	EB
Distribution Force	-	0.8	0.2	-
Fixed End Moment	-	+800	-1000 +200	
Carried over moment	+400	-	-	
Final moment	+400	+800	-800	

$$\text{Mid span moment due to loading} = \frac{120 \times 10 \times 10}{8} = 1500 kNm$$

Bending moment diagram drawn on tension side is shown if Fig.12.20.

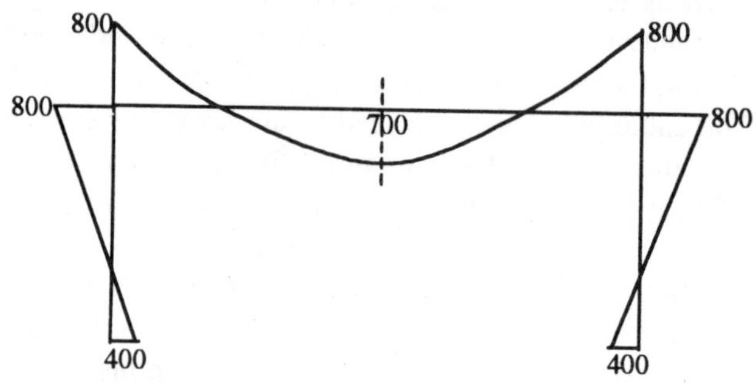

Fig 12.20

Four possible cable profile are chosen, three based equivalent loaidng and fourth one based on resultant bending moment diagram shown in Figure. The profiles are shown in Fig 12.21.

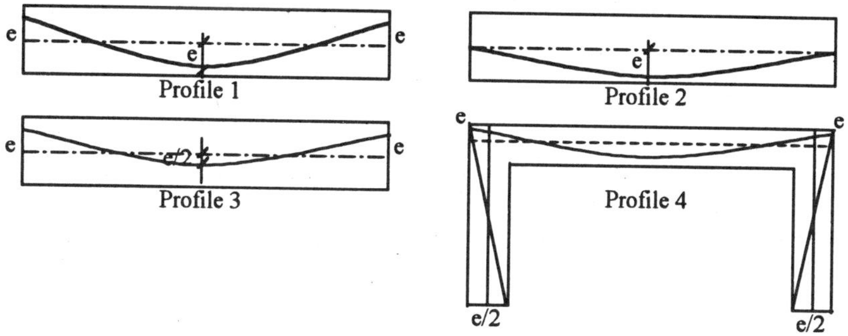

Fig. 12.21

Profile 1 is based on equivalent loading. Hence it is only on the beam and it is parabolic with negative eccentricity at ends and possitive eccentricity at centre. The eccentricity is the same based on the bending moment diagram.

Profile 2 is also based on equivalent loading. It is possible with zero eccentricity at ends and with a positive eccentricity of 'e' at the centre based purely on loading.

Profile 3 is Similar the profile 1 only with a reduced eccentricity of e/2 at mid span of beam.

Profile 4 is based on the final bending moment diagram caused by the external loading. The profile is parabolic with equal negative eccentricities of e at ends and positive eccentricity of 'e' at mid span. The columns are also proposed to be prestressed with straight cables with outward eccentricity of 'e' at top and inward eccentricity of e/2 at bottom fixed end.

As discussed earlier each cable profile with a given prestressing force 'P' is analysed and the bending moment diagram drawn

Profile1: It is the same as given in example 12.4. The Bending moment Diagram is shown below again.

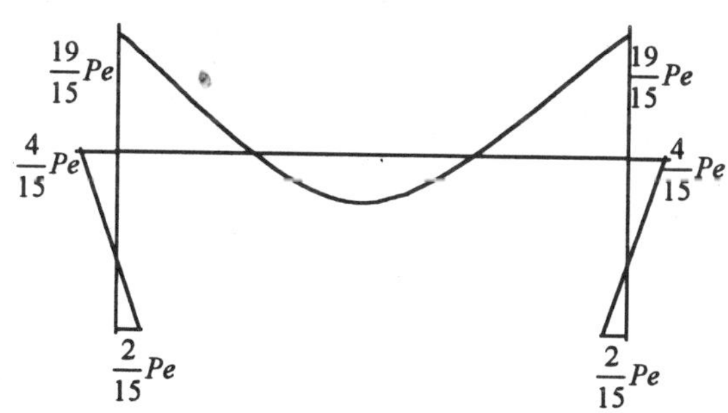

Fig. 12.22

Profile 2: The fixed end moments exist only in the beam portion

$$M^{F}_{PBC} = M^{o}_{PBC} + M^{l}_{PBC}$$

Since M^{l}_{PBC} alone causes unbalanced moment that alone is considered.

$$M^{l}_{PBC} = -[\frac{2}{3}e]P^{(-)} = \frac{2}{3} Pe(bottomtension)$$

Clockwise at B towards C

Anticlockwise at C towards B.

As in the case of example 12.4 only half the portal is considered for moment distribution.

Distribution Table				
Joint	A	B		E
Member	AB	BA	BE	EB
Distribution Factor	-	0.8	0.2	-
Fixed End Moment Distribution moment	-	-	$\frac{2}{3}Pe$	-
Carried over moment	-	$-\frac{8}{15}Pe$	$-\frac{2}{15}Pe$	-
Final moment	$-\frac{4}{15}Pe$	$-\frac{4}{15}Pe$	$-\frac{2}{15}Pe$	$+\frac{5}{15}P$

The balanced final moment of the indeterminate component should be added to determinate component of M_P^o This is presented in the tabular form with due sign conversation (bottom tension positive and top tension negative)

Profile 2							
Joint	A	B		E	C		D
	AB	BA	BC		CB	CD	DC
M_P^I	0	0	$\frac{2}{3}Pe$		$\frac{2}{3}Pe$	0	0
M_P^I after redistribu -tion	$-\frac{4}{15}Pe$	$\frac{8}{15}Pe$	$\frac{8}{15}Pe$	$\frac{8}{15}Pe$	$\frac{8}{15}Pe$	$\frac{8}{15}Pe$	$-\frac{4}{15}Pe$
M_P^o	0	0	0	$-\frac{15}{15}P$	0	0	0
M_P	$-\frac{4}{15}Pe$	$\frac{8}{15}Pe$	$\frac{8}{15}Pe$	$-\frac{7}{15}P$	$\frac{8}{15}Pe$	$\frac{8}{15}Pe$	$-\frac{4}{15}Pe$

$$M'_{BA} = -\left[\tfrac{2}{3}.e\right]V^{(-)}$$

Final bending moment diagram on the frame due to the effect of prestressing as per profile 2 is shown in Fig.12.23 below.

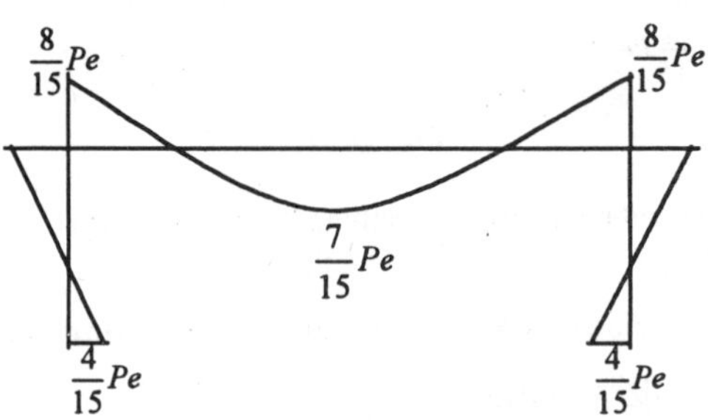

Fig. 12.23

Profile-3:

The fixed end moments exist only in the beam portion

$M^{F}_{BC} = M^{0}_{PBC} + M^{1}_{PBC}$

Since M^{I}_{PBC} alone causes unbalanced moment that alone is considered for distribution.

$$M^{I}_{PBC} = -[\tfrac{2}{3}\cdot\tfrac{3}{2}e - e]P^{(-)} = 0$$

The indeterminate moment due to the profile on beam portion is zero. That means the profile is a concordant profile. There need be no distribution. Then the final moment details are presented in tabular form.

Profile 3							
Joint	A	B		E	C		D
	AB	BA	BC		CB	CD	DC
MP	0	0	0		0	0	0
MP after redistribu -tion	0	0	0	0	0	0	0
MP^{0}	0	0	*Pe*	$-\frac{Pe}{2}$	*Pe*	0	0
MP	0	0	*Pe*	$-\frac{Pe}{2}$	*Pe*	0	0

The final bending moment diagram on the frame due to the effect of prestressing as per profile-3 is shown in Fig.12.24 below

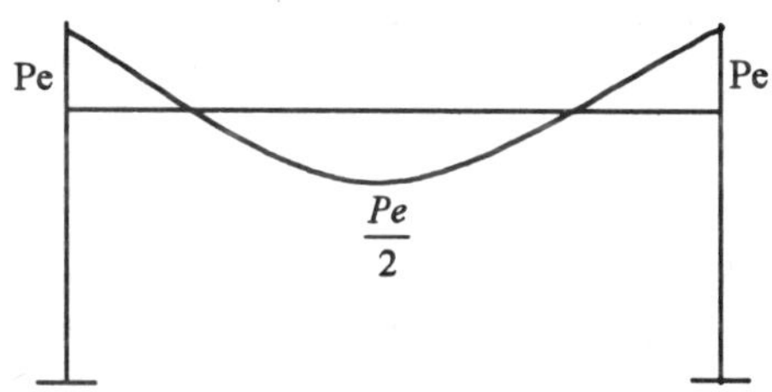

Fig. 12.24

The bending moment diagram is drawn on compression side.

Profile - 4

This profile runs both on beam element and column elements. Hence there are fixed end moments both on columns and beams. But both the geometry of the frame and the prestressing is symmetrical and hence half the portal alone is considered for moment distribution.

The fixed end moment = $M^{F}_{P} = M^{0}_{P} + M^{1}_{P}$

Since the indeterminate component alone causes unbalanced moment, that alone is considered for distribution.

$$M^{I}_{PBC} = -[P^{(-)}\frac{(+e)}{2}] = [\frac{+Pe}{2}]$$

$M^{I}_{PBC} = -P^{(-)}(-e) = -Pe$ (outward tension)

$$M^{IS}_{FBC} = -P[\frac{2}{3}.2e - e] = \frac{+P.e}{2} \text{ (bottom tension)}$$

Moment distribution table				
Joints	A	B		E
Member	AB	BA	BE	EB
Distribution Force		0.8	0.2	
M'_P	$\frac{Pe}{2}$	Pe	$\frac{Pe}{3}$	-
Distribution Moment		$\frac{-4Pe}{3}.\frac{4}{5}$	$\frac{-4Pe}{3}.\frac{1}{5}$	-
Carried over moment	$-\frac{8}{15}Pe$	-	-	-
Final Moment	$-\frac{0.5}{15}Pe$	$-\frac{Pe}{15}$	$+\frac{Pe}{15}$	-

The balanced final moment of the indeterminate component should be algebrically added to determinate component M^o_P This is presented in the tabular form with sign convention. (/bottom tension positive and top tension negative).

Profile 4							
	A	B		E	C		D
Member	AB	BA	BC		CB	CD	DC
M'_P	$\frac{Pe}{2}$	$-Pe$	$+\frac{Pe}{3}$		$+\frac{Pe}{3}$	$-Pe$	$\frac{Pe}{2}$
M'_P after redistribu -tion	$-\frac{0.5Pe}{15}$	$\frac{Pe}{15}$	$\frac{Pe}{15}$	$\frac{Pe}{15}$	$\frac{Pe}{15}$	$\frac{Pe}{15}$	$-\frac{0.5Pe}{15}$
M^o_P	$-\frac{7.5Pe}{15}$	Pe	Pe	$-Pe$	Pe	Pe	$-\frac{7.5Pe}{15}$
M_P	$-\frac{8}{15}Pe$	$\frac{16}{15}Pe$	$\frac{16}{15}Pe$	$-\frac{14}{15}P$	$\frac{16}{15}Pe$	$\frac{16}{15}Pe$	$-\frac{8}{15}Pe$

The final bending moment diagram on the frame due to the effect of prestressing as per profile 4 is shown in Fig.12.25 below.

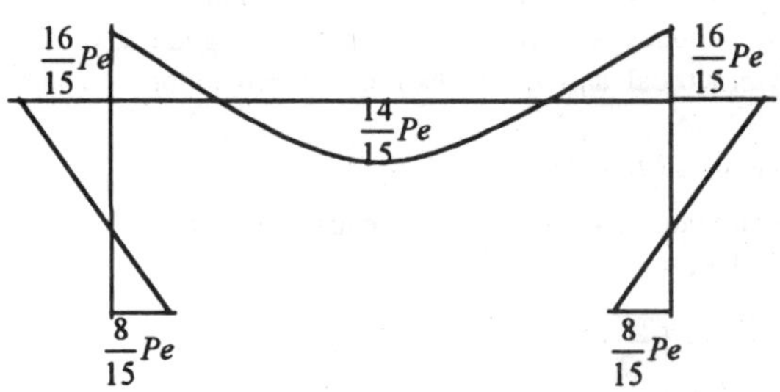

Fig. 12.25

12.7 Comparision for Choice for Best Profile

By comparing the bending moment diagrams generated by various profiling 1,2,3,4 shown in Figures 12.21, 12.22, 12.23, 12.24, with that caused by loading shown in Figures 12.19, it is seen that profiles 1,2 and 4 lead to bending moment diagram matching with that caused by loading diagram. Of these, the profile 2, and 4 lead to bending moments of salient points which are proportionate to the bending moments caused by external loading. Hence profile 1 is eliminated. Profile 4 leads prestressing force requires only half that is required in profile two (2) to resist the bending moment caused by external loads, i.e., In case of profile 4, $\frac{16}{15}P.e = 800kNm$ .i.e. $P.e. = \frac{800}{2}\times\frac{15}{16}$. In case of profile 2, $\frac{8}{15}Pe = 800kNm$, $Pe = \frac{800}{e}\times\frac{15}{8}kN$

Hence it looks profile 4 is preferable. In profile 4, total length of cable requried is with more than 20 m to cover the beam (10m) and 2 columns (10 m) i.e. in profile 4, where as in case of profile 2, it is only 10 m. Through the area of steel is halved, the length is doubled in profle 4. Hence the volume of steel remains the same for both cases approximately. But in case of profile 4, the operations required and the external anchorages required are more. Further, anchoring cables both in the column and the beam at the junction may lead to stress concentration leading to structural problems. Hence profile 2 is adjusted as the best profile for design for the problem given in 12.5. The quantum steel depends on the permissible eccentricity and section dimensions of the beam. This has to be separately worked out.

12.8 Effect of Time Dependent Deformation on Indeterminate Prestressed Concrete Structures

It is well known that any imposed deformation such as (sinking of supports, lack of fit etc.,) on an indeterminate structure leads to additional stresses on the structure. Creep and shrinkage also cause deformation on structures. But the effect of creep and shrinkage deformations on prestressed concrete indeterminate structures are different. While creep deformation is caused by an applied load, shrinkage deformation is due to evaporation of water and natural environmental conditions which have nothing to do with any applied load.

The applied load in an indeterminate structure can be in the form of prestress or an external load. ***Creep causes losses in prestress, which has to be separetely dealt with.*** It is known that an indeterminate structure can be analysed in two units namely a corresponding determinate structure with the given loading and the same determinate structure with the indeterminate forces as external loading.

If there is a change in the stress/strain resultants caused by creep deformation on the determinate structure with the given loading there will be similar changes in stresses/strain caused by creep deformation on the same structure due to the indeterminate load applied as external forces. The solution for an indeterminate structure by consistant deformation methods is defined by a compatibility condition $\Delta_{10} + X_1\delta_{11} = 0$ where Δ_{10} is the deformation caused by the loads on the determinate structure and δ_{11} is the deformation caused at the same place in the same direction on the same determinate structure by the in determinant component namely X_1. If there is change in the Δ_{10} leading to $K\Delta_1$ due to creep effect, it will be the same in δ_1 also and by the same amount leading to $K\delta_{11}$. Hence the compatibility conditions remain unaltered and the value of the indeterminate force X_1 remains unchanged. This can be proved with greater details by taking specific cases. ***In conclusion it is stated that in an indeterminate structure, while creep causes losses in prestressing, it does cause any change in the Indeterminate components of the structure caused by prestressing*** **[12].**

The same is not true about shrinkage. If the shrinkage deformation is expressed by ($\varepsilon_{s\alpha}.L$), the corresponding deformation introduces, stress resultants in any indeterminate structure. This can be evaluated by compatibility equation $\Delta_{10} + X_1\delta_{11} = \varepsilon_{s\alpha}.L$. This is the normal procedure in any indeterminate structure. This internal stress developed by shrinkage strains also causes creep. Hence Δ_{10} and also δ_{11} will have a changed value due to creep and the changes in Δ_{10} and δ_{11} are proportional to each other. But $\epsilon_{s\infty}$ is a constant and that does not get altered with creep. Hence in the compatibility equation, out of the three terms, two terms get altered proportionately but not the third term. Hence the solution for X_1 is altered. ***Hence it is proved that shrinkage strain in an indeterminate structure causes a change in the stress resultants*** **[11]**

Further the creep varies with time. Since there is variation in the creep coefficients (φ) with time, the resultant stresses caused by shrinkage strain (with creep effect) also varies with time. This has been eloborately analysed by Trost and a general expression has been derived as.

$\varepsilon_{c(t)} = \dfrac{\sigma_{cpo}}{E_c}(1+\varphi_\infty) + \dfrac{\sigma_{cp1} - \sigma_{cpo}}{E_c}(1+\rho\varphi_\infty) + \varepsilon_{s\infty}$ This has beam explained in art 5.3.2. If one were to analyse the effect of shrinkage and creep caused by shrinkage in indeterminate prestressed concrete structures, in this expression the strain due to shrinkage alone shall remain and there is no other stress. In such a case $\sigma_{cpo} = 0$ where σ_{cp1} is the new stress developed in concrete due to shrinkage. There is no strain in concrete other than shrinkage strain.

Hence $\varepsilon_{s\alpha}$ becomes equal to $\dfrac{\sigma_{cp1}}{E_c}(1+\rho\varphi_\infty) = \varepsilon_{s\infty}$ and $\sigma_{cp1} = \dfrac{E_c \in_{s\infty}}{(1+\rho\varphi_\infty)}$.

Presently concrete stress is expressed as $E_c.\varepsilon_{s\infty}$ whereas in reality, it is $\dfrac{E_c.\in_{s\infty}}{1+\rho\varphi_\infty}$. The actual stress in concrete is less because of creep effect. The creep effect caused by stresses introduced due to shrinkage reduces the resultant stress by a coefficient $\frac{1}{1+\rho\varphi_{\infty}}$. This can be proved for all cases.

The reduction coefficient of $\frac{1}{1+\rho\varphi_{\infty}}$ is given in Table 12.3 for various values of creep coefficient. The same can be interpolated or extrapolated within certain limits.

Table 12.3

φ_{∞}	1,5	2	3	4	5
$\frac{1}{1+\rho\varphi_{\infty}}$	0,4	0,333	0,25	0,2	0,167

References

1. R.B.B. MOORMAN, (1952) **"Equivalent Load Method for Analyzing Prestressed Concrete Structures,"** J. Am. Concrete Institute Jan. (Proc., Vol. 48), pp. 405-416.
2. LIN .T.Y. and Ned H. BURNS (1981) **"Design of Prestressed Concrete Structures"**, Third edition John Wiley & Sons Inc., New York.
3. LEONHARDT. F, (1964) **"Prestressed Concrete Design and Construction"**, Wilhelm Ernst and Sohn, Berlin,.New York .
4. F.S. ROSTASY, (1962) **"Connections in Precast Concrete Structures-Continuity in Double-T Floor Construction,"** J. Prestressed Concrete Institute Aug. 1962.
5. LIN, T.Y., (1960) **"Prestressed Concrete Structures,"** John Wiley and Sons, Inc. pp 288-292.
6. VISVESVARAYA, H.C. and N. RAGHAVENDRA, (1970) **"Continuity in prestressed concrete construction,"** Seminar on Problems of prestressing, Madras, pp I-91-94.
7. BAKER, J.F., M.R. HORNE, and J. HEYMAN, (1956) **"The Steel Skeleton."** Vol.2. Cambridge University Press, London.
8. PRAGER, W. and P.G. HODGE, (1951) **"Theory of perfectly Plastic Solids,"** John Wiley and Sons Inc., New York.
9. H. CROSS and N.D. MORGAN (1932) **"Continuous Frames of Reinforced Concrete,"** John Wiley & Sons, New York.
10. T.Y. LIN, (1937) **"A Direct Method of Moment Distrubution,"** Trans. Am. Soc. C.E., pp. 561-605.
11. WALTHER RANE, (1972) Vorlesungen uber **spannbeton** - Ausge wahlte Kapital - Lehrstulilfur Massivbau, University of Sluttgart.
12. SAEED-Un-DIN, K. (Nov. 1958) **"The Effect of Creep Upon Redundant Reactions in Continuous Prestressed Concrete Beam,"** Mag. Concrete Research, 10, 109.
13. PARME, A.L. and PARIS, G.H. (1951) ***"Designing for Continuity in Prestressed Concrete Structures,"*** Journal of the American Concrete Institute, 23,45-64 sept. 1951.

13. Cylindrical Structures in Prestressed Concrete – Pipes and Tanks

13.1 General

Liquid carrying structures, a pipe or a tank should be impervious against leakage of liquid from inside to outside and also against ingression of external materials inside. Earlier Steel or cast iron were used for pipes and tanks. In the first half of twentieth century, concrete was introduced as a structural material for large diameter pipes and tanks. Well made concrete provides an impermeable protection against leakage and also against ingression of materials from outside. But environmental causes like shrinkage or changes in temperature lead to tensile stresses in concrete, which bring down the impermeability of the concrete even if cracks are not developed. Concrete becomes more vulnerable for impermeability if the dimensions become large. In such cases, introduction of a precompression may serve the purpose of keeping the impermeability intact. For water flowing or retaining structures like pipes and water tanks, it is known that circular cross-section is the most suitable and practical cross-section. When such circular cross-sections are under internal or external pressure, hoop forces are developed through the thickness of the element. While hoop compression, caused by external loading, can be resisted well by concrete, hoop tension created by internal pressure cannot be resisted by concrete because of its inherent weakness in tension. This hoop tension is a function of internal pressure and diameter. Hence pipes and tanks of higher diameter or with higher internal pressure, require prestressing. (Introduction of precompression) to resist the hoop tension. This prestressing is done in circular direction around the periphery of the tank or pipe.

The level to which hoop tension is developed is a function of applied pressure, which is constant in a pipe and is a variant in a tank. Based on the hoop tension developed, quantum of prestressing is decided. Prestressing cannot be introduced for full length of pipe or full height of tank in one stretch. Prestressing process over the length of pipe or height of the tank causes curvature along the length in the pipe and along the height in tank leading to flexural forces. Because of these flexural forces cracks may develop around the perimeter at frequent internals which is not desirable. To avoid this, longitudinal prestressing is also called for in tanks and pipes. Though the structural behaviour is similar in tanks and pipes, their boundary conditions, bring in variation in the analysis. Hence pipes and tanks are separately dealtwith in this chapter. Relevant statutory regulations are also separately presented.

13.2 Prestressed Concrete Pipes

Prestressed concrete pipes are ideally suited for pressure range of 0.4 to 2 N/mm^2. For this range, cast iron, or mild steel pipes are not economical and RCC pipes are not suitable. Technique of prestressing was first applied successfully as early as 1930. However, further developments were severely hindered during the second world war. The first commercial application of steel cylinder prestressed concrete pipe was made in the year 1942. Since then numerous pipelines have been installed throughout the world using steel cylinder prestressed concrete pipes. During the same period studies were undertaken on the economics of production of prestressed concrete pipe with all concrete cores i.e. without steel cylinder. In 1944, the first non-cylinder prestressed concrete pipeline was installed. Since then various techniques have been developed by different manufacturers, in the manufacture of PSC pipes.

13.3 Types of Prestressed Concrete Pipes

There are mainly two types of prestressed concrete pipes in use today, namely the cylinder type and the non-cylinder type. Cylinder type has a steel cylinder core over which concrete is shot creeted and prestressed Non-cylinder type has only a prestressed concrete pipe. The former is even now predominant in the United States and in North America.

13.3.1 Cylinder type

In cylinder type there are two basic designs. The first design was introduced in 1942 in U.S.A. and is termed as the prestressed concrete lined cylinder type pipe. They are found suitable upto 1200 mm dia. It consists of a concrete lined steel cylinder with steel joint/rings welded to their ends, joint/bell wrapped with a helix of highly stressed wire and coated with a dense cement mortar to protect the wire. A typical cylinder type prestressed concrete pipe is shown in Fig.13.1. In 1950, a second type of prestressed cylinder concrete pipe was developed. In this, the steel cylinder with joint rings is embeded in the concrete core so that the stressed wire helix is in contact with concrete rather than with the steel cylinder. The protective coating is either cement mortar or concrete. These are found suitable for diameter upto 3600 mm. This cylinder type of pipes is not relevant in this chapter and hence not discussed further.

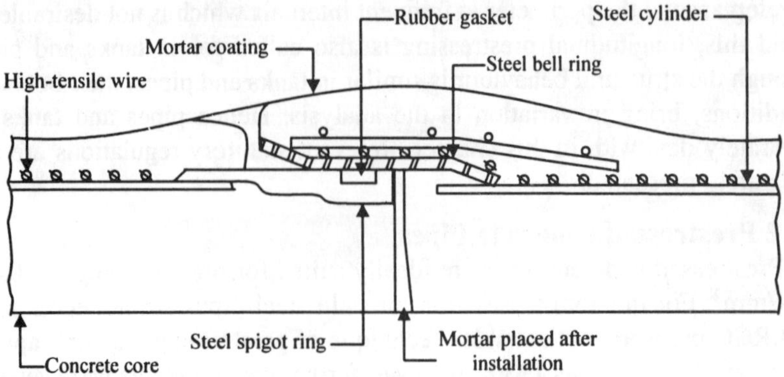

Fig 13.1 Longitudinal section through joint of prestressed-concrete cylinder pipe

13. 3.2 Non-Cylinder Type

There are two basic designs of non-cylinder pipes, viz. composite or wire wound and non-composite. The composite design consists wrapping of high tensile steel wire under controlled tension around a core pipe of concrete, which has previously been precompressed, longitudinally by high tensile wires. A cover coat of mortar or concrete is applied afterwards to protect the high tensile steel wire wrapping. Figure 13.2 shows a longitudinal section of a non cylinder type of a pipe.

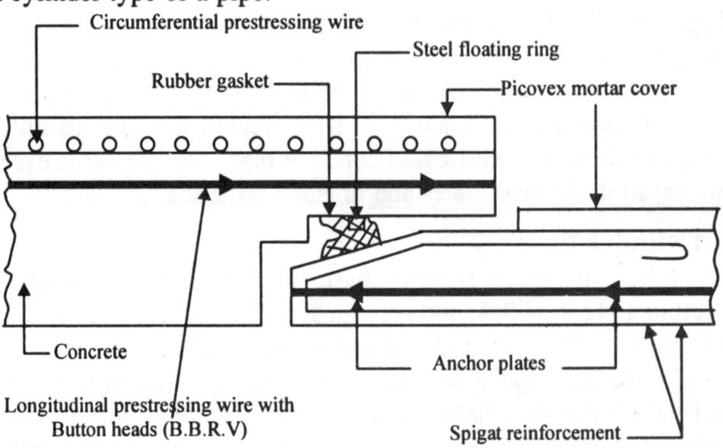

Fig 13. 2. Typical Longitudinal Section of Non-Cylinder Prestressed Concrete Pipe

In non-composite type, a cage of high tensile wire is laid slack in the mould and the pipe is concreted in a single operation over the entire thickness of the wall. Immediately after concreting the mould is raidally expanded. This radial expansion of the mould core ensures that the circumferential reinforcement embeded in green concrete is expanded and tensioned. After concrete has set, expansion of the mould core is reversed and as a result the concrete is prestressed by compression caused by elasticity of the tensioned steel. Since the pipe wall has only one layer this prestressing includes also the concrete outside the reinforcement.

Non-cylinder prestressed concrete pipes are suitable for all types of water supply schemes. It is specially suitable where shell for the cylinder is not available. Under normal conditions, this is economical for medium heads. Tapping under pressure was considered impracticable with this type of pipe. But now it is possible due to improved methods of cutting concrete cores. Replacement of this type of pipe can also be done quickly due to the development of special devices for cutting concrete. Repairs to these pipes on the spot, to some extent, can be made using epoxy. The joint of this pipe are made with flexible rubber rings. Two types of rubber ring joints are commonly used. One is with circular chord (hollow or solid) either rolled or fixed gasket and the other is with a 'V' section gasket. Both the joints have been successfully used for pressures upto 2N/mm^2, since all exposed surfaces are made of concrete, no mortar is required to be placed in the joint recaps. Shear strength of the joint is considerably increased by the compression induced by the longitudinal prestress. The rubber and concrete joint effectively isolates the steel in each pipe from that of its neighbour so that the pipeline is electrically discontinuous. The possibility of long time current pick-up and electrolytic corrosion is thus eliminated. This is particularly important in these days of increasing potential hazard of interference from adjacent cathodically protected metallic pipelines of cables. Even in each individual pipe, the circumferential prestressed wires are separated from the longitudinal prestressed wires by a layer of concrete. Only selected aggregates and skilled manufacturing techniques produce optimum quality, concrete which is required in the manufacture of pipes. Pipes manufactured using high quality concrete have the following properties. (a) impermeability even at high pressures. (b) high structural strength and (c) high resistance against attack by deleterious substances and therefore less friction.

13.4 Unique Properties of Non-Cylinder Prestressed Pipes

- The behaviour of prestressed concrete pipe when hydraulic shock occurs is worthy of mention. Under hydraulic shock, excessive pressure developed brings about bursting the pipe material. Normally, such a burst pipe has to be replaced, but this does not apply to prestressed concrete pipe. When the pressure drops, cracks in the pipe close and the pipe becomes water-tight as before. The pipe then undergoes a self healing process due to subsequent calcification.
- Under normal working pressure, the pipe wall is always in a state of compression, a condition most favorable for impermeability.
- Under internal hydrostatic pressure, the mortar coat acts integrally with the core pipe. The ability of the cover coat to act integrally with

the core is due to the fact that the coat is also pre-compressed to some extent due to creep of concrete and loss of stress in wire winding.

13.5 Advantages of Prestressed Concrete Pipes

13.5.1 Economy

Installed cost is generally less than that for pipes of other materials such as C.I. or steel for ductile iron. Considerable additional economy can be achieved by specifying design pressure in increments according to the variation in working pressure along the pipeline. The differences of cost on cast iron pipe increases considerably as the diameter of the pipe increases. Prestressed concrete pipes resist external load to a much higher degree than the steel pipes. The rubber ring joint is extremely simple for fixing and reliable. Laying of PSC pipe is faster. It eliminates expansion joints due to its flexibility. The effectiveness and durability of this joint has been proved by 40 years of experience under a wide range of job conditions. The flexible joint allows small rotation and deformation without allowing for any leakage path.

13.5.2 Local Manufacture

For large requirements of pipes it is often more economical to install a special manufacturing plant near the site of the pipeline than to transport the pipe to the site from an existing plant. Establishment of such a factory brings important benefits to the local community by consumption of local raw materials and services and in providing employment.

13.5.3. Economical use of Raw Material

In the design of prestressed concrete pipes, both concrete and steel are utilised to their maximum possible extent; and therefore, overall economy can be effected in the consumption of steel and cement.

13.5.4 Hydraulic properties

Prestressed concrete pipes posses excellent hydraulic properties and resistance to corrosion similar to RCC pipes. Hydraulic flow co-efficient is not affected with age, like in the case of metal pipes.

13.5.5 Durability

Prestressed concrete pipe is essentially concrete pipe, containing a very small percentage of steel compared to its total weight. If this pipe is protected by a durable coat, there is no reason why its life should not be as much as, if not more, than the life of a RCC pipe. Expert's estimate that the life of a prestressed concrete pipe is generally more than 100 years when buried in non-corrosive soil environment.

13.5.6. Service Performance

The service record of these pipes since 1944 is very satisfactory, although a few failures of earlier pipelines were reported. These are mainly due to the use of oil tempered wire, the use of calcium chloride as additive for achieving early strength and the use of different cement for the core and coat. If proper materials are used, service life of these pipes could be around 100 years with least maintenance.

13.6. Loads on Pipes

13.6.1 External Loads

External loads are primarily earth loads. Estimation of earth loads acting on pipes in trenches or under embankments was studied by many researchers. The most commonly used theory today is that of Prof. Marston of the Iowa University. According to his theory, a pipe in trench installation carries only a part of the total vertical load of the overlying backfill, while the rest is carried through friction by the side of the trench. Consequently, the vertical earth load on the pipe is dependent on the friction characteristics and the unit weight of the fill material (and not directly on its compaction) the load increases with increase of the trench width at the level of the pipe crown. It is therefore, of great importance to keep the width of the trench as small as possible. Horizontal earth load has in general a favourable effect on the pipe since it relieves some of the stress created by the vertical load. The lateral deformation for concrete pipes under earth load is generally so small that no passive reaction of the backfill on the side of the pipe can be expected. However active earth pressure, should be assumed to act horizontally on both sides of the pipes.

A commonly used assumption which covers also the influence of lateral earth pressure is that of a bulb shaped distribution of the earth pressure over the perimeter of the pipe. The reaction of the supporting ground is also assumed to have a bulb shape over the central angle. ($\pi-\alpha$) where is the central angle subtended by bedding.

The angle of bedding α depends on the preparation of the bottom of the trench and can be assumed equal to 90° under moderately good conditions of placing. Total vertical (↓) component is caused by earth load from above. Total vertical (↑) Component of reaction is the Earth pressure from below and friction from soil from the sides.

These loads result in normal force and bending moments expressed in the form of C_1W and C_2WD is where D mean diameter. W represents the load per unit length due to earth fill (W_F), Live Load (W_L) weight of fluid and self weight (W_P). The salient positions of criticality are the crown, side and invert as shown in Fig.13.3. The value of C_1 and C_2 are available in specialised literature on soil mechanics and also IS 783 [1]. These values are given in Table 13.1 as reported by Pundit and Gupta [2] These moments and normal forces are more used to check the safety of the structure than for designing the prestressed concrete pipes.

Table 13.1 Values of Coefficinets C_1 and C_2 (2)

α_b	Type of loading	At crown(C)		At Side (S)		At Invert (I)	
		C_1	C_2	C_1	C_2	C_1	C_2
30°	W_L or W_E	0.487	0.031	-0.633	-0.043	-0.376	0.087
	W_F	0.237	0.040	0.058	0.050	0.352	0.097
	W_P	0.078	0.040	-0.302	-0.050	-0.125	0.097
60°	W_L or W_E	-0.425	0.033	-0.576	-0.043	-0.345	0.073
	W_P W_F	0.230	0.038	0.059	-0.048	0.310	0.078
		0.071	0.038	-0.302	-0.048	-0.168	0.078
90°	W_L or W_E	-0.382	-0.035	-0.539	-0.045	-0.324	0.063
	W_F	0.220	0.035	0.062	-0.044	0.272	0.061
	W_P	0.061	0.035	-0.295	-0.044	-0.207	0.061
120°	W_L or W_E	-0.351	0.037	-0.516	-0.046	-0.316	0.055
	W_F	0.205	0.032	0.065	-0.039	0.233	0.046
	W_P	0.046	-0.032	-0.282	-0.077	-0.245	0.046
150°	W_L or W_E	-0.331	0.040	-0.504	-0.047	-0.313	0.049
	W_F	0.185	0.027	0.067	-0.032	0.196	0.033
	W_P	0.026	0.027	-0.267	-0.032	-0.282	0.033
180°	W_L or W_E	-0.318	0.043	-0.500	-0.048	-0.318	0.044
	W_F	0.159	0.022	0.068	-0.024	0.159	0.022
	W_P	0.0	0.022	-0.250	-0.024	-0.318	0.022

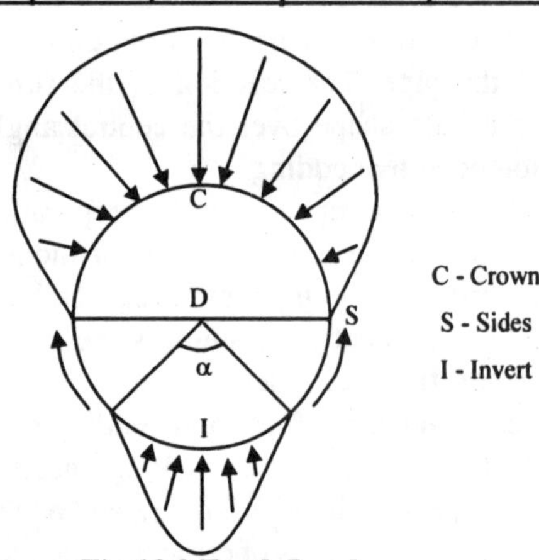

Fig 13.3 Earth Load

13. 7 Design of Prestressed Concrete Pipes

Design of prestressed concrete pipes is made as per specifications IS 784[3]. American Water Work Association (AWWA)[4] also gives some design specifications. BS 4265 lays down specifications for manufacture and testing of prestressed pressure pipes.

13.7.1 Design specifications as per IS 784

- The minimum grade of concrete should be M45 for core concrete, the minimum concrete cover on all reinforcement in any finished pipes should be 12 mm.
- The circumferential prestressing steel should be protected against corrosion due to ingress on water by the outer cover coating of cement mortar or concrete or any approved material with a minimum thickness of 18 mm.
- Outer protective coatings should be applied within 16 hours on circumferential wire winding.
- The tensile stress in cover coating due to any reason shall not cross the modulus of rupture of coating material.
- The cover coatings can be reduced to 15 mm if a precompression is available.
- Inner protective coatings in the form of non cementitious material like epoxy, bitumen etc., should be used after checking the basic material properties for its ingredients. There should not be any element which is injuries to health.
- Longitudinal prestressing may be done using high tensile wires anchored in permenant anchorages. This joint portion at each end should transfer pre-compression to the core concrete. The spacing shall not be greater than twice the thickness or 150 mm which ever is greater. Transfer of forces shall not happen before the core concrete attains a strength 0.55 f_{ck} or 1.8 times induced the longitudinal prestressing.

Curtis and Cowans [5] have suggested recommendations for evaluating longitudinal prestressing steel area as follows.

$$A_t = \frac{2\pi t_c(0.5D + 0.5t)(0.275 f_{cp} - f_{ct})}{f_{pl}}$$

t = core concrete shell thickness

f_{pl} = stress in longitudinal steel at the time of wire winding

f_{cp} = average circumferential stress in core wall on account of circumferential prestressing.

f_{ct} = Permissible tensile stress in core concrete at the time of wire winding.

Doanides [6] has also made suggestions for longitudinal prestress at the time of wire winding as follows.

- Provision may be made to cover a possible maximum tension of $0.6f_{cp}$ at the spigot in the longitudinal direction where f_{cp} is a compressive stress generated in the circumferential direction at the time of transfer. The tension may get reduced to 0.355 f_{cp} during service load stage.

- The longitudinal prestressing is limited so that the compressive stress in concrete does not cross 0.8 $\sqrt{f_{ci}}$ at the time of transfer and 0.5 $\sqrt{f_{ck}}$ at service load stage.
- Circumferential Prestressing can be made only after concrete reaches a minimum strength of 32 Mpa. The compression caused by circumferential prestressing shall not exceed 55% of strength of concrete at the time of transfer of prestress. The clear spacing of successive turns of circumferential prestressing should be less than (a) diameter of wire, (b) 1.5 times the maximum nominal size of the aggregate used for outer coat concrete (c) 5 mm, and whichever is the greater.
- Whenever the tensile stress is generated, the permissible stress can be limited direct tensile stress of concrete which could be assumed 0.6 times the flexural tensile stresses as per Neville [7]. The permissible circumferential stress and longitudinal stress in core concrete under different load combinations are given in Table 13.2 and Table 13.3 respectively.

Table 13.2 Permissible Circumferential Stresses in Core Concrete

S.No.	Load combination	Permissible stress (Mpa)	
		Tension	Compression
1.	Operating conditions (self weight + weight of water +earth fill load + working pressure +final prestress)	0	$0.33f_{ck}$
2.	Operating conditions plus surge of water hammer pressure.	$0.56\sqrt{f_{ck}}$	$0.4 f_{ck}$
3.	Operating conditions plus live load (with impact)	$0.56\sqrt{f_{ck}}$	$0.4 f_{ck}$
4.*	Hydrostatic proof test pressure + self weight of water + final prestress.	$0.56\sqrt{f_{ck}}$	$0.4 f_{ck}$
5.	Self weight + initial prestress.	0	$0.50 f_{ck}$

* The hydrostatic proof test pressure should be taken equal to 1.5 times the working pressure or working plus allowance for surge, whichever is greater.

Table 13.3 Permissible longitudinal stresses in core concrete

S. No.	Load combination	Permissible stress (Mpa)	
		Tension	Compression
1.	Longitudinal prestress after losses		Minimum residual compression, $f_{tc} = IMPa$
2.	Initial prestress less the loss of longitudinal prestress at the time of wire winding.	$0.56\sqrt{f_{ck}}$	$0.5\,f_{ck}$
3. .*	Longitudinal prestress after losses plus beam action	$0.34\sqrt{f_{ck}}$	$0.5\,f_{ck}$
4.*	Handling before winding	0.67 Mpa	$0.5\,f_{ck}$

* To compute the longitudinal stresses due to beam action caused by uneven bearing, the pipe should be assumed to span its full length and support a total load equal to the self weight of the pipe, the weight of water in the pipe and an earth load equal to 2.2 times the diameter of the pipe (in meters) inkN per linear metre.

13.8 Design of Prestressed Concrete Non Cylinder Pipes

The design philosophy is that there should be no tensile stress generated at any stage in the pipe both the circumferential direction or along the length of the pipe and if at all, a residual compression of 1 Mpa has to be maintained. The circular pipes with free ends are subjected to hoop tension due to water pressure inside. These hoop tensile stresses are to be nullified by the hoop compression generated by circumferential prestressing and if necessary some amount of residual hoop compression may be available in concrete. The prestress on the outer pheriphery of the pipes cannot be effected at one stretch all together. Progressive winding of wires arround the pipe length leads to reduction of diameter at places where prestressing is already effected and hence there is curvature introduced along the length of pipe leading to flexural tensile stresses. There may be longitudinal flexural tensile stresses also due to bending of the pipe as a whole during to handling or external loads like over burden sort etc. To keep the pipes cracks free prestressing is normally resorted to. The design is based on the structural behaviour explained.

IS 784 [3], defines different stages for which the design has to satisfy the stress limitations on the materials. They are

- Circumferential prestressing with or without longitudinal prestressing. Bending of pipes with or without longitudinal prestressing.

- Condition in which a pipe is supported by saddles at extreme points with full water load but zero hydrostatic pressure,
- Full working pressure conforming to the limit state of serviceability, and
- The first crack stage corresponding to the limit state of local damage.

In addition, it is also necessary to examine the stage of bursting or failure of pipes corresponding to the limit state of collapse, mainly to ensure a desirable load factor against collapse.

The design output consists of arriving at the thickness of the shell, prestressing force in the circumferential direction or the diameter of prestressing wire, their spacing and stress level, prestressing force in the longitudinal direction or the diameter of prestressing wire, their spacing and stress level. The data available are, the internal pressure to the resisted, and the diameter of pipe, flow conditions, the material properties of concrete and steel, the residual compression required in concrete. Fig. 13.4 gives the details of the forces in a slice of pipe.

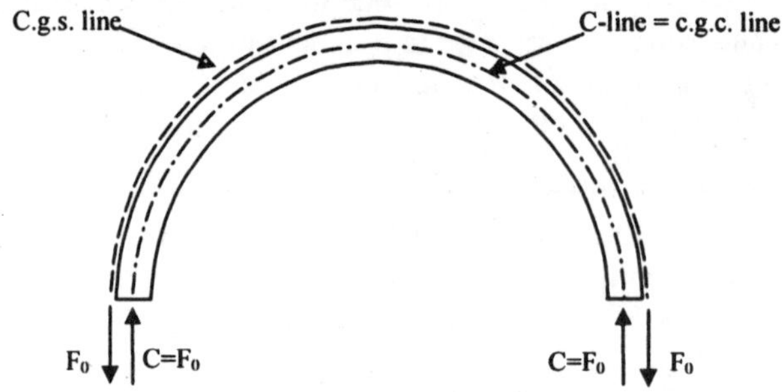

(a) Due to Pressers F_0

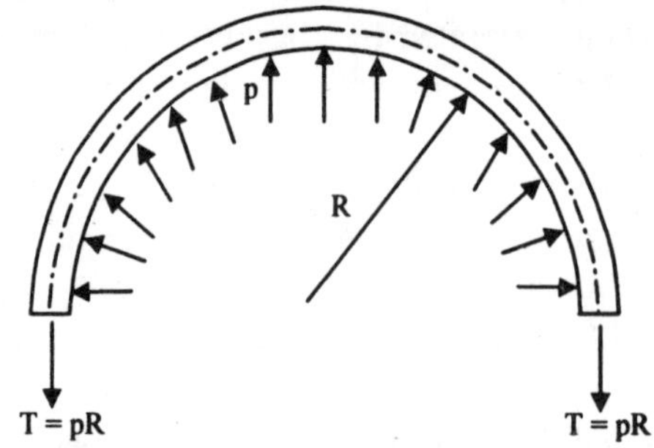

(b) Due to Internal Pressure p

Fig 13. 4 Forces in a horizontal slice of pipe (half slice as freebody)

The design water pressure inside the pipe p_w

Diameter of pipe D

Thickness of pipe t

Permissible compression stress in concrete at transfer $\overline{f_{ct}}$

Minimum compression stress to be available as residual compression $\overline{f_{c\,min}}$

The hoop tensile force on the pipe due to

Water pressure $\dfrac{p_W.D}{2}$

Hoop tensile stress in concrete $\dfrac{p_W D}{2t}$

Hoop tensile stress generated should be less than the available hoop compressive stress generated by prestressing. The available compression stress due to prestress after losses = $\eta\overline{f_{ct}} - f_{cmin}$

Hence $\dfrac{p_w D}{2t} < \eta\overline{f_{ct}} - f_{cmin}$

$$t \geq \left[\frac{p_w D}{2}\right]\frac{1}{\eta\overline{f}_{ct} - f_{c\min}}$$

$$t \geq T_w.\frac{1}{\eta\overline{f}_{ct} - f_{c\min}}$$

Where T_w is the Hoop tension and 't' is the thickness of core concrete shell. The minimum residual compressive stress is defined by statutory authorities in different ways. I.S 784 specified this value as $0.2 f_{ck}$. This is too much on the conservative side. British and American codes permit even tension to a limited level at service load stages. The service load stage is a combination

of load stages. ACI 344-2R permits even tension of 0.3 $\sqrt{f_{cy}}$ N/mm^2.

Hence the prestress in the circumferential direction should create a compressive stress of $\overline{f_{ct}}$ at transfer in such a way there is a residual compression after the pipe is subjected to internal of p_w

If 't' is the actual thickness of the shell provided.

$$f_{ct} = \frac{T_w}{\eta.t} + \frac{f_{cmin}}{\eta}$$

If the prestressing force transmitted in one meter length of the pipe is P. P =[1000 × 2t.f_{ct}] where t is in mm, f_{ct} is in N/mm^2, P is in N/m.
If it is achieved by wires of diameter 'd' through 'n' number of turns, and A_s is the area of cross section of steel, f_s is stress in steel wire at transfer.

$$A_s.f_s = P = 1000 \times 2 \times t.f_{ct}$$

$$A_s.f_s = \frac{\pi d^2}{4}.2.n$$

$$\frac{\pi d^2}{4}.2.n = 2000.t.f_{ct}$$

$$n = \frac{4000.t.f_{ct}}{\pi d^2.f_s}$$

There will be elastic deformation of concrete while transferring prestress and hence there will be loss in prestress in steel to that extent f_{ste} is the effective stress in steel. The corresponding f_{sti} (initial stress) has to be higher to take care of elastic shortening less.

$$\frac{f_{sti} - f_{ste}}{E_s} = \frac{f_{ci}}{E_c} = \frac{A_s f_{ste}}{2000t}.\frac{1}{E_c}$$

$$\frac{E_s}{E_c} = m; \quad \frac{A_s}{2000t} = \mu \text{ (percentage steel)}$$

$$\frac{f_{ct}}{E_c} = \mu.\frac{f_{ste}}{E_c}$$

$$\frac{f_{sti} - f_{ste}}{f_{ct}} = \frac{E_s}{E_c} = m$$

$$\frac{f_{sti} - f_{ste}}{\mu.f_{ste}} = m$$

$$\frac{f_{sti} - f_{ste}}{f_{ct}} = \frac{E_s}{E_c} = m$$

$$f_{sti} - f_{ste} = \text{m}.\mu f_{ste}$$

$$f_{sti} = (1 + \text{m}\mu) f_{ste}$$

For prestressed concrete non-cylindrical pipe μ varies 0.5 to 1% and the ratio of modulli of elasticity of high strength concrete and high tensile steel varies from 5 to 8, f_{sti} could also include other losses due to relaxation, creep and shrinkage.

The longitudinal tensile stresses could be generated due to handling of pipe as a beam, due to unsupported length when laid in soil, with or without water load, and during prestressing. Since prestressing is transferred only section after section for the prestressing stages, solution can be found out using the influence lines, drawn based on the principle of beam on elastic foundations as explained in art 13.12. Anyhow the field condition are only approximated in any design or analysis.

Example 13.1

Design a prestressed concrete non cylindrical pipe of internal diameter of 500mm to carry water with a pressure of 1N/mm^2. The concrete used has 14N/mm^2 as permissible compressive stress at transfer. Steel wire of 3mm diameter is available and it can be stressed to a level of 1000N/mm^2 .The prestressing losses may be assumed to be 20% and a loss coefficient of 0.8 may be assumed. Barring other loads and their combinations, for water pressure load, the pipe concrete should have a residual compressive stress 0.7 N/mm^2 at service load conditions. Evaluate also the test pressure that the pipe can with stand if concrete can be stressed to a level of 0.7 N/mm^2 in tension. For prestressing, steel, E modulus may be assumed to be 250kN/m^2 and that of concrete is 35kN/m^2.

Solution

Pipe diameter	500mm
Internal prestressing water	1N/mm^2
Permissible compressive stress	14 N/mm^2
Residual compression required	0.7 N/mm^2

Hence thickness of pipe 't' $\quad t = [\dfrac{1.0 \times \frac{500}{2}}{0.8 \times 14 - 0.7}] \approx 24\text{mm}$

Adopt a thickness of 25 mm

Actual compressive stress in concrete = f_{ct}

$$f_{ct} = \frac{\frac{500}{2}}{0.8 \times 25} + \frac{0.7}{0.8} = 13.4 \text{ N / mm}^2$$

Assuming 3mm steel wire to be used with a final stress level of 1000N/mm^2 the number of turns/meter length of pipe length will be

$$\frac{4000 \times 13.4 \times 25}{\pi \times 9 \times 1000} = 48 \text{ turn / m. length}$$

$$\text{Spacing} = \frac{1000}{48} = 21mm$$

If p_w is the test pressure immediately after transfer of prestress ($\eta = 1$)

$$f_{ct} = \frac{p_{W.}D}{2t} + \frac{f_{t,min}}{\eta}$$

$$P_W = \frac{2t\beta}{D}(f_{ct} - \frac{f_t \min}{\eta})$$

$$P_W = \frac{2t\beta}{D}(f_{ct} - \frac{f_t \min}{\eta})$$

If f_{si} is the winding stress. m = 6, $\mu = 0.14$

f_{sc} = [1+6×0.00014] 1000 = 1084 N/mm^2

According to Curties and Cowans recommendations

$$A_l = \frac{2\pi t_c(\frac{D}{2} + 0.5t_c)(0.275\overline{f_{cp} - f_{ci}})}{f_{pi}}$$

$$\frac{2\pi(25)(250+2.5)(0.275\times 13.4-0)}{1000} = 151.868 \text{ Sq. mm.}$$

4mm wires spaced at 125 mm along the perimeter will give an area of 152 Sq.mm and it would be adequate to provide the longitudinal prestressing.

Example 13.2

A prestressed concrete pipe is to be designed to withstand a liquid pressure of 1.6N/mm^2 internally. The diameter of the pipe is 1200mm and a shell thickness of 100 mm is proposed. The maximum permissible compressive stress in concrete is 16N/mm^2 at transfer. A residual compression of 2N/mm^2 is expected to be maintained at service load. The possible losses in prestress can be expected to be 20%. 4 mm and 7 mm high-tension wires are available. The maximum level to which they can be subjected to is 1000 N/mm^2 at the time of prestress. Permissible longitudinal tensile stress is 0.8 $\sqrt{f_{ci}}$. The strength of concrete at transfer is 50 N/mm^2. Check the safety of the designed pipe against longitudinal stresses and bursting as per IS 784 – specifications.

Solution

Internal dia 1200 mm

Thickness 100 mm

Length of one pipe unit 6 m

Permissible compressive stress in concrete at

transfer of prestress 16 N/mm^2

Permissible tensile stress 2 N/mm^2

Strength of concrete at the time of transfer of prestress 50 N/mm^2

Hightensile wires available 7mm, 5mm, 4mm

Permissible prestressing level in steel 1000 N/mm^2

Loss factor on wires .0.8

Assuming the shell thickness of 100 mm, the hoop compressive stress to be generated due to prestress

$$= \frac{N_d}{\eta t} + \frac{f_c min}{\eta}$$

$$= \frac{1.6\times 1200}{2\times 0.8\times 100} + \frac{2}{0.8} = 12 + 2.5 = 14.5\text{N} / \text{mm}^2$$

If the internal pressure is not there, there will be hoop compressive stress of 14.5 N/mm^2 which is less than 16 N/mm^2

Assuming 5 mm wires to be used for circumferential prestressing and the level of prestress is 1000 N/mm^2 .

The number of windings/metre length 'n'

$$n = \frac{4000\times 100\times 14.5}{\pi\times 5^2\times 1000} = 73.89$$

$$\text{Pitch} = \frac{1000}{73.89} = 13.53\ mm$$

Due to circumferential compression of $f_{cp'}$ = 14.5 N/mm^2, the transient tensions developed at spigot end in the longitudinal direction as per Doanides [6] =0.6 f_{cp}

= 0.6 × 14.5 = 8.7 *N/mm*^{2}

Maximum tensile stress presented in the longitudinal direction

$= 0.8\sqrt{50} = 5.61\text{N} / \text{mm}^2$

External compression to be given through longitudinal prestressing

= 8.7 – 5.6 = 3.1 N/mm^2

Cross section area of pipe = (π × 1.4 × 0.10) = 0.44 Sq.m

Force required = ((π × 1.4 × 0.1 ×10^6) × 3.1 = 13.63 kN

Using 7mm wires at 1000 N/mm^2 as prestressing steel

No. of wires of 7 mm dia required $= 13.63 \times \dfrac{1000}{1000} \times \dfrac{1}{38.465} = 35.42$

Adopt 7 mm at 115 mm c/c

Flexural stresses has to be checked for conditions specified in I.S.784. Then stresses are expected to be developed during handling or due to eventual condition that pipe is not fully resting in soil and resting only at the end due to enlarged diameter.

Self weight = (1.30 × 0.10 × 24 × 3.14) 9.8 kN/m

Water load on the pipe $= \dfrac{\pi}{4} \times 1.2^2 \times 1000 = 36.36\text{kN} / \text{mR}$

Total load = 9.8 + 36.36 = 46.16kN/m

Bending moment caused by this load $= 46.16 \times 46.16 \times \dfrac{6^2}{8} = 207.72 kNm$

Modulus of section of the pipe section $= \dfrac{\pi}{64}\left(1.4^4 - 1.20^4\right) = 0.0866\text{m}^3$

Maximum bending stress generated $= \dfrac{207.72}{0.6} = 2398.6 = 2.398 kN / mm^2$

Longitudinal prestress available = 3.1N/mm^2

Net Compression available after is nullifying the tension = 3.1 – 2.399

= 0.7 N/mm^2

13.9 Prestressed Concrete Water Tanks

As in the case of pipes, or any water carrying or retaining structure, it is of primary importance that water tanks also should be impervious to the penetration of liquid either from internal leakage or external seepage due to rains. Based on this requirement the restrictions are laid on liquid retaining structures in different codes of practice. The Indian Standards Institution lays out the relevant conditions in Code I.S.3370 — 1965 [8] in four different parts. It demands that the design of liquid retaining structures have to be based on no cracking in concrete due to tensile stresses may be caused by i) liquid pressure from inside, ii) earth pressure from outside incase of underground structures, iii) the interaction between the reinforcement and concrete during drying shrinkage (iv) the boundary conditions at foundations and/or other parts of structure (v) the differential temperature conditions at the faces due to the large thickness of mass concrete. I.S. code also limits the tensile stresses based on cracking at both faces, irrespective of loading in cases of members less than 225 mm thick. Hence it becomes absolutely necessary either to introduce certain factors which could bring in precompression in desired places (where cracking is expected) by some means or to reduce the possibility of causes of occurrences of cracks. The later possibility involves lot of troubles and hence the first approach is normally preferred and the same is achieved by prestressing. Accordingly prestressing has become an automatic choice in case of large liquid retaining structures to keep them water tight, leak free and who keep the dimensions with in acceptable limits.

13.10 Choice of Shape of Tanks

Structures of different shapes and forms are being chosen depending on various factors like (i) availability of space (ii) aesthetic considerations (iii) design and construction feasibility etc. A cylindrical structure is one of the simplest form which can carry tensile stresses more by membrane action. Hence it possesses considerable advantages of economy. The design of cylindrical tank is also simple avoiding biaxial stresses when compared to rectangular or any prismatic shape. The scope of this chapter is limited to only prestressed concrete cylindrical pipes and tanks. A typical sketch of a prestressed concrete service reservoir with economical dimensional proportion is given in Fig 13.5. While choosing the dimension for a tank the proportions of various parameters may be used as given in 13.5 and explained in Table 13.9 later.

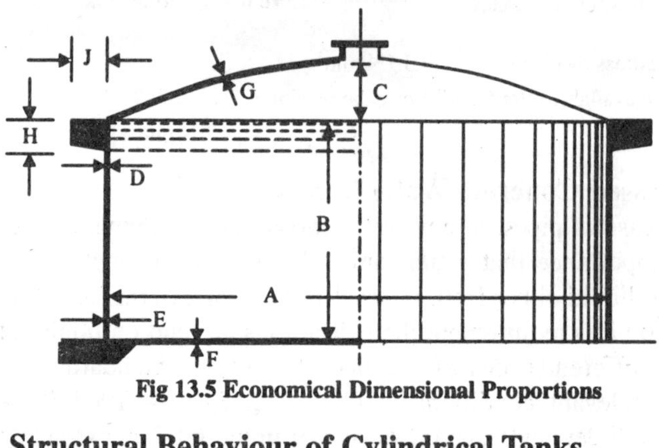

Fig 13.5 Economical Dimensional Proportions

13.11 Structural Behaviour of Cylindrical Tanks

In case of cylindrical tanks, whose walls are free to deform at top and bottom, the walls should expand circumferentially due to radial pressure of liquid unlike in pipes, this is a varying pressure over the depth. With walls having uniform thickness, the extension will increase linearly from top to bottom as shown in dotted line in the Fig.13.6. If the walls are having restrained end conditions the expansion and angular movement of wall at that point will be controlled. This causes a bending moment (M_0) and a radial force (Q) at the restrained ends. This will automatically have some effect on the stress distribution at different points on the structure i.e. Hoop tension along the height of the wall will also be affected at different points. Fig.13.6 illustrates the behaviour of tanks whose walls are monolithic with rigid base and free at top. Radial force Q at support will nullify the deflection at point B and the bending moment M_0 will resist the angular deformation at B due to load and the radial force (shear) Q. Even if the end condition at B is different, equilibrium and compatibility equations at that point can be written and the unknown forces could be solved. Considering the wall of the tank to be cut vertically in several strips of unit width along the pheriphery, it is noted that under the action of hydrostatic pressure each of these strips will tends to bend outwards as vetical beams. Hence the deflection of all the strips at different heights above the base must be equal. As these strips deflect outwards, the circumference of the tank increases. Since the strips are monolithically combined with each other the increase in circumference of the tank due to deflection cannot take place without causing hoop tension in the wall. If 'T' is the hoop force/unit height it will give rise to inward radial force 2T/D. This can be represented as p_h. This varies from top to bottom and depends on the level 'h' from top. If the net outward deflection at a section is y. the diameter of the tank at the section increases by 2y and causes a circumferential strain 2y/D. If 't' is thickness of well, hoop tensile force, T can be expressed as

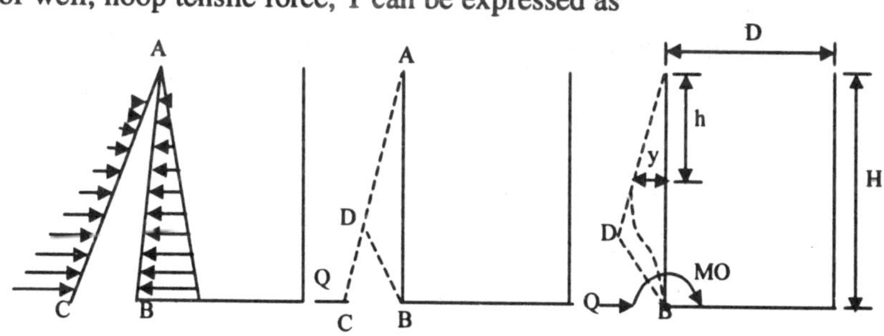

Fig 13.6 Structural Behavior of Tanks

$$T = \frac{2y}{D} E(1 \times t); p_h = \frac{T}{D/2} = \frac{4yE.t}{D^2}$$

In each strip there will be an outward force = W_h. Therefore net force causing deflection $p_b = W_h - p_h$.The action of wall is similar to that of a two way slab where the load is in two directions. The distribution of water pressure will cause hoop tension and beam action. It will be such that the increase in circumference due to hoop force is the same as the increase in circumference due to outward deflection due to beam action. The relationship between deflection, bending moment and load in the beam is given by

$$EI\frac{d^4y}{dh^4} = p_b$$ where I is moment of inertia per unit width of tank wall.

$$EI\frac{d^4y}{dh^4} = W_h - p_h$$

$$EI\frac{d^4y}{dh^4} - W_h \frac{4yEt}{D^2} = 0$$

If W_h is represented as P the expression can be written as

$$EI\frac{d^4y}{dh^4} + y\frac{Et}{R^2} - P = 0$$

Where 'y' is the deflection or change in radius inwards due to radial pressure P. The circumferential strain due to $P = y/R$.

$$\frac{y}{R} = \frac{2\pi(R-y) - 2R}{2\pi R}$$; Hoop stress (compression) = $\frac{(PR)}{t}$

Hoop strain = $\frac{PR}{Et}(-ve)$

Equating circumferential strain, to Hoop Strain

$$\frac{y}{R} = \frac{PR}{Et}; \frac{P}{y} = \frac{Et}{R^2}$$

Force per unit deflection is defined as k.

$$k = \frac{Et}{R^2}$$

Substituting k for $\frac{Et}{R^2}$ in the basic equation of equilibrium

$$EI.\frac{d^4y}{dh^4} + k.y = P$$

$$\frac{d^4y}{dh^4} + \frac{k}{EI} = \frac{P}{EI}$$

$$y = \frac{P}{4EI\beta^4} + e^{+\beta h}(C_1 \cos \beta h + C_2 \sin \beta h)$$

$$+ e^{-\beta h}(C_3 \cos \beta h + C_4 \sin \beta h) \quad where \quad \beta = \sqrt[4]{\frac{k}{EI}}$$

The solution for this equation is

$$y = \frac{W_h}{4EI\beta^4} + e^{\beta h}(A \sin \beta h + B \cos \beta h) + e^{-\beta h}(C \sin \beta h + D \cos \beta h)$$

Table 13.4

Sl. No.	End Conditions	Boundary conditions
1.	Fixity at bottom top free	$h = 0, \frac{d^2y}{dh^2} = 0, \frac{d^3y}{dh^3} = 0$ $\text{h} = \text{H}, \frac{dy}{dx} = 0, y = 0$
2.	Hinged bottom, top free	$h = 0, \frac{d^2y}{dx^2} = 0, \frac{d^3y}{dh^3} = 0$ $\text{h} = \text{H}, \frac{d^2y}{dh^2} = 0, y = 0$
3.	Hinged bottom, top hinged to the dome.	$h = 0, \frac{d^2y}{dh^2} = 0, y = y(dome)$ $\text{h} = \text{H}, \frac{d^2y}{dh^2} = 0, y = 0$
4.	Semi sliding with hinged end at bottom, hinged at top.	$h = 0, \frac{d^2y}{dh^2} = 0, y = y(dome)$ $\text{h} = \text{H}, \frac{d^2y}{dh^2} = 0, y = y_2$

From this expression, the inward radial force p_h at any point and bending moment at any level can be solved. The constants of the above equation are evaluated using the boundary conditions at top and bottom. The boundary conditions for various end conditions are mathematically expressed as shown in Table 13.4.

It could seen that degree of restraint imposed at the junction between the wall and the base and the between the wall and top cover will have an over riding influence on the design. To overcome the effect of tensile stresses due to hoop forces and bending action a precompression is introduced corresponding to hoop tension and flexural tension by prestressing the wall circumferentially and vertically Theoretical solution for several combination of boundary conditions at the top and at bottom, with or without varying thickness of the side well has been given by Timoshenko [9], W.S. Gray [10] and many others. A typical case of the tank with fixed base and free top is shown in Fig. 13.7 for the benefit of readers.

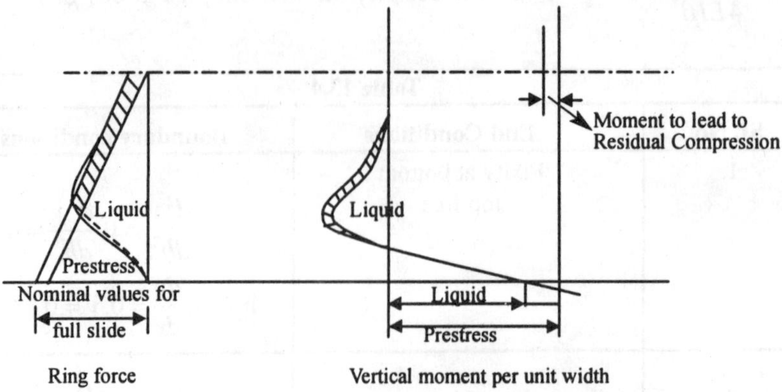

Fig 13.7. Ring Forces and Vertical Moment (Fixed base)

13.12 Design Effects of Prestressing.

The object of prestressing in a cylindrical tank is to apply opposing load to liquid pressure which will introduce ring compression and flexural Stresses neutralising the effect of liquid forces. By this arrangement of balancing the liquid forces by prestressing, the concrete walls are placed under minimum stress condition when the tank is full. Thus the danger of cracking of concrete is greatly reduced. As an additional precaution against seepage from tank, the circumferential prestressing, is done in such a way that there is a residual compression in concrete when the tank is full. Depending on edge conditions, there may be flexural stresses on the vertical section of the wall which are counteracted by opposite type of stresses introduced by vertical prestressing. In order to avoid development of crack in the wall of the tank when the tank is empty, the vertical prestresssing is so arranged that there is a residual tension less than the modulus of rupture value for concrete.

The vertical prestress could very well be increased to have zero tension or residual compression. But this may lead to a possibility of compressive stress at outer face exceeding the limit when tank is empty. In short, it may be remembered that horizontal prestress leads to flexural forces caused by introduction of edge restraints. Variation of types of restraints at the wall foot and or at top modify the induced effects like the ring tension and vertical moment for no change in the liquid pressure or prestressing force. Behaviour of cylindrical tanks with different boundary condition at the base and at the top and the mathematical formulation are given by Crezy [11] in his book on water tanks. I.S.337-(part IV) –1965 give the details of hoop tension at various heights and the bending moment for different ends conditions of water tanks. They are also given in Table 13.5;13.6, 13.7, 13.8.

13.13 Design for Flexure in Vertical Plane

These coefficients for hoop forces and bending moments are given when the tank is full. In non-prestressed water tanks the actual stresses on the wall (both Hoop and flexural) are zero when the tank is empty. When the tank is filled with water slowly it starts acting and stresses are getting developed. When the tank is partially full, the hoop stresses developed will be lower than that when the tank is full and hence the hoop stresses need not be checked up against limiting values. But elastic curve in the vertical plane is not similar to elastic curve when the tank is full and the flexural stresses have to be checked up at various stages. But it is not done in practice since the stresses are expected to be with in permissible limits because of large thickness of wall necessited by uncracked conditions. But in case of prestressing, two points are to be kept in mind. Cracking is avoided by prestressing and hence wall thickness assumed is very small. Secondly, the prestressing is done from the bottom. The prestressing force required the bottom for tank full condtions are very high. This is transmitted at the bottom even when the tank is empty. Such high concentrated prestressing force only at bottom leads to an elastic curve which causes high flexural stresses.

Hence various stages of prestressing have to be considered for checking for stresses and deciding about the vertical prestressing. Influence lines are drawn for effect of prestressing at various sections for hoop forces and flexural forces. From the influence lines, the maximum hoop compression, flexural stresses, the sections at which they occur the stages of prestressing etc when they occur, etc., are decided The details for drawing influence line for moments in vertical plane are discussed in at 13.12.

But in the simplified approach, for the design, only final stage is considered and prestressing forces required the vertical direction to keep any section with in the limiting stresses is decided. Here too the bending moment created by circular prestressing is evaluated proportionately from such bending moment caused by water forces from inside. This is only as approximate method.

Table 13.5.

$\frac{H^2}{Dt}$	Coefficients at point 0.1H	0.2H	0.3H	0.4H	0.5H	0.6H	0.7H	0.8H	0.9H	1.0H
0.4	+.0005	+.0014	+.0021	+.0007	–.0042	–.0150	–.0302	–.0529	–.0816	–.1205
0.8	+.0011	+.0037	+.0063	+.0080	+.0070	+.0023	–.0068	–.0224	–.0465	–.0795
1.2	+.0012	+.0042	+.0077	+.0103	+.0112	+.0090	+.0022	–.0108	–.0311	–.0602
1.6	+.0011	+.0041	+.0075	+.0107	+.0121	+.0111	+.0058	–.0051	–.0232	–.0505
2.0	+.0010	+.0035	+.0068	+.0099	+.0120	+.0115	+.0075	–.0021	–.0185	–.0436
3.0	+.0006	+.0024	+.0047	+.0071	+.0090	+.0097	+.0077	+.0012	–.0119	–.0333
4.0	+.0003	+.0015	+.0028	+.0047	+.0066	+.0077	+.0069	+.0023	–.0080	–.0268
5.0	+.0002	+.0008	+.0016	+.0029	+.0046	+.0059	+.0059	+.0028	–.0058	–.0222
6.0	+.0001	+.0003	+.0008	+.0019	+.0032	+.0046	+.0051	+.0029	–.0041	–.0187
8.0	.0000	+.0001	+.0002	+.0008	+.0016	+.0028	+.0038	+.0029	–.0022	–.0146
10.0	.0000	.0000	+.0001	+.0004	+.0007	+.0019	+.0029	+.0028	–.0012	–.0122
12.0	.0000	–.0001	+.0001	+.0002	+.0003	+.0013	+.0023	+.0026	–.0005	–.0104
14.0	.0000	.0000	.0000	.0000	.0001	+.0008	+.0019	+.0023	+.0001	–.0090
16.0	.0000	.0000	–.0001	–.0002	–.0001	+.0004	+.0013	+.0019	+.0001	–.0079

	Coefficients at points .80H	.85H	.90H	.95H	1.00H
20.0	+.0015	+.0014	+.0005	–.0018	–.0063
24.0	+.0012	+.0012	+.0007	–.0013	–.0053
32.0	+.0007	+.0003	+.0007	–.0008	–.0040
40.0	+.0002	+.0005	+.0005	–.0005	–.0032
48.0	.0000	+.0001	+.0005	–.0003	–.0026
56.0	.0000	.0000	+.0004	–.0001	–.0023

Moments in Cylindrical Walls - Fixed Base Free at Top (IS : 3370-Part IV)

Moment M_w = (Coefficient) × (wH^3) kNm/m

Positive sign indicates tension at the outside face

Table 13.6.

$\frac{H^2}{Dt}$	Coefficients at point									
	0.0H	**0.1H**	**0.2H**	**0.3H**	**0.4H**	**0.5H**	**0.6H**	**0.7H**	**0.8H**	**0.9H**
0.4	+0.149	+0.134	+0.120	+0.101	+0.082	+0.066	+0.049	+0.029	+0.014	+0.004
0.8	+0.263	+0.239	+0.215	+0.190	+0.160	+0.130	+0.096	+0.063	+0.034	+0.010
1.2	+0.283	+0.271	+0.254	+0.234	+0.209	+0.180	+0.142	+0.099	+0.054	+0.016
1.6	+0.265	+0.268	+0.268	+0.266	+0.250	+0.226	+0.185	+0.134	+0.075	+0.023
2.0	+0.234	+0.251	+0.273	+0.285	+0.285	+0.274	+0.232	+0.172	+0.104	+0.031
3.0	+0.134	+0.203	+0.267	+0.322	+0.357	+0.362	+0.330	+0.262	+0.157	+0.052
4.0	+0.067	+0.164	+0.256	+0.339	+0.403	+0.429	+0.409	+0.334	+0.210	+0.073
5.0	+0.025	+0.137	+0.245	+0.346	+0.428	+0.477	+0.469	+0.398	+0.259	+0.092
6.0	+0.018	+0.119	+0.234	+0.344	+0.441	+0.504	+0.514	+0.447	+0.301	+0.112
8.0	–0.011	+0.104	+0.218	+0.335	+0.443	+0.534	+0.575	+0.530	+0.381	+0.151
10.0	–0.011	+0.098	+0.208	+0.323	+0.437	+0.542	+0.608	+0.589	+0.440	+0.179
12.0	–0.005	+0.097	+0.202	+0.312	+0.429	+0.543	+0.628	+0.633	+0.494	+0.211
14.0	–0.002	+0.098	+0.200	+0.306	+0.420	+0.539	+0.639	+0.666	+0.541	+0.241
16.0	–0.000	+0.099	+0.199	+0.304	+0.412	+0.531	+0.641	+0.687	+0.582	+0.265

	Coefficients at points				
	.75H	**.80H**	**.85H**	**.90H**	**95H**
20.0	+0.716	+0.654	+0.520	+0.325	+0.115
24.0	+0.746	+0.702	+0.577	+0.372	+0.137
32.0	+0.782	+0.768	+0.663	+0.459	+0.182
40.0	+0.800	+0.805	+0.731	+0.530	+0.217
48.0	+0.791	+0.828	+0.785	+0.593	+0.254
56.0	+0.763	+0.838	+0.824	+0.536	+0.285

Ring Tension Cylindrical Walls - Fixed Base Free at Top (IS : 3370-Part IV)

Ring tension N_d = (Coefficient) (wHR) kN/m

Table 13.7

$\frac{H^2}{Dt}$	Coefficients at point									
	0.1H	0.2H	0.3H	0.4H	0.5H	0.6H	0.7H	0.8H	0.9H	1.0H
0.4	+.0020	+.0072	+.0151	+.0230	+.0301	+.0348	+.0357	+.0312	+.0197	0
0.8	+.0019	+.0064	+.0133	+.0207	+.0271	+.0319	+.0329	+.0292	+.0187	0
1.2	+.0016	+.0058	+.0111	+.0177	+.0237	+.0280	+.0296	+.0263	+.0171	0
1.6	+.0012	+.0044	+.0091	+.0145	+.0195	+.0236	+.0255	+.0232	+.0155	0
2.0	+.0006	+.0033	+.0073	+.0114	+.0158	+.0199	+.0218	+.0205	+.0145	0
3.0	+.0004	+.0018	+.0040	+.0063	+.0092	+.0127	+.0252	+.0153	+.0111	0
4.0	+.0001	+.0007	+.0016	+.0033	+.0057	+.0083	+.0109	+.0118	+.0092	0
5.0	.0000	+.0001	+.0006	+.0016	+.0034	+.0057	+.0080	+.0094	+.0078	0
6.0	.0000	.0000	+.0002	.0000	+.0019	+.0059	+.0062	+.0078	+.0068	0
8.0	.0000	.0001	–.0002	.0000	+.0007	+.0020	+.0038	+.0057	+.0054	0
10.0	.0000	.0000	–.0002	–.0001	+.0002	+.0011	+.0025	+.0053	+.0045	0
12.0	.0000	.0000	–.0001	.0002	.0000	+.0005	+.0017	+.0032	+.0039	0
14.0	.0000	.0000	.0000	–.0001	.0001	.0000	+.0012	+.0026	+.0033	0
16.0	.0000	.0000	.0000	–.0001	–.0002	–.0004	+.0008	+.0022	+.0029	0

	Coefficients at points				
	.75H	.80H	.85H	.90H	95H
20.0	+.0008	+.0014	+.0020	+.0024	+.0020
24.0	+.0005	+.0010	+.0015	+.0020	+.0017
32.0	.0000	+.0005	+.0009	+.0014	+.0013
40.0	.0000	+.0003	+.0006	+.0011	+.0011
48.0	.0000	+.0001	+.0004	+.0008	+.0010
56.0	.0000	.0000	+.0003	+.0007	+.0008

Moments in Cylindrical Walls - Hinged Base Free at Top (IS : 3370 - Part IV)
Moment M_w = (Coefficient) (wH^3)kN m/m
Positive sign indicates tension at the outside.

Table 13.8

$\frac{H^2}{Dt}$	Coefficients at point									
	0.0H	0.1H	0.2H	0.3H	0.4H	0.5H	0.6H	0.7H	0.8H	0.9H
0.4	+0.474	+0.440	+0.395	+0.352	+0.305	+0.264	+0.215	+0.165	+0.111	+0.057
0.8	+0.423	+0.402	+0.381	+0.358	+0.330	+0.297	+0.249	+0.202	+0.145	+0.076
1.2	+0.350	+0355.	+0.361	+0.362	+0.358	+0.343	+0.309	+0.256	+0.186	+0.098
1.6	+0.271	+0.303	+0.341	+0.369	+0.385	+0.385	+0.362	+0.314	+0.233	+0.124
2.0	+0.205	+0.260	+0.321	+0.373	+0.411	+0.434	+0.419	+0.369	+0.280	+0.151
3.0	+0.074	+0.179	+0.381	+0.375	+0.449	+0.506	+0.519	+0.479	+0.375	+0.210
4.0	+0.017	+0.137	+0.253	+0.367	+0.469	+0.545	+0.579	+0.553	+0.447	+0.256
5.0	–0.008	+0.114	+0.235	+0.356	+0.469	+0.562	+0.617	+0.606	+0.503	+0.294
6.0	–0.011	+0.103	+0.223	+0.343	+0.463	+0.566	+0.639	+0.643	+0.547	+0.327
8.0	–0.015	+0.096	+0.208	+0.324	+0.443	+0.564	+0.661	+0.697	+0.621	+0.386
10.0	–0.008	+0.095	+0.200	+0.311	+0.428	+0.552	+0.666	+0.730	+0.678	+0.433
12.0	–0.002	+0.097	+0.197	+0.302	+0.417	+0.541	+0.664	+0.750	+0.720	+0.477
14.0	0.000	+0.098	+0.197	+0.299	+0.408	+0.531	+0.659	+0.761	+0.752	+0.513
16.0	+0.002	+0.100	+0.198	+0.299	+0.403	+0.521	+0.650	+0.764	+0.776	+0.543

	Coefficients at points				
	.75H	.80H	.85H	.90H	95H
20.0	+0.812	+0.817	+0.755	+0.603	+0.344
24.0	+0.816	+0.839	+0.793	+0.647	+0.377
32.0	+0.814	+0.861	+0.847	+0.721	+0.436
40.0	+0.802	+0.866	+0.880	+0.778	+0.483
48.0	+0.791	+0.854	+0.900	+0.802	+0.427
56.0	+0.781	+0.859	+0.911	+0.852	+0.563

Ring Tension in Cylindrical Walls - Hinged Base
Free at Top (IS : 3370-Part IV)
Ring tension N_d = (Coefficient) (wHR) kN/m
Positive sign indicates tension

Example 13.3

. A Cylindrical prestressed water tank of internal diameter of 28m to store water to a depth of 8 m is to be designed with M40 concrete. High tensile steel of 5 mm, 7mm and 8mm diameter are available for prestressing. The permissible compressive stress in concrete is limited to 12 N/mm^2 at transfer of prestress at any other stage. During service, a residual compression of 1 N/mm^2 is desired in concrete. High tensile steel can be stressed upto 1000N/mm^2. 20% loss in prestress is expected due to varying causes during the service stages. The base slab of the tank and sidewalls are monolithically cast and hence the base may be assumed to be fixed and top free.

Solution

For the required storage of 8m depth, the average thickness of wall of 150 mm is tentatively assumed for a prestressed concrete tank. The maximum hoop tension that could be developed in this tank and the maximum positive and negative moment along the vertical plane are taken from I.S.3370-1965 (part IV).[Table 13.5 13.6]

H=8m, D=28m, t=0.15m, $\eta = 0.8$,

$$\frac{H^2}{Dt} = \frac{64}{28 \times 0.15} = 15.2$$

ρ_w= 10H = 10 × 8 = 80kN/mm^2

Maximum hoop tension = 0.67 W.H.R.

= 0.67×10×8×14

= 750.4 kN/m

= 750 N/mm

Maximum negative bending moment = .009×W.H^3 kNm/m

= .009×10×8^3

= 46.08kNm/m = 46080Nmm/mm

Minimum wall thickness to satisfy the stress requirement of residual compression = t

$$t = \frac{\rho_w}{\eta f_{ct} - f_{\min}} = \frac{750}{0.8 \times 2 - 1} = \frac{750}{86} = 87.2$$

Thickness available after providing for cover for vertical cables = 150-30=120 mm

Required circumferential prestress $= \left[\frac{750}{0.8 \times 140} + \frac{1}{0.8}\right] = 6.96 + 1.25$

= 8.21N/mm^2

5 mm wire is used for Circumferential prestress. The spacing of 5mm wires along the height

$$= S = \frac{2 \times P_w}{P_w} \cdot \frac{f_s \cdot A_d}{f_c D.t}$$

$$= \frac{2 \times 750 \times 1000 \times 20}{0.08 \times 8.21 \times 28 \times 10^3 \times 140}$$

= 11.6 mm

No. of turns/m .height of the tank $= \frac{1000}{11.6} = 86.2$

The spacing of these wires may be increased as prestressing moves towards the top of the tank. The maximum spacing (minimum force) may be at a height 0.1H from top. In between the spacing can be linearly increased because the water pressure which introduces hoop tension increases linearly towards bottom

Ring Tension at 0.1H from top

= .098 × 10 × 8.0 × 14 = 109.76kN/mm = 10976N/mm

Prestressing required $= f_c$

$$= \frac{109.76}{0.8 \times 140} + \frac{1}{0.8} = 0.98 + 1.25 = 2.23\text{N/mm}^2$$

Spacing of 5mm wire to introduce 2.23 N/mm^2 of hoop compression

$$= \frac{2 \times 109.76 \times 1000 \times 20}{0.8 \times 2.23 \times 28 \times 10^3 \times 140} = 62.3 \text{ mm}$$

No. of wires/metre height of the tankwall $= \frac{1000}{62.3} = 16.2$

Spacing of wires may increased from 11.0mm at bottom to 62mm at top linearly. Since maximum bending moments and hoop tension generated by water pressure is calculated using coefficients, the maximum bending caused by hoop compression due to circular prestressing is evaluated proportionately.

Maximum radial pressure due to prestress

$$= \frac{2.f_s.f_s}{s.D} = \frac{2 \times 20 \times 1000}{11.6 \times 28 \times 10^3} = 0.1231 \text{ N/mm}^2$$

Maximum bending moment in the vertical plane due to the radial pressure caused by prestressing.

$$= M_w \times \frac{W_t}{W_P} = 46080 \times \frac{0.1231}{0.08} = 78905.6\text{Nmm/mm}$$

$= 78.905 \times 10^6$ *Nmm / mm*

The section modulus of the tank thickness

$$Z = 1000 \times \frac{150^2}{6} = 375 \times 10^4 \text{mm}^3$$

The compressive stress generated

$$f_z = \frac{f_c\text{min}}{0.8} + \frac{78.905 \times 10^6}{375 \times 10^4} = 22.06\text{N/mm}^2 > 12.\text{N/mm}^2$$

Vertical prestressing required causes very high compressive stresses. This demands a revision in the thickness of the tank wall. The thickness is increased to 250mm. The circumferential prestress required also gets changed.

Required circumferential prestress $= \frac{750}{0.8 \times 240} + \frac{1}{0.8}$

= 3.91+1.25 = 5.16N/mm^2

Spacing of 5 mm wire of used for circumferential pressers

$$s = \frac{2 \times 750 \times 1000 \times 20}{0.08 \times 5.16 \times 28 \times 10^3 \times 240} \quad 10.8 \text{ mm}$$

No. of turms/m. height of tank $= \frac{1000}{10.8} = 92.5$

Prestressing required at top 0.14 $= \frac{109.76}{0.8 \times 240} + \frac{1}{0.8} = 0.572 + 1.25$

=1.822N/mm^2

Spacing of 5 mm wire to introduce 1.822N/mm^2 compression

$$= \frac{2 \times 109.76 \times 1000 \times 20}{0.08 \times 1.822 \times 28 \times 10^3 \times 240} = 44.8\text{mm}$$

No. of wires/m length at top = $\dfrac{10000}{448} = 22.32$

Spacing of wires may be increased from 9.25 mm from bottom to 44.8 mm at top linearly

Maximum radial pressure at bottom due to pressure = $\dfrac{2A_s.f_s}{sD}$

$$= \frac{2 \times 1000 \times 20}{9.25 \times 28 \times 10^3} = 0.1544 \text{ N/mm}^2$$

Maximum bending moment in vertical plane due to radial pressure

caused by prestressing $= 46080 \times \dfrac{0.1544}{0.08} = 88.93 \times 10^6\ Nmm/m$

Modulus of section $Z = 1000 \times \dfrac{250^2}{6} = 1041.6 \times 10^4 \text{ mm}^3$

Compressive stress generated $= \dfrac{1}{8} + \dfrac{88.93 \times 10^6}{1041.6 \times 10^4} = 1.25 + 0.54$

$= 9.76 \text{ N/mm}^2 < 12\text{N/mm}^2$

Vertical prestressing force required = 9.76 × 1000 × 250 = 2440 × 10^3N

= 2440kN

Force on 7 mm wire for limiting stresses value of 1000/mm^2 = 38.465kN

No. of wire/m of perimeter $= \dfrac{2440}{38.465} = 63.44$

Hence provide 7 ply 12.7mm strand which can give a force of 125 kN

No. of strand required will be $= \dfrac{2440}{125} = 19.52$

Provide 12.7 mm 7 ply strand at 50 mm spacing along the perimeter for vertical prestressing. I.S.code gives guideline for vertical prestress as to generate 0.3 f_c in concrete .0.3 × 9.76 = 2.928N/mm

Vertical prestress required = 2.928 × 1000 × 250 = 732 × 103N = 732kN

But 2440 kN of force is provided which more than adequate.

Ultimate tensile force at the base = 92.5×1500×20 = 2775. P_w = 750kN

Load factor $= \dfrac{2775}{750} = 3.7$

Nominal 0.2% of untensioned steel may be provided both circumferencially and vertically. Spacing of 20 cm may be adopted. 8mm to 12mm dia base may be used

Direct tensile strength of concrete $(0.7\sqrt{f_{ck}})0.6 = 2.65\text{N/m}^2$

Cracking load = 1000 × 250 [0.8 × 5.16 + 2.65]10^{-3}

= 1694.5kN

Factor of safety against cracking $= \dfrac{1694.5}{750} = 2.26$

Hence the design is acceptable. It may be noted that the Hoop Tension and bending moment due to water pressure as well as that to prestress are evaluated using coeff. given in I.S. Code. Hence only the maximum and minimum values are considered. A more rigorous analysis and design can be made as for drawing influence lines as could be seen from next illustrative example.

13.14 Influence line for Moments inVertical plane

Influence line for bending moments on the vertical plane for every cross section at intervals of 0.1H are drawn, by applying unit concentrated force at those sections for the given boundary condition. The equation of equilibrium for an assumed hoop compression deflection (y) caused by an applied unit force around the circumference is written as

$$EI\frac{d^4y}{dh^4} + y\frac{Et}{R^2} - P = 0$$

$$y = \frac{P}{4EI\beta^4} + e^{+\beta h}(C_1 \cos\beta h + C_2 \sin\beta h)$$

$$+e^{-\beta h}(C_3 \cos\beta h + C_4 \sin\beta h), \text{where } \beta = \sqrt[4]{\frac{k}{EI}}$$

Now the edge conditions have to be applied and the constants have to be evaluated. It could be seen from calculation that the complementary solution of the equation contains two parts, which give the disturbances due to edge conditions from two ends. In case of tall tanks, the disturbances due to edge conditions of the farther edge can be neglected without much loss of accuracy. Hence the solution can be simplified as

$$y = f(h) + e^{-\beta h} (C_3 \cos \beta h + C_4 \sin \beta h)$$

For the purpose of explaining the solution, a structure assumed to be at bottom with hinged end. The wall is considered as semi infinite beam on elastic foundation. The solution of basic equation is arrived by the principle of super position.

Referring to the Fig.13.8, conditions hinged end is shown as combination of three systems shown in the fig In the actual condition, there is no moment or deflection at point A (wall foot) Fig.13.9 shows an infinitely long beam subjected to concentrated load P. Due to load P, there will be bending moment (M_1) shear force (Q) and deflection (y_1) at wall foot. First the moment at A is nullified by applying equivalent and opposing moment ($-M_1$) at wall foot. This will introduce additional deflection y_2 at the wall foot. Then a shear force Q is applied nullify the deflection (y_1+y_2). The values of M_1, y_1, y_2 etc can be expressed as follows

$$M_1 = \frac{P}{4\beta}(e^{-\beta l})(Cos\beta l - sin\,\beta l)$$

$$y_1 = \frac{PB}{2k}(e^{\beta l})(Cos\beta l + sin\,\beta l)$$

$$y_2 = \frac{wMl}{K}\beta$$

$$Q_1 = \frac{P}{4}e^{-\beta l}(Cos\beta l + sin\,\beta l) + M_1\beta$$

Considering any section at a distance h from top

$$M_x = \frac{P}{4\beta}e^{-\beta s}(Cos\beta s - Sin\beta s) - M_1 2e^{-\beta(H-h)}$$

$$[Cos\beta(H-h) + Sin\beta(H-h)] + \frac{Q_1}{B}e^{-\beta(H-h)}Sin\beta(H-h); \text{where s} = \text{H} - \text{l} - \text{h}$$

A computer programme is used to compute M_x at sections with x varying from 0.05 H to 0.95 H increase being 0.05H. l is also varied from 0.05 H to 0.5H. P is taken as 100 kg. (unit load) and the influence line for moments at various sections are plotted. One such influence line is shown in Fig.13.9

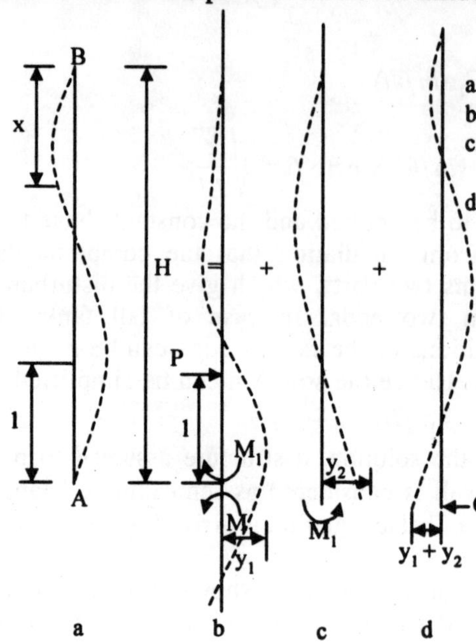

Fig 13.8

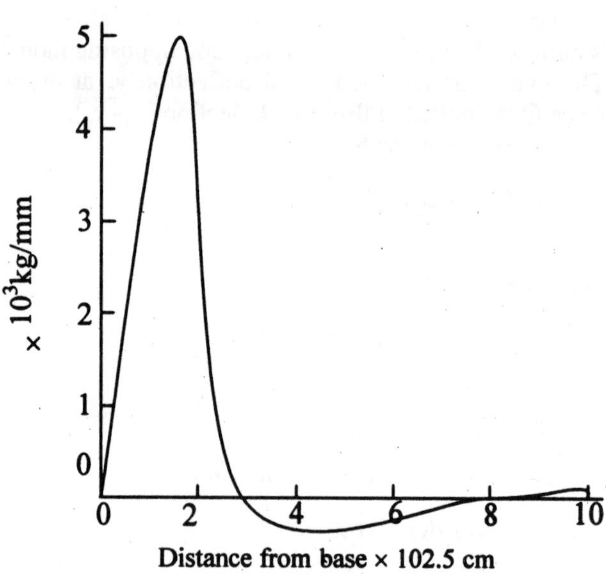

Fig 13.9 Influence Line for Bending Moment at Section at 0.1 H due to Unit Radial Force

13. 15 Design Procedure

The procedure for design of a water tank is explained stepwise on the following lines.

Step 1: The dimensions of the tank based on storage capacity required; space available etc may be first arrived at. The economic proportion of the diameter to the height shall be of the order of 2 to 3:1 for large reservoirs. Approximate thickness of the wall to be provided may also be assumed. Table 13.9 will provide some guidelines for this purpose.

Step 2: Then the system and type of prestressing to be adopted may be fixed. Circular prestressing is a type where the prestressing is done continuously over different windings (layers). Normally circular prestressing is chosen. But rarely other types of prestressing are also chosen for which anchorages are provided in specially erected plaster on the circular wall itself.

Step 3: Effective prestressing force that could be achieved with one cable pulling is evaluated.

Step 4: Hoop tensile forces generated due to water load is evaluated. The ring force diagram on a vertical section of the tank is then plotted. The spacing of wire based on this ring force diagram and effective prestress that could be transferred by one cable is then decided.

Step 5: The flexural moment (vertical)/generated due to cable tension, is evaluated using influence diagram as explained earlier.

Step 6: On similar lines the moment (vertical) due to liquid forces is evaluated using influence diagram as explained earlier

Step 7: Ring forces at different levels, moment due to water pressure, moment due to prestress (positive and negative) are tabulated and design moments are arrived at each section.

Step 8: The vertical prestressing required is calculated. Otherwise wall is to be designed for vertical moments as a R.C.C. structure.

Step 9: If the circular prestressing is not adopted and if segmental prestressing is done along the circular direction, plastress for anchoring the cables have to be designed.

13.16 Design Example

A typical design example is worked out for proper understanding of the procedure. Only minimum calculations are presented. The example shown is meant to explain only the design calculations of prestressing and its side effects on tank. The problem chosen is a field problem. The actual unit adopted in the field design is C.G.S units and the same is adopted here also.

Table 13.9 Ref Fig.13.5.

Capacity in m³	Dimension in m								
	A	B	C	D	E	F	G	H	I
378	12.50	3.15	1.56	0.12	0.05	0.05	0.05	0.20	0.15
945	16..90	4.30	2.11	0.12	0.12	0.05	0.05	0.22	0.15
1890	21..35	5.35	2.67	0.12	0.12	0.05	0.05	0.30	0.17
2835	24.40	6.10	3.05	0.12	0.15	0.05	0.05	0.36	0.19
3780	26.95	6.70	3.36	0.12	0.18	0.05	0.05	0.38	0.22
5670	30.80	7.80	3.86	0.12	0.23	0.05	0.05	0.43	0.25
7570	33.85	8.55	4.23	0.12	0.24	0.05	0.06	0.48	0.27
9450	36.40	9.15	4.55	0.22	0.26	0.05	0.06	0.51	0.31
18900	46.00	11.45	5.75	0.22	0.44	0.05	0.10	0.69	0.38
37800	57..90	14.50	7.24	0.22	0.74	0.05	0.11	0.89	0.49

Example 13.4

Design a prestressed concrete balancing reservoir for a water supply scheme for a city. The following shall be the details of the reservoir.

Ground level	82.9 m
Floor level	79.9 m
Maximum water level	99.4 m
Over flow pipe level	100.4 m
Top level of wall	101.4 m
Volume of tank	10000 m^3

Water tank has to be covered with a dome of rise 5m.

Solution

Dimensions of the tank

A cylindrical water tank is proposed

Storage height = 99.4 – 79.9 = 19.5m.

Diameter required $=\sqrt{\dfrac{10{,}000}{\pi \times 19.5}} = 25.6$

A tank of 26m-diameter and 20.5 m height is proposed. This takes care of overflow height (level) also. 40-cm wall thickness is proposed. M 350 concrete is proposed. Stresses: Permissible stresses are as per I.S. 1343/1980.

Analysis

Stresses due to Water Pressure

Considering a vertical strip of unit width, water pressure per unit area at any level "h" from water surface is given by where the applied water pressure is resisted by radial pressure Ph due to hoop tension and bending forces pb. The deflection "y" at any height is given by. $y = \dfrac{wh}{4EI\beta^4}$

$+ \; e^{\beta h}(A\sin\beta H + B\cos\beta H) + e^{-\beta h}(C\sin\beta H + D\cos\beta H)$

Applying the boundary conditions at top.

i.e. At h = 0, y = 0 $\dfrac{d^2y}{dh^2} = 0$ we get D = B, C = A

Rearranging the above equation with D = -B and C = A.

$$y = \frac{wh}{4EI\beta^4} + A\sin\beta H(e^{-\beta h} + e^{+\beta h}) + B\cos\beta H.(e^{-\beta h} - e^{+\beta h})$$

Applying boundary conditions at bottom of the tank.

h = H, y = 0 and $\dfrac{d^2y}{dh^2} = 0$ we get two simultaneous equations. They are used for solving A and B and y is expressed in proper form.

Hoop tension T/Unit height = $\dfrac{2.y.E.t}{D.}$

Bending moment/unit width/unit height = $\dfrac{d^2y}{dh^2}.EI$

The details are evaluated using facilities of computer and the results are alone presented here

The hoop tension is plotted in fig.13.10 along the height of tank. The corresponding required prestressing cables and spacing are presented in Table 13.10

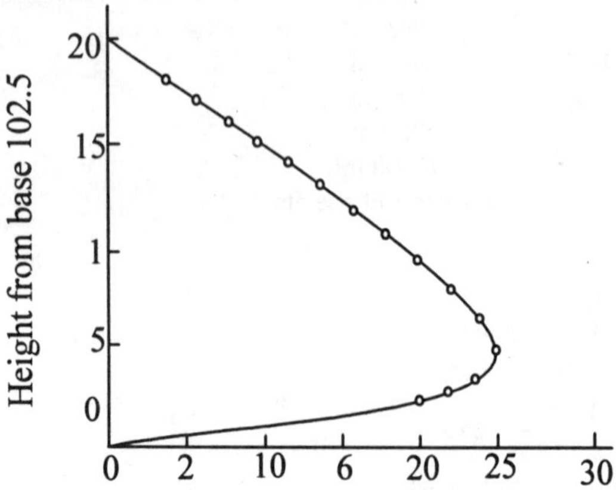

Fig 13.10

Vertical Bending Moment Stress due to Circumferencial Cable Tension

The procedure and the expression for calculation of bending moment due to cable tension is discussed in Art.13.12. The bending moment positive and negative are calculated at any height h from top for cable position 1 from bottom is given by

$$Mx = \frac{P}{4\beta} e^{-\beta s} (Cos\beta s - Sin\beta s)$$
$$- M_1 e^{-\beta (H-h)} (Cos\beta (H-h) + Sin\beta (H-h)$$
$$+ \frac{Q_1}{B} e^{-\beta (H-h)} Sin\beta (H-h)$$

where s = (H-1-h),M_1 and Q_1 are given by art 13.12. The computed values of M_x are also given in Table 13.11.

Calculation of Losses

It is proposed to use segmental prestressing around the tank with plaster at 60^0. The cable will be running around for 120^0 and anchored. This type leads to heavy losses in prestress due to thickness.

Friction of Losses

Diameter of the tank at C.L. of the cable – 26.84m.

Cable length $\frac{2\pi}{3} \times 26.84 = 28.1m$

Prestressing force to = 6 × 120 × 10^2 = 72,000kg = 72 tons.

Prestressing force after friction losses to $e^{-(\mu\alpha + k.x)}$ is assumed to be 0.85 To at farther end

Average force in cable $\frac{1-0.85}{2} To = 0.925 To$ = 66.5 tons.

Table 13.10			
Cable No.	**Distance from bottom (cm)**	**Cable No.**	**Distance from bottom (cm)**
1	65	25	745
2	105	26	775
3	145	27	805
4	185	28	835
5	210	29	870
6	235	30	905
7	260	31	940
8	285	32	975
9	310	33	1010
10	335	34	1045
11	360	35	1080
12	385	36	1115
13	410	37	1165
14	435	38	1215
15	460	39	1265
16	485	40	1315
17	510	41	1365
18	535	42	1415
19	565	43	1495
20	595	44	1575
21	625	45	1655
22	655	46	1735
23	685	47	1845
24	715	48	1955

Table 13.11 Tension at inner face

Section	Distance from base (cm)	Moment (kg.m)					Design moment kg.m.
		Due to water	Due to cable tension				
			Max. positive B.M.	Max. Negative B.M.	B.M. at End of prestressing		
02.5	-9497		6565	--	3975	-5523	6565
2.	205	-8726	8927	--	5421	-3305	8827
3.	307.5	-5108	6896	--	3171	-1937	6896
4.	410	-2021	6584	-1400	3406	-1400	6584
5.	512.5	-292	4675	-2640	1858	-2640	4675
6.	615	350	3256	-3732	171	-3732	-3256
7.	717.5	411	2884	-3993	246	-3995	2884
8.	820	270	6032	-3892	3122	-3892	6032
9.	922.5	123	2114	-3697	118	-3697	2114
10.	1025	30	1761	-3595	-101	-3595	1761
11.	1125-5	-10	2032	-3048	392	-3048	2032
12.	1230	-19	862	-3161	-548	-3161	862
13.	1332.5	-14	1107	-2821	-141	-2821	1107
14.	1435	-7	885	-2617	-267	-2617	885
15.	1537	-2	528	-2267	-403	-2267	528
16.	1640	0	677	-1916	159	1916	677
17.	1742-5	0	571	-1670	529	-1670	571
18.	1845	0	1013	-1432	1013	-1432	1013
19.	1947.5	0	13.17	-873	1390	-873	1370

Creep and Shrinkage losses

Ultimate creep strain due to a sustained units stress is given as.

$$C = 1.75\left(\frac{1.25-h}{0.25}\right)\sqrt{\frac{400}{F_c}}\frac{(Fc)}{Fr} \times 10-6cm/cm/kg/cm^2$$

h = 0.75 (humid atmosphere is assumed), Fc = 350 kg/cm^2 , Fr = (Assumed to be equal to F_c)

On substitution of proper values C works out to be 3.74×10^{-6}

F_c,Max hoop force in the wall = 235.67 t/m.

(obtained from the plot of the hoop force variation as obtained from the computer output)

Assume 10% immediate loss in prestress = 23.57

Assume 20% loss due to time dependent effect

Radial prestress just after initial loss should $= \dfrac{235.67-23.57}{0.8}$ = 324T

$$f_{rsp} = \frac{324 \times 1000}{100 \times 40} = 81kg/cm^2$$

Average stress in steel immediately after stressing operation, f_{sp} = 11000 kg/cm^2

Residual shrinkage strain,

$$\in_{shr} = \frac{2.0 \times 10^{-4}}{\log(T-2)} = \frac{2.0 \times 10^{-4}}{\log 3} = 1.354 \times 10^{-4}$$

E_C = 18,000 × 350 = 3.37×10^5 kg/cm^2

$E_s = 2.1 \times 10^6$ kg/cm^2, $m = E_s/E_c = 6.25$

Loss of prestress due to creep and shrinkage.

$$= E_s \frac{c \times frsp + \in_{shr}}{+ + m - \dfrac{frsp}{fsp.}(1.05CE_c)}$$

$$-2.1 \times 10^5 \left[\frac{(3.74 \times 10^{-6} \times 81) + 1.354 \times 10^{-4}}{1 + 6.23 - \dfrac{81}{11000}(1 + 0.5 \times 3.74 \times 10^{-6} \times 3.37 \times 10^5)}\right]$$

= 855kg/cm^2

No relaxation loss is accounted for because of the use of stress relieved wires. Assuming 6% loss due to successive prestressing.

$$\text{Losses} = \frac{6}{100} \times 11000 = 660kg/cm^2$$

Final prestress = 11,000 – (855 + 660)=9485 kg/cm^2

0.6Fs = 0.6 × 15000 = 9000kg/cm^2

Which is less than the final prestresses. Hence the initial prestress in the cable may be reduced to give a final prestress 0.6 F_s.

Initial prestress required = $\dfrac{12000 \times 9000}{9485} = 11{,}400 \text{kg}/\text{cm}^2$

Final prestressing force in cable = 9 × 6 = 54 t.

The plot of the hoop force is used to locate the cable positions. The moment caused in the vertical plane by the horizontal pressers is given in Table 13.11. They are got from influence line for various sections for bending moments.

Design for Moment

To avoid cracks mild steel reinforcement is used with a reduced allowable stress of 1000 kg/cm^2 For M 350 concrete f_c = 115kg/cm^2.

$\sigma_{st} = 1000 kg / cm^2$ in the water face.

=1250 kg/cm^2 in the face away from water.

m = 8 Water face : j = 0.84, Q = 23.2; other face: j = 0.86, Q = 21.

Vertical reinforcement in the inner face should be based on maximum bending moment. Section-2 has got the maximum bending moment causing inside tension. The bending moment is 8927kg.m. The thickness required for the max.

moment = $\dfrac{8927}{21} = 20.6cm$

The reinforcement required is $A_{st} \dfrac{8927 \times 100}{1000 \times 0.86 \times 36.7} = 28.2cm^2$

Using 16 mm dia rods, spacing = 7.15 cm or say 7 cms. This reinforcement is provided in the inner face for a height of 5 m from the bottom. For height of 5m to 10m the maximum design bending moment to be used is 6032kg.m. Ast required = 6.65cm^2

16mm rods at 28cm c/c will be adequate at 10m.

For height 10m to 15m from bottom the design bending moment is assessed to be 2032 kg/m and the reinforcement required is 6.65cm^2 16mm rods at 28cm C/c would satisfy the requirement. For outer face the entire height is divided into 2 zones namely upto 5m for base and further height beyond 5m upto the top. Design bending moment for the bottom 5m is 5523kg.m. and the the reinforcement required is 18.1cm^2 20mm rods at 10cm spacing be adequate to cover the bending moment and for the height above the bending moment to be latred for is 3995 Kg.m. 20mm dia rods at 20cm spacing will be adquate.

Design of Pilaster

Maximum force/cable = 66.5t.

Permissible shear stress = 10 kg/cm^2 (M 350)

Try a pilaster with dimensions 200 × 30 cm
Permissible stress in shear stirrups.

σ_{sq} = 1760 *kg /mm²* (*deformedbars*)

If P_q is the % of shear reinforcement.

$$\frac{Q}{bjd} = P_q.\sigma_{sq} = 10$$

P_q = 0.00268 i.e. 0.268% A_{st} =11.2 cm²

Provide 8 legged 12 mm dia stirrups at a spacing of 10 cm for a distance 30 cm on either side of the anchorage.

Beyond that the spacing may be increased.

13.17 Construction Problems

The successful design and construction of liquid retaining structure demands two basic conditions. 1. It must be safe against applied water loads with a possible overload factor. 2, It must be leak proof.

The first point can be taken care of in design office to very great extent. Construction will affect the safety factor only nominally. The second one is purely a construction al feature, which depends on choice of method types of joints provided, choice of prestressing, type of prestressing etc.

13.17.1. Choice of Method of Construction

The walls of the tanks of large capacity will be constructed with one two types of form works. I) Fixed form work, ii) moving or sliding form work. For efficient concreting the form work should not be fixed for large height at one streych.

Moreover economy in form work is achieved by choosing sliding or moving form work. This needs efficient skilled labour and equipment. Otherwise there will be number of recess in the concrete particularly at the joint. In case of large tanks the movable or sliding form work may also be provided with bottoned backing frame. If such frame is not adopted, a true cylindrical form cannot be achieved. Any warping of steek will cause secondary effects, which may lead to cracking. So maintaining proper form is an absolute necessity in water tanks.

13.17.2 Fixing Prestressing Tendons

Depending on the type of prestressing such as normal or circular prestressing, tendons have to be arranged. If normal post tensioning method is chosen it should be decided as in how many sectors the tendons are prestressed per layer. Hull rohrs are kept in the tank walls while casting. The anchoring is always done at pilasters specially constructed for the same. In all cases it is essential that tendons are maintained in smooth unbroken lines without kinks. This problem is greatly simplified if a support mesh of mild steel cast inside the wall which can be used to keep the cables in position. If circular prestressing is adopted, anchoring should be perfect. After the winding a thin layer of concrete should be gunited with rich mix to provide a cover for the prestressing steel.

13.17.3 Joints and Joining Materials

In cylindrical tanks there will be a number of joints of different types and purpose. They are of three basic types, namely "construction" "contraction" and "movement" joints. The first will naturally occur at any interruption in the continuity of work. The others are introduced in predetermined positions in the work as a deliberate function of the design. They are either required to deal with long term plastic movements arising from changes in the applied load or temperature.

13.17.4 Construction joints

These joints occur where the break in construction is involved. The aim shall be to have the strength at the joint as near as possible to that of monolithic construction. Elaborate rebates are possibly to be avoided. Shear key to assist shear transfer and bonding may nominally be provided. The surface is chiped and roughened. Lowest 25 to 30 cm shall be cast along with base slab if fixity is to be achieved.

13.17.5 Contraction joints

Shrinkage movements and consequent cracking in the base slab and wall which follow from drying out and hardening may cause cracking. By arranging contraction joints the form and position cracking could be avoided. These joints shall have continuous running of steel with (gap) break in concrete castings. These gaps are filled with plastic seal or bituminous materials. In case of base slab there will be a bottom mass concrete which could cover the joint from being exposed to soil. In case of walls, the joints are staggered at different levels and joining materials have to be carefully chosen.

13.17.6 Movement joints

As the name implies, movement joints generally are designed to accommodate variation in relative position of different parts of the structure. In prestressed concrete water tanks, these joints are mostly at the ends. Sealing material have to be chosen so that it is not much susceptible for atmospheric variations. Metal plates may be used for achieving sliding conditions.

Reference

1. IS: 783-1985 (1986) **"Code of Practice for Laying of Concrete pipes (First Revision)'**, Indian Standard Institution, Manak Bhavan, New Delhi – 110 002.
2. G.S.PANDIT and S.P. GUPTA (1993) **"Prestressed Concrete"** - CBS Publishers & Distributors 4596/1A, 11 Darya Ganj, New Delhi - 110 002 (INDIA) First Edition.

3. IS: 784-1959, (1958) **Indian Standard Specification for Prestressed Concrete Pipes,** Indian Standards Institution, New Delhi, pp 1-6.
4. C-30/64, A.W.W.A., (1964) **Specification for Prestressed Concrete Pressure Pipe, Steel Cylinder Type for Water and Other Liquids** American Waterworks Association.
5. CURTIS, A.R. and H.J.COWAN, (1954) **"Design of longitudinal cables in circumferentially would prestressed tanks,"** Magazine of Concrete Research, Vol.5, March 1954, pp 123-126.
6. DOANIDES, P.J., (1972) **"Some notes on the prestressed concrete pipes of the Veeranam aquaeduct for the water supply of Madras",**International Symposium on Prestressed Concrete Pipes, Poles, Pressure Vessels and Sleepers, Proceedings, Vol.2, Madras pp p/3-1 to p/3-40.
7. A.M. NEVILLE (1996) **"Properties of Concrete",** Edison Wesslley – Longman Publishers Limited, Ivth edition.
8. IS: 3370-IV, (1974) **"Indian Standard Code of Practice for Concrete Structures for the Storage of Liquids",** Part IV-Design Tables (Third reprint), Indian Standards Institution, May 1974, pp 14-49.
9. TIMOSHENKO, S. and WOINOWSKY KRIEGER, S., (1959) **"Theory of Plates and Shells",** McGraw-Hill, New York.
10. W.S. GRAY (1960) **"Reinforced Concrete Reservoirs and Tanks'** Fourth Edition Published by the Cement and Concrete Association 52, Grosvenor Garden, London SWIW OAQ.
11. CREASY, L.R. (1961) **"Prestressed Concrete Cylindrical Tanks'**, John Wiley Publication, New York.

14 Partial Prestressing (Principles, Analysis, and Design Concepts)

14.0 General

Prestressing of structures is resorted to overcome the effect of tensile stresses in concrete which may affect the performance of the structure during service load conditions. When a prestressed concrete structure is subjected to flexure or any other type of external force causing tensile stresses on concrete, initial precompression on concrete due to prestresssing first gets nullified. Hence tensile stresses introduced by external loading, if it is not greater than precompression, does not lead to any distress like cracking or reduction of stiffness. The performance during service load is not affected. The level to which the concrete has to be precompressed, depends upon the most commonly occurring liveload and service load on the structure. But there are stipulations by statutory or non statutory authorities that the concrete structure should remain uncracked under worst service loads for which it is designed. Normally the worst service loads defined for the structure for the design purposes may rarely occur during the life time of the structure. If concrete is prestressed for such a rare service load stress(tension), the initial compression that has to be given to the concrete is very high and that leaves the concrete ever in compression which causes creep on concrete. The purpose of prestressing as indicated in chapter 1 is to keep the structure under **no stress conditions or under optimum stress condtions for the maximum period of the life time**. If the prestressing is done for the rare services load, it is not under optimally stressed conditions or under no stress conditions during the maximum span of its time. This is harmful for the structure. For example, bridges are designed for heavy military vehicle loads and in an area, which is far interior from the borders of the country. The movements of military vehicles in such a place is rare and if the structure is prestressed for such military vehicle loads, the structure is unnecessarily subjected to high compression and that is harmful for the structure and it also making the structure costly. In such structures, some amount of tensile stresses may be permitted for such unusual loading, which may rarely occur and even if occurs and it is only temporary. If the structure is with its elastic limits, no damage is done to the structure. Based on these phenomena only, the structures are classified in to four types in most of the international codes. They are fully Prestressed members without any allowance for flexural tension under maximum service load, (Type 1); Limited prestressed members where small tensile stresses not leading to cracks are permitted under service loads, (Type 2); Partially prestressed structures which are prestressed not for full service load conditions which may occur rarely, but prestressing is done only for normally occuring service loads so that no tension is developed under normal service load, (Type 3); and Only reinforced concrete where there is no prestressing, and cracking is permitted at the service loads (Type 4). This type of classification has also been discussed in chapter 1. Whatever be the type to which the structure belongs, it has to be designed for the limit state of collapse with sufficient factor of safety against collapse. Reinforcement interms of high tensile steel as well as additional untenioned steel have to be provided so that the system or the section has got enough capacity to resist the ultimate moment under limit state of collapse. While the partial prestressing is done for service load, its capacity under limit state of collapse, should be more than what is called for.

Partially prestressed structure, may have no tension under normal service load but may have tension and cracks under extreme services load which may be of momentary nature and get closed as soon as soon as the extreme loads are removed. This can be achieved in two ways. In the first approach, the prestressing given by using high tensile steel, but stressing it only to the lower level of capacity say $0.4f_{spu}$ as against the permitted level of stress in steel say $0.6\ f_{spu}$ at service loads. The additional facility of tensile stress in the high tensile steel is made use as a normal reinforcement in a R.C structure for carrying the additional service load moment, which may corne rarely, and later for carrying the ultimate moment. This type structure where the prestressed steel is only partially stressed and not to its permissible capacity is referred as partially prestressed concrete structure.

Since the material, high tensile steel, is costlier than ordinary steel, in the second type of structure high tensile steel is stressed to its permitted level of stress at service loads say $0.6f_{pu}$ and untensioned steel is provided to cater for the additional service load (rare) moment and later both tensioned and untensioned steel participate to carry the ultimate moment at ordinary reinforcement. Untensioned steel is used to increase the tensile capacity of the tension chord. This type of structural system is called Prestressed reinforced concrete structure and it is more practical.

In any structure, untensioned steel have to be provided for hanging shear reinforcement, to support anchorage reinforcement to take care of non structural cracking etc., These untensioned reinforcements serve also to increase moment carrying capacity of the structure.

In this chapter, analysis and design of flexural members as partially prestressed members is discussed and a few terms of definition as described in national and international codes are also eloborated. It may give the first insight in the area of partial prestressing. It is worthwhile for the readers to understand the subject, so that it could be extended in various places where statutory regulations provide for variation in stress limits under different laoding stages. In this way it is possible to provide the stress free or optimally stressed structure under normally occuring serivice loads and also bring in economy in the structure. I.S. 1343 does not provide for such a facility for structures built in India, but it has given four types of structures as detailed earlier.

14.1 Case for Partial Prestressing

F.Leonhardt, [1] suggests that principal tensile stresses can not be fully avoided in any structure. In his opinion fully uncracked structure is only a myth. Hence he advocates partial prestressing in many cases and states that partial prestressing improves the behaviour of concrete structures:

- by decreasing creep effect and corresponding time dependent deformations;
- by controlling cracks and crack widths as per designer's requirements.
- by having desirable stress amplitudes to keep the structures within the allowable range of fatigue strength of both materials in the case of very frequent load cycles or of oscillations.

According to the definition of full prestressing, no tensile stresses are permitted to occur anywhere with structure. But this is hardly possible to achieve in practice and it has now been realised that even "full" prestressing cannot prevent all tensile stresses in concrete. Principal tensile stresses always occur in shear zones of beam like structures, which can be neutralized only with vertical prestressing, or diagonal prestressing. Similarly tensile stresses also occur due to bond between tendons and concrete, due to anchor forces of tendons, due to internal or external restraint by shrinkage, due to temperature or differential creep, due to differences in stiffnesses between aggregates and cement mortar, etc. Therefore, since long, one has relaxed the definition of full prestressing (without any cracks) under the compulsion of practical and economic considerations. This means, in case of full prestressing, though no flexural tensile stresses are permitted, other tensile stresses like the principal tensile stresses in the web-zones are permitted. Undoubtedly, this appears to be illogical because, firstly the flexural tensile stresses, if seen physically, are also principal tensile stresses only in constant bending moment zone, and secondly the diagonal shear cracks occurring in the webs are undoubtedly more damaging than fine flexural cracks. It has, therefore, been now generally accepted that in many cases, partial prestressing is a far better and often more economical. A safe solution could be achieved, using partial prestressing.

14.2 Partial Prestressing and the Codes of Practice

From the very beginning the German Code DIN 4227 [2] permitted "limited" prestressing in which tensile stresses up to the limit of cracking were permitted. Partial prestressing was recognised by the professionalists when the new CEB-FIP Recommendations [3] were presented at the 6th FIP congress in Prague in 1970. These Recommendations, for the first time, recognised a full range between reinforced concrete and fully prestressed concrete. According to the CEB-FIP classifications, 4 types of classifications of structures as mentioned earlier emerged out.

The Swiss Code SIA Norm 162 [4], and British Code B.S. 8110 [5] permits partial prestressing in structures. Of course in U.S.S.R. partial prestressing has been practiced from early years of 20th century.

In Great Britain P.W. Ables [6] was able to introduce partial prestressing when the first report on prestressed concrete by a committee of the Institution of Structural Engineers classified prestressed concrete into three types with the type -3 structures having visible hair cracks under frequently occurring working loads.

The Indian Code I.S.1343-1980 [7] has recognised partial prestressing by including provision for intermediate degrees of prestressing, through it does not permit to use the same in structural practice.

In general, partial prestressing serves to have (i) a better stress distribution during services at load stage, (ii) a greater economy because of use of untensioned steel (Fe 415) which is less costly to resist the failure moments, (iii) reasonably good control on cracks and crack width because of higher percentage of reinforcements, and (iv) a good ductility with facilities for rotations capacities when required.

The level of partially prestressing is defined in many ways by many researches or many codes. A few terms used to define partial stressing are prestressing index, partial prestressing ratio or degree of prestress. But what should be rememebered is consistency of the terminology and understanding of the partial prestressing in the entire procedure of the design and detailing. Detailing of partial prestressed concrete structures is very important specially with respect to cracks and crackwidths.

14.3 Definitions for Partial Prestressing

Partial prestressing is defined in many ways by various authors and as adopted in various codes. They are briefly described in a consolidated form by Pandit and Gupta [8] and presented here

14.3.1 Prestressing index (PI)

Prestressing index is defined as the ratio of ultimate strength of prestressing steel to that of the entire steel in the tension chord. It may be expressed by the expression.

$$PI = \frac{A_p f_{pu}}{A_p f_{pu} + A_s f_y} \qquad \text{..... (14.1)}$$

where A_p = area of prestressing steel
A_s = area of untensioned reinforcing steel
f_{pu} = ultimate strength of prestressing steel
f_y = yield strength of untensioned reinforcing steel.

Thurlimann[8] proposed a similar definition for prestressing index.

$$PI = \frac{A_p f_{py}}{A_p f_{py} + A_s f_y} \quad \text{..... (14.2)}$$

in which f_{py} is the yield stress of prestressing steel corresponding to 0.2 percent residual or permanent strain.

14.3.2 Partial Prestressing Ratio (PPR)

Partial prestressing ratio is another expression for the measure of partial prestressing. It is defined by the equation,

$$PPR = \frac{M_{up}}{M_u} \quad \text{..... (14.3)}$$

where M_{up} = ultimate moment capacity corresponding to prestressing steel alone and M_u = ultimate moment capacity due to total tension reinforcement.

In a steel controlled design, where the steel reaches to limiting first and then concrete reaches to its limiting strain, when the internal lever arms for the two types of steel do not differ substantially, the value of PPR is close to the value of PI.

14.3.3 Degree of Prestress (DP)

In the definition of Degree of prestress, the decompression moment M_{de} has also been identified as a measure of partial prestress. The degree of partial prestress is defined as

$$DP = \frac{M_{de}}{M_w}, \text{ when } M_w \text{ is the moment under sterlice loads} \quad \text{..... (14.3)}$$

14.4 Analysis

Since the structure is allowed to have cracks under extreme service loads calculations of crackwidth and crack spacing also play a role for checking the servicebility of the structure for the extreme services loads. For precracking conditions, the structures must be visualised as a composite structure with the effective cross section properties inclusive of the effect of steel. Accordingly the level of the centroidal axis has to be decided and the stress in various materials, have to be calculated for the stage of decompression. The structures gets into the post cracking stage, the neutral axis or the axis of zero stress has to be considered for evaluating strain levels in concrete and steel. Hence for precracking and post cracking, separate analysis have to be carried out.

14.4.1 Precracking Stage

Any partially prestressed concrete beam is expected to have cracks during its service load specially at higher range of service loads which will be temporary in nature. But under regularly occurring service load, it is not supposed to have cracks. Hence even under service loads, there are two stages for the structure, namely a cracked condition and an uncracked condition. In case of uncracked condition (namely precracking stage) the sections subjected to flexure may be analysed using linear elastic theory. The section may be visualised as a composite structure comprising of concrete, prestressing steel, and untensioned reinforceing steel. Considering appropriate modulli of elasticity for prestressing steel, reinforcement steel and concrete, the transformed section properties can be evaluated. The position of the centroidal axis of the beam can be identified as y_{it} from top fibre. The composite sectional properties can be defined as A_i, I_i, Z_{ib}, Z_{it}, y_{ib}, y_{it}. The stress in steel at any stage evaluated as initial stresses due to prestressing minus losses in stress in steel due to immediate and time dependent losses in prestress plus additional stresses due to curvature created by service load. Similarly the stress in concrete at any level may be evaluated as stress in concrete due to prestressing at that level after losses plus the stress generated due to curvature, caused by external service load.

The elastic stresses in precracking stage gets modified due to creep effect and the stress in untensioned reinforcement steel is increased by a factor $(1+\phi)$ due to creep of concrete where ϕ is creep coefficient. The effect of prestressing in untensioned steel is compressive. The effect of shrinkage also introduces compressive stresses in untensioned steel.

14.4.2 Post Cracking Stage

As mentioned earlier, partially prestressed members may crack even, during service load, specially at higher service load. At this time it is mandatory that stresses in concrete and steel must be within limits and the crack width also should be within the permissible limits. These cracks may close down, as soon as the extra service load is removed of the element. The depth of the neutral axis in the partially prestressing concrete section can be determined by considering that as a cracked section and as a eccentrically loaded column with large eccentricities. It may be assumed that the distribution of the strains are linear and the change (addition) in the force in prestressing steel due to curvature caused by applied moment is negligible and can be ignored.

Considering the case of a flanged section with neutral axis falling in the web as shown in the figure 14.1 centre of compression coincides with the prestressing tendons just at the time of prestresing. As the external load inclusive of self weight moment is applied on to the beam, the centre of compression moves away are from the centre of tension. If the moment is of sagging type, the movement of centre of compression is upwards towards the compression zone and the distance between the centre of tension and the centre of compression forms lever arm for resisting the external moment. **Considering only the applied moment due to dead load + external load and the corresponding stresses in steel and concrete, equilibrium of forces and moments are written and the position of neutral axis for this applied moment evaluated.** Then using the assumption of linear strain distribution the strain in reinforceing steel and in concrete at any level is evaluated. From the strains, the stress in the material are evaluated. For arriving at the final stresses, initial stresses due to applied prestress is to be added.

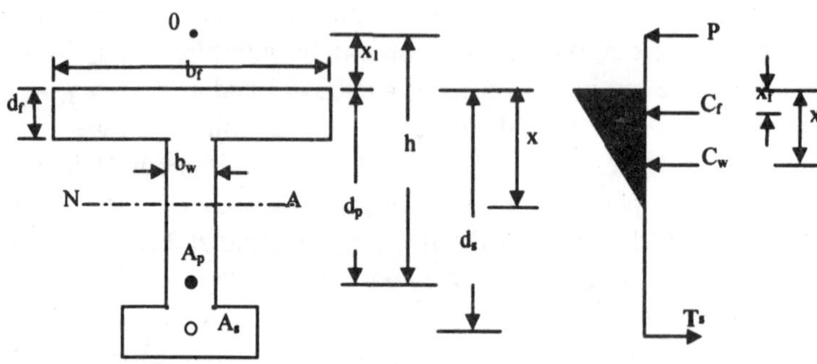

Fig 14.1Equilibrum of Forces due to service load Moment only

Referring to Fig.14.1 the upward shift 'h' of centre of compression due to application of moment M_w due to loads (self weight in applied loads) can be expressed as

$$h = \frac{M_w}{P} = d_p + x_1 = d_p + k_p d_s \qquad .....(14.5)$$

where k_p is a constant All distances are referred to be base distance "d_s', the distance of untensioned steel from top surface with a coefficient. As the moment is applied, the section rotates about an axis called the neutral axis for that moment. The stresses above the neutral axis are compressive and those below are tensile. With the applied prestressing force on steel and the stresses on concrete, the section is in equilibrium. The external moment introduces new stresses in concrete and steel and keeps the section in equilibrium. The increase in steel stress in prestressing steel is assumed negligible. Hence the tensile force due to service loads is assumed to be taken only by untensioned steel with a force T_s. The compressive stress in concrete due to this moment (M_w) be f_c. The depth of neutral axis is at 'x' from top defined as [$k.d_s$]. Thickness of flange of beam $d_f = k_1 d_s$ Ratio of web width to flange width be namely $b_w/b_f = k_2$. Now the equilibrium of the moment (applied) with internal stresses is considered. Since the prestress and corresponding stress in concrete are in self equilibrium the force in high tensile steel and stresses

in concrete due to prestress are not considered.
Then the equilibrium of moments can be written as
$C_f(k_pd_s + x_f) + C_w(k_p d_s + x_w) - T_s(k_p d_s + d_s) = 0$ .....(14.6)

where the compressive force in flange is C_f, compressive force in web is C_w and tensile force in untensioned reinforcing steel is T_s. They may be expressed in terms of the extreme fibre stress in concrete at top fibre **due to external moment alone and not due to prestressing. Due to prestressing, the fibre stress at top fibre is assumed to be zero.**

Referring to Fig.14.1. The compressive force in top flange can be written as

$$C_f = [b_f . f_c k_1 d_s] - [\frac{f}{kd_s}.k_1.d_s.b_f.\frac{1}{2}k_1 d_s]$$

$$= b_f . f_c d_s k_1 [1 - \frac{k_1 d_s}{2kd_s}]$$

$$= [b_f f_c d_s \frac{(2k - k_1)}{2k}.k_1]. \qquad ....(14.7)$$

The force in the web can be written as

$$C_w = \frac{f_c}{kds}(k - k_1)ds\frac{(k - k_1)ds}{2}k_2 b_f$$

$$= f_c b_f d_c k_s [\frac{(k - k_1)^2}{2k}.k_2] \qquad ....(14.8)$$

The force in the untensioned reinforcement can be written as

$$T_s = f_c A_s = \frac{m(1-k)}{k} f_c p_s b_f d_s \qquad ....(14.9)$$

where $k = \frac{x}{d_s}$

$$k_1 = \frac{d_f}{d_s}$$

$$k_2 = \frac{d_w}{d_f}$$

$p_s = \frac{A_s}{b_f d_s}$ proportion of untensioned steel with respect to cross section. The distances x_1 and x_w of the compressive forces C_f amd C_w from the top face may be expressed as

$$x_f = k_3 d_3 = \left(\frac{3k - 2k_1}{2k - k_1}\right)\frac{k_1}{3}d_s$$

Substitution of C_f, C_w, T_s, x_f and x_w leads to the following cubic equation in k_1

$$k^3 + 3k_p k^2 (\frac{k_5 + k_6}{k_2} - k_s) + \left(k_7 - \frac{k_7 + k_6}{k_2}\right) = 0 \qquad .....(14.10)$$

where $k_5 = 3k_1 (k_1 + 2k_p)$
$k_6 = 6mps (1 + k_p)$
$k_7 = k_1^2 (2k_1 + 3k_p)$

Solving for equation 14.10 for k, the position of neutral axis is determined. The extreme fibre stress f_c may be determined from the force equation in which the internal and external forces are equated to each other.

$$P = C_f + C_w - T_s \qquad .....(14.11)$$

Substituting for C_f, C_w and T_s in equation 14.11,

$$f_c = \frac{p}{b_f d_s}\left[\frac{2k}{k_1(2k - k_1) + k_2(k - k_1)^2 - 2mp_s(1 - k)}\right] \qquad .....(14.12)$$

Knowing f_c, the stress in untensioned steel may be determined from using the expression,

$$f_s = m\left(\frac{1 - k}{k}\right)f_c \qquad .....(14.13)$$

An iterative graphical procedure for the determination of the stress f_s in untensioned steel has been given by Levi described by Guyon [9].

Example 14.1

A partially prestressed concrete beam has a cross section as shown in Fig14.2. It has a prestressing steel of 1500 mm^2 which has 0.2% proof stress of 1250 Mpa and a failure capacity of 1500 Mpa. The steel is prestressed with an effective stress level of 1000 N/mm^2. It is also reinforced with untensioned steel with an area of 1000 mm^2 which has a limiting stress level of 415 Mpa. The centre of compression at failure could be assumed to be at 0.175 m from top. The service load moment to be resisted is 1MNM. Determine the values of pressing index, partial prestressing ratio, Degree of prestress etc.

Solution:

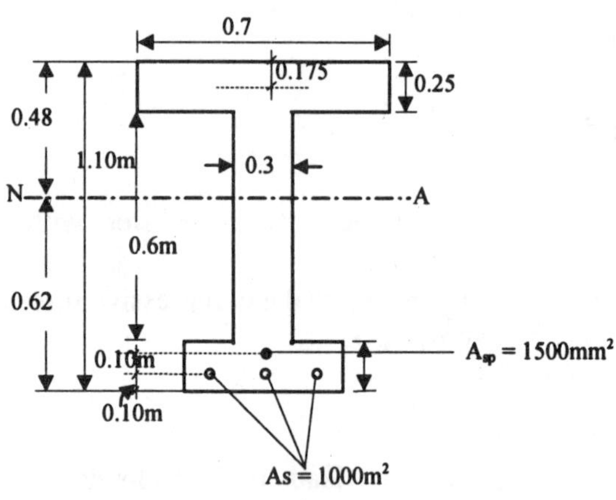

Fig. 14.2

Prestressing Index

The prestressing Index, $PI = \dfrac{A_{sp} f_{pu}}{A_{sp} f_{pu} + A_s f_y}$

$$= \frac{1500 \times 1500}{1500 \times 1500 + 1000 \times 415} = 0.844$$

The prestressing index as per. Thurlimann [10],

$$PI = \frac{A_{sp} f_{py}}{A_{sp} f_{py} + A_s f_p} = \frac{1500 \times 1250}{1500 \times 1250 + 1000 \times 415} = 0.818$$

ii) Partial Prestressing Ratio:

The partial prestressing ratio is defined as $\dfrac{M_{up}}{M_u}$ by a Naaman [11].

$$PPR = \frac{1500 \times 1400 \times 0.725}{1500 \times 1400 \times 0.725 + 1000 \times 415 \times 0.825}$$

iii) Degree of Prestress:

$A = 0.7 \times 0.25 + 0.3 \times 0.6 + 0.25 \times 0.4 = 0.455\ mm^2$.

Depth of neutral axis from bottom fibre,

$$= \frac{(0.7 \times 0.25) \times (0.975) + (0.3 \times 0.6) \times (0.55) + (0.4 \times 0.25) \times (0.125)}{(0.7 \times 0.25) + (0.3 \times 0.6) + (0.4 \times 0.25)}$$

$$= \frac{0.170 + 0.99 + 0.0125}{0.175 + 0.18 + 0.1} = 0.620$$

$$I = \frac{1}{12} \times 0.7 \times 0.25^3 + 0.7 \times 0.25 \times (0.48 - 0.125)^2 +$$

$$\frac{1}{12} \times 0.3 \times 0.6^3 + 0.3 \times 0.6 \times (0.55 - 0.48)^2 +$$

$$\frac{1}{12} \times 0.3 \times 0.25^3 + 0.4 \times 0.25 \times (0.62 - 0.125)^2$$

$$= 0.0537 m^4$$

Intensity of prestress at bottom $= \dfrac{-1.5}{0.455} - \dfrac{1.5 \times 0.42 \times 0.62}{0.0537} = -10.57$ Mpa.

Hence the decompression moment,

$$M_{de} = -10.57 \times \frac{0.0537}{0.62} = -0.9154 Nm$$

The degree of prestress $= \dfrac{M_{de}}{M_w} = \dfrac{0.9154}{1.0} = 0.9154$

Example 14.2

A partially prestressed concrete beam is rectangualr in cross section with 0.4m width and 1.2m depth. It is prestressed with a force of 1800kN using 2000mm² of HT Steel located at 400 mm from soffit. It is also having 1500mm² of untensioned steel located at 150mm from soffit and compression steel of 800 mm² at 150mm from top. Evaluate the stresses in steel just after transfer, at the stage of decompression and the limit state of cracking. Assume the modular ratio between concrete and steel as 6 and flexural tensile capacity of concrete as 7N/mm².

Solution:

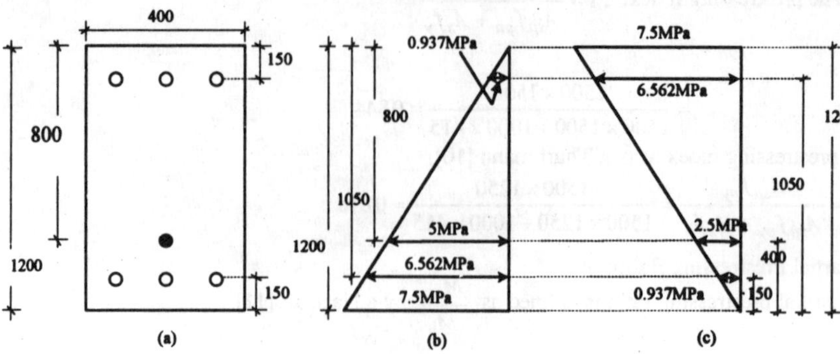

Fig. 14.3

The sectional properties of gross concrete section are

$A = 0.4 \times 1.2 = 0.48\text{m}^2$

$Z = \frac{1}{6} \times 0.4 \times 12^2 = 0.096m^2$

Intensity of stress due to prestress at the top fibre,

$$f_t = \frac{-1.8}{0.48} + \frac{1.8 \times 0.2}{0.096} = 0\,Mpa$$

Intensity of stress due to prestress at bottom fibre,

$$f_b = \frac{-1.8}{0.48} - \frac{1.8 \times 0.2}{0.096} = -7.5\,Mpa$$

(i) No-load condition (Just after transfer)

The stress diagram for concrete at no-load condition or just after transfer is shown in Fig.14.3. With modular ratio m = 6, the stress in untensioned steel is given by the stress in concrete at that level multiplied by modular ratio.,

Hence $f_s = f_{cps} * 6$(modular ratio) where f_{cps} is the stress in concrete at steel level

The stress in untensioned steel can be calculated from similar triangle

$$= \frac{7.5}{1.2} \times 1.05 \times 6 = 6.562 \times 6 = 39.372 \text{ N/mm}^2$$

Similarly the stress in compression steel is given by

$$= \frac{7.5}{1.2} \times 1.05 \times 6 = 0.937 \times 6N/mm^2 = 5.622 \text{ N/mm}^2$$

Stress in prestressing steel,at transfer $f_{sp} = \dfrac{1.8}{2000 \times 10^{-6}} = 900\,Mpa$

(ii) At stage of decompression

In order to cause of decompression at the bottom fibre, the applied load must create a tensile stresses of 7.5 Mpa at the extreme bottom fibre. Because it is a symmetrical uncracked section, the moment which causes 7.5 N/mm^2 tension at bottom fibre shall also cause 7.5 N/mm^2 compression at top fibre.

Hence the stress at the top fibre at the stage of decompression is 7.5N/mm^2 compression. (since the stress at top fibre due to prestress is zero).

The stress diagram at this stage is shown in Fig.14.3c

From the similar triangles, the stresses at various levels are

$$\frac{7.5}{1.2}=\frac{y_1}{1.05};y_1=6.5625Mpa$$

$$\frac{7.5}{1.2}=\frac{y_2}{1.4};y_2=2.5Mpa$$

$$\frac{7.5}{1.2}=\frac{y_3}{0.15};y_3=0.937Mpa$$

Hence the steel stresses at this stage are

$f_s = -0.937 \times 6 = -5.622\ Mpa$

$fs' = y_1 \times m = -6.5625 \times 6 = -39.375\ Mpa$

$f_{sp} = 900 + (5 - 2.5) \times 6 = 915\ Mpa.$

Note that the stresses in concrete at the level of prestersssing steel changes from –5Mpa at no load condition to –2.5 Mpa at the stage of decompression. Hence the stress in prestressing steel increases by (5 – 2.5) × 6 = 15 Mpa.

Limit State of Cracking

At this stage, the applied load must cause a tensile stress equal to (7..5 + f_{cr}) = 7.5+7=14.5Mpa at the bottom fibre. The same stress will be caused at top fibre. At that time top stress will be 14.5N/mm^2 compression and at bottom 7 N/mm^2 tension because first 7.5 Mpa of precompression has tobe nullified by external moment. The stress pattern is given in Fig. 14.4.

The stresses in untensioned steel can be calculated as $f_s = (f_{cps} .m)$

$$=\left(\frac{0.2407 \times 7}{0.3907}\right) \times 6 = 25.87\text{Mpa.}$$

Similarly, stress in compression steel

$$f_s' = -\left(\frac{0.6593 \times 14.5}{0.8093}\right) \times 6 = -70.872Mpa$$

Stress in H.T. steel $f_p = 900 + (5 - 0.1667) \times 6 = 928.99Mpa$

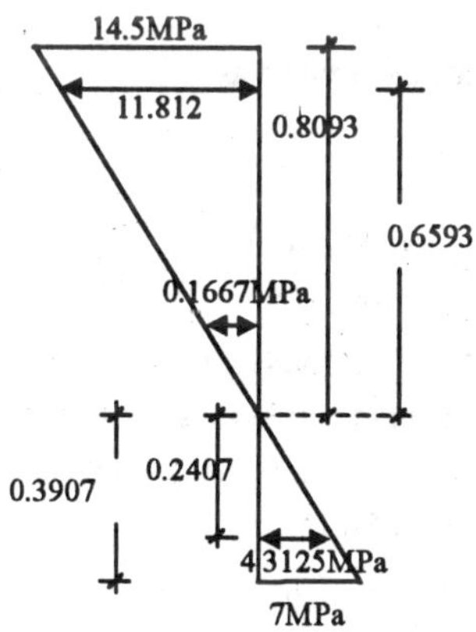

Example 14.3

Partially prestressed concrete beam having a top flange of 800mm, 200mm, a web of 700mm × 250 mm and a bottom flange of 600mm × 400mm is prestressed with an effective prestress of 1400 kN applied at 1.0m from top. It has also untensioned reinforcement of 4000 mm^2 located at 1.08 m from top. It is subjected to sagging bending moment of 1800 kNm. Assume a modular ratio of 6. Evaluate the stress in concrete at top fibre and stresses in untensioned steel. The increase in stress in prestressing steel due to external moment may be assumed as neglible and the initial stress in concrete due to prestress at top = 0.

Solution:

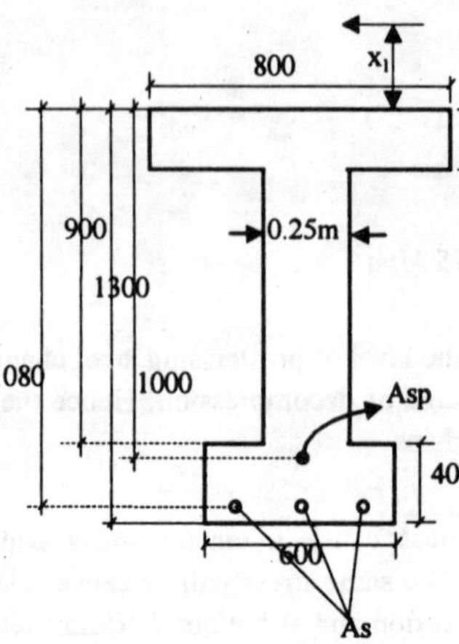

Fig 14.5

Bending moment =1800kNm.

Effective prestress =1400kN at a distance of 1m from top, m = 6 and A_s = 4000 mm² The shift of centre of compression from the location of prestressing steel

$$h = \frac{M_w}{P} = \frac{1800}{1400} = 1.2857\text{m}$$

Referring to Fig.14.5, and 14.4.2 and related expressions $x_1 = k_p d_s = 1.2857 - 1 = 0.2857\text{m}$

$$\text{Or } k_p = \frac{0.2857}{1.08} = 0.2645$$

The other constants in Eq.14.10 are determined as

$$k_1 = \frac{df}{d_s} = \frac{0.2}{1.08} = 0.185$$

$$k_2 = \frac{b_w}{b_f} = \frac{0.2}{0.8} = 0.312 \; P_s = \frac{A_s}{b_f d_s} = \frac{4000 \times 10^{-6}}{0.8 \times 1.08} = 0.004629$$

$k_5 = 3k_1 + 2k_p = 3 \times 0.185(0.185 \times 2 \times 0.2645) = 0.3962$

$k_6 = 6mP_s(1 + k_p) = 6 \times 6 \times 0.004629\,(1 + 0.2645) = 0.2107$

$k_7 = k_1^2(2k_1 + 3k_p) = 0.185^2\,(2 \times 0.185 + 3 \times 0.2645) = 0.0398$

Substituting the values of constants in Eq.14.10

$$k^3 + 3k_p k^2\left(\frac{k_5 + k_6}{k_2} - k_5\right) + \left(k_7 - \frac{k_7 + k_6}{k_2}\right) = 0$$

$$k^3 + 3 \times 0.2645 \times k^2 + \left(\frac{0.3962 + 0.2107}{0.312} - 0.3962\right)\text{k} +$$

$$\left(0.0398 - \frac{0.0398 + 0.2107}{0.312}\right) = 0$$

$k^3 + 0.7935k^2 + 1.5489k - 0.7963 = 0$ or $k = 0.383$

Depth of neutral axis, $x = kd_s = 0.383 \times 1.08 = 0.41$m

Hence from Eq. 14.12 and 14.13 the stresses in concrete and steel are

$$f_c = \frac{P}{b_f d_s}\left[\frac{2k}{k_1(2k-k_1)+k_2(k-k_1)^2-2mp_s(1-k)}\right]$$

$$= \frac{1.4}{0.8 \times 1.08}\left[\frac{2 \times 0.383}{0.185(2 \times 0.383 - 0.185) + 0.312(0.383 - 0.185)^2 - 2 \times 6 \times 0.004629(1-0.383)}\right]$$

$= 14.52$ Mpa

$$f_s = m\left(\frac{1-k}{k}\right)f_c$$

$$= 6\left(\frac{1-0.383}{0.383}\right) \times 14.52 = 140.34\text{Mpa.}$$

Note: The actual stress in concrete at any level is the stress cuased by prestressed plus the stress caused by external momentt with neutral at axis at 0.383 d_s. Since the stress caused by prestressed at top fibre is given as zero, the actual stress at top fibre = 14.52 Mpa.

14 .5 Design

The design approach for partially prestressed concrete has been explained very clearly by Walther R [12]. When a prestressed element is subjected to tension, the force required to get to the stage of decompression can be expressed as

$P^* = \varepsilon^*_{sp} E_{sp} A_{sp}$ and P^* is also equal to $P_\infty + \frac{f_{cpa}}{E_c}.E_{sp}.A_{sp}$. The force required to get to stage of decompression is equal to the prestressing force existent in the cross section plus the force required to release the initial compression in concrete $f_{cp\infty}$ due to prestress. $f_{cp\infty}$ is the stress in concrete after all loses due to creep shrinkage etc (Ref. Fig. 14.6).

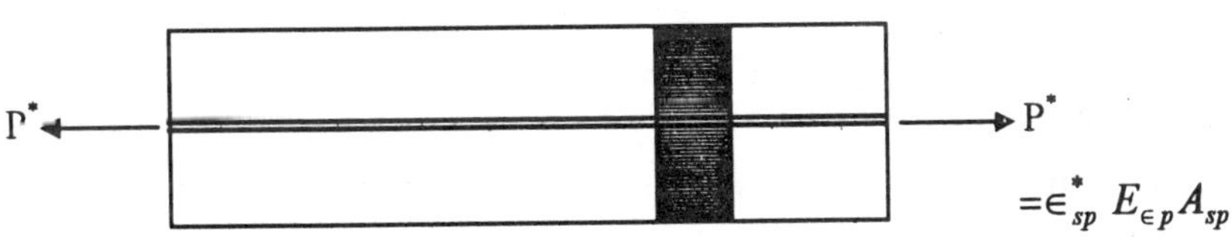

$$P^* = P_\infty + \frac{f_{cp\infty}}{Ec} E_{sp} A_{sp}$$

Fig. 14.6

Referring the stress strain curve of the High tensile steel which is tensioned to a strain level of ε_{sp}, looses certain elongation due to elastic shortening of concrete and reaches a state of ε^*_{sp} and strain gets further reduced to $\varepsilon_{sp\infty}$ because of creep deformation and other time dependent deformation as shown in Fig. 14.7. But creep and other time dependent deformations do not occur immediately. Hence when external load starts

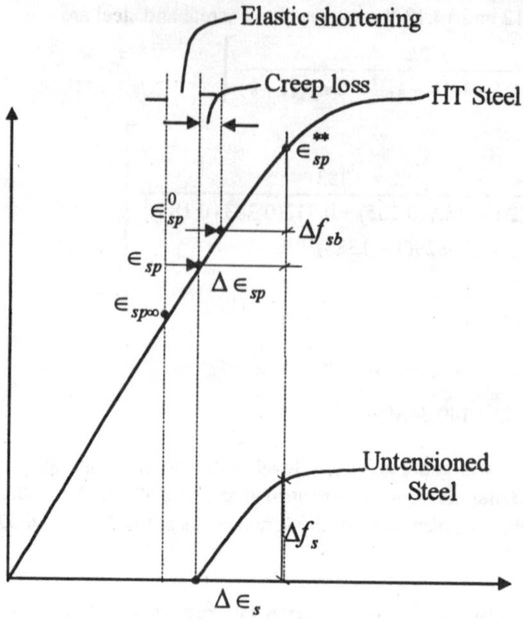

Fig 14.7

acting on the structure, after the stage of decompression, the high level steel gets further elongation by $\Delta\varepsilon_{sp}$ and there is a corresponding increase in stress. Δf_{sp}. If untensioned steel is present, the untensioned steel also gets the same stress Δf assuming the untensioned steel to be in same level of location. Fig 14.7 gives the details of the strains and stresses in high tensile steel and untensioned steel of a partially prestressed concrete beam.

Hence at service loads, the total strain in high tensile steel is expressed as

$$\varepsilon_{sp}^{**} = \varepsilon_{sp}^{*} + \Delta\varepsilon_{sp}$$

and the corresponding force in tension chord (in high tensile steel and untensioned steel) $P^{**} = (f_{sp}^{*} + \Delta f_{sp})A_{sp} + \Delta f_s . A_s$. Here it is assumed that prestressing steel and untensioned steel jointly resist the load.

In any design, the permissible stresses at service load and the limiting stresses at the time of limit state of collapse are defined, based on strain criteria.In a partially prestressed concrete beam, it is desirable that the permissible stresses in untensioned steel and the permissible stress in high tensile steel are reached simultaneously under service load. In such a case the stress level on the prestressing steel during prestressing should be equal to $f_{sp(permissible)} - f_{cs(permissile)}$ so that both steel may reach to their limiting stress simultaneously at service load. Just after transfer of prestress, there is tension in high tensile steel and compression in untensioned steel corresponding to compressive strain in concrete at this level. As the load is applied, a stage of decompression is realised and at the point of time, the untensioned steel also gets to zero stress. Then as the load is increased, tension occurs both in untensioned steel and high tensile steel, corresponding to strain in concrete at steel level, which is very small. But as soon as the load crosses the load causing the stage of decompression, there is an increase in stress in both high tensile steel and untensioned steel. This corresponds to again modular ratio times the stress in concrete at steel level. In the untensioned steel, there will be a pre compression due to prestress which alone gets nullified. After the load crosses the load causing stage of cracking the rate of increase of stress in both steel is higher and the stress increase in both the steel is one and the same if they are located at the same level. Both of them reach the limiting service load stress level at service load together. With this basis, the design procedure is explained. The flexural element provided with sufficient high tensile steel, is stressed to a level which is less than the permissible stress level at service load by an amount equal to the permissible stress level in the untentioned steel. As and when the load is applied, the initial compression gets nullified and there is a small increase in the tension of high tensile steel at the stage of decompression. The force in the high tensile steel is equal to the force which was existent at the time of force transfer, plus the force required to release the elastic compressive stress in concrete expressed as P^*. Once this force is applied on the tensile steel, the element reaches the stage of decompression and any further curvature of the element will increase the tensile force in the high tensile steel and also in the untentioned steel. If they are at the same level, the stress increase will be the same. While there is a stress increase from zero in the untentioned steel after the stage of decompression, the stress increases from high tensile steel is from f_{sp}^*. At the service load when cracking is permitted, the total tensile force in the tension chord is given by the force in the untentioned steel caused by the service load from the stage of decompression and the force in the tension steel is given as P^* plus the force in the tension steel after the stage of decompression. A strain distribution pattern is assumed at the service load. From the strain distribution pattern the stress in concrete at the top fibre and the level of neutral axis is known. Using these values, the total compressive force is evaluated. Since position of the neutral axis is known the lever arm is also known and the total moment resisted for this assumed strain distribution is evaluated. If this total moment resisted is not equal to the moment applied at service load, the strain distribution can be altered. It may be noted that the strain at the top concrete fibre should not lead to a stress more than the permissible stress allowed in concrete. By iteration, the strain distribution pattern with top fibre concrete stress not exceeding the stress in concrete at service load and the position of the neutral axis is arrived at to generate a moment equal to service load moment. For this strain distribution pattern in compressive zone, the total compression force is evaluated and the total compressive force must be equal to total tensile force. The total tensile force in high tensile steel is known as equal to the total force at the stage of decompression plus the additional tensile force due to the stress got from the strain profile. This total tensile force in high tensile steel may not match with the total compressive force which would normally be the case. The additional tensile force required has to be provided by the untensioned steel. Since the stress level in the untensioned steel is also known, the area in the untensioned steel is evaluated. Now the concrete cross section, the area steel by high tensile as well as untensioned steel, their positions are known, the design should be checked for limit stage of collapse. Assuming that the steel initiates the collapse, the total tensile force in the tension chord (that provided by high tensile steel and untensioned steel) multiplied by lever arm will be the moment carrying capacity. If this is found to be inadequate, additional untensioned steel may be added to have higher moment resisting capacity. This is illustrated in the design example 14.4.

Example 14.4.

Design a rectangular beam of 800 × 1200 to resist the total moment of 450 tonne meters at service load.(a) as a fully prestressed concrete beam. (b) as a limited prestressed concrete beam wherein a tension of 3N/mm² is permitted at service loads and (c) as a partially prestressed concrete beam the steel being prestressed lower than its permissible stress at service load and untensioned steel used for carrying service loads. But it should be prestressed to cover 300kNm. High tensile steel of capacity 1,600N/mm², untensioned steel of capacity 415N/mm² are available. The maximum permissible stress at service loads in high tensile steel is 960 N/mm² and untensioned steel is 240N/mm². Maximum compressive stress in concrete at service load =20N/mm².

Solution

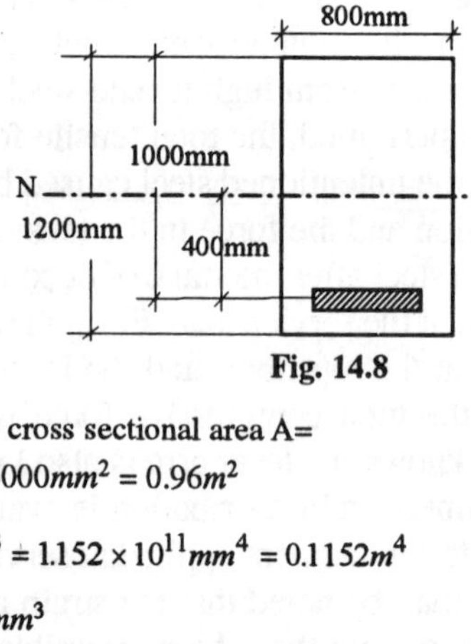

Fig. 14.8

Reffering to Fig.14.8 cross sectional area A=

A = 800 × 1200 = 96000mm^2 = 0.96m^2

$$I = \frac{1}{12} \times 800 \times 1200^3 = 1.152 \times 10^{11} mm^4 = 0.1152 m^4$$

$Z_b = Z_t = 192 \times 10^6\ mm^3$

If the service load moment to be resisted 450 tm.

Case (I): (fully prestressed concrete beam). The beam shall not have tensile stress at any level at any stage upto service load stage considering the serivice load stage, the bottom to be stress should be reistricted to zero stress. Let the prestress to be given be P. The bottom fibre stress at service load stage =

$$\frac{P}{96 \times 10^4} + \frac{P \times 400}{192 \times 10^6} - \frac{450 \times 10^7}{192 \times 10^6} = 0$$

solving for P, $P = 7.5 \times 10^6 N$

Stress at top fibre at service load

$$\frac{7.5\times10^6}{96\times10^4}-\frac{7.5\times400}{192\times10^6}+\frac{450\times10^7}{192\times10^6}=15.624<20N/mm^2$$

Case (ii) Limited prestressed concrete beam.

The required prestressed should limit the bottom fibre stress not be exceed a tensile stress of $3N/mm^2$.

This can be expressed as

$$\frac{P}{96\times10^4}+\frac{P\times400}{192\times10^6}-\frac{450\times10^7}{192\times10^6}=-$$

Solving for P, $P = 6.54 \times 10^6 N$

Then the stress at top fibre during service load

$$=\frac{6.54\times10^6}{96\times10^4}-\frac{6.54\times10^6\times400}{192\times10^6}+\frac{450\times10^7}{192\times10^6}=16.624N/mm^2$$

Case (iii) Partially prestressed concrete beam. The beam should be able to resist 300 ton.m without having any tension at the bottom. At service load moment, of 450 ton, it can have cracks, of limited, crack width. The prestressing force to be provided can be arrived at using the expression.

$$\frac{P}{96\times10^4}+\frac{P\times400}{192\times10^6}-\frac{300\times10^7}{192\times10^6}=0$$

Solving for P, $P = 5\times10^6 N$

Then stress at top fibre

$$=\frac{5\times10^6}{96\times10^4}-\frac{5\times10^6\times400}{192\times10^6}+\frac{300\times10^7}{192\times10^6}=10.417\text{ N/mm}^2$$

$P_\infty = 5 \times 10^6$

Stress at steel level due to prestressing alone

$$\frac{5\times10^6}{96\times10^4}+\frac{5\times10^6\times400}{1.152\times10^{11}}\times400=12.15\ \text{N/mm}^2$$

$E_{sp} = 2 \times 10^5\ N/mm^2$

$E_s = 2 \times 10^5\ N/mm^2$

$E_{concrete} = 3.0 \times 10^4\ N/mm^2$

Limiting stress at failure in prestressing steel = $1600\ N/mm^2$

Permissible stress limit at service load = $960\ N/mm^2$

Permissible stress in untensioned steel at service load = $240\ N/mm^2$

Stress at which H.T Steel can be stressed

after elastic shortening (960-240) = $720\ N/mm^2$

Area of high tensile steel required $=\dfrac{5\times10^6}{720}=6944.5mm^2$

Total tensile force at stage of decompression

$= 5 \times 10^6 + 12.152 \times 6.66 \times 6944.5$ = $5.562 \times 10^6 N$

Final moment at service load is 450m(4500kNm)

This moment should be resisted by a couple with compression in concrete and tension in high tensile steel and untenioned steel. The section is cracked.

Assuming a Strain distribution as shown in Fig. 14.9

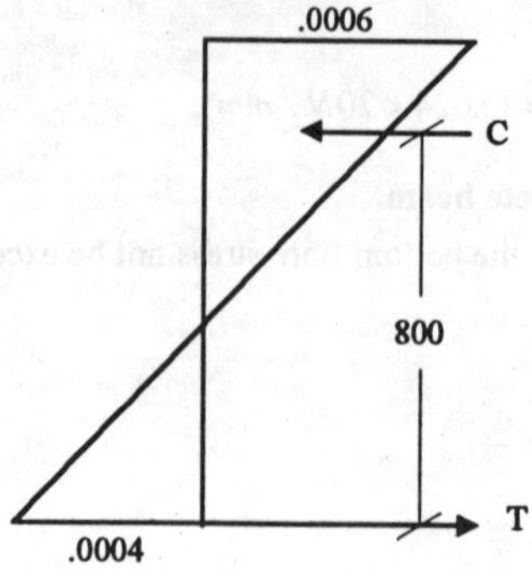

Fig. 14.9

Compressive Stress at the top fibre $= 6 \times 10^{-4} \times 3 \times 10^4 = 18\text{N/mm}^2 < 20\text{N/mm}^2$

Tensile stress in steel fibre $= .0004 \times 2 \times 10^5 = 80\text{N/mm}^2$

Total compression $= \dfrac{18}{2} \times 0.6 \times 0.8 \times 10^6 = 4320\text{kN}$

Lever arm $= 0.8$ m

Moment $= 4320 \times 0.8 = 3456$ kNm = 345.6 ton metre.

Moment required to resisted = 450 kNm

Hence assumed a strain distribution as .0006 at top and .0004 at steel level is not correct.

Assuming a revised strain distribution of .00066 at top and .0022 at steel fibre level as shown in Fig.14.10.

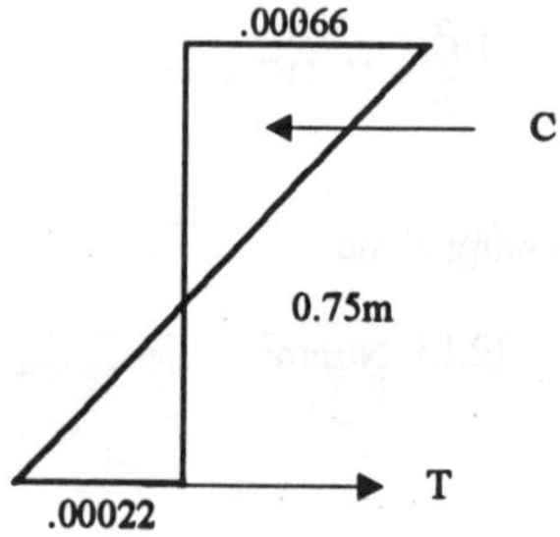

Fig. 14.10

Compressive stress at top $= .00066 \times 3\times10^4 = 20\text{ N/mm}^2$

Tensile stress in steel level $= .00022 \times 2 \times 10^5 = 44\text{N/mm}^2$

Tensile stress in steel fibre $= .00022 \times 2 \times 10^5 = 44\text{N/mm}^2$

Total compression $= \dfrac{20}{2} \times 0.75 \times 0.8 \times 10^6 = 6000\text{kN}$

Lever arm = 0.75m

Moment $= 6000 \times 0.75 = 4500$ kNm.

Moment required to be resisted = 4500 kNm

Tensile stress in high tensile steel $= .00022 \times 2 \times 10^5 = 44\text{N/mm}^2$

A_{sp} = 6944.5

Additional tensile stress $= 6944.5 \times 44 = 305.558$kN

Total tensile force in high tensile steel = 500+30.558 = 530.558 tons

Total compression = 600 tons

Additional tensile to be provided by untensioned steel (600 – 530.558)

= 69.442 tons

Untensioned steel required $= \dfrac{69.442 \times 10^4}{240} = 2893\ mm^2$

25mm dia bars 6 numbers will give an area = $2940\ mm^2$

The section is reinforced with 24 Nos. of 12 × 7 straind and 6 numbers of 25mmφ F_e 415 steel.

14.6 Cracks and Crack Width

The number of formulae have been proposed in by various authors as detailed below. Since cracks are permitted to occur during service loads and width of cracks are defined with a maximum limit, procedure for evaluating of crack width and spacing is mandatory for partial prestressed members. The details of research works done and the formulation of expression for crackwidth and spacing are presented here for information the usage of these expressions for practical cases have to be done with caution depending on situation, type of structure and its purpose, the loadaing, the locations etc.

Bennott and Chandrasekhar [13] reported extensive tests on partially prestressed beams using 7 mm diameter hard drawn crimped wires as prestressing steel and deformed bars or strands as untensioned steel and derived an expression for mean crack width w_m as follows:

$$w_m = (K_1\varepsilon_s + K_2\delta\varepsilon_s)C_s$$

where K_1 = Constant which is equal to3.8 for deformed bars and strands and 5 for crimped wires.

K_2 = Constant which is equal to 20 for deformed bars and strands and 30 for crimped wires.

ε_s = strain in untensioned steel occurring after decompression of concrete at the level of tendons.

$\delta\varepsilon_s$ = change of strain in untensioned steel occurring between decompression at the level of tension face and decompression at the level of tendons.

C_s = minimum concrete cover over untensioned steel.

As an alternative, Bennett and Chandrasekar [13] also proposed a simpler expression for crack width as

$$w_m = K_3 f_{ft} C_s$$

where K_s = Constant which is equal to 0.000435, 0.000725 and 0.001160 mm^2/N for deformed bars, strands and wire reinforcement respectively.

f_{ft} = ficticious or hypothetical tensile stress in concrete at the tension face in Mpa.

The CEB-FIP [14] recommedned the following general equation for the maximum crackwidth w_{max} *in mm*

$$w_{max} = \left(\delta f_s - \frac{KE_s}{P_s}\right) \times 1$$

Where δf_s increase of stress in Mpa in untensioned steel beyond the stage of decompression at the level of untensioned steel.

K is constant which is related to the nature of bond and loading conditions (Table 14.1)

P_s is percentage of untensioned steel as defined in Table 14.1 but not less than the specified minimum value, P_s(min).

Table 14.1

Type of member	P_s	P_s(min)	KE in Mpa
(i) Rectangular and T-. section in simple bending	$\frac{A_s}{b_w d_s}$	0.010	0.37
(ii) Rectangular and T section in compression and bending	$\frac{A_s}{b_w(d_s - x)}$	0.016	0.60
(iii) Ties	$\frac{A_s}{\text{total area}}$	0.040	1.50
(iv) Beams with bottom flange	$\frac{A_s}{\text{Area of bottom flange}}$	0.040	1.50

The CEB-FIP model code [15] recommends the following simplified version.

(a) For non repetitive loading

$$w_{max} = (\delta f_s - 40) \times 10^{-3} mm$$

(b) For repetitive loading.

$$w_{max} = f_s \times 10^{-3} mm$$

Beeby and Taylor[16] proposed the following expression for crackwidth in partially prestressed elements.

$$w_m = \frac{1.75 f_{ft}(D - x)}{E_c}$$

Where D = overall depth of the beam

X = depth of neutral axis computed for a cracked section subject to compression and bending, f_{ft} is ficticious or hypothetical formula stress in concrete at tension face in Mpa.

Nawy and Huang[17] proposed the following equations for pretensioned and bonded post-tensioned beams for crackwidth.

$$w_{max} = 8.32 \times 10^{-6} \frac{A_{ct}(\delta f_p)}{\epsilon_{ut}}$$

where A_{ct} effective area of concrete in uniform tension which may be taken as the area between soffit and a layer located at a clear distance c above the topmost layer of tension reinforcement in which c is the clear concrete cover below the lowermost layer.

δf_p = net change in stress in Mpa in prestressing steel beyond the stage of decompression.

$\in_{ut}$ = sum of perimeters of all tension steel, w_{nax} gives the maximum crack width at the centroid of tension steel. The crack width at soffit, (w) may be taken as w_{max} multiplied by the ratio of distances of soffit and centroid of tension steel measured from neutral axis.

Example 14.5

Determine the crack width in case of example 14.3 at the time of applied bending moment of 1.8MNm. High tensile steel is in the form of strands and the untensioned steel is provided using deformed bars Fig.14.5.

Solution

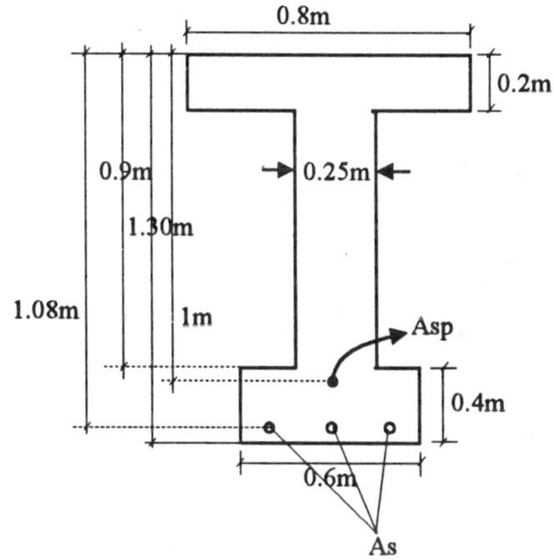

Fig 14.5

The crack width has to be calculated at the stage of loading at which the external bending moment M = 1.8MNm and the stress in concrete and steel at top fibre and in the tensile zone respectively are f_c =14.52 Mpa and f_s =140.34 Mpa. This has been calculated in the example 14.3.

For calculation of crack width, number of methods have been discussed. Here the method suggested by Bennott and Chandrasekar is made use.

The mean crackwidth wm can be expressed by the general equation

$$w_m = (k_1\varepsilon_s + k_2\delta\varepsilon_s)C_s$$

where $\in_s = \dfrac{f_s - f_{so}}{E_s'}$

in which f_{so} is the stress in untensioned steel at the stage of decompression at the level of tendons.

$$f_c \quad -\frac{1.400}{0.575} - \frac{1.400 \times 0.345 \times 0.345}{0.1096} + \frac{1.256 \times 0.645}{0.1096} = 3.43\text{Mpa}$$

Hence the stress at the level of untensioned steel is

$\frac{3.43 \times 0.4}{0.62} = 2.212$. It follows that

$f_{so} = m \times 2.212 = 6 \times 2.212 = 13.27 Mpa$

$$\varepsilon_s = \frac{140.34 - 13.27}{2 \times 10^5} = 0.000635$$

The change of strain in untensioned steel $\delta\varepsilon_s$ may be expressed as

$\delta\varepsilon_s = \frac{M'_{dc} - M_{dc}}{E_c I_g} \times (d_s - d_t)$ in which M_{dc} is the moment required to cause decompression at the bottom fibre. The intensity of prestress at bottom fibre,

$$f_{pc} = -\frac{1.400}{0.575} - \frac{1.400 \times 0.345 \times 0.345}{0.1096} = -5.277 Mpa$$

Hence, $M_{dc} = \frac{5.277 \times 0.1096}{0.645} = 0.896 MN.m$

$$\delta\varepsilon_s = \frac{(1.256 - 0.896)0.43}{\left(\frac{2 \times 10^5}{6}\right) \times 0.1096} = 0.000042$$

Taking $K_1 = 3.8$ and $K_2 = 20$, the mean crack width in accordance with $w_m = (K_1\varepsilon_s + K_2\delta\varepsilon_s)C_s$ $= (3.8 \times 0.0000635 + 20 \times 0.000042) \times 100 = 0.3253$ mm

In the simplified equation $W_m = K_3 f_{ft} C_s$, the fictitious tensile stress at bottom fibre,

$$f_{ft} = f_{PB} + \frac{My_B}{I_g} = -5277 + \frac{1.8 \times 0.645}{0.1096} = 5.316 Mpa$$

Taking $K_3 = 0.000435$, the mean crack width in accordance with equation above $w_m = 0.000435 \times 5.316 \times 100 = 0.231$ mm

Reference

1. LEONHARDT F (1977) **Recommendations for the degree of prestressing in prestressing concrete structures,** FIP Notes 69, July-August 1977, pp. 9-14.
2. DIN 4227 (1979) **Spannbeton; Bauteile aus Normalbeton mit beschriankter oder voller Vorspannung - Teil I Teil weiser Vorspannung** - II Augabe Dezember 1979.
3. COMITE EUROPEAN DU BETON. (1970) **"Federation Internationale de la precontrainte: International Richtlinien zur Berechnung and Ausfuhrung von Betonbauwerken.** Veroffentilicht zum 6." FIP - Kongreb, Prag.
4. SIA Norm 1962 **Betonbauten - Schweizerischer Ingenieur und Architekten** -Verein Postfach,8039 Zurich.
5. BS 8110 - 1985 **"British Standard for Structural use of Concrete"** British Standards Institution, London WIA 2BS.
6. ABELES. P.W., (1952) **The use of high strength steel in ordinary reinforced and Prestressed concrete beams,** Preliminary Publications, Fourth Congress, International Association for Bridge and Structural Engineers. Also supplement to above, Final Report, Fourth Congress, 1953.
7. IS 1343 - 1980 **Indian Standard "Code of Practice for Prestressed Concrete",** Bureau of Indian Standards, New Delhi.
8. G S PANDIT and S P GUPTA (1993) **"Prestressed Concrete"** CBS publishers and Distributors, 4596/1A Darya Ganj, New Delhi - 110 002 (India).

9. GUYON.Y. (1974) **Limit state design of prestressed concrete, Vol I.,** Applied Science Publishers (London).
10. THURLIAMAN B., (1971) **A case of partial prestressing, Bericht No.41,** ETH, Zurich.
11. NAAMAN, A.E. and SIRIAKSORN, A., (1979) Serviciability based design of paritally prestressed concrete beams, Journal of the prestressed Concrete Institute, Vol.24, No.2, March - April 1979. Pp.64-89.
12. *WALTHER R* - (1972) **Ausgewahlte Kaptel in Spannbeton bau**. Vorlesungschrift - Universitat - Stuttgart ".
13. BENNET, E.W. (1971) and CHANDRASEKHAR, C.S., **Calculation of the width of cracks in class 3 prestressed beams,** Proceedings of the Institution of Civil Engineers (London), Vol I.49, pp.333-46.
14. CEB - FIP (1984) **Recommendations for Practical Design of Reinforced and Prestressed Concrete Structures,** Thomas Telford, London.
15. **CEB /FIP Model Code of Concrete Structures (MC 78)** (1978), Committee Euro-International du Beton, Paris.
16. BEBBY A W and TAYLOR H.P.J., (1970) **Cracking in partially prestressed members,** Sixth Congress FIP Prague, Prelimnery Publications Concrete Society (London).
17. NAWY, E.G. and CHIANG J.Y., (1980) **Serviceability behaviour of postensioned beams** Journal of the Prestressed Concrete Institute, Vol.25, No.1, January-February pp. 74-95.

15 Other Structures

15.0 General

Behavior of prestressed concrete planar structures, and cylindrical structures, their analysis and design aspects have been considered in the previous chapters. Behavior of structures in space can not be considered as super imposition of number of planar structures, because of interactive nature of behavior. Slabs, Folded plates, Shells, Domes, are typical of examples of such structures. A few points in each of these structures specially with respect to prestressing and its effects in their design are discussed only briefly in this chapter for the benefit of readers. Full details of various methods of analysis as well as design are not dealt in this chapter.

15.1 Flat Slabs

Simply supported slabs, when subjected to axi-symetrical loading, bends with a curvature of that of a bowl in the centre portion. The curvature adjusts itself at the ends based on the boundary conditions. A circular slab is the best shape for resisting the force by bending. If there is a ring beam around the boundary, the slab behaves either simply supported or fixed at the ring beam. If fixity is provided, there is a negative curvature for a small distance from the support. But such circular slabs are not very common in regular constructions but for in water tanks and lobbies of auditorium etc. In regular constructions, square or rectangular slabs are very commonly adopted. For long spans, the thickness of the slab becomes large and sometime simple beam bending theory may not become applicable for such thick slabs. For dealing with such cases, flat slabs concepts were developed. Flat slabs are supported at the corners by columns with and without column heads. Such slabs are also prestressed to make them thin. Because of the structural system, there is a indirect support to the slab system along the lines joining the columns, which is seen clearly seen in the yield lines formed at the time of failure. Hence the analysis of flat slabs is done with a idealisation of end strips, column strips and middle strips as shown in Fig.15.1. The column strips have large negative bending moment and smaller quantity of positive bending moment whereas the middle strips have large positive bending moment with a limited quantity of negative bending moment. To take care of the bending moment in orthogonal directions with a limited thickness, prestressing is resorted to. A practical and optimal design for prestressing leads to a pattern with more number of cables in column strips with less number of cables at larger spacings in the middle strips. The typical cable profiles for a flat slab is shown in Fig.15.2. ACI - AISC joint committee 423[1,2] on prestressed concrete structures covers the analysis and applications of postensioning of slabs. There are number of softwares like ADAPT [3] which have been developed based on the procedure given in ACI-AISC committee report. Analysis and design is done both for strength and serviceability requirements. The philosophy of analysis and design are, the same as for any other structure, considering all possible loadings and their combinations. Stresses in the materials are checked to be with in their limits at the service load stage and resisting capacity is checked against the factored load carrying capacity. Deformations, camber, cracks and crackwidths are calculated at service load stage considering appropriate section properties. The preliminary analysis for given loading is done with the principles structural mechanics. The flat slab is divided into equivalent frames in orthogonal directions. The frames are idealised using the column strip and middle strip portions as the top boom elements and the same columns are used for frames in both directions.

The top boom will have different moments of inertia along its length namely the middle portion, and the column head portion. The bending moment diagram in orthogonal direction on the frames are drawn. The boom bending moment in orthogonal frames are used for designing the required prestressing force and cable profile in that direction. Then this prestressing profile is converted in to the equivalent loads as suggested by Alamy[4] and the analysis is redone by equivalent frame method as developed by Vanderbilt and Corley [5]. The stresses and deformations are evaluated. They are checked against the permissible limits. Additional untentioned reinforcements are provided for controlling cracks and crack widths and for limit state of collapse.

As an approximate method, half width on either side of the panel can be considered as the boom portion of the frame in each direction and frame analysis carried out. That would mean, there are only column strips and no middle strips. Even while working out shiftness an equivalent depth of slab is considered taking both the thickness of slab in the middle strip portion and at column head portion and the analysis is done. Illustrative example 15.1. describes this approximate method clearly.

Prestressing flat slabs are commonly used for multistoried buildings and they lead to a facility of having one more floor, for every 14 floors, since the longitudinal and transverse beams are avoided and that beam depth is saved at each storey height.

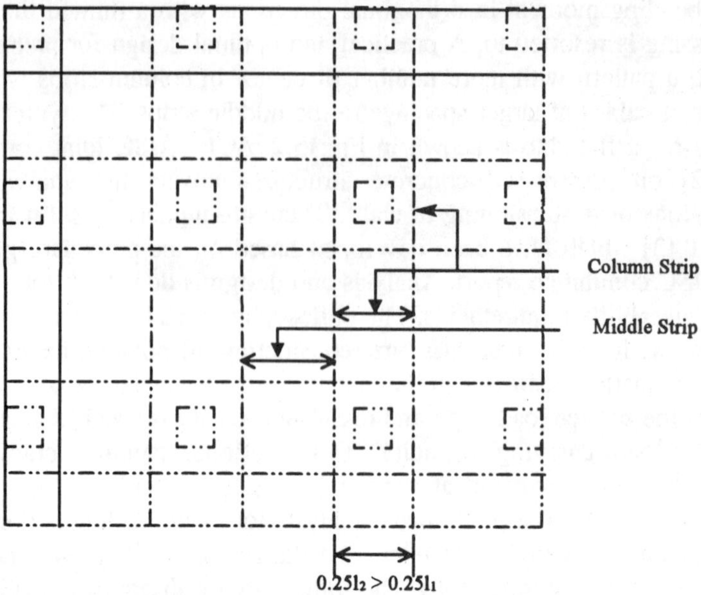

15.2 Two Sides Supported Slabs

The slabs used in bridges are supported only on 2 sides and free on the other two sides with or without stiffners at ends. Large span slab bridges can also be prestressed with cables running along the traffic directions and they may have parabolic profiles or straight profiles. Analysis of such slabs is done using principle of super position of 2 cases namely analysis of the slabs for the loading and the analysis for prestressing. The bending moments got from the analysis of slabs for loading are used to arrive at the prestressing, whereas in reality, they act together. While the loading leads to bending moment in orthogonal directions, the prestressing is decided based on bending moment only in the traffic directions and any prestressing given in the traffic direction affects the bending moment in the width direction also which is not being looked into. Analysis, considering prestressing forces as internal forces along with the loading definitely results in different bending moment diagrams and principal stress trajectories from those got by principle of super position adopted till date. The present day design method is conservative in the traffic direction but does not meet the requirements in width directions. But because of the redistribution effect the structure may not collapse.

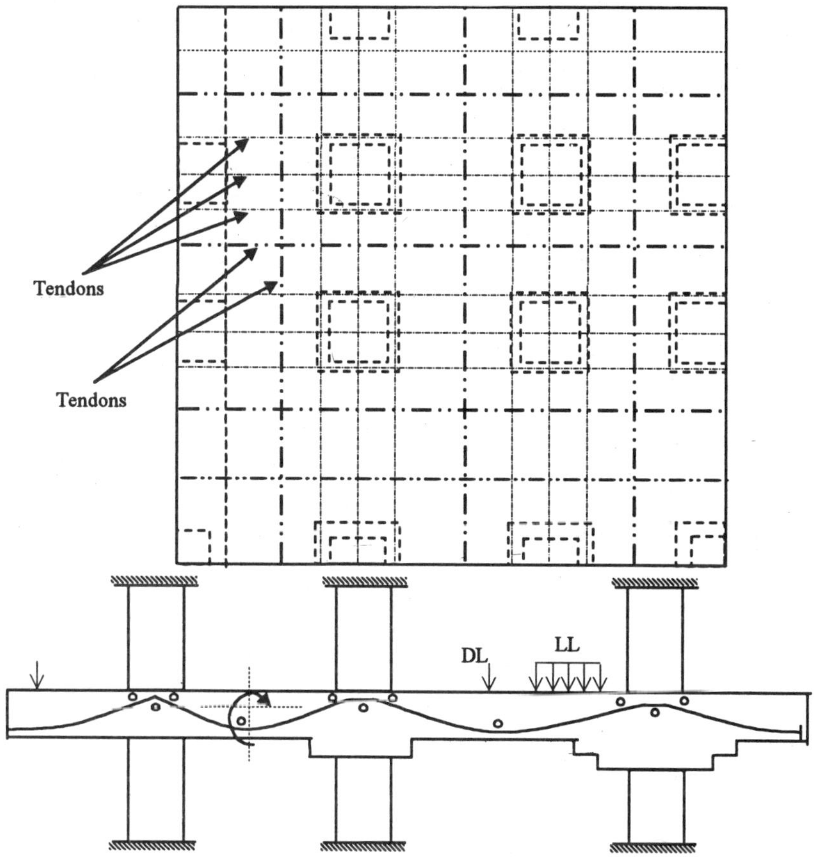

Fig 15.2 a.b. Plan Elevation of a Post-Tensioned Slab

Example 15.1

A typical floor of a multistoried building consists of a flat slab with the slab thickness of 150 mm supported on columns at 6 m intervals with drop panels of 2.01 m. At the drop panel the slab thickness is 250 mm. Columns are of size 500 × 500 mm. Design a typical internal panel of this floor as a flat slab prestressed in both the directions. A superimposed dead load of 1kN/m^2, floor finished plaster load of 1.4 kN/m^2 and live load of 2 kN/m^2 may be assumed.

Solution

Panel Size: 6000 × 6000 (internal)

Spans are as shown in Figure 15.3

Overall slab depth outside drop panels = 150 mm

and at drop panels = 250 mm

Drop panels size = 2010 mm × 2010 mm.

All columns are 500 mm × 500 mm

Loading

Dead load of slab = 24 × 0.15 = 3.6 kN/m^2

Superimposed dead load partitions = 1.00 kN/m^2

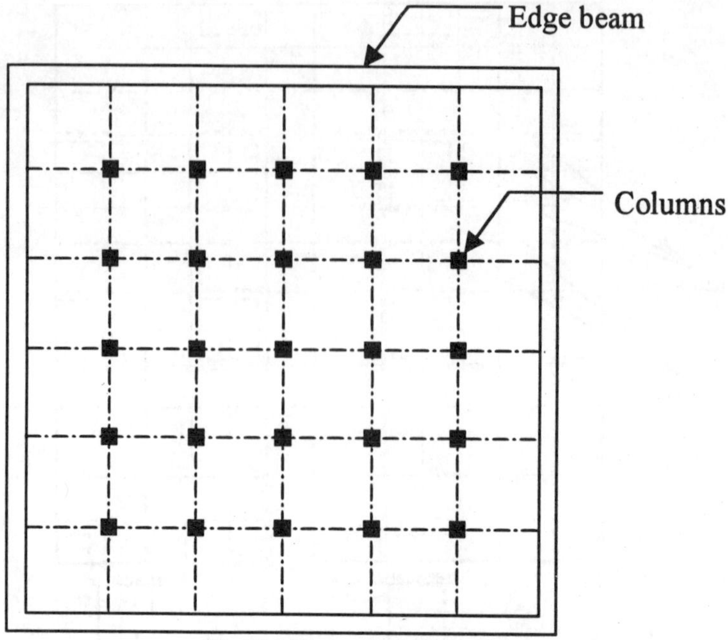

Fig 15.3 Plan of Full Floor Slab

Floor finish, plaster, suspended cealing etc = 1.40 kN/m^2

Total dead load = 6.0 kN/m^2

Live load = 2.0 kN/m^2

Total load = 8.0 kN/m^2

Extra dead load on drop panel = 24 × 0.1 = 2.4 kN/m^2

i.e. 2.4 × 1.005 = 2.412 kN/m width in each span.

Geometry

In order to take drop panels into account an equivalent slab thickness is calculated.

The properties of slab

Taking moment of areas of unit strip of slab in one direction about the top surface the depth of centred of the combined unit is calculated.

Depth of centroidal axis is 97.9 mm from top

$\bar{y} = 97.9$ mm

Moment of inertia of the equivalent slab is calculated on the 2nd moment of area about the centroidal axis of the individual areas of regular slab portion and column head portion of the slab

Moment of inertia (I) = 4.42 × 10^9 mm^4

Area of cross section at support = 1101000 mm^2

$Z_t = 4.52 \times 10^7$ m^3

$Z_b = 2.91 \times 10^7$ m^3

Equivalent depth (h_c) = $\sqrt[3]{\frac{1.2 \times I}{6000}} = 206.8mm$

At midspan, Area = 900000 mm^2

$Z_t = 2.25 \times 10^7$ mm^3

$Z_b = 2.25 \times 10^7$ m^3

From standard tables, stiffness factors, carryover factors, F.E.M. factors are obtained for haunched beams shown in Fig.15.6.

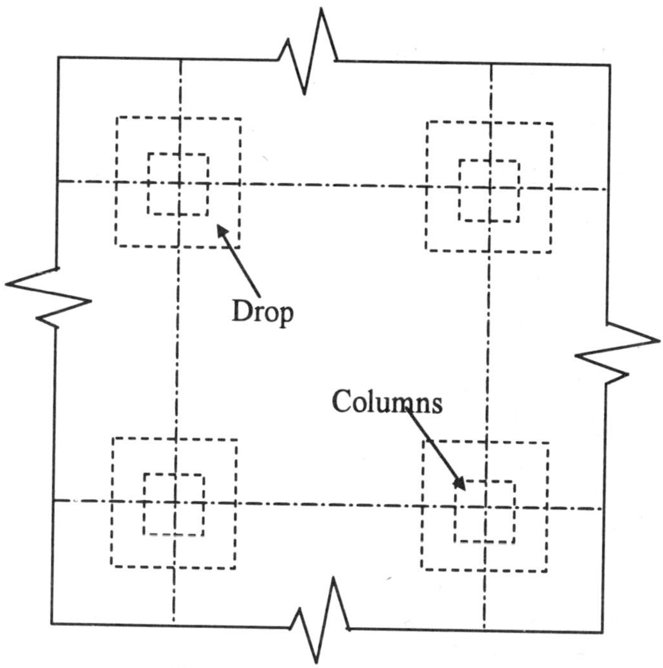

Fig 15.4 Plan of Typical Internal Panel

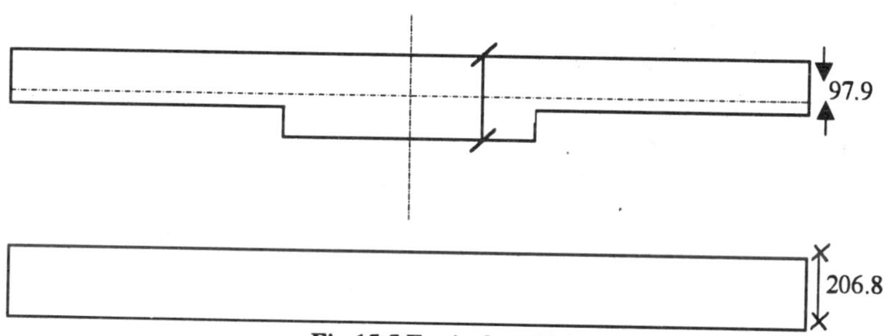

Fig 15.5 Equivalent Slab

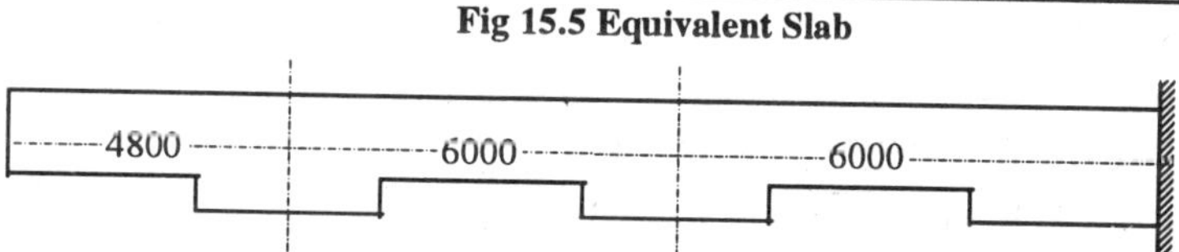

Fig 15.6 Slab for Stiffness Factor

Stiffness and carry over factors are designated as C and K respectively

Span AB : $C_{AB} = 0.677$; $K_{AB} = 4.42$

$C_{BA} = 0.469$; $K_{BA} = 6.37$

For other spans: $C_{BC} = C_{CB} = C_{CD} = C_{DC} = 0.619$

$K_{BC} = K_{CB} = K_{CD} = K_{DC} = 6.771$

Relative stiffness: [E being same for all members]

At A, Upper column = $\dfrac{4 \times I}{L} = 5952380$

$$\text{Lower column} = \frac{4 \times I}{L} = 59523380$$

$$\text{Span AB} = \frac{K_{AB} \times I}{L} = 1553910$$

At B, Upper Column & Lower Column same as above.

$$\text{Span BA} = \frac{K_{BA} \times I}{L} = 2239460$$

$$\text{Span BC} = \frac{K_{BC} \times I}{L} = 1904340$$

At C, Upper column and Lower column same as above

$$\text{Span CB} = \frac{K_{CB} \times I}{L} = 1904340$$

$$\text{Span CD} = \frac{K_{CB} \times I}{L} = 1904340$$

Distribution factors

Distribution factors for the slab members are only calculated as these are only relevant for the study.

At joint A →
- Lower Column
- Upper Column (U)
- Slab AB=0.115

At joint B →
- U
- BA = 0.140
- BC = 0.119
- L

At joint C →
- U
- CB = 0.12
- CD = 0.121
- L

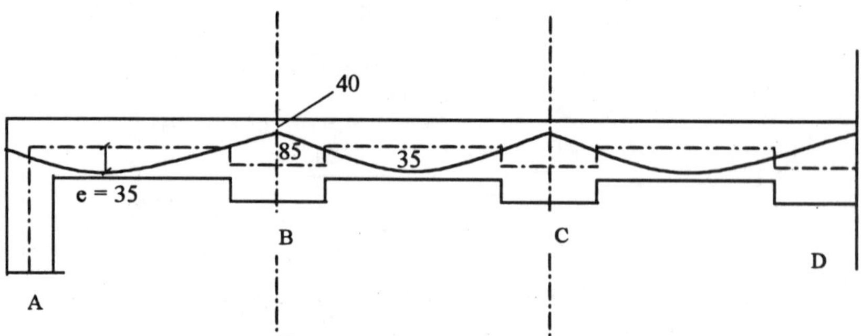

Fig 15.7 Idealized Tendon Profile

The cable profile shown in Fig.15.7 is idealized in the sense that the sharp kinks of the cables over support centre lines as shown will be actually smooth summit curves. However for theoretical calculations of upward forces, the idealized profile will be considered. The inaccuracy is within limits of approximations as accepted.

Clear over as recommended by IS:1343 = 30 mm

Effective cover upto centre line of strands=40 mm

Effective drape (a) for

$$\text{Span AB} = \frac{1}{2}(e_1 + e_3) + e_2 = 77.5\ mm$$

$$\text{Span BC,CD} = \frac{1}{2}(e_3 + e_3) + e_4 = 120\ mm$$

Friction losses

For a typical internal span, average deviation per metre

$$\alpha' = 16 \times \frac{\textit{total drape of tendons}}{L^2}$$

Total drape = 150 -40 - 40 = 70 mm

$\mu = 0.2/rad$

k = 0.0040/m

$P_x = P_o e^{-(\mu\theta + kx)}$

Where P_o is the force at anchorage and P_x at a distance of x meters from anchorages.

At D i.e., at a distance of (2.8 × 6) m. from anchorages i.e., from A

$P_D = 0.842 \times P_0$; m= $\dfrac{(P_0 - P_D)}{2.8 \times 6000}$ in kN/mm

where m is reinforcement ratio

Now with a slip of 5 mm (D)

A_{sp} = strand area of 98.7 mm^2 and E_s = 195 kN/mm^2

Slip loss is felt upto a distance of 'x', where $x = \sqrt{\dfrac{\Delta . A_{sp} . E_s}{m}}$

while arriving at the cable forces, the following two restraints are to be kept in mind.

The jacking force Po < 80% of breaking force for strand i.e, < 0.8 × 18.4 kN i.e < 147.2 kN and the maximum force after locking, i.e., the initial force at a distance 'x' < 70% of 184 kN i.e., < 128.8 kN

Now, from trials, P_o = 139.2 kN

m= 0.00131 kN/mm

x= 8579 mm

P_{max}= 128 kN

Both the restraints are satisfied.

Long term Losses

Relaxation of steel = $\dfrac{59}{1000} \times 98.7 = 5.82\ kN$ for 66% average stress from IS 1343–1980

Shrinkage = .0002 × 195 × 98.7 = 3.85 kN for transfer of force after 8 days. Assuming E for concrete as 3.12 × 10^4 N/mm

Assuming average compressive stress of 2.5 N/mm^2 on safeside

$$\text{creep loss} = \frac{2.2 \times 2.5 \times 195 \times 98.7}{3.12 \times 10^4} = 3.39\text{kN}$$

Total loss = 13.06 kN

For an average stress level of 66.6% of UTS, this loss= $\frac{13.06 \times 100}{0.66 \times 184} = 10.66\%$

Residual force = 89.34% of initial force.

The force flow is shown in Fig. 15.8.

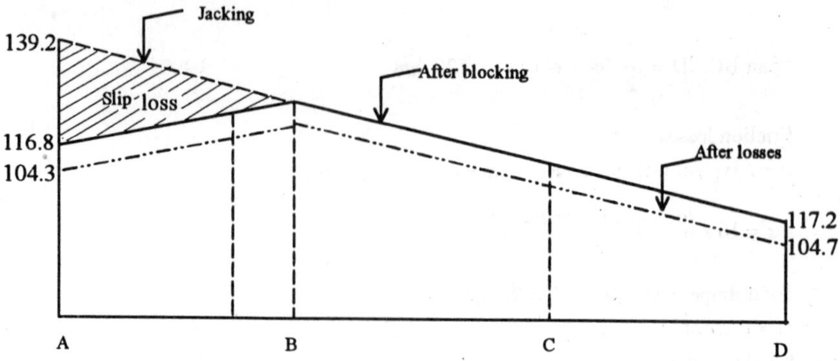

Fig 15.8 Force Profile Per Tendon

Average tendon force

Span AB = 107.1 kN

Span BC = 112.5 kN

Span CD = 108.2 kN

Balancing 50% of total load i.e. = 8/2 = 4 kN/m^2

The load of 4 kN/m^2 should be balanced by P.e.

$$\text{Thus for span AB } P = \frac{4 \times 4.8 \times 4.8}{8 \times 77.5 \times 10^{-3}} = 148.6 \text{ kN/m}$$

Similarly for span CD = 150.0 kN/m

Number of strands required in AB = 148.6/107.1 = 1.387/m

And number required in CD = 150.0/108.2 = 1.386/m

6 × 2 strands (13 mm) will be adequate to provide this force

Unbalanced UDL per span = Total load -(No. of strands $\frac{wL^2}{8}$)

For span AB = 0.234 kN/m^2

For span BC = 0 kN/m^2 These are the permanent loads

For span CD = 0.226 kN/m^2

F.E.M s for the above UDL and also for live load are calculated from standard charts for haunched beams.

Permanent loads

$$M_{AB} = 6 \times 0705 \times 0.234 \times 4.8^2 + 6 \times 0013 \times 2.4 \times 4.8^2 = 2.71 kNm$$

$$M_{BA} = 6 \times 1126 \times 0.234 \times 4.8^2 + 6 \times 0171 \times 2.4 \times 4.8^2 = 9.31 kNm$$

$$M_{BC} = M_{CB} = 6 \times .0955 \times 0 \times 6^2 + 6 \times 0146 \times 2.4 \times 6^2 = 7.58 kNm$$

$$M_{CD} = M_{DC} = 6 \times .0955 \times 0.226 \times 6^2 + 6 \times 0146 \times 2.4 \times 6^2 = 12.24 kNm$$

Live load

$M_{AB} = 6 \times .0705 \times 2 \times 4.8^2 = 19.49$ kNm

$M_{BA} = 6 \times .1126 \times 2 \times 4.8^2 = 31.13$kNm

$M_{BC} = M_{CB} = M_{CD} = M_{DC} = 6 \times .0955 \times 2 \times 6^2 = 41.26$kNm

Permanent load case and a few critical live load cases are investigated. After usual moment distribution the following values of bending moments are obtained.

Case I Permanent load (All moments are in kNm)

Sign convention: internal moment clock wise is positive

$M_{AB} = 2.50$

$M_{BA} = -9.21$

$M_{CB} = -8.04$

$M_{CD} = 11.66$

$M_{DC} = -12.59$

Span moments AB = 1.79

-do- BC = –0.55

-do- CD = 1.19

Case 2. Live load on spans AB and BC

$M_{BC} = 42.96$ kNm

Case 3: Live load on spans AB and CD (All moments are in kNm)

$M_{AB} = 19.04$

$M_{BA} = -27.67$

$M_{CD} = 35.98$

$M_{DC} = -44.35$

Span moment AB = 11.21

- do - CD = 13.84

Case 4: Live load on span BC (All the moments are in kNm)

Span moment BC= 15.00

Stresses at serviceability limit state:

At support $A_c = 1101000$ mm^2 ;$Z_t = 4.52 \times 10^7$ mm^3; $Z_b = 2.91\ 10^7$ mm^3

At midspan $A_c = 900000$mm^2; $Z_t = Z_b = 2.25 \times 10^7$ mm^3

Sagging moments are considered positive

Table 15.1

Location	Design moment (M) (kNm)	Stresses at	P/A_c(N/mm^2)	M/Z(N/mm^2)	Resultant stress (N/mm^2)
A	-21.54	Top	1.39	–0.96	0.43
		Bottom	1.39	0.96	2.35
Span AB	13.00	Top	1.43	0.58	2.01
		Bottom	1.43	–0.58	0.85
B	-50.47	Top	1.20	–1.12	0.08
		Bottom	1.20	1.74	2.94
Span BC Or CD	15.02	Top	1.44	0.67	2.11
		Bottom	1.44	–0.67	0.77
C or D	-56.94	Top	1.14	-1.26	-0.12
		Bottom	1.14	1.96	3.10

Permissble tensile stress = -0.45 $\sqrt{30}$ = -2.46 N/mm^2

Permissible compressive stress = 0.24 × 30 = 7.20 N/mm^2

Stresses are safe everywhere.

Stresses at transfer

This check is made immediately after blocking.

The dead load of slab and prestress before losses are the only two loads acting at this stage.

Unbalanced UDL in a span = $\left(\text{concrete weight} - \dfrac{P.a.8}{L^2}\right)$

Span AB = $3.6 - \dfrac{12}{6} \times \dfrac{1071 \times .0775 \times 8}{.8934 \times 4.8^2} = -2.85 kN/m^2$

Similarly on span BC = -3.12 kN/m^2

Span CD = -2.86kN/m^2

In addition, launch dead loads will act downwards as before. The FEM's are as under,

M_{AB}= 27.38 kN.m

M_{BA} = 38.75 kNm

M_{BC} = M_{CB} = 56.69 kNm

M_{CD} = M_{DC} = 51.48kNm.

After moment distribution, the moments are as under, [clockwise moment positive]

M_{AB} = -23.18kNm   Span momAB=-12.55kNm

M_{BA} = 43.15 kNm   Span mom BC = -20.72 kNm

M_{BC} = -55.16 kNm   Span mom CD = -18.40 kNm

M_{CB} = 57.21 kNm   [Sagging moment are considered positive]

M_{CD} = -52.27 kNm

M_{DC} = 51.09 kNm

Table 15.2

Location	B.M. (kNm)	Stresses at	P/Ac(N/m/^{2})	M/Z(N/mm^2)	Resultant stress (N/mm^2)
A	23.18	Top Bottom	1.56 1.56	1.03 -1.03	2.59 0.53
Span AB	-12.55	Top Bottom	1.60 1.60	-0.56 0.56	1.04 2.16
B	55.16	Top Bottom	1.34 1.34	1.22 -1.90	2.56 -0.56
Span beared	-20.72	Top Bottom	1.68 1.68	-0.72 0.72	0.76 2.60
C or D	57.21	Top Bottom	1.36 1.36	1.27 -1.97	2.63 -0.61

For 7 day old concrete (f_{ci} =20 N/mm^2)

Permissible stress in tension = $-0.45\ \sqrt{20} = -2.01\ N/mm^2$

and in compression = $0.24 \times 20 = 4.8\ N/mm^2$

Stresses are safe everywhere.

Ultimate moment analysis

Moments of resistance (M) at mid section A, d =75 *mm*

From table 11 of IS:1343, $\frac{A_p t_p}{b_d f_{ck}} = \frac{2 \times 184000}{6000 \times 75 \times 30} = 0.163$

$\frac{fpu}{0.87 fp}$; fpu =0.87×.987×1864=1600N/mm^2

x_u=0.341×75=25.6*mm*

M=1600×12×98.7 (75-0.42) ×25.6×10^{-6} =122 *kNm*

Similarly for midspan M=190*kNm*

Interior supports M = 381*kNm*

Now external ultimate moment as per IS:1343 = 1.5(DL+LL).

To obtain the actual dead load moment. This should be added to the moments caused by the upward prestressing effect to the serviceability design moments tabulated earlier.

Thus the external ultimate moments are (after calculations)

At external support A= -103.9 *kNm*

At midspan of BC or CD = 81.1 *kNm*

At interior support B or C or D =-273.8 *kNm*

Ultimate shear analysis

At internal columns, shear force (including the enhancement factor of 1.25 for column moment effects)

V_{ult} = 1.25 × 36 × 8 × 1.5 = 540, Vertical component of one strands tendons = Psinθ where θ = $(\frac{4 \times drape}{Span})$

Minimum average P over interior span = 108.2 kN/strand

Drape for internal columns = 120 mm

Vertical component = 108.2 sin $(\frac{4 \times 120}{6000})$ =8.66 kN/strand

No. of strands crossing the column or within L/2 of column in both directions taken, together = 4 + 2 = 6

Therefore total vertical component = 2 × 6 × 8.66 = 103.9 kN

Residual shear = 436.1 kN

Assuming a column head of 500 mm×500mm,

Critical perimeter = 4 × 500 + 2π × 1.5 × 375 = 4356 *mm*

Equivalat side is evaluated as [500 + 2 × 375] = 1250

(This allows for some curvature of perimeter line. Other use the perimeter = 4 × 1250 = 5000)

For bonded directions,

$\rho_{ps} = \frac{4 \times 98.7}{1250 \times 210} = 0.0015;$

$\rho_s = \frac{678(612)}{1250 \times 210} = 0.0026\ \rho^* = \frac{1864}{410} \times .0015 + .0026 = .0094$

Similarly, in opposite direction,

ρ_{ps}=0.00075; ρ_s= 0.0026; ρ^* = 0.0060;

$$Vc(av) = \frac{1}{2}[(0.0094 \times 30)^{1/3} + (.0060 \times 30)^{1/3}] = 0.61N/mm^2$$

ε_s=1.1

Shear resistance =1.1 × 0.61 × 4356 × 0.21=614kN>436.1kN

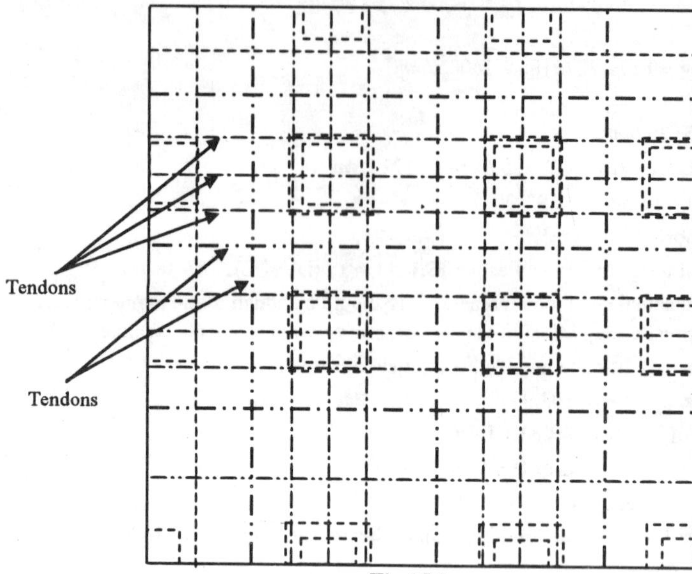

Fig 15.9

Tendons

1. 12 strands (6×2)@200c/c in internal column bands
2. 12 strands (6×2)@1050 c/c in internal panels and 10 strands (5×2) @ 1010c/c in external panels.

Reinforcement.

1. 8 No. T 12@ 200 in each direction over internal columns at top
2. 3 No. T 12 @ 200 along each edge at top.
3. T 10 hair pin bars @ 250 c/c.

15.3 Skew Slabs

Bridge decks slabs may be supported in such a way that the supports are not orthogonal to the traffic directions. They are called skew bridge decks and the behaviour of skew bridge decks are different from that of right angle bridge decks. For such bridge decks the principal trajectories (forces flow lines) are shown in Fig15.10 which are used by designers for arriving at the preliminary design, the prestressing cables and their positioning. After the preliminary design is over, it is mandatory that analysis has to be done once again for design forces and final designs made. The possible profiles for prestressing for such a type of skew slab are shown in the Fig.15.11. Positioning of prestressing also affect the force flow leading to torsion and uplift forces at corners calling for additional reinforcement at these places.

The behaviour of skew slabs are not amenable for closed form solutions and hence numerical and discrete approaches like finite element method,

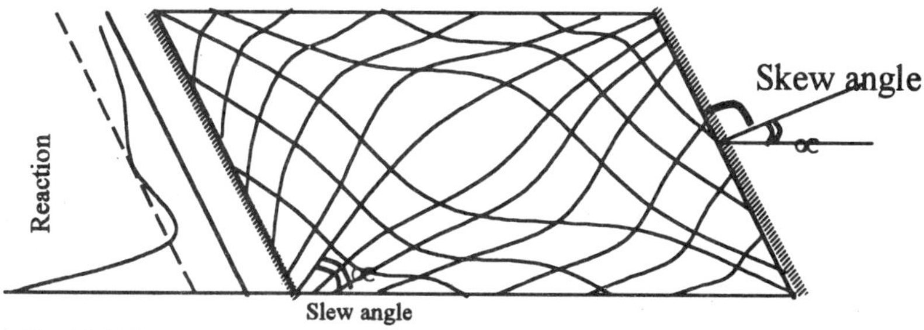

Fig 15.10 Principal Moment Trajectories and Distribution of Reaction for Skew Slabls

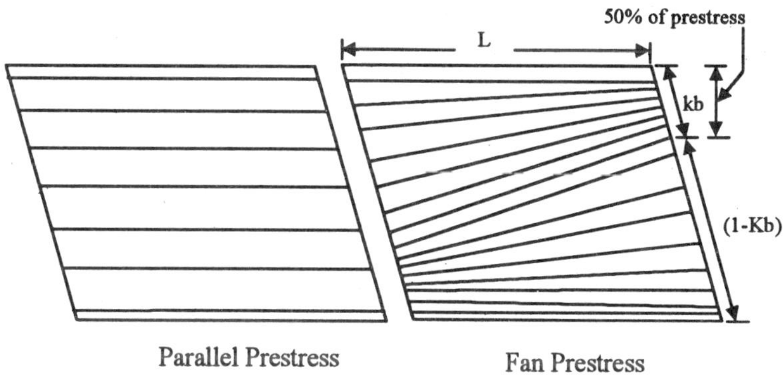

Fig 15.11 Paralles and Fan Type of Prestress for Skew Slabs

finite strip method etc are made use of. One such analysis has also been validated using experimental work by Robinson[6], Naruka Ohmura, and Yammato[7] have also studied the behaviour of skew bridge deck experimentally and suggested that the primary bending happens between obtuse angled corner and the reactions along the support line is not uniform.

The reactions are more concentrated at obtused along the corners. Rusch and Hergenroder [8] have drawn influence lines in the form of contours for bending moment at critical sections for different skew angles and different aspect ratios. The same is in the presented form of Table by Rusch [9]. Such charts and tables are used for design of skew slabs in profession. Prestressing brings in equivalent loads/equivalent moments various cross sections and modifies stress resultants.

15.3.1 Design of Skew Slabs

For Skew angle less than 15°, skew slabs are analyzed and designed as right angled slabs without much offence to load carrying characteristics. If the angle of skew is greater than 15^0 to the traffic direction and if the $\frac{L}{B}$ ratio is more than 4, the two triangular portions of the slab at ends as shown in figure 15.12b. will alone have the problem of the skew and in-between portion behaves as right-angled slab. This could be seen from the yield

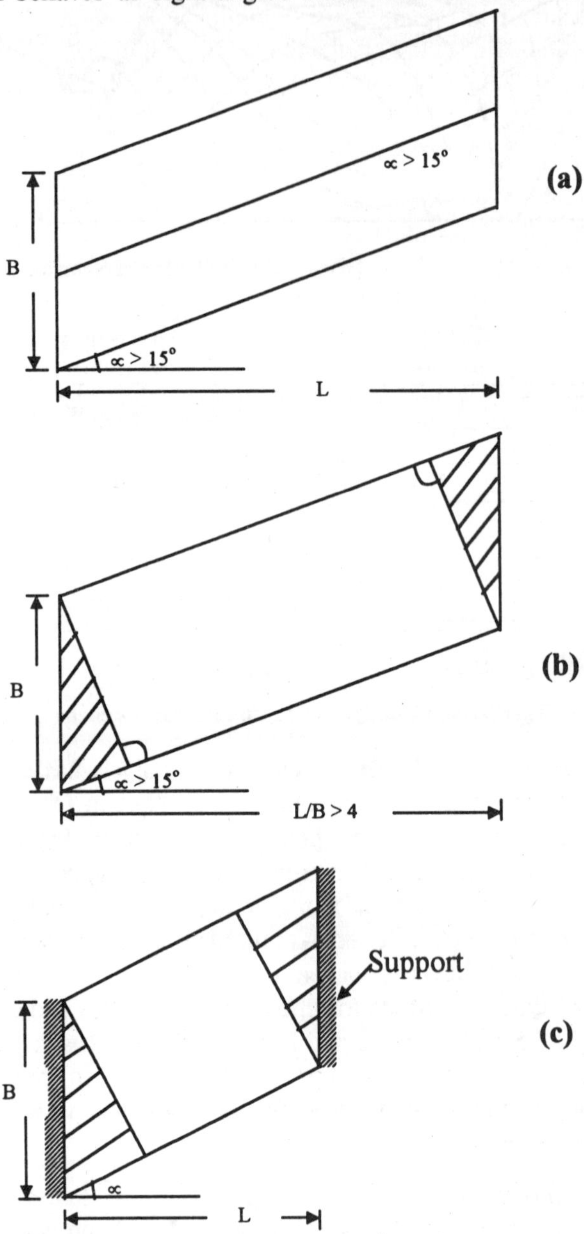

Fig 15.12

pattern and also from the force flow lines. Force flow lines run nearly parallel to the free edges in this zone. Hence the reinforcement in this zone can be decided based on longitudinal bending for this length and a nominal 20 % of reinforcement in the transverse direction. For the two end triangular portions the same longitudinal steel will be extended up to support. Similar transverse steel will also provided. In addition 15 % of the longitudinal steel will be provided at the corners both at the top and bottom and in the direction which will be normal to the line joining the diagonal corners of the slab. This is meant to take care of the torsional moments, at the skew end.

For slabs when the skew angle is more than 15° and the $\frac{L}{B}$ ratio > 1.5< 4 (Fig.15.12c) a detailed analysis has to be done. The present day methods of analysis are using a grillage modeled out of the slab or finite element analysis.

15.3.2 Grillage Analysis

If elastic analysis of skew slab is done using a grillage model the torsional stiffness for the individual girders has to be considered. The grillage may form either a skew mesh or orthogonal mesh, the former is generally adopted for low skew angles only. The results will bc acceptable only if the spacing of the grillage is very close. A spacing more than half the depth of the longitudinal grillage beam itself is generally not acceptable. If such modeling is done with closer spacing than half the depth of longitudinal beams, the beam could be assumed to represent a portion of the width of the deck and have the same torsional stiffnessof the slab, both in longitudinal and transverse direction. Reinforcements arrived for beams, if distributed over the width of the slab will be adequate in the longitudinal and transverse direction. The torsional moments on the individual beams have to be converted to equivalent moments and shear. Provisions have to be made for these equivalent moments and shear both in the longitudinal and transverse beam and then distributed over the representative width of slab in each direction

Here also 15% of longitudinal beam reinforcement have to be provided at the corneas diagonally both at top and bottom put together.

If the spacing of the grillage becomes larger than what is given above, the torsional stiffnesses of the individual beams do not represent the torsional stiffnesses of the deck for the equivalent length within acceptable limits. Hence the provisions made for additional transverse and longitudinal bending moment will not be adequate.

15.3.3 Finite Element Analysis

Finite element analysis is onc of the best suited methods available as on date. The choice of elements, the aspect ratio etc., have to be carefully arrived at. If quadrilateral elements are used, the triangular portion at both the ends of the slab have to be modeled with triangular elements The stress resultants got from the finite element analysis which would be in the form of moments and stresses have to be converted in to equivalent compressive and tensile forces and the reinforcement have to be provided for the same. Provision of reinforcement in the longitudinal and transverse directions and also diagonal reinforcement at the corners could be made in the similar lines as explained earlier. Design of the bearings have to be done taking reactive forces got out of the appropriate springs depicting correctly the proposed bearings in the mathematical model.

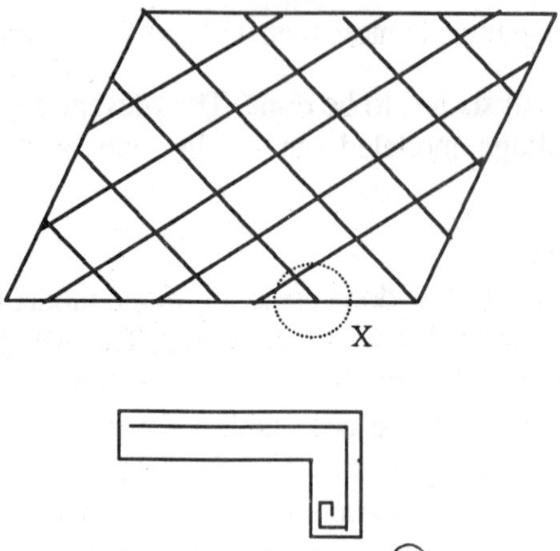

Fig 15.13

Higher bearing reactions occur in the region of obtuse angle corners. The resilience of the bearings has a considerable effect on the distribution of bearing reactiions and finite element or boundary element modelling has to be done accordingly. Use of highly resistant rubber / elastomiser bearings reduces the obtuse angle corner reactions but principal moments increase in the midspan region as contrary to use of rigid bearings.

The provision of reinforcement can be optimized by choosing the direction of reinforcement both in the two directions as the direction of the line joining obtuse angled corners and the direction normal to it. The area of reinforcement to be provided can be decided by the bending moment in that direction, considering the line joining the obtuse angled corners as the span. These reinforcements which take care of main bending moment have to have sufficient anchorages or bond length at the free end of the reinforcement. Free ends of such specially angled reinforcement may be along the free edge of the slab. To provide sufficient anchorages therein, the free edges of the slab will be stiffened by a beam or reinforcement. The slab main reinforcements meeting at that point will be tied to the stiffeners arrangement so that anchorages are provided as shown in Fig.15.13. Here also 15% of longitudinal beam reinforcement have to be provided diagonally at the corners both at top & bottom. This can be provided as a mesh form in two directions within a distance of L/5 from the corner. The total volume or reinforcements used by this approach will be less than the other method of detailing explained earlier. But the difficulty of preparing grill work makes this choice of detailing as impracticable.

Having got the philosophy of the design and the provision of reinforcement for the design forces in the case of a reinforced skew slabs, the layout of prestressing, in case of prestressed slabs can easily be identified as fan type of prestressing with more concentration at the obtuse angled corners and more distributed spacing at the acute angled corners as shown in Fig 15.11b.

15.3.4 Prestressing of Skew Slabs

Wittfoht [10] has reported the effect of different layouts of prestressing tendons based on the concentrations of bending moment under loadings. Extensive research work has also been done at I.I.T.Madras, India on the behaviour of skew plates both experimentally and theoretically and reported (11,12). The following results got out of this literature are useful in the design of skew slabs.

- Longitudinal bending of skew slab bridge decks is primarily between the obtuse angled corners for a definite band width.
- At free edge maximum moment also occurs nearer to the obtuse corner rather than at centre.
- Maximum longitudinal moment and also the deflection reduce with increase in skew angle for fixed side-ratios.
- But the transverse moment becomes more pronounced with increase of skew angle and this should be kept in view while designing the reinforcement and prestressing.
- The fact that more of moment is thrown towards the obtuse angled corner shows the greater necessity of higher amount of prestresss there and hence calls for fan-shaped arrangement of cables converging towards these corners.
- By prestressing the skew slabs, the stiffness of the slabs is increased and hence deformations during service loading are reduced.
- Normally in skew slabs, the points of maximum deflection are nearer the obtuse angled corners. This shift of points of maximum deflection towards obtuse angled corners is more if the skew is more.
- With a fan shaped layout for prestressing tendons, the points of maximum deflection due to prestressing alone, along the free edges shift away from obtuse angled corners.
- Providing prestressing, and providing it in a fan shaped layout counteracts the effects due to live load very effectively.

- At service load, in case of prestressed concrete skew slabs with fan shaped tendons, the deflections are small, and more uniform, A symmetrical deflection profile is possible even along the free edge.
- The deformation in transverse direction in case of a skew slab with prestress tendons laid out in a fan shape is more or less uniform leading to better distribution of transverse moments.
- There is more even distribution of reaction along the support line in case of prestressed skew slabs with fan shaped layout of tendons.
- The choice of fan ratio, which would be most effective, depends on the skew angle and the span to width ratio for guidance. This can be represented by an expression $k = 0.33 \cot \alpha$. where α is the skew angle, k is the ratio in which the width is to be divided so that 50% of prestressing is to be provided on earlier zone. The value of k is limited to 0.5. This formula is valid for $60° > a > 15°$ (Ref.Fig.15.11).
- Though there is significant improvement in the behaviour of prestressed skew slabs with prestressing tendons laid out in a fan shape at service stage, there is very little improvement in the ultimate load carrying capacity of the slab because of redistribution effect at failure stage.

15.4 Folded Plates

Folded plates is widely used as roofs for a variety of structures such as assembly halls, gymnisia and auditoria, where large column free areas are required. These are form resistant, stressed skin structures where the external forces are resisted also by their geometrical form to a large extent, and by the membrane stresses which is the most efficient form of mobilising the capacity of the structural elements.

With increase in the span the longitudinal tensile stresses and tranverse bending moments increase. Prestressing becomes inevitable in such long span roofs, to counter the large tensile stresses in concrete. Also the high transverse bending moment requires the provision of intermediate transverse stiffners to reduce the same.

15.4.1 Structural Action of Folded Plates

The structural behaviour of folded plates is characterised by"beam", "slab" and "plate" actions. These structures resist the system of transverse loads by the beam action of the full cross section in the longitudinal direction, by the in-plane (plate) action in the longitudinal direction and by the slab and plate action in the transverse and longitudinal directions with the fold lines acting as elastic supports. Loads normal to each plate cause it to bend transversely between the ridges as a continuous slab (slab action). The plate functions as one-way slab and longitudinal slab action is ignored because of the assumption that the length of each plate is more than twice its width. The bending of the slab transmits normal loads at the fold lines as reactions (ridge loads).These ridge loads are resolved into components in the planes of the plates meeting at that ridge. These loads parallel to the plates are then transmitted to the supports by the beam action. The plate functions as a beam (shallow or deep depending on the dimensions) between its supports to transmit loads inits plane to those supports. Because of its large depth and small thickness, each plate offers considerable resistance to bending. This plate action explains the remarkable rigidity of folded plate construction. The structural behavioural aspects is shown in Fig.15.14.

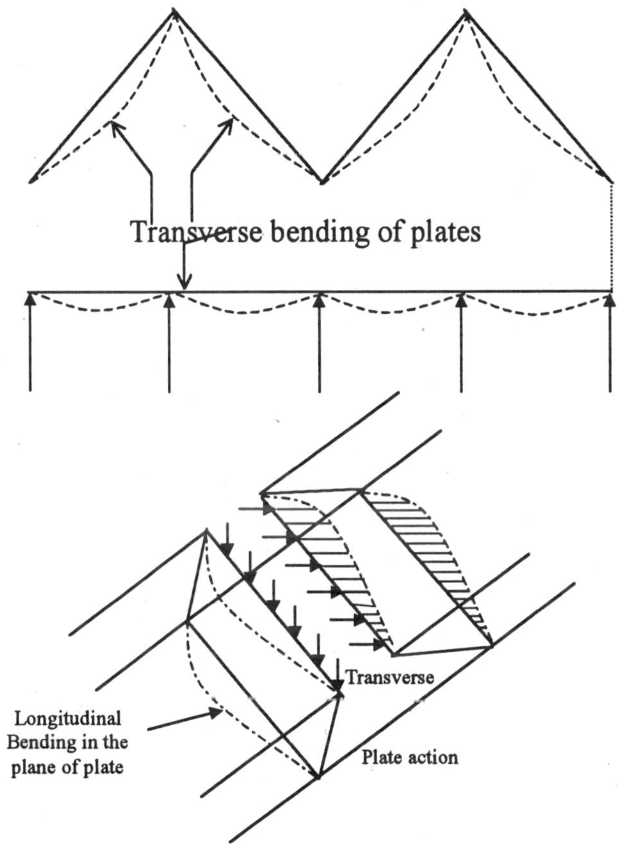

Fig. 15.14. Behaviour of Folded Plates

The magnitude and distribution of the moments and the inplane stresses are influenced by the deformation of the folds relative to one another. The increase in the span of the folded plate and the relative sinking of folds in the transverse direction leads to large transverse bending moments. With increase in the span, the longitudinal tensile stresses also increase.

15.4.2 Need For Prestressing

With increasing spans, the beam action becomes more predominant and the enitre cross section behaves as a beam. Then the longitudinal tension assumes very high values demading the use of large quantities of steel and also greater thickness to accommodate the steel. Hence, the design becomes uneconomical. Moreover, placing this steel without undue congestion presents practical difficulties. In long spans, splicing tension bars is also a problem. Laps are not effective in tension members and welding becomes necessary. These problems can be overcome by prestressing as in the case of beams. The choice of the placement of the tendons, of course has to be done carefully to the best advantage with respect to structural behaviour. By prestressing (introducing precompression at appropriate places) the longitudinal tensile stresses due to dead load or other loads are covered to a large extent. Hence use of untensioned steel becomes minimum. The entire concrete cross section being in compression and being uncracked, becomes effective in resisting structural forces. The other advantages of prestressing are reduction in the transverse bending moments and the control of deflection with the limitation.

15.4.3 Analysis and Design

The task committee on Folded plate construction of ASCE(13) has suggested different methods of analysis for analysing folded plates for the design forces. They are beam method, the folded plate theory neglecting or including relative joints displacement, and elasticity method.

While the beam method consists only longitudinal action, the folded plate method consists of beam analysis in the longitudinal direction with the cross section of the folded plate as a cross section of the beam and a plate analysis of individual plates forming the folds in the folded plate. The individual plates are analysed for transverse bending and also for longitudinal bending. These plates are considered as supported at diaphragms if diaphragms are existing and their deformation along with the fold lines evaluated. These deformation are introduced as junction forces to be included in the transverse analysis considering the folds as a frame. With the advent of computers, these operations have been made mathematical and mechanical and a direct stiffness method for the folded plate has been suggested by DeFries-Skene and Scordiles [14]. The other authors who have developed procedures for analysis are Gaffar[15], Sympson[16] and Witney[17].

Modern software packages like SAP or NOSTRAN are using finite elements for the analysis of such folded plate. In these packages, beam element, plate shell elements and shell elements are used with proper degrees of freedom depending on type of elements chosen.

For long spans prestressing reduces the longitudinal bending moment due to upward equivalent load. Diaphragm stiffeners or ties at frequent interval help in reducing the design forces in longitudinal behaviour. Prestressing also helps in the serviceability behaviour. The location and the profiling of the cables plays a major role. Sitapathi Rao[18], Rajagopalan and Geetha[19,20] have done work and reported some salient aspects regarding the use of folded plates, and prestressing the same for use as roof of structures. The salient aspects on the behaviour and effect of prestressing in the behaviour of folded plates could be listed as below.

- In the case of large-span folded plate roof structures, provision of edge beams and intermediate transverse stiffeners are essential and lead to reduction in both longitudinal tensile stresses and the design transverse bending moment.
- With prestress in the edge beams, both the longitudinal stress and the transverse bending moments are further reduced. Prestressing the folds of plates is not effective.
- Provisions of prestress at the junctions of folds in the tension zone is effective and they are counter productive if done at compression zone.
- Ties or solid diaphragms are more efficient as transverse stiffeners than ribs. They are equally efficient in reducing the transverse bending moment.
- Provision of intermediate transverse stiffeners in the form of ties/solid diaphragms only in the end panels is as efficient as stiffeners (ties/solid diaphragms) provided in all the panels.
- More number of transverse stiffeners results in a uniform distribution of the transverse bending moment. Increase in the number of stiffeners reduces the design transverse bending moment. However, there is a limit beyond which any increase in the number of intermediate transverse stiffeners does not give any advantage.
- Ties provided in the end panels only at every quarter span section is preferred for multiple folded roofs.

15.5 Cylindrical Shell Structures

Similar to folded plates, cylindrical shell roofs are also very commonly used to cover the large column free areas. These roof structures are economical because of their form resistant nature also They are aesthetically pleasing because of their curvature. Cylindrical shell has the facility of curvature namely the form resisting nature in the transverse direction. In the longitudinal directions it behaves as a beam with a cross sectional shape of the arch. The structural behaviour of long cylindrical shells is characterised by a beam action in the longitudinal direction and a arch action (shell action) in the transverse direction. The arch action will lead to a horizontal trust at the end which has to be contained by stiffness at the end either by an edge beam or by a continuity of the shell unit. Of course, at the extreme end barrel, there should be an edge beam. Because of the arch action in the transverse direction the thickness of the shell is minimum and the weight of the shell gets reduced. The full rise of the arch helps in providing a lever arm for longitudinal bending which keeps the total tension and compression as small as possible. The longitudinal tension can be taken by reinforcement or can be taken by prestressing. These prestressing cables are provided in the edge beams or at the bottom portion of the shell. The cables can also be profiled within the edge beam depth or a to some extent in the sloping portion of the shell. If it is in the sloping surface of the shell there is also lateral eccentricity for the cables. Marshell [21] and Haas [22] have done experimental investigation on shell structures. Typical arrangements of prestressing steel in Cylindrical shells are shown in Fig.15.15.

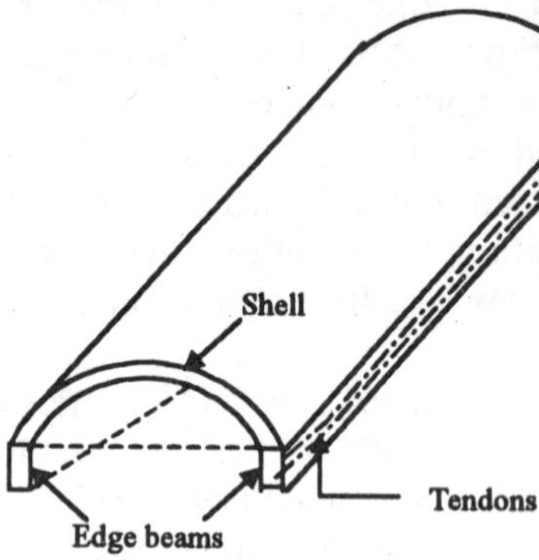

Fig 15.15. Arrangement of Tendons in Edge Beams of Long Shells

15.6 Doubly Curved Shell Structures

The concept of curvature resisting the external load is developed even the longitudinal direction. This has lead to doubly curved shells. The hyperbolic paraboloid shells are outcome of such philosophy. The prestressing for hyperbolic-parbolied shells which can be got by ruled surfaces can also be prestressed with straight cables as shown in Fig.15.16. The analysis and design of such shells is considered beyond the scope of this book. There are theories developed by Jenkins [23] and Schorer Tottenham[24] using which the stress resultants, can be found out. This stress resultants are later converted in the longitudinal and transverse bending for which prestressing can be provided. Approximate solution are also developed by Mc. Namee[25]. Typical prestress pattern is shown in Fig 15.16.

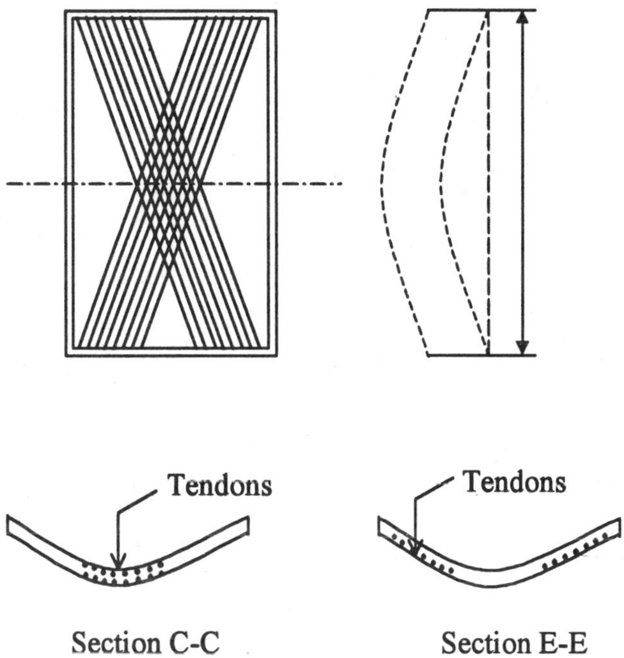

Fig 15.16 Arrangement of High Tensile Wires in Pretensioned Hyperboloid

References

1. **ACI-AISC joints committee 421 on concrete slabs,** (1983) American Concrete Institute Detroit, Michigan.
2. **ACI-AISC joint Committee 423 on prestressed concrete slabs** (1983) American Concrete Institute Detroit, Michigan.
3. ADAPT, (1990) **Theory Bijan, Florian And Associates 1733**, wood side Road, Red wood City California 94061 U.S.A.
4. AALAMI B.O., (1990) **Load balancing and A comprehensive solution for posttensioning,** ACI structural journal V.87 No.6 November-December 1990.
5. VANDERBILT M.D and CORLEY W.G., (1983) **Frame analysis of concrete buildings,** concrete International December 1983 P.P.33-43.
6. ROBINSON, K.E. (1959) **The behaviour of simply supported skew bridge under concentragted loads C and CA** publications, London, Research Ref. 8-. U.K.
7. NARUKA, M., OHMURA, H., and YAMAMATO . T., (1961) **"On model tests of skew girder bridges"** International Association of Bridge and structural Engineering Zurich, Vol.21, . P.P. 193-215.
8. RUSCH.H, and HERGENRODER A., (1964) **Eintlussfelder Momente schief winkliger Platten,** Selbrstverlag MPA, Bauswesen der T.H.
9. RUSCH.H, (1967) **Berechnungs tafeln fur, schiefwinklege Fahrbahnplatten Von Strassen-brucken,** DAfst, 166 Wilhelm Ernst and Sohn Berlin.
10. WITTFOHT H., (1962) Consideration on the theory and practice of prestressed skew slab bridges, Technical Report Translation No.100, Cement and Concrete Association London
11. DASH J.K. (1984) **Analysis of skew plates with special reference to Bridge decks,** Thesis submitted for the award of the degree of Doctor of philosophy to .Indian Institute of Technology Madras, India.

12. CHAKRAPANI R.V., (1988) **Experimental study on the behaviour of prestressed skew slab used as bridge decks.** Thesis submitted for the award of degree of Master of Science to Indian Institute of Technlogy Madras, India.
13. **"Phase I Report on Folded Plate Construction",** (1963) by the Task Committee on Folded Plate Construction, Journal of the Structural Division, ASCE, Vol.89, No.ST6, Proc. Paper 3741, - December 1963, pp.365-406.
14. DEFRIES-SKENE, A., and SCORDELIS, A.C., (1964) **"Direct Stiffness Solution of Folded Plates"** Journal of the Structural Division, ASCE, Vol.90, No.ST4, Proc. Paper 3994, August 1964, pp 15-47.
15. GAAFAR,L., (1954) **"Hipped Plate Analysis considering joint displacement,"** Transactions, ASCE, Vol.119.
16. SIMPSON,H., (1958) **"Design of folded plate roofs,"** Journal of the Structural Division, ASCE, Vol.84, Proc. Paper 1508, - January 1958.
17. WHITNEY , C.S., B.G. ANDERSON and H. BIRNBAUM, (1959) **Reinforced concrete folded plate construction',** Journal of the structural Division, Proc. ASCE Vol.85, No. ST 8, pp 15-43.
18. SITAPATI RAO.P (1975) **"Analytical Investigation on transversely stiffened folded plates,"** Thesis submitted for the award of the degree of Doctor of Philosophy to I.I.T., Madras.
19. GEETHA S (1994) **"Behaviour of reinforced and prestressed concrete folded plate structures".** Thesis submitted for the award of the degree of the Master of Science to I.I.T., Madras,- May 1994.
20. GEETHA S and RAJAGOPALAN N (1995) **"Behaviour of reinforced and prestressed concrete folded plate structures"** Indian Concrete Journal, May -October 1995 p.p.589-594.
21. MARSHALL W.T., (1954) **"The elimination of moments in shell roofs by prestressing,"** Proceedings of the Institution of Civil Engineers, Vol.3, Part III, April 1954, p.p 276-282
22. HAAS A M (1957) **"Prestressed shells and precast units,"** A-19, Proceedings of the World Conference on Prestressed concrete, San Fransisco, July, 1957.
23. JENKINS R.S., (1947) **"Theory and design of cylindrical shell structures,"** The O.N. Arup Group of Consulting Engineers, Bulletin 1, London.
24. TOTTENHAM H., (1963) "Simplified method of design for cylindrical shell roofs," The structural Division, Proceedings of the American Society of Civil Engineers, ST5, October 1963.
25. MC NAMEE, (1954) **"Existing methods for the analysis of concrete shell roofs"** Proceedings of the Symposium on Concrete Shell Roof Construction, Cement and Concrete Association, London.

Subject Index